Amorphous-
Nanocrystalline Alloys

Amorphous-Nanocrystalline Alloys

A.M. Glezer
N.A. Shurygina

CISP

CRC Press
Taylor & Francis Group
Boca Raton London New York

CRC Press is an imprint of the
Taylor & Francis Group, an **informa** business

CRC Press
Taylor & Francis Group
6000 Broken Sound Parkway NW, Suite 300
Boca Raton, FL 33487-2742

Printed on acid-free paper

International Standard Book Number-13: 978-1-138-50237-6 (Hardback)

Visit the Taylor & Francis Web site at
http://www.taylorandfrancis.com

and the CRC Press Web site at
http://www.crcpress.com

Contents

Foreword vii

1.	**Amorphous metallic materials**	**1**
1.1.	Production methods	2
1.2.	Special features of the structure	5
1.2.1.	Structural relaxation	14
1.2.2.	Defects in amorphous alloys	16
1.3.	The mechanical properties	22
1.3.1.	Methods of investigating the mechanical properties	22
1.3.2.	Strength	30
1.3.3.	Plasticity and plastic deformation mechanisms	40
1.3.4.	Fracture	58
1.3.5.	Temper brittleness	72
1.4.	Magnetic properties	84
1.4.1.	Ferromagnetism and ferrimagnetism of amorphous metals	84
1.4.2.	Magnetically elastic phenomena in amorphous alloys	92
2,	**Nanocrystalline alloys**	**103**
2.1.	Classification of nanocrystalline alloys	103
2.2.	Methods for obtaining bulk nanocrystalline materials	105
2.3.	Structure	110
2.4.	Physical and mechanical properties	115
2.4.1.	Theoretical strength and theoretical hardness	117
2.4.2.	The Hall–Petch relation and its anomaly	119
2.4.3.	Structural mechanisms of plastic deformation.	124
2.4.4.	Destruction	130
2.4.5.	Magnetic properties	133
3	**Amorphous–nanocrystalline alloys**	**141**
3.1.	Methods for the preparation of amorphous–crystalline materials	141
3.1.1.	Melt quenching	141
3.1.2.	Controlled crystallisation	144
3.1.3.	Deformation effect	145
3.1.3.1.	Volumetric strain. Terminology	146
3.1.3.2.	Surface treatment	150

3.1.4. Compacting of powders 153
3.1.5. Pulsed treatment 160
3.1.5.1. Photon treatment 160
3.1.5.2. Laser treatment 161
3.1.6. Production of thin films 162
3.1.7. Ion implantation 167
3.2. The structure of alloys 170
3.2.1. Special features of the transition from the amorphous to crystalline state under thermal effects 171
3.2.2. Features of the transition of an amorphous state to a crystalline state under deformation effects 222
3.2.3. Structure when compacting powders 272
3.2.4. Structure under pulsed light annealing 274
3.2.5. Changes of the structure under the action of pulsed laser radiation 276
3.2.6. Structure of amorphous–crystalline films 277
3.3. Mechanical properties 314
3.3.1. Mechanical properties of amorphous alloys in early stages of crystallisation 316
3.3.1.1. Influence of the size of nanocrystalline particles on strength 334
3.3.1.2. Structural mechanisms of plastic deformation 340
3.3.1.3. Plasticity 354
3.3.2. Mechanical properties in the late stages of crystallisation 358
3.3.3. Mechanical properties during quenching from a melt at a critical rate 373
3.3.4. Mechanical properties under deformation effects 379
3.3.5. Mechanical properties after pulsed treatment 389
3.3.6. Mechanical properties of amorphous–nanocrystalline films 391
3.4. Magnetic properties 394
3.4.1. The theory of magnetism in nanocrystals with a strong intergranular interaction 398
3.4.2. Magnetic properties of Fe–Si–Nb–Cu–B alloys (Finemet). 405
3.4.3. Magnetic properties of alloys after MPD 415
3.4.4. Magnetic properties of Fe–M–B–Cu (Nanoperm) and (Fe, Co)–M–B–Cu (Thermoperm) alloys 421
3.4.5. Magnetic properties of Fe–Nd–B alloys 425
3.4.6. Magnetic properties of films 436

Index **441**

Foreword

Amorphous–nanocrystalline alloys are a new class of materials born at the turn of the 20th and 21st centuries as a result of the rapid development of new technologies and, in particular, nanotechnology (various methods for obtaining amorphous and nanocrystalline powders and films, compacting, melt quenching, megaplastic deformation, implantation, laser , Plasma and other methods of high-energy effects, etc.). They arose at the intersection of intensive research and development of promising amorphous and nanocrystalline materials. At the same time, in terms of the level of physical and mechanical properties, two-phase amorphous–nanocrystalline materials in some cases exceed the properties of both nanocrystalline and amorphous materials, thereby creating a noticeable synergistic effect. The unusual nature of materials with an amorphous–nanocrystalline structure consists, first of all, in the fact that the structural–phase components of such a two-phase system radically differ among themselves in the nature of the atomic structure. Indeed, the crystalline phase has an ordered atomic structure at large distances and is characterized by the presence of translational symmetry. The atomic structure of the amorphous phase is, on the contrary, disordered and devoid of translational symmetry of long-range crystalline order, possessing only a pronounced topological and compositional short-range order in several coordination spheres. Such a "unity of opposites" is not limited to the above. In the table below, we summarized some essential differences between the amorphous and crystalline state of a solid

The situation becomes even more unusual if a nanocrystalline phase is used instead of the usual crystalline structural component, which, among other things, has an additional complex of unusual properties. Essentially, amorphous–nanocrystalline materials can equally be considered as natural amorphous–nanocrystalline

Table F.1. Comparative characteristics of the crystalline and amorphous states of metallic solids

Crystalline state	Amorphous state
Long-range atomic order Translational symmetry	Short-range atomic (topological and compositional) order No translational symmetry
High values of elasticity moduli	Low values of elasticity moduli
Structural anisotropy exists	Absence of structural anisotropy
Dislocation mode of plastic deformation	Non-dislocation mode of plastic deformation
Strain hardening	No strain hardening
High thermodynamic stability	Low thermodynamic stability

composites that possess physico–mechanical properties that are important for practical use.

In this monograph, the methods of obtaining amorphous–crystalline materials (quenching from a melt, controlled crystallization, deformation effect, pulsed (photon, laser and ultrasonic) treatment, thin film deposition, ion implantation) are considered successively. Detailed information is given on the structural features of the transition from the amorphous phase state to the nanocrystalline state under thermal and deformation effects. Theoretical and experimental studies are analyzed in which the mechanisms of plastic deformation and the features of the emerging physico–mechanical properties are described. Areas of practical application of amorphous–nanocrystalline alloys are considered.

We found it possible to build the content of the book in such a way that first of all the main characteristics of the main structural components–amorphous metallic materials and further–nanocrystalline ones–were first examined in it. It is only at the final stage of the presentation of the material that the structure and physico–mechanical properties of amorphous–nanocrystalline composite materials are described in detail.

The authors of the book are aware of the fact that this monograph is not devoid of shortcomings both in its form and in its content. Perhaps we did not mention or did not discuss a number of important and interesting studies on the topic under discussion. In advance, we apologize to their authors and note that this was done not for malicious intent, but only for reasons of limited edition volume.

1
Amorphous metallic materials

Amorphous metallic alloys (AMA) are a new class of metallic materials characterised by the unique combination of the magnetic, electrophysical, mechanical and corrosion properties. Recently, the AMA have been used on an increasing scale in aerospace, electronic and electrical engineering industries as magnetically soft materials in the cores of transformers and high-sensitivity sensors, such as brazing alloys, catalysts, and corrosion-resisting constructional materials.

The amorphisation of the melt is based on cooling at a relatively high rate in order to prevent the occurrence of the crystallisation processes and, consequently, the self-freezing of the disordered configuration of the atoms. Glass transition takes place quite easily in some well-known groups of non-metallic materials (silicates, polymers, oxides, etc). In these substances the nature of the strongly directed interatomic bonds imposes considerable restrictions on the speed with which the atomic or molecular interaction should take place for the formation of the thermodynamically equilibrium crystalline state. The metallic melts, not having directional bonds, are characterised by the high rate of atomic rearrangement even in the case of the high degrees of supercooling below the equilibrium glass transition temperature T_g [1.1].

It was assumed over a large number of years that it is not possible to transform the metallic melts to the amorphous state by quenching. In this case, it is sufficient to mention one of the best-known Russian publications 'The Physical Encyclopedic Dictionary', published in 1962. On page 31 in volume 2 in the section 'Quenching' we may read the following:' The rate of crystallisation of the metals and metallic alloys is too high in order to be able to produce them in the glassy state'.

Starting in the 60s, there were first timid and then more and more confident reports in both the USSR and abroad according to which the ejection of a liquid droplet onto a heat-conducting substrate produces non-crystalline phases in the metallic alloys. The pioneers of these investigations were P. Dyuvez and I.S. Miroshnichenko. In spite of the obvious scientific significance of these results, they were met with little enthusiasm, since the 'pyatachki', a product of the so-called 'Dyuvez gun', of little use for serious physical research, seemed more exotic than the subject of a deep study of the structure and physico-mechanical properties of amorphous alloys. Only in the late 1960s, when a group of Japanese scientists, led by C. Masumoto, used the method of spinning to obtain amorphous alloys, an 'amorphous boom' broke out. Melt quenching on a rapidly spinning disc or other similar methods of melt quenching made it possible to obtain reproducible structural structures that are suitable for large-scale studies, which, as it turned out, have a unique combination of physico-chemical and mechanical properties.

1.1. Production methods

At the present time there are several methods of producing metallic materials with the amorphous structure which can be divided into three large groups: 1) gas phase deposition of metals; 2) the introduction of defects into the metallic crystal; 3) solidification of liquid metal [1.2].

The methods included in the first group include the following: vacuum spraying, sputtering and chemical reactions in the gas phase. The second group includes the methods of irradiation the crystal surface with particles, the effect of the shockwave, and a number of other methods. The third group includes different methods of quenching from the liquid state. A detailed description of the methods of preparation of the amorphous alloys and their physical properties have been published in [1.3].

Without mentioning the significance of all the above described methods of amorphisation, it is important to note the controlling role of the method of melt quenching which can be used to produce amorphous alloys in the form of ribbon or wire (sometimes thick) specimens, separated from the crystal substrate, in very wide ranges of the compositions and physical and mechanical properties.

In cooling over a relatively long period of time producing the thermodynamically equilibrium state of the liquid, the melt solidifies at the solidification temperature T_m. However, at a high cooling rate the liquid does not solidify even in undercooling below T_m. The

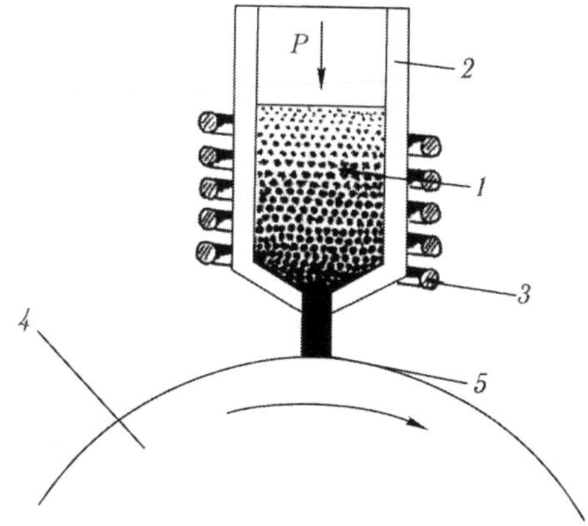

Fig. 1.1. Diagram of equipment for melt quenching by the spinning method: 1) the melt, 2) the crucible, 3) the induction coil, 4) the cooling disc, 5) the ribbon [1. 2].

liquid in this state is referred to as the supercooled liquid. Further, if the cooling rate is maintained relatively high, the liquid does not transform into the crystal, the structure of the liquid remains without change to relatively low temperatures but, in the end, the liquid still does not solidify. The supercooled liquid solidifies at the temperature referred to as the glass transition temperature T_g.

There are several methods of carrying out superfast melt quenching, but the most widely used method is the spinning method in which the melt is supplied under pressure on a rapidly spinning disc [1. 4]. The diagram of the experimental setup is shown in Figure 1.1. As a result, ribbons with a thickness of 20 to 100 μm are produced, and the structure of the ribbons depends on the composition of the alloy and the cooling rate.

The common principle of all systems for quenching from the melt is that the melt 1, supplied into the crucible 2, rapidly solidifies using the induction coil 3, spreads in a scene layer on the surface of the spinning cooling disc 4 and transforms to the final product in the form of the thin ribbon 5. At a constant composition of the alloy the cooling rate depends on the thickness of the melt and the characteristics of the cooling disc. For different alloys, the cooling rate also depends on the properties of the melt (heat conductivity, specific heat, viscosity, density). In addition, an important factor is the coefficient of heat transfer between the melt and the cooler which is in contact with the melt.

At the present time, there are more than 200 amorphous binary, ternary and multicomponent systems of metallic alloys which can be divided into two groups: the amorphous alloys of the metal–metal type, and the amorphous alloys of the metal–metalloid type. Successes in Russian science in the production of these materials, and also the examination of the structure and properties have been described extensively in the publications [1.5–1.12]. It is useful to provide a general classification of the metallic systems, produced in the amorphous state [1.13].

The first group of the amorphised systems includes the alloys of the transition metal type of the groups VIIB and VIII of the periodic system of elements or the noble metal of the group IB combined with the metalloid (B, C, Si, P). These alloys are most important for practice because they have high magnetic and mechanical characteristics.

This applies especially to the amorphous alloys based on iron, cobalt and nickel which represent the basis of the high-strength magnetically soft amorphous materials. As a result of adding alloying elements, the susceptibility of these alloys to amorphisation may be greatly increased, i.e., the critical quenching rate, required for producing the amorphous state, is reduced.

The second group of the amorphous alloys includes the system consisting of the transition metals of the groups IVB, VB or VIB, combined with the metals included in the composition of the amorphous alloys of the first group. Suitable examples are Cu–Ti, Ni–Nb and (Co, Ni, Fe)–Zr alloys. Here we can also include the alloys whose composition contains the metals of the platinum group which are usually characterised by the very high-temperature of transition to the crystalline state (higher than 1000 K).

The third group of the amorphous alloys includes the system consisting of the transition metal of the groups VIIB and VIII or a noble metal of group IB and metals – lanthanides. Suitable examples are La–Au and Gd–Fe alloys.

The fourth group of the amorphised alloys contains binary and multicomponent alloys, consisting of alkali-earth elements and some metals. Typical examples: Ca–Al, Mg–Zn and Be–Zr.

The fifth group of the amorphous alloys includes a relatively small group based on metals – actinides (U–Co, Np–Ga, Pu–Ni). Finally, another class of the amorphous metallic systems, consisting of pure metals: amorphous nickel, molybdenum and a number of other metals [1.14], was reported recently.

1.2. Special features of the structure

The amorphous state of the solid body has been the subject of extensive research because of its unusual properties and the structure. In particular, these alloys are characterised by the absence of the crystal lattice and the long-range order (there is no correlation between the atoms over long distances), but the short-range order is retained (correlations are found at two or three coordination spheres [1.15]) which influences the properties of these alloys. If the topological short-range order, describing the degree of local ordering by the type of the crystal, has no analogue in the conventional crystals, the chemical (compositional) short-range order, describing the tendency of the atoms to be surrounded by the atoms of a specific type, is very similar to that which is present almost always in the multicomponent crystals. In addition, the methods of quantitative description of the compositional short-range order in the amorphous systems [1.16] and of the short-range order in the crystals [1.17] are basically identical.

The main difficulty is the method of describing the structure of the amorphous state. This is associated with the absence of translational elements of symmetry and the concept of the elementary cell. Combined with the low efficiency of the methods based on the interaction of the solid body with the electromagnetic radiation of different types of waves (neutrons, X-rays, electrons) this deprives the researcher of usual crystallographic terms and concepts and also the powerful tools of structural analysis. Since the amorphous state of the solid body reflects extensively the structure of the liquid, the description of its structure must be based on taking into account the fluctuations of density, local environment and chemical composition. All these factors result undoubtedly in the probability and statistical nature of the description of the structure.

The absence of the strict order in the distribution of the atoms in the amorphous state does not mean that the distribution of the atoms and molecules in their structure is chaotic. The diffraction methods of investigating the structure indicate the presence of the short-range order in the distribution of the atoms in the amorphous solid bodies (as in the liquids). A number of structural models of amorphous alloys have been proposed and they can be divided into two large groups [1.10]: the first group of the models is based on the quasi-liquid description of the structure using a continuous network of chaotically distributed atoms; the second group of the models is based on the quasi-crystalline description of the structure in the form of crystals characterised by a high density of defects of different types.

Attempts for the theoretical description of the structure of the amorphous solids are associated with considerable difficulties. If the structure of the disordered systems which are in thermodynamically equilibria (for example, gas or liquid) can be described using the distribution functions in one-, two- or multiparticle approximations, there are certain difficulties in the theoretical description of the structure of the non-equilibrium systems. At the present time, the special distribution of the atoms in the amorphous solids is evaluated on the basis of several structural models based on rational physical assumptions or using the results of analysis of special features of the distribution of atoms in the actual amorphous metallic alloys determined by the method of integral Fourier transform of the measured intensity of scattering of the X-rays (or neutrons).

The structure of the amorphous solids is described by the radial distribution function (RDF) and other functions associated with it [1.18, 1.19]. The radial distribution function is the most important characteristic and for a system consisting of atoms of the same type is determined by the following equation

$$W(r) = 4\pi r^2 \rho(r), \tag{1.1}$$

where r is the distance from the fixed atom; $\rho(r)$ is the function of the atomic distribution order or the pairing function of the atomic distribution, determined by the following equation

$$\rho(r) = \lim \sum_{l \neq m}^{N} \delta\left[r - (r_l - r_m)\right], \tag{1.2}$$

which has the following physical meaning: if the coordination system is combined with the centre of one of the particles (m is fixed), the product $\rho(r)dV$ is the mean number of the particles in the element with the volume dV which is characterised by the radius-vector r. Here $\delta(r)$ is the Dirac delta function, N is the number of atoms, and r_1 and r_m denote the coordinates of the centres of gravity of the atoms.

The function $W(r)$ is the number of atoms in a spherical cell with radius r and the thickness of the layer equal to unity. This function is equal to 0 for the values of r smaller than the diameter of the sphere of the atom, and with increasing distance r the correlation between the particles gradually attenuates, and at $r \to \infty$ the functions $\rho(r)$ tends to the mean value $\rho_0 = N/V$, where N is the number of the particles, V is the volume. The radial distribution function is characterised by the large first peak, corresponding to the first nearest neighbours, and by gradually expanding peaks with lower intensity which correspond to the second, third, etc. neighbour particles around the selected atom.

In many cases, function $W(r)$ is replaced by the reduced radial distribution function

$$G(r) = 4\pi r[\rho(r) - \rho_0] \tag{1.3}$$

and by the pairing function of the radial distribution

$$g(r) = \rho(r)/\rho_0. \tag{1.4}$$

The radial distribution function and the reduced radial distribution function are linked by a simple relationship

$$W(r) = rG(r) + 4\pi r^2 \rho_0. \tag{1.5}$$

When describing the amorphous solid bodies consisting of atoms of n different elements, the total RDF is determined by the equation

$$W(r) = 4\pi r^2 \sum_{i=1}^{n} \sum_{j=1}^{n} w_{ij} \rho_{ij}(r), \tag{1.6}$$

The partial functions of the atomic distribution $\rho_{ij}(r)$ are included in this distribution with specific weight factors w_{ij}. The partial atomic distribution functions $\rho_{ij}(r)$ $(i, j = 1, 2, \ldots, n)$ represent the average density of the particles of the j-th type in the element with the volume dV with the coordinate r, if the point with the coordinate $r = 0$ contains a particle of the i-th type.

The results of the experimental investigation of the scattering of the X-rays, electrons or neutrons can be used to determine only their Fourier image, not the distribution functions. For example, the intensity of coherent scattering of the X-rays by a group of atoms of the same kind, determined by the experiments, is expressed by the equation

$$I_N(K) = N|f(K)|^2 \int_0^V q_N(r) \exp(-iKr)\, dV, \tag{1.7}$$

here N is the number of atoms; $q_N(r)$ is the structural factor, describing the interference of the waves scattered by the atoms of the substance in the direction of the receiver of the radiation; $f(K)$ is the atomic factor, i.e., the value determining the scattering of the X-rays by the isolated atom; $K = k - k'$ and k and k' are the wave vectors for the incident and reflected rays, respectively. The absolute value of K is determined by the wavelength λ of the X-ray radiation employed and the scattering angle Θ (the angle 2Θ is the angle between the incident and reflected rays, i.e., between the vectors k and k'):

$$K = 4\pi\sin(\Theta/\lambda). \qquad (1.8)$$

Instead of the scattering intensity, experiments are often carried out using the interference function referred to as the structural factor:

$$J(K) = \frac{I_N(K)}{N|f(K)|^2}. \qquad (1.9)$$

Using the Fourier transform of the equations (1.7) and (1.9), it is possible to determine the reduced radial distribution function:

$$G(r) = 4\pi r[\rho(r) - \rho_0] = \frac{2}{\pi}\int_0^\infty [J(K) - 1]\sin(Kr)K\,dK. \qquad (1.10)$$

The expression below the integral

$$K[J(K) - 1] = F(K), \qquad (1.11)$$

is referred to as the reduced interference function.

The intensity of coherent scattering, measured by experiments for the amorphous multicomponent systems, can be written in the following form:

$$I_N(K) = \sum_{i=1}^{n}\sum_{j=1}^{n} N_i f_i(K) f_j^*(K) \frac{4\pi}{K}\int_0^\infty [q_{ij}(r) - \rho_{0,j}]\sin(Kr) r\,dr, \qquad (1.12)$$

and the total interference function

$$J(K) = 1 + \sum_{i=1}^{n}\sum_{j=1}^{n} w_{ij}(K)\frac{4\pi}{K}\int_0^\infty \left[\frac{\rho_{ij}(r)}{C_j} - \rho_0\right]\sin(Kr) r\,dr, \qquad (1.13)$$

where $\rho_{0,j} = (N_j/N)\rho_0$, $\rho_{i,j}$ are the partial atomic distribution functions; $f^*(\mathrm{K})$ is the function complexly conjugate with the function $f(K)$;

$w_{ij}(K)$ are the weight multipliers, determined by the relationship

$$w_{ij}(K) = C_i C_j f_i(K) f_j^*(K) / \left|\langle f(K) \rangle\right|. \tag{1.14}$$

In the last equation, C_i and C_j are the coefficients equal to the concentrations of the individual components; $\langle f(K) \rangle$ is the atomic factor averaged out with respect to the structure. The individual terms and expressions (1.13) without the weight multipliers are referred to as the partial interference functions.

As in the case of the single-component systems, the total reduced radial distribution function can also be introduced

$$G(r) = 4\pi r \sum_{i=1}^{n} \sum_{j=1}^{n} w_{ij} \left[\frac{\rho_{ij}(r)}{C_j} - \rho_0 \right] \tag{1.15}$$

or

$$G(r) = \frac{2}{\pi} \int_0^{\infty} [J(K) - 1] \sin(Kr) K \, dK. \tag{1.16}$$

Since the scattering intensity $I(K)$ can be measured in experiments only to the maximum values $K_{max} \approx 100$–200 nm^{-1} (at high values of K the value $I(K)$ is very small), then in the calculation of the radial distribution function we face additional difficulties. To overcome these difficulties partially, it is recommended to use the approximate equation [1.18]

$$G(r) = \frac{2}{\pi} \int_0^{K_{max}} [J(K) - 1] \exp\left(bK^2\right) \sin(Kr) K \, dK. \tag{1.17}$$

The typical values b = (0.05–0.2) nm. To prevent scattering intensity on the small angles associated with the structural defects, the lower integration limit is introduced into the relationship (1.17); this limit corresponds to the minimum values of $K_{min} \approx 10$–20 nm^{-1}.

All the previously mentioned functions are used to examine the structure of the amorphous solids. Each function has its own advantages and shortcomings. However, the most important functions are $J(K)$, $W(r)$ and $G(r)$ and also the paired distribution function $g(r)$. The experimental functions can be used to determine the short-range order parameters. For example, the area below the first peak of the radial distribution function of the single-component of the system can be used to determine the mean coordination number, i.e., the number of the nearest neighbours:

$$Z = \int_0^{r_{min}} \rho(r) 4\pi r^2 \, dr, \tag{1.18}$$

and taking into account the position of the first maximum it is possible to determine the mean atomic spacing of the given structure. Here r_{min} is the coordinate of the first minimum of the radial distribution function.

The calculation of the short-range order parameters for the multi-component system is more complicated because they are described using the sum of the partial functions of the atomic distribution $\rho_{ij}(r)$ with their weight multipliers $w_{ij}(K)$ showing the relative contribution of the individual components of the total interference function of the total reduced radial distribution function. In the case of binary amorphous alloys with three independent investigations of the scattering intensity of the X-rays (at different wavelengths) it is possible to determine all three partial interference functions of the paired distribution $J_{11}(K)$, $J_{12}(K)$, $J_{13}(K)$ [1.19] and other functions introduced above. If the experimental results can be used to determine, for example, the partial atomic distribution functions $\rho_{ij}(r)$, it is then also possible to evaluate the partial coordination number Z_{ij} which is the number of the nearest atoms of the j-th type around the atom of the i-th type, i.e.

$$Z_{ij} = \int_0^{r_{min}} 4\pi r^2 \rho_{ij}(r) dr, \tag{1.19}$$

Here r_{min} is the coordinate for each of the functions $4\pi r^2 \rho_{ij}(r)$. Since the atomic distribution function $\rho_{ij}(r)$ is the number of the atoms (of the i-th and j-th type) per unit volume at the distance r from the atom, it can be determined as follows

$$\rho_i(r) = \sum_j \rho_{ij}(r). \tag{1.20}$$

The number of the nearest adjacent atoms around the atom of the i-th type is equal to

$$Z_i = \sum_j Z_{ij}. \tag{1.21}$$

The mean coordination number for the alloy can also be determined using the equation (1.18), where $\rho(r)$ is the total atomic distribution function.

Initially the structure of some simple amorphous solids was described by analogy with the liquid (within the framework of the 'quasi-liquid' model) using the Bernal model proposed at the time for describing the structure of simple liquids. The model is based on the chaotic dense packing of the rigid spheres. By analogy with the liquid, the structure of some simple amorphous solids may be represented by five polyhedral configurations (clusters): tetrahedron, octahedron, trigonal prism, Archimedes antiprism, and the tetragonal dodecahedron.

However, the computer simulation methods, used actively for producing the quasi-liquid structure (the methods of successive annexation methods of collective rearrangement) did not make it possible to produce the structure of chaotically close-packed rigid spheres of the same density as that observed in the experiments. Further modernisation of the model (the application of 'soft' spheres, governed by the paired interatomic potentials of the Leonard–Jones type, instead of the 'rigid' spheres) resulted in a considerable improvement of the agreement between the theory and experiment. In addition to the Bernal polyhedrons, considerable successes have been achieved using the Voronoi polyhedrons. However, it is not yet clear whether a specific type of polyhedron has an unambiguous relationship with the atomic structure to be described. In addition to this, any displacement of the atoms or distortion of the regions of the structure results in changes in the type of the Voronoi polyhedron [1.8].

In the final analysis, the amorphous single-component structure can be regarded as an ensemble of the distorted octahedrons and tetrahedrons existing in a simple close-packed structure; the sequence of alternation of the configurations of the tetrahedrons and octahedrons can be described on the basis of the paired correlation functions [1.20].

The attempts to solve the problem of the two-component amorphous systems within the framework of the quasi-liquid models have resulted in the conclusions by investigators regarding the accuracy of the stereochemical assumptions proposed for the first time in [1.21]. The stereochemical approach is based on the description of the amorphous structure by specific structural elements consisting of the central atom A and the atoms B surrounding the

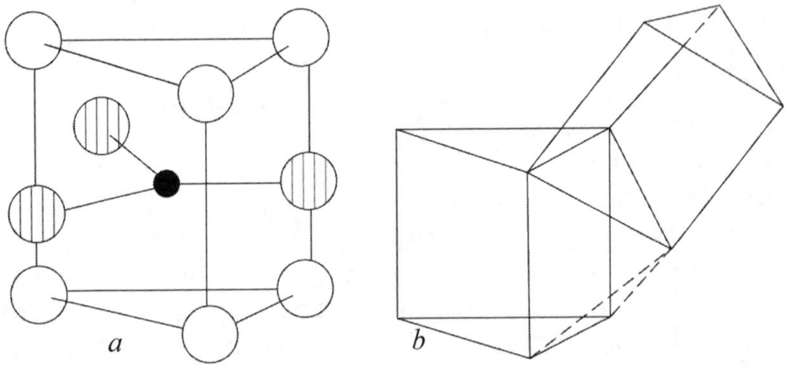

Fig. 1.2. *a*) the trigonal-prismatic nanocluster used for describing the structure of the two-component amorphous alloy of the metal–metalloid type: *b*) the diagram of 'contact' of the nanoclusters; ● – the metalloid atom; ○ – the metal atoms; ⦶ – the atoms of the metal distributed in the second coordination sphere.

central atom which form together some coordination cell (for example, in the form of a trigonal prism) (Fig. 1.2). The most accurate description of the structure can be achieved using the coordination cell whose symmetry is identical with that realised in the crystalline phase, formed in the same binary system. It is natural to assume that these structural elements existed in the initial melt in the form of associates and in supercooling of the melt they were inherited by the metallic glass.

For the majority of the amorphous systems of the metal–metal and, in particular, metal–metalloid type, the quasi-crystalline model provides an accurate description of the structure. The binary alloys of different composition are often regarded as the 'two-phase mixture' of close-packed regions of pure metal and the regions with the close-packed structure typical of the basic metallic nanocluster. In more complicated cases (for example, for multi-component systems) it is necessary to use polycluster models [1.22] in which the amorphous matrix is formed by several types of nanoclusters each of which is locally ordered and separated from the neighbours by intercluster boundaries in the form of planar defects consisting of two-dimensional monolayers with the imperfect local loading of the atoms.

We will describe briefly a different approach to explaining the structure of the amorphous state – the pseudo-crystalline model. The development of theoretical considerations regarding the structure of the metallic glasses resulted in the construction in the 80s of models describing the structure of the amorphous state bicrystals with a high density of defects of different types. The defective crystal, containing dislocations is not adequate to the structure of metallic glass [1.23], even if the density of dislocations exceeds 10^{12} cm^{-2}, The regular filling of the crystals with the defects of the disclination type is capable of transforming the crystal to the structural state similar to amorphous. The disclinations theory shows [1.24] that the dislocations cause only Cartan torsion in the crystal without changing its metrics. Therefore, the crystal is studied in the Euclidean space, and the dislocations are linear defects of the structure of the crystal. However, if the crystal contains disclinations, they greatly change its metrics, i.e., the crystal should be regarded as a crystal in space with the Riemann–Christoffel curvature as functions of the tensor of the disclination density, and the disclinations are linear defects of such a structure. A similar structure of the models of the structure of the amorphous alloys produced from polytypes in the distorted space is used widely, mostly for calculating the electronic properties. For example, it was shown [1.25] that the close-packed non-crystalline structure can be produced by imaging the polytypes figures from the distorted space in the Euclidean space. This imaging is obtained by introducing a grid of disclination lines transforming the space curvature to zero. As an example, investigations were carried out [1.25] of the structure generated by the imaging of the polytype {3, 3, 5} on a three-dimensional sphere to the Euclidean space. The resultant structure can be classified by the type of the grid of disclination lines. As regards the metallic glasses produced by melt quenching, the disclination model is not very accurate because it is not clear how the crystals with a very high disclination density form in the process of superfast cooling of the liquid phase. Evidently, the disclination assumptions are suitable for describing the amorphous state formed as a result of high plastic deformation of crystalline intermetallic compounds [1.26]. The disclinations in these models appear by the natural mechanism during the disordered filling of the conventional three-dimensional space with structural units in the form of regular tetrahedrons or polyhedrons representing the geometrical models of the elementary atomic clusters.

Although the ensemble of the chaotically oriented microcrystals or nanocrystals has no translational symmetry at large distances, it has been shown [1.23] that its radial distribution function differs in principle from the same characteristic of the amorphous state. The fact that the microcrystalline model is not suitable in this case reflects the fundamental difference in the nature of the topological short-range order of the amorphous and crystalline phases: polytetrahedral in the first case and crystalline (with the elements of the translational symmetry) in the second case. At the same time, microcrystalline approaches to describing the amorphous state have proved to be very durable. This is associated primarily with the fact that the X-ray diffraction patterns of the nanocrystalline objects are very similar to the X-ray diagram of the amorphous alloys.

1.2.1. Structural relaxation

The relaxation process in the amorphous alloys has many common features with the well examined processes of relaxation in the amorphous polymers and oxides but in the case of the amorphous alloys it has a considerably stronger effect also on the large number of physical and structural parameters [1.6, 1.11]. In principle, it is not rational to discuss the properties of the amorphous alloys without taking into account the relaxation parameters because the measured properties can prove to be strongly dependent on the degree of structural relaxation of the specific state.

Homogeneous relaxation, referred to more frequently as structural relaxation (SR), takes place homogeneously throughout the entire volume of the specimen without affecting its amorphous state. Structural relaxation is accompanied by changes of the short-range order leading to a very small decrease of the degree of non-equilibrium of the glass. The unstable atomic configurations, formed at the moment of amorphisation during quenching, change to stable configurations as a result of small atomic displacements. Consequently, the density of the amorphous matrix increases as a result of partial annihilation and removal of the excess free volume [1.27]. It is important to note that the displacement of the atoms during structural relaxation is smaller than the atomic distances and takes place only in local regions. The magnitude of the heat of transformation to the stable phase which may be used as a measure of this non-equilibrium nature changes only slightly in this case.

Structural relaxation is accompanied by changes of many physical properties of the amorphous alloys: specific heat, density, electrical resistance, internal friction, elastic constants, hardness, magnetic characteristics (the Curie temperature changes, magnetic anisotropy is induced), corrosion resistance, etc.

Several models have been proposed for describing the structural relaxation processes. They can be divided into two groups: 1) the models of the activation energy spectrum or *AES* (*activation energy spectrum*)-models; 2) The model proposed by van den Beukel et al.

In the first model, it is assumed that the structural relaxation is caused by the local atomic rearrangement in the amorphous material taking place with different relaxation times (activation energies). The fundamentals of the model were developed in the studies by W. Primak [1.28, 1.29] and were applied later to the relaxation processes in glasses [1.30–1.32]. It is assumed that the activation energy of these processes is distributed in a continuous smooth spectrum. The rate of variation of the physical property is proportional to the rate of variation of the density of 'kinetic processes'.

The second model [1.33–1.35] uses the approach to describing structural relaxation is based on the short-range order classification, proposed by Egami [1.36]. It is assumed that the relaxation in the first stage takes place by compositional (chemical) short-range ordering. This contribution is described efficiently using the AES model and is a reversible process, taking place with the activation energy spectrum from 150 to 250 kJ/mole. The chemical ordering is relatively fast and after it is completed the topological short-range ordering becomes controlling. The topological relaxation is described by the Spaepen free volume model [1.37–1.39] with the unique activation energy of approximately 250 kJ/mole and is an irreversible process. It should be stressed that the main assumption in the Spaepen model is that the heterogeneous structure of the glass results in the formation of regions with a free volume excessive in relation to the 'ideal' structure – 'relaxation centres'. Thermally activated atom displacements can take place in these regions and causes the redistribution of the free volume inside the material and also its partial transfer to the free surface.

The Spaepen free volume model is now used only very rarely for describing the mechanical behaviour of metallic alloys because of obvious shortcomings (the direct determination of the magnitude of the excess free volume is not possible, the unique activation energy of relaxation processes must be considered). The Van den Beukel model has also been criticised [1.40] because it is very complicated as regards the chemical and topological ordering independent of each other and taking place at different times. The model of the activation energy spectrum is used most frequently at present. In particular, a model of the directional structural relaxation, i.e., the relaxation oriented by the external stress, described in [1.41–1.44], is especially promising.

Nevertheless, the changes of the majority of the physical properties of the metallic alloys during annealing have been studied in detail and described by both the model of 'topological and chemical ordering', proposed by Van den Beukel [1.45–1.40] and by the model of the activation energy spectrum [1.31, 1.41–1.44, 1.49, 1.50].

1.2.2. Defects in amorphous alloys

To understand the role of a defect in a specific process it is necessary to consider most of all the state of the structure free from defects. Comparison of the state with and without defects can be carried out in terms of the topological properties or stress fields. The majority of defects typical of crystals lose their specific features in the amorphous state. Nevertheless, the results of a large number of experiments carried out to investigate the structure-sensitive properties of the amorphous alloys show that the structural defects can also exist in the amorphous alloys. The deviations from the low-energy equilibrium state in the structure of amorphous materials may be described by the increase of the density of these defects. Because of this, we are not capable of regarding as a defect some small ordered region in the amorphous matrix although the small disordered regions in the crystal can be regarded as clusters of elementary defects.

Several attempts have been made to present a generalised definition of the structural defects in amorphous solids [1.51]. On the one hand, a model of the ideal amorphous structure has been developed and subsequently defects were introduced into the structure on the basis of the purely geometrical considerations, by analogy with the procedure applied to the crystal [1.52]. This was followed by measurement of the resultant displacements which are usually very large. This definition of the defects, using the assumptions of the local deformation, is applicable only to covalent amorphous solid bodies [1.52].

There is another approach to describing defects in amorphous alloys. In this approach, the configuration of the individual atoms is considered and defects as a collectivized phenomenon are ignored [1.53].

Some investigators subdivide the defects in the amorphous alloys into internal and external. The former are typical of the material even after deep relaxation and the latter annihilate during the relaxation changes in the structure. Since it is quite difficult to obtain by experiments any detailed information about different types of defects and their distribution, these experiments are carried out by computer modelling methods and their evolution during different external defects [1.54]. In addition to this, these calculations can evidently help in understanding the amorphous state.

The defects in the amorphous alloys can be divided into point defects, microscopically elongated and macroscopic. The main point defects, existing in the amorphous matrix, are [1.52]: broken bonds; irregular bonds; pairs with the changed valency; the atoms with a small stress field (quasi-vacancies); the atoms with a large stress field (quasi-implanted atoms).

The elongated defects may include: quasi-vacancy dislocations; quasi-implanted dislocations; the boundaries between two amorphous phases; intercluster boundaries.

The macroscopic defects include pores, cracks and other macro-imperfections.

Figure 1.3 shows the main point defects which can be found in the amorphous solids. An important source for the formation of structural defects in the amorphous alloys is the free volume determined by the high coefficient of expansion of the liquid.

In [1.55] it is assumed that the cooperative molecular variations are formed from time to time by the pores which are sufficiently large in order to realise molecular 'jumps'. At high supercooling the formation of these pores and the corresponding displacement of the molecules become extremely difficult and lead to the formation of an amorphous structure. At the same time, Cohen and Turnbull [1.56] assumed that the migration of the atoms in a system consisting of rigid spheres becomes possible only in the presence of a cavity which is larger than the critical size. It was found that the free volume is statistically distributed in the amorphous matrix without any changes of free energy during its redistribution within the same amount. Thus, the free volume can be regarded either in the form of cavities of a given size or as a formation continuously distributed in the matrix.

In [1.57] the authors formulated the assumption of the free volume as regions with reduced density. They are characterised by a specific size distribution (from fractions of the atomic radius to hundreds of nanometres), which depends on the production

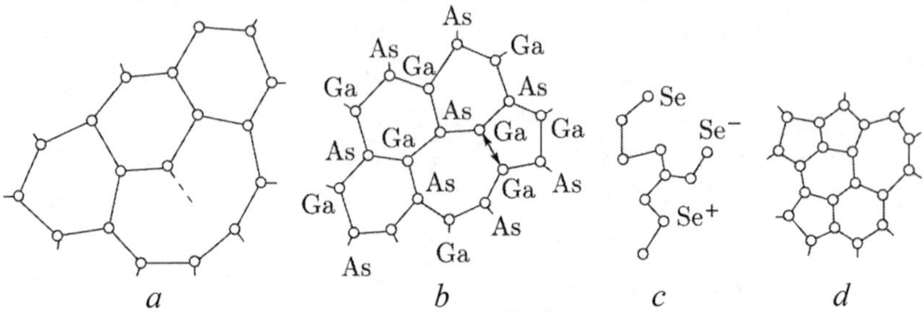

Fig. 1.3. Point defects which may exist in amorphous alloys: *a* – broken bonds; *b* – irregular bonds; *c* – pairs with changed valency; *d* – quasi-vacancies (free volume).

conditions, composition, heat treatment conditions and a number of other factors. In the existing continuous spectrum of the dimensions of the free volume regions the different dimensional fractions of these defects should have greatly different activation energies of migration mechanisms. In comparison with other structural models of the defects of the amorphous state, a significant advantage of the free volume is that it is simple and informative. This theory can be used for theoretical and sometimes also experimental investigation of the evolution of free volume regions from their nucleation (from regions of rarefaction in the melt through the transformation of the shape and redistribution in the amorphous matrix volume during quenching) to the change of the morphology and size distribution of the parameters and under different external effects (including thermal effects).

Irrespective of the nature and origin of the free volume, its role in the processes of relaxation, deformation, mass transfer and many other processes in the supercooled liquid is controlling.

The investigation of the structure of the amorphous alloys by small-angle scattering of the X-rays is in fact the only method capable of providing direct information on the shape of defects and their size distribution. Comparative analysis of the experimental data, obtained for the $Fe_{82.5}B_{17.5}$ basic alloy and the alloys produced by alloying this alloy with 0.01 at.% Sb, Ce or Nb shows that [1.58] the addition of the surface-active active elements greatly influences the total number and size distribution of the quenched defects. The quantitative analysis of the data for the amorphous alloys is complicated by the fact that the shape of the defects (as shown by the analysis of the forces acting during solidification) changes in the cross section and may depend strongly on the combination of the parameters of the produced amorphous alloys. Therefore, the examination of the effect of microalloying with the surface- active elements by the methods of small-angle scattering of the X-rays was carried out on the alloys produced at the same values of the main parameters of production of the amorphous ribbons.

The special procedural features of recording make it possible to obtain the almost complete information on the inhomogeneities with the size smaller than 200 nm. These inhomogeneities are non-equiaxal, are elongated and slightly flattened along the normal to the surface of the ribbon (along the Z axis). The observed ellipsoidal defects are oriented by their major axes in the direction normal to the direction of rotation of the disc in the plane of the ribbon (along the Y axis). This is indicated by the anisotropy of the scattering profile, observed in inclined small-angle recording and obtained for an iron-based amorphous alloy (Fig. 1.4).

Computer simulation experiments were carried out. The results show that the vacancies and vacancy clusters, introduced into the single-component three-dimensional amorphous structure at the absolute zero temperature and at higher temperatures are stable in most cases. They have either the spherical or cylindrical but not flat shape [1.59]. In this study it was shown that in the presence of external loading the vacancies can act as centres of the nucleation of the pores and microcracks.

The presence or absence of dislocations in the amorphous structure is a relatively disputable question [1.60]. In particular, in a number of theory it is proposed to investigate the amorphous state as the crystalline state with the density of the dislocations higher than some critical value (of the order of 10^{14} cm^{-2}). Evidently, in this case, the dislocations internal belong in the amorphous state and cannot disappear during structural relaxation.

Egami et al [1.61] attempted to describe the structural defects in the amorphous alloys from a different, more general viewpoint, in the form of sources of internal stresses and a specific type of the local

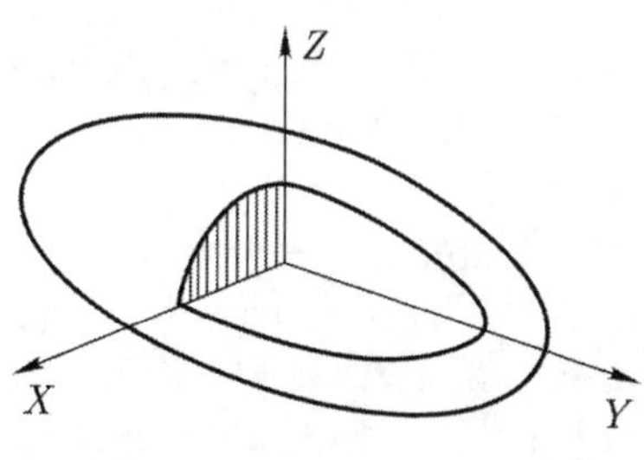

Fig. 1.4. Anisotropy of the shape of the regions of the free volume, determined by small-angle scattering of the X-rays, the X axis coincides with the direction of the major axis of the ribbon, Z axis – with the normal to the surface of the ribbon, produced by melt quenching [1.56].

atomic symmetry. In this approach, it is not necessary to consider the ideal structure, and all the quantitative characteristics are related to the appropriate equilibrium states, but not necessarily that with the lowest energy of the atomic structure.

In [1.62] the computer model proposed in [1.63] for α-iron was used to show that the spectrum of the distribution of local pressure P in the amorphous matrix can be used to determine the degree of structural relaxation: the narrow spectrum corresponds to the efficiently relaxed state, and a wider spectrum to the non-relaxed state. The variation of the distribution of P in relaxation can be described by the redistribution of structural defects, determined by the level of stresses on the atomic level. The recombination of the defects of the p- and n-type explains the variation of the density and the main changes of the radial distribution function. The shear defects of the τ-type are not sensitive to structural relaxation.

The correspondence between the theoretical, experimental and simulation results for the structural relaxation of the amorphous alloys makes it possible to determine the number of defects in the actual materials. The results show that, for example, annealing of the $Fe_{40}Ni_{40}P_{14}B_6$ alloy at 350°C for 0.5 h results in the recombination of approximately 10% of the defects present in the structure [1.52].

In [1.64] A.S. Bakay proposed a polycluster model of the structure of the amorphous state. The polycluster structures are formed by the clusters of the atoms where each atom is locally ordered. The cluster boundaries, investigated in the model, represent planar defects and consist of the two-dimensional single layers with the imperfect local ordering of the atoms. These boundaries contain a large number of the pseudo-locations and pseudo-implanted atoms and are responsible for the diffusion and mechanical properties of the polycluster structures.

The presence of the local order in the main part of the volume makes it possible to determine the structure of the cluster boundaries – the regions of disruption of the local order. In contrast to the intergranular boundaries, in the polycrystalline aggregates described by the grain-boundary dislocations, the boundaries between the clusters are not connected and, in fact, consist of the point defects of different types which join to form complexes.

In the framework of the proposed polycluster model, the defects are characterised as the regions of the highest degree of disordering. They may contain both point and elongated defects. In addition to the regular vacancies and interstitial atoms, they also contain partial point defects (the analogue of the free volume), the cluster boundaries and one-dimensional defects – like the dislocations in crystals – the edges of the boundaries and their sections, filled with the partial point defects [1.64]. In addition to this, the clusters may contain regions of localisation of high compression, tension or shear – the analogues of the defects of the *p*, *n*- and τ-types.

A special feature of the polycluster structure is the two-dimensional form of the cluster boundaries. The relaxation process of such a structure takes place in the direction of improvement of the local and short-range orders inside the crystals and further in the direction of establishment of the long-range order (i.e., crystallisation). In the stage preceding crystallisation, the regions with the high density of the cluster boundaries can be arranged with the formation of large clusters which are considerably larger than the atomic spacing. In the large number of the amorphous structures, the polycrystalline structures occupy an intermediate position as regards the degree of structural imperfection and the excess of the free energy between the polycrystalline structures and the system of close-packed atoms. Therefore, the polycrystalline structure can be found in the stage of structural relaxation in the transition from the truly amorphous structure to a crystalline one.

We will attempt to compare briefly the polycluster model with the free volume model. In the framework of the free volume model, all the displacements of the atoms under the effect of thermal fluctuations and also stress or under the effect of tunnelling take place as in the polyclusters in the areas of localisation of the vacancy cavities with a relatively large volume. However, the presence of these atoms and their local structure are not linked with the existence of the local ordering. The distribution of the free volume is spatially homogeneous, and the clusters of the cells, containing the excess free volume are three-dimensional and relatively large.

The free volume, existing in the polyclusters, is observed in the form of partial vacancies in the areas of mismatch of the boundaries and in the form of vacancies inside the clusters. The exchange of the free volume in the polyclusters is prevented until the local regular clusters can be rearranged by the dislocations, leading to crystallisation.

1.3. The mechanical properties

1.3.1. Methods of investigating the mechanical properties

The unique nature of the mechanical properties of the amorphous metallic materials is based on the combination of high strength and sufficiently high ductility. When examining the strength of the amorphous materials it was found that the strength is close to theoretical.

As regards the mechanical properties, the amorphous alloys are greatly superior not only to the inorganic glasses but also compete efficiently with the widely used constructional steels and alloys in which only one parameter: either strength or ductility, is high.

However, when obtaining the information on the mechanical characteristics of the amorphous alloys, it is necessary to take into account a number of the special features which impose certain and in some cases quite considerable restrictions on the specific methods of reliable analysis of the properties [1.65]:

1. High absolute values of the elasticity limit, yield strength and tensile strength;

2. The unstable nature of the flow under active loading, associated with the absence of strain hardening;

3. The presence of two completely different plastic flow mechanisms manifested by the temperature and strain rate, and also the scheme of the stress state in the deformation zone;

4. The small ratio of the thickness to the width of the measured ribbon specimens, produced by melt quenching;

5. The quality of the surface of the ribbon (different state of the contact side, adjacent to the drum in the manufacturing process, and the free side;

6. The strong effect on the mechanical properties of the 'prior history' of the investigated specimens (the production parameters, heat treatment), and also small amount of the impurities.

Several methods of mechanical tests have been proposed for investigating the mechanical properties of the amorphous ribbon specimens produced by melt quenching. Each method will be examined in detail.

1. Uniaxial tensile

The uniaxial tensile test is used widely because in principle it can provide a large amount of information on both the strength and ductility properties of the amorphous alloys [1.66–1.68].

The typical deformation curve shows the main relationships governing the mechanical behaviour of the amorphous ribbons: high tensile and yield strength at almost complete absence of strain hardening and small (but not zero) macroscopic ductility. However, the method of tensile testing the ribbon specimens is characterised by a number of shortcomings some of which cannot be removed. Firstly, it is the plastic deformation energy in tensile loading, comparable with the elastic energy concentrated in the conventional testing machines and resulting in the catastrophic fracture in testing [1.69]. The construction of the loading systems with higher rigidity especially for testing amorphous ribbons makes it possible to determine more accurately a number of special features of plastic deformation of the amorphous alloys. Secondly, in the majority of cases it is difficult to obtain reliable information in this loading method of the ribbon specimens because of the non-uniform thickness and width, uneven edges, longitudinal warping [1.70, 1.71].

Different methods of treating the edges of the ribbons – mechanical and chemical polishing, cutting to produce the special shape of the specimens in the form of a 'blade', 'eight', therefore careful centring in the testing machine reduce the scatter of the values of strength from specimen to specimen to 5% [1.72, 1.73].

Uniaxial tensile loading is also associated with difficulties when it is attempted to determine the dependence of the yield strength of the amorphous ribbons on the heat treatment temperature. This is due to the fact that when reaching a specific temperature, the specimens are greatly embrittled – the material fails by brittle fracture without reaching the yield limit [1.74].

2. Dynamic methods of exciting bending oscillations

To investigate the mechanical properties and obtain the absolute characteristics of the stress relaxation in the amorphous alloys in the form of thin ribbons, it is recommended to use a deformation machine of the spring type [1.75–1.77]. Using an electrode and a sonic generator, the specimen is excited during deformation at the resonance frequency and carries out bending oscillations like a spring. This is accompanied by the modulation of the high-frequency oscillations by the low-frequency oscillations of the sample.

The excitation in the investigated specimen of the bending oscillations (longitudinal or transverse) at the given frequency f [1.78, 1.79] makes it possible to calculate the elastic characteristics – the Young modulus E and the shear modulus G from the ratio

$(E, G) = Af^2$, where A is a constant defined by the method of excitation of the elastic oscillations and the geometry of the sample.

The Young modulus of the thin ribbons can be measured by exciting in them the bending resonant oscillations by the electrostatic method. The value of E is calculated by the equation:

$$E = \frac{48\pi^2 l^2 \rho f^2}{m^4 h^2}, \qquad (1.22)$$

where $m = 1.8751$, h is the thickness, t is the length, ρ is the density of the specimen, f is the resonance frequency [1.80]. The main error in the determination of E by this method is associated with the non-uniform thickness of the sample. However, the analysis results show that the relative error in the determination of E does not exceed $\pm 5\%$.

The authors of [1.81] determined G of the amorphous ribbons by the torsional oscillations, excited in the ribbon specimen using an inertia system in the form of a disc secured at the free end of the sample. The shear modulus was calculated from the equation

$$G = \frac{I\left(f^2 - f_0^2\right)}{1/3bh\left(l - 0.63h/b\right)}, \qquad (1.23)$$

here f and f_0 is the oscillation frequency of the initial system with and without the sample, respectively and $f_0 \ll f$, I is the inertia moment of the specimen around the axis of rotation, b is the width, h is the thickness and l is the length of the sample. The error in the determination of G by this method was $\pm 10\%$.

3. The internal friction method. Torsional deformation

It is well-known that the internal friction method is highly sensitive to the structure and is used for examining structural relaxation, the transition to the glassy state and crystallisation of the amorphous alloys [1.82, 1.83]. In a number of investigations, the inelasticity is studied by the stress relaxation method in which the samples, secured by the cantilever joint, are excited electrostatically in the bending oscillation mode [1.79, 1.84]. The internal friction is calculated from the following equation

$$Q^{-1} = k/N, \qquad (1.24)$$

here N is the number of freely damping oscillations of the sample, corresponding to a specific decrease of their amplitude; k is the apparatus constant.

At present, the internal friction, stress relaxation and also the restoration of the shape of the amorphous alloys are studied extensively using the method with torsional deformation [1.85–1.87].

To investigate the restoration of the shape, the samples are heated to the required temperature T_{an}, plastically deformed at T_{an} and cooled in the loaded state to room temperature. Subsequently, they are taken out from the loading device and the plastic torsional angle is determined. The determined angle is then used to determine the residual strain γ_0. Subsequently, the change of the torsional angle, caused by the effect of internal stresses, is determined during loading the freely suspended sample. This change is subsequently converted to the restoration strain γ_r [1.87].

4. The bend test

To obtain information on the elastic properties of the amorphous ribbons, experiments are carried out using the bend test method. However, in this case, and also in uniaxial tensile loading, the mechanical properties are highly sensitive to the geometry and quality of the sample surface.

The simplest and most readily available method is the one associated with the fixing of the ribbon samples in toroidal devices of different diameter. Recording the residual strain after fixing, calculations are carried out to determine the degree of deformation for the given diameter of the device using an approximate equation.

The method of bending tests of amorphous alloys in the form of ribbons 30–50 μm wide was examined in the greatest detail and most efficiently in [1.88]. The ribbon is placed between two parallel plates of the testing machine (Fig. 1.5).

The sample is then compressed with different strain rates and the *F–D* diagram is recorded; here *F* is the load and *D* is the distance between the plates. It is assumed that the sample has the form of a thin sheet and the bent area of the ribbon forms a cylindrical surface, with the ends of the ribbon tightly contacting the sheets. In addition to this, the external load *F* is uniformly distributed in the width of the ribbon, and along the X_1 axis the load rapidly decreases to almost zero over a short section, starting from point *L*.

The bend test in these experiments is also used to determine the ductility of ribbons of amorphous alloys [1.74, 1.89, 1.90]. There is a simple expression for the ductility parameter ε_f characterising the critical degree of deformation in the surface layer of the ribbon at the moment of fracture [1.89]

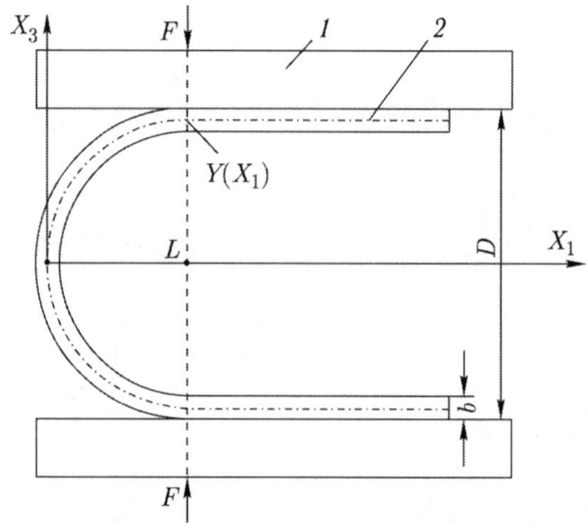

Fig. 1.5. The bend test of a thin ribbon sample: 1) loading plates; 2) the ribbon.

$$\varepsilon_f = b/(D-b), \qquad (1.25)$$

here b is the width of the ribbon and D is the distance between the compressing plates at the moment of fracture.

Examining the expression (1.25) it can be seen that in the absolutely ductile state of the analysed sample when the plates come into contact and the ribbon is folded without fracture, $D = 2b$ and $\varepsilon_f = 1$. If the material is brittle, the embrittlement increases with increase of D. Thus, in transition to absolute brittleness $D \to \infty$ and $\varepsilon_f \to 0$. In practice, experiments of this type are usually carried out in the clamps of a micrometre. It should be taken into account that at every moment of time the bent ribbon should have the strictly cylindrical surface and it is therefore necessary to carry out a large number of experiments in order to prevent the error which may form in this situation.

5. The microindentation method

This method is used widely for determining the mechanical characteristics of hard materials, including thin amorphous ribbons [1.91–1.93] and is based on the Vickers indentation method [1.94].

As shown in [1.95], an important role in the microhardness measurements is played by the relationship between the thickness of the ribbon specimen of the amorphous alloy and the indentation diameter, i.e., the penetration depth of the indentor. In the range of the load and thickness of the ribbon, satisfying the ratio $b/d > 1.6$, where b is the ribbon thickness; d is the indentation diameter, the microhardness values (HV) remain constant and correspond to the true values of the measured parameter. At $b/d < 1.6$ traces of deformation in the form of 'rays' appear on the ribbon side opposite to the indentation. This indicates the 'puncturing' of the indentation. It should be noted that within the framework of this restriction ($b/d > 1.6$) the substrate material does not influence the value of HV in the measurement error range [1.95]. Taking into account the geometry of the indentor pyramid, characterised by the tip angle of 136°, it may be concluded that the thickness of the investigated material should be at least 12 times greater than the maximum permissible penetration depth of the indentor.

The microhardness values greatly differ in indentation of the contact surface (adjacent to the cooling disc in melt quenching) and the free surface of the ribbons [1.96]. To prevent the effect of the surface layer, the difference in the two surfaces of the ribbon, and also the possible 'puncturing' effects it is necessary to use the method of measuring the microhardness in the middle region of the end surface of the ribbon specimens (Fig. 1.6). Measurements are taken after embedding the sample in a compound, grinding and polishing [1.97, 1.98]. However, this method also has limitations, associated with the maximum load and minimum thickness of the sample.

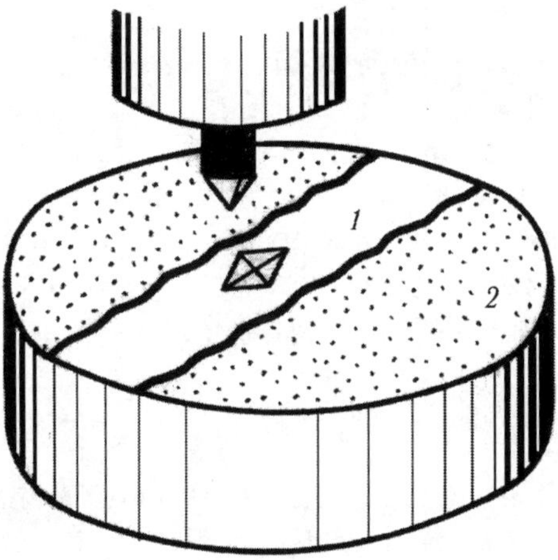

Fig. 1.6. Indentation on the end surface of a ribbon specimen of an amorphous alloy: 1) specimen; 2) the compound.

The indentation of the thin ribbons of the amorphous alloys is accompanied unavoidably by a problem with the selection of the substrate which should, on the one side, ensure tight contact with the investigated sample and, on the other side, cause a minimum error in the experimental results. The effect of the substrate is often ignored and, consequently, the error in the interpretation of the indentation results is larger.

A method of evaluating strength and ductility on the basis of microhardness values was proposed in [1.99]. In the case of the amorphous alloys there is a distinctive linear relationship between hardness and strength: $HV = K\sigma_T$ ($K = 3.2$), as confirmed experimentally by the simultaneous measurement of HV and σ_T using independent methods (uniaxial tension, bending) of iron- and cobalt-based alloys with different composition. Consequently, it is possible to determine the values of tensile strength σ_B and yield strength σ_T with the accuracy of ~10% on the basis of 30–40 indentations for each point using the measured values of HV. This is especially important in cases in which similar estimates using other methods of mechanical testing are not possible (for example, after heat treatment causing embrittlement of the amorphous alloys).

The method for determining the ductility of thin strips of metallic alloys, proposed by the authors of [1.90] and also based on the indentation on the substrate can be used for testing in microregions. The method can be used to record changes in the ductility of the ribbon specimens in the temperature range of transition of the amorphous alloys from the amorphous to crystalline state.

An important advantage of the microindentation method is that it can be used for tests not causing catastrophic macrofracture of the sample. In addition to this, local loading by the Vickers method taking into account special features of the ribbon specimens of the amorphous alloys can be used to evaluate an important quantitative fracture toughness criterion – cracking resistance K_{1c} [1.100], using the semi-empirical relationship [1.101]:

$$K_{1c} = A(E/HV)^{1/2}\, P/C^{3/2}, \tag{1.26}$$

here $A = 0.016$ is the proportionality coefficient; E is the Young modulus; HV is Vickers microhardness; P is the critical load for the formation of radial cracks; C is the length of the radial crack.

The traditional methods of evaluating K_{1c} – three- and four-point bending of the notched specimen, off-centre tensile loading, double torsion and others [1.102], require time-consuming machining operations, special testing equipment, a large number of specimens of complicated shape (with the cut-out layers, holes, notches). The conventional methods of determination of cracking resistance cannot be used for small and thin specimens. Therefore, the microindentation in this sense is a promising mechanical testing method.

6. The tensile test

The mechanical tensile test is used for the determination of special features of fracture, fracture toughness and more accurate values of σ_T [1.103]. The fracture tests are carried out by transverse shear and also tensile loading of flat specimens with a sharp notch or with a central orifice [1.104, 1.105]. Force F is applied to the edges of the notched strip, as shown in Fig. 1.7. The tests are carried out with uniaxial tension recording the load–elongation curve.

The low deformation rate results in the process of steady fracture in which the crack propagates at a constant rate and in the limit at a constant load. High deformation rates are characterised by the 'start–stop' type fracture in which crack propagation takes place in 'jumps' at the appropriate 'teeth' on the fracture curve.

The fracture energy Γ in tests of this type can be described by the equation

$$\Gamma = 2F/t, \tag{1.27}$$

here F is the force required for catastrophic fracture; t is the ribbon thickness.

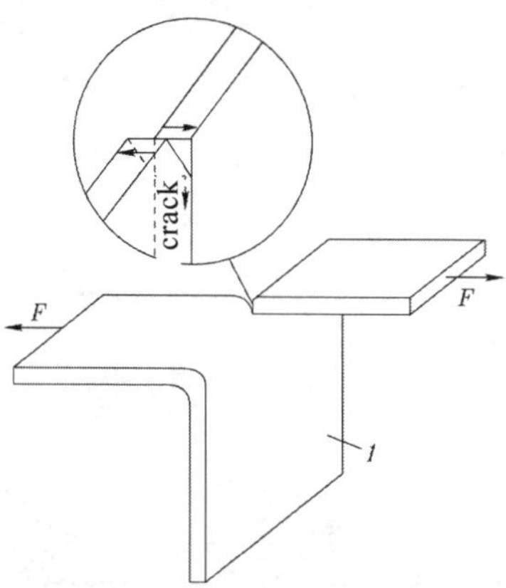

Fig. 1.7. The fracture tests of a strip specimen with a notch (1) for determining impact toughness and the yield strength [1.104].

This equation reflects the fact that the entire work, carried out by the force F, is used for the fracture process: for the formation of the material in the vicinity of the base of the main crack and for the formation of new interfacial surfaces. The typical feature of this test method is the formation of a shear crack. Using the fracture energy, it is possible to estimate the toughness parameters of the amorphous alloy.

As noted previously, the following relationship is fulfilled in this case:

$$\Gamma = \gamma_p + \gamma_s, \tag{1.28}$$

where γ_s is the work used for the formation of a new surface; γ_p is the work used for deformation at the base of the main crack.

Assuming that the plane stress shear fracture forms if there is no strain hardening, the critical work, required for the creation of the unstable fracture conditions can be determined by the following equation

$$\gamma_{pc} = \frac{G_c}{2} = \frac{\sigma_T}{2}\varepsilon_f t k_\alpha, \tag{1.29}$$

here k_α is the stress concentration factor; G_c is the crack opening force.

Equation (1.29) can be used to determine the value of the ideal yield strength σ_T, and using the well-known Irwin equation [1.106] the impact toughness:

$$K_{1c} = (\mu G_c)^{1/2} \tag{1.30}$$

The method of fracture of the notched ribbon specimens [1.105] can be used to estimate not only the rate dependence of the energy and nature of fracture but also the temperature dependence of the fracture energy Γ: the value of Γ increases quite rapidly in transition to the range of the negative test temperatures.

1.3.2. Strength

A unique phenomenon was observed when examining the strength of amorphous alloys: the strength of these alloys is close to theoretical, i.e., $\sigma_T \approx G/30$. If it is assumed that the real strength of the crystals does not correspond to the theoretical strength because of the presence of dislocations, the strength of the amorphous alloys, close to theoretical, is associated with the fact that there are no dislocation defects in the non-crystalline structure or, at least, their Peierls barrier is very large. The high value of the ratio σ_T/G is obtained in the amorphous alloys on the one hand higher in comparison with the value of σ_T of the crystals above, on the other hand, with the lower (by approximately 30–50%) value of G. The low elasticity modulus values (Table 1.1) reflect in fact the absence of ordering of internal atomic shifts which is typical of the amorphous state and which take place during the passage of an elastic pulse through the amorphous matrix. The highest shear modulus of the amorphous alloys of the metal–metalloid type was recorded for the materials containing boron as metalloid atoms (Table 1.1)

The amorphous alloys can be regarded as an ideal elastoplastic material with very low strain hardening. Consequently, the tensile strength of the amorphous alloy is equal to the yield strength [1.107].

1. Effect of composition and parameters of superfast quenching
The values of the elastic modulus and strength of the amorphous alloys depend strongly on the chemical nature of the components. For example, for the amorphous alloys with the dominance of the simple metalloid (for example, phosphorus), the change of the type of atoms of the metallic base increases the values of the Young modulus E in the following sequence: Pd – Ni → Co → Fe [1.108]. The type of the metalloid atoms is also very important, and this is associated evidently with the electronic structure of the amorphous alloys.

Attention should be given to the following relationship: in alloying the amorphous alloy with the metallic element, the Young modulus increases (or decreases) in accordance with the sequence of the increase (or decrease) of the crystallisation temperature of the amorphous alloy. This is clearly indicated by the example of the same Zr–Si alloy, doped with a number of transition metals (Fig. 1.8). The value of T_k increases in the same sequence: Ti→Nb→V→Ta→Mo [1.109].

The strength and hardness of the amorphous alloys increase with the increase of the difference of the numbers of the groups or numbers of the periods of the main metallic components and other elements of the amorphous alloy (Fig. 1.9).

At the same time, the tendency for embrittlement as a result of the increase of the strength of the bonds between the atoms of the elements included in the composition of the alloy [1.110] becomes stronger. The strength of the amorphous alloys based on nickel

Table 1.1. Values of the Young modulus *E*, shear modulus *G*, the bulk elasticity modulus *B* and the Poisson coefficient ν of some amorphous alloys

Alloy	*E*, GPa	*G*, GPa	*B*, GPa	ν
$Fe_{80}B_{20}$	170	65	140	0.30
$Fe_{75}P_{15}C_{10}$	150	–	–	–
$Fe_{78}B_{10}Si_{12}$	120	–	–	–
$Fe_{75}P_{16}B_{6}A_{13}$	130	–	–	–
$Fe_{75}P_{16}C_{5}A_{13}Si_{2}$	85	58	–	–
$Fe_{74}P_{16}B_{7}A_{13}$	100	–	–	–
$Ni_{76}P_{24}$	95	35	110	–
$Ni_{78}Si_{10}B_{12}$	80	–	–	–
$Ni_{74}P_{16}B_{7}Al_{3}$	86	–	–	–
$Ni_{49}Fe_{29}P_{14}B_{6}A_{12}$	130	54	–	–
$Ni_{36}Fe_{32}Cr_{14}P_{12}$	140	–	–	–
$Co_{85}P_{15}$	120	–	–	–
$Co_{74}Fe_{6}B_{20}$	179	68	166	0.32
$Co_{73}Si_{15}B_{12}$	90	–	–	–
$Pd_{80}Si_{20}$	68	35	180	0.40
$Pd_{77.5}Cu_{6}Si_{16.5}$	90	32	168	0.40
$Pd_{77.5}Ag_{6}Si_{16.5}$	85	30	164	0.40
$Pd_{77.5}Ni_{6}Si_{16.5}$	90	32	170	0.40
$Pd_{64}Ni_{16}P_{20}$	92	33	166	0.40
$Pd_{64}Fe_{16}P_{20}$	93	33	162	0.40
$Pd_{60}Ni_{15}P_{25}$	96	34	202	0.42
$Cu_{50}Zr_{50}$	132	–	–	–
$Nb_{50}Ni_{50}$	100	–	–	–
$Ti_{85}Si_{15}$	90	–	–	–
$Zr_{85}Si_{15}$	81	–	–	–
$Zr_{55}Mo_{30}Si_{15}$	125	–	–	–

increases with increase of the concentration of carbon, chromium, molybdenum or tungsten, reaching the maximum value in the Ni–Cr–Mo–C or Ni–Cr–W–C systems [1.111]. The strength of the Fe–B system increases with increase of the boron content up to 25 at.% [1.112].

In the amorphous alloys based on the Fe–Ni–B system the strength increases with increase of the iron and boron content. The replacement of boron with silicon results in a decrease of the strength

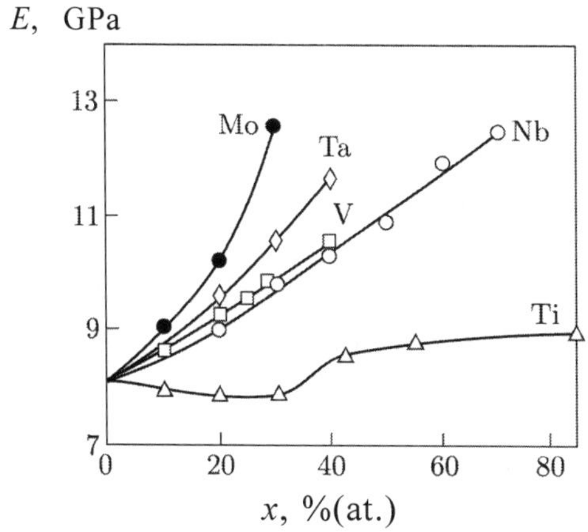

Fig. 1.8. Effect of alloying elements on the value of the Young modulus of the amorphous alloys $Zr_{85-x}Me_xSi_{15}$ (here *Me* is Ti, V, Nb, Ta and Mo).

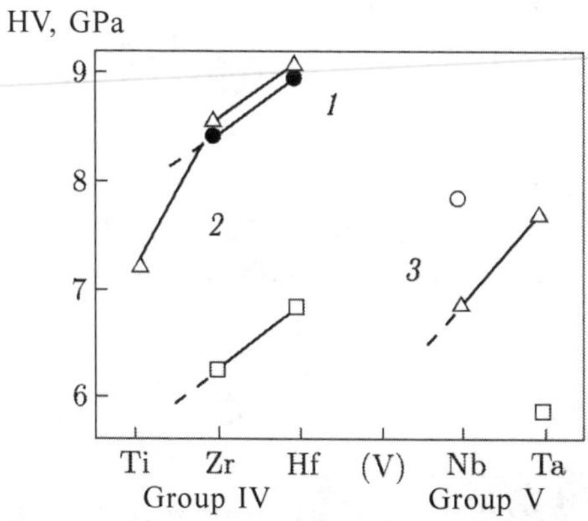

Fig. 1.9. Variation of hardness for the $Fe_{85}Me_{12}B_6$ (1), $Co_{82}Me_{12}B_6$ (2) and $Ni_{82}Me_{12}B_6$ (3) alloys (*Me* – Ti, Zr, Hf, Hb, Ta) in dependence on the position of the alloying element in the periodic table of elements.

[1.113]. In the $Fe_xNi_{80-x}P_{14}B_6$ alloys the hardness reached 9000 MPa, and in the $Fe_xNi_{80-x}B_{20}$ systems it was equal to 13 000 MPa [1.114].

An important parameter for the strength and other characteristics is the quenching rate from the melt in the range of the same interval in which the total amorphisation takes place. For the spinning process (the quenching of the melt jet on a rapidly spinning cooling disc) the quenching rate can be qualitatively evaluated on the basis of the thickness of the ribbon of the amorphous alloy: as the thickness of the ribbon increases, the effective quenching rate from the melt decreases.

The strength of the amorphous alloys also changes in dependence on the spinning conditions. The quenching rate and the superheating temperature strongly influence also the type of short-range order in the initial (quenched) condition which in turn determines a number of the mechanical and physical properties of the amorphous alloys. In addition to this, the production parameters also have a strong effect on the nature of variation of the mechanical and physical properties and subsequent processing of the amorphous alloys, in particular, in heat treatment [1.115].

2. Effect of temperature and strain rate

At relatively high temperature corresponding to inhomogeneous deformation, there is no dependence of σ_T on temperature (or, more accurately, the dependence of σ_T in this temperature range is identical with the dependence of *G*) (Fig. 1.10) [1.9]. In the homogeneous deformation range (at temperatures higher than $0.8T_k$) there is a strong temperature dependence of σ_T leading to a large decrease of the strength of the amorphous alloys at temperatures close to T_k (Fig. 1.10).

Completely different results are obtained in the range of low temperatures. As shown by the results in [1.116], the microhardness of the amorphous alloys based on iron, cobalt and Fe–Ni shows a relatively strong temperature dependence at temperatures lower than 200 K. As indicated by Fig. 1.11, the value of *HV* rapidly decreases when the temperature is reduced down to the boiling point of liquid nitrogen (77 K).

The rate dependence of the deformation stresses has been studied less extensively. Figure 1.12 shows the dependence of σ_B of two amorphous alloys on the strain rate in uniaxial tension. It may be seen that in the rate range 10^{-5}–$2\cdot10^{-4}$ s^{-1} the value of σ_B for the $Fe_{40}Ni_{40}P_{14}B_6$ alloy is almost completely independent of the deformation rate, and at a higher rate it starts to rapidly decrease. At the same time, for the $Fe_5Co_{75}Si_{13}C_7$ alloy the value of parameter σ_B decreases monotonically with increase of the deformation rate [1.117].

When discussing the question relating to the effect of temperature and strain rate on the strength of the amorphous alloys, attention must also be given to the results obtained in [1.73]. The excess free volume (EFV) is regarded as a structural defect, and the removal of this defect does not lead to any change of the nature of the symmetry and topological characteristics of the amorphous state. However,

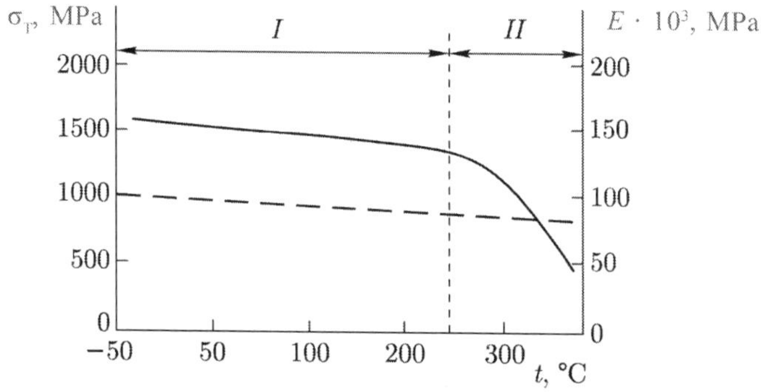

Fig. 1.10. Temperature dependence of the Young modulus and yield strength for the $Pd_{78}Cu_6Si_6$ alloy in the range of higher temperatures: I) the region of inhomogeneous deformation; II) the region of homogeneous deformation.

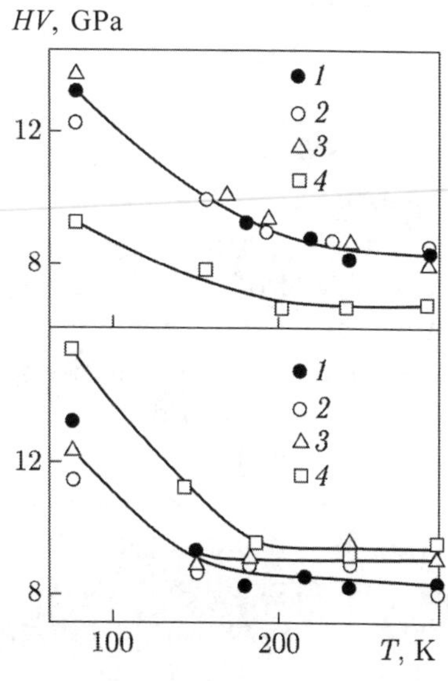

Fig. 1.11. Temperature dependence of the hardness of the amorphous alloys based on iron (*a*) and cobalt (*b*) in the range 77–300 K: *a* – $Fe_{84}B_{16}$ (1); $Fe_{63}Co_{20}B_{17}$ (2); $Fe_{70}Ni_8Si_{10}B_{12}$ (3); $Fe_{40}Ni_{36}Mo_4B_{18}$ (4); b – $Co_{68}Fe_5Si_{15}B_{12}$ (1); $Co_{60}Fe_5Ni_{10}Si_{10}B_{15}$ (2); $Co_{67}Fe_4Cr_7Si_8B_{14}$ (3).

this mobile compound part of the free volume may be responsible for the structural rearrangement in the change of the physical and mechanical properties of the amorphous alloys in the process of structural relaxation and, possibly, the early stage of crystallisation. The authors have shown that both the intensity and the slope of the curve decrease, indicating the decrease of the volume of the scattered micropores. Figures 1.13 and 1.14 show the data for the variation of the volume of this fraction of the pores in dependence on the annealing temperature and time.

It may be seen (Fig. 1.13) that the volume of the micropores rapidly decreases in the temperature range 450–500 K and the time

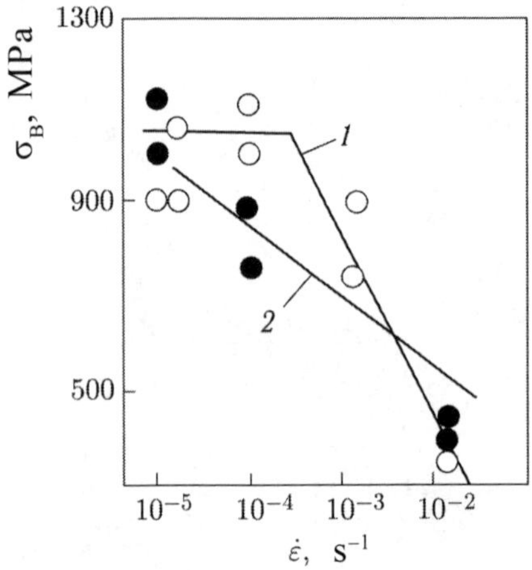

Fig. 1.12. Dependence of σ_B on the strain rate for the $Fe_{40}Ni_{40}P_{14}B_6$ (1) and $Co_{75}Fe_5Si_{13}C_7$ (2) alloys.

10^3–10^4 s (the volume is approximately halved). Thus, annealing at relatively moderate temperatures (below the glass transition temperature) results in a large decrease of the volume of the micropores and, therefore, the rate of low-angle scattering.

Figure 1.15 shows the temperature dependence of the strength of the $Fe_{77}Ni_1Si_9B_{13}$ alloy. It may be seen that the dependence is quite complicated and can be conventionally divided into three sections.

In two regions (18–100°C) and (200–350°C) the strength decreases linearly (or quasi-linearly) with increasing temperature. The temperature range from 100–150 to 200°C is characterised by the anomalous dependence of strength. When part of the EFV (excess free volume) is removed by long-term annealing at elevated temperature or under the effect of high pressure, the strength decreases linearly with increasing temperature and there is no anomalous temperature dependence of strength; the strength increases in the entire investigated temperature range. Two conclusions can be drawn on the basis of the results:

1. The decrease of the EFV as a result of the application of high (0.7 GPa) pressure or annealing at 240°C for 5.5 h increases the strength in the entire investigated temperature range.

2. The anomalous increase of strength in the tests with a constant loading rate may be associated with the fact that loading activates the process of removal of part of the EFV already at 150–200°C; the latter also results in an anomalous increase of strength at these temperatures.

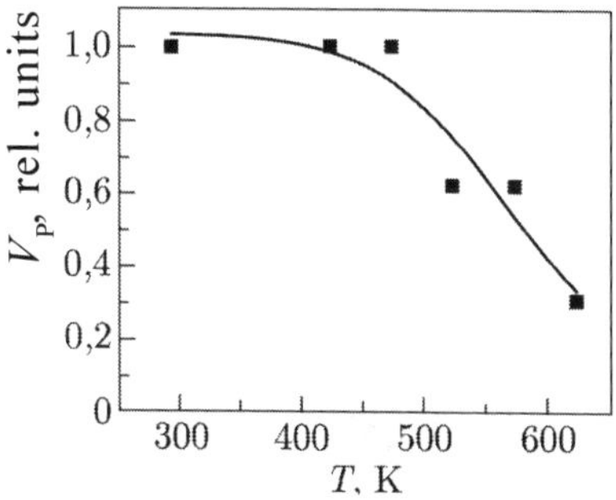

Fig. 1.13. The curve of the isochronous (t = 1800 s) annealing of pores in the $Fe_{77}Ni_1Si_9B_{13}$ alloy.

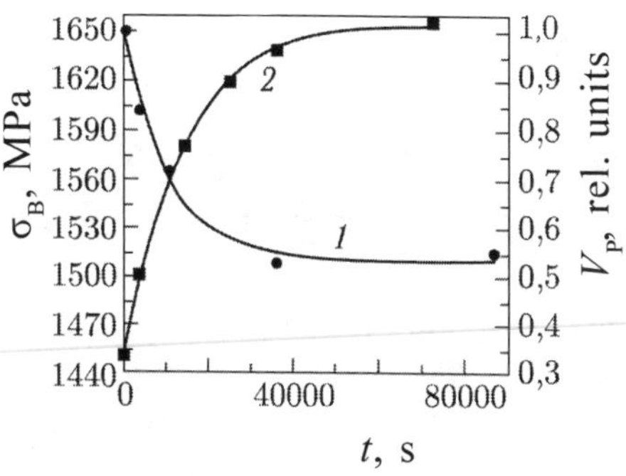

Fig. 1.14. The curve of isothermal (240°C) annealing of pores in the $Fe_{77}Ni_1Si_9B_{13}$ (1) alloy and the change of the fracture strength of the specimens after annealing (2) [1.73].

On the basis of these results it can be concluded that the excess free volume, localised in the form of micropores, has a strong effect on the strength characteristics of the amorphous alloys, and the effect of porosity on strength is relatively general for a wide range of the materials [1.118].

In [1.27] it was reported that the application of external hydrostatic pressure increases the strength and microhardness of amorphous alloys on average by 7–9%. For example, in the samples of $Co_{60}Fe_{10}Si_{15}B_{10}$ alloy the strength at 18 and 100°C was 1500 and 1400 MPa, respectively, and the strength of the same samples after the effect of a pressure of 1 GPa at these temperatures increased to 1610 and 1520 MPa.

3. Effect of structural relaxation on the properties of amorphous alloys

Structural relaxation has a strong effect on many properties of the amorphous alloys. The strength characteristics are an exception in this sense. Changes in the structure, caused by low-temperature annealing of the amorphous alloys, are associated with the atomic rearrangement without diffusion over long distances. This process

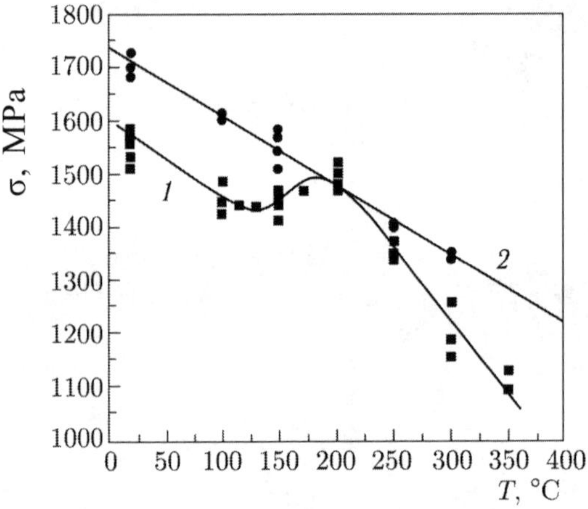

Fig. 1.15. Temperature dependence of the samples of the $Fe_{77}Ni_1Si_9B_{13}$ alloy. The initial specimens (1) and the specimens subjected to the effect of a pressure of 0.7 GPa at 18°C (2) [1.73].

is accompanied by the removal of residual stresses and annihilation of the excess free volume,

The most characteristic cases of changes of the strength or hardness in dependence on the preliminary annealing temperature in the structural relaxation stage are shown in Fig. 1.16. The curves in Figs. 1.16 *a* and *b* are usually recorded for the amorphous alloys of the metal–metalloid type, and the curve in Fig. 1.16 *c* in the metal–metal amorphous alloys.

The amorphous alloys of the Fe–Ni–P and Fe–Ni–P–B systems show two extremes in the dependence of microhardness (strength) on annealing temperature [1.57] (Fig. 1.16 *a*), referred to as the effects of low-temperature and high-temperature hardening, respectively. The low-temperature hardening effect (LTHE) for the amorphous alloys of different composition is observed to different degrees in the temperature range 50–150°C, whereas the high-temperature hardening effect (HTHE) is distinctive in all amorphous alloys of the metal–metalloid system. The correlation of the temperature range of manifestation of the HTHE with the crystallisation temperature of a specific alloys has also been observed. The experimental results show that the magnitude of the peak and the temperature range of the manifestation of HTHE depend strongly on the chemical composition of the alloy, the production parameters of the alloy and the duration of the thermal effect. For example, the excess pressure of the jet of the melt which does not have any significant effect on the value

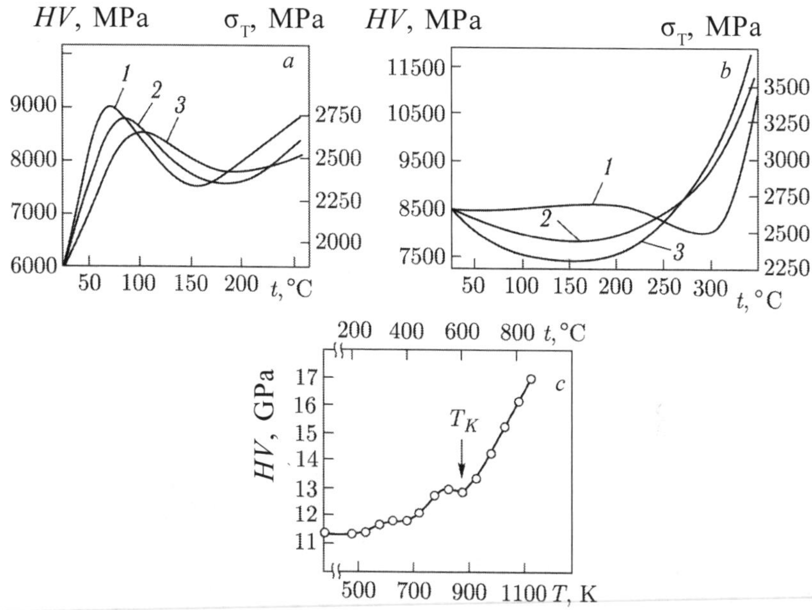

Fig. 1.16. Typical dependences of hardness on the temperature of preliminary annealing for the Fe–Ni–P (*a*), Fe–P (*b*) and Ni–Ta (*c*) amorphous alloys. Annealing time, h: 1) 10; 2) 1; 3) 0.1.

of HV or σ_T has, in the initial condition, a strong effect on the temperature of isothermal annealing, corresponding to the maximum of the LTHE for each alloy, and also on the magnitude of the peak. The increase of excess pressure at any annealing time increases the minimum annealing temperature required for the occurrence of LTHE and also increases the maximum increase of the strength as a result of the LTHE in relation to the initial condition of the amorphous alloy (this effect is most distinctive at the minimum annealing time). The temperature range of the manifestation and magnitude of the HTHE are influenced only slightly by the excess pressure of the melt jet. It is important to note the opposite (in comparison with the LTHE) nature of the effect of the excess pressure on the magnitude of the HTIHE: with increase of the excess pressure the temperature range of manifestation of the HTHE is displaced in the direction of low temperatures. In addition to this, it should be noted that the increase of the excess pressure limits the effect of the duration of isothermal annealing on the maximum increase of the strength in relation to the initial condition in the case of both LTHE and HTHE.

Evidently, deformation in the elastic range of the amorphous alloy in the initial condition stimulates structural relaxation at room

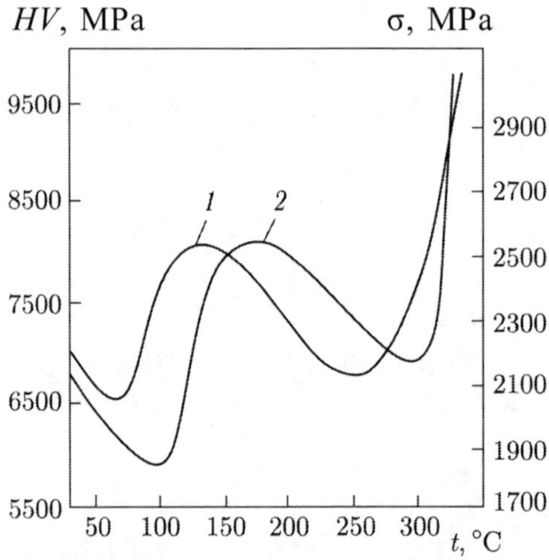

Fig. 1.17. Effect of preliminary elastic deformation on the characteristics of low-temperature hardening of the $Fe_{65.5}Ni_{17}P_{17.5}$ amorphous alloy: 1) initial condition; 2) after elastic deformation by uniaxial tensile loading at room temperature.

temperature to a considerably greater degree than the diffusion processes. This results in the annihilation of part of the defects which provide a certain contribution to the process of heterogeneous plastic flow, as indicated by the displacement of the range of occurrence of LTHE to higher temperatures (Fig. 1.17).

1.3.3. Plasticity and plastic deformation mechanisms

Special features of the amorphous metallic state include the presence of the Newtonian viscosity in a wide temperature range because the amorphous alloys by analogy with the crystalline materials are plastically deformed at the critical shear stress.

One of the important properties of the amorphous alloys is their plastic flow capacity [1.1]. Usually, the plastic deformation is regarded as synonymous with the nucleation, multiplication and annihilation of dislocations moving in the solid. However, in the amorphous solid bodies there is no translational symmetry and, consequently, no dislocations in the classic concept of the nature of these defects. Consequently, the amorphous solid should be regarded as absolutely brittle. In particular, this situation is characteristic of the inorganic glasses, although they show features of very slight plastic flow. However, plastic deformation does take place in the amorphous alloys, resulting in an anomalously high strength which is found in the non-crystalline solids if brittle fracture of the solids is prevented at stresses considerably lower than the yield limit. The capacity of amorphous alloys for plastic flow (by which they differ from other amorphous solids) is undoubtedly associated with the collectivised metallic nature of the atomic bonds at which the processes of collective atomic displacements can take place quite easily [1.6].

As in the crystals, the deformation in the amorphous alloys can be divided into elastic, inelastic and plastic, depending on the degree of reversibility with time. Elastic deformation is completely and instantaneously reversible after removing the load, inelastic deformation is completely reversible with time and, finally, plastic deformation is irreversible with time after removing the external load.

Two types of plastic flow are found in the amorphous alloys depending on temperature: homogeneous deformation and inhomogeneous deformation (Fig. 1.18).

The critical temperature of transition from one type of plastic flow to another (the equicohesion temperature T_e) depends, for the specific composition of the amorphous alloy, on the strain rate. The transition with increase of the temperature to the Newton viscous flow can be explained by the strong temperature dependence of the value σ_T for the viscous flow which below the critical temperature T_e also exceeds the value of σ_T for the inhomogeneous flow. In turn, the strong dependence of σ_T for the viscous flow is determined by the large reduction of the shear viscosity of the amorphous systems in the range of T_g [1.99].

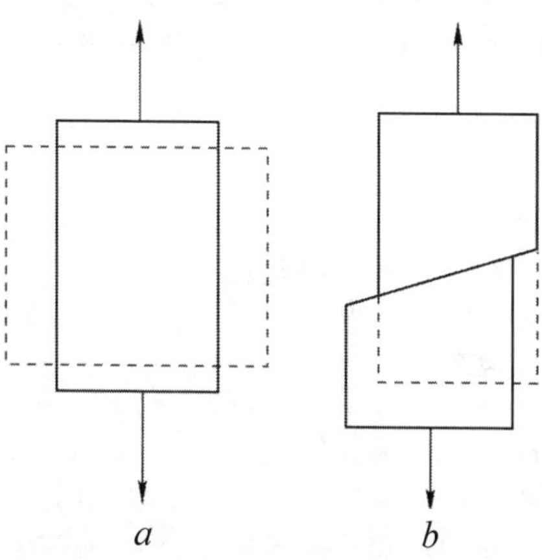

Fig. 1.18. The diagram explaining homogeneous (*a*) and inhomogeneous (*b*) plastic deformation of the sample.

Homogeneous deformation is observed in the amorphous alloys at high temperatures (slightly lower than T_g or higher) and at low values of the applied stress ($\tau < G/100$). In homogeneous plastic deformation, each element of the solid body undergoes plastic shape changes because the homogeneously loaded specimen is subjected to homogeneous deformation (Fig. 1.18 *a*).

Usually, the investigations of the homogeneous flow are carried out in the creep or stress relaxation experiments. At low applied stresses, the homogeneous flow is governed by the law of flow of the Newtonian liquid: $\dot{\gamma} \approx \tau$. In this case, the shear viscosity can be determined using the equation

$$\eta = \tau\dot{\gamma}, \qquad (1.31)$$

where τ is the applied shear stress; $\dot{\gamma}$ is the strain rate.

The homogeneous deformation in the amorphous alloys greatly differs from the viscous flow in the crystals. The main difference is that the stage of steady-state creep is not reached in the amorphous alloys and the strain rate decreases continuously with time. At the same time, the value of the effective viscosity, calculated for each moment of time using equation (1.31), continuously increases. The important result is the fact that a similar increase of viscosity takes place with time in accordance with a linear law, in both the freshly-quenched and relaxed states [1.119].

The degree of plastic deformation which can be realised in the homogeneous flow is in principle not limited. In the case of the accurate selection of the temperature, the values of the applied stress and the conditions of preliminary heat treatment of the amorphous alloys of the given composition it is possible to obtain the macroscopic degree of deformation of the order of hundreds of percent.

At present time, there are several laws indicating that in certain conditions some amorphous alloys become superplastic [1.120–1.122]. The superplastic effects in the alloys of the Ti–Ni–Cu [1.121] and Pd–Cu–Si [1.122] systems were observed in mechanical tensile tests and thermal expansion, respectively. The experimental results show that the superplasticity is observed in the glass transition temperature range. In these experiments, a qualitative model of the observed effect was proposed which can be described as follows. The glass transition process in heating is accompanied by the 'defrosting' of the structural defects – the areas in which the excess free volume is greater than the specific critical value [1.123]. These deffects are responsible for the elementary shear acts, and the large increase of the mobility of these defects in the glass transition range determines the decrease of the shear viscosity of the material and, consequently, the increase of plasticity.

The inhomogeneous flow is observed at low values ($T < 0.8T_g$) and high applied stresses ($\tau > G/50$). This type of flow is observed in tensile loading, compression, rolling, drawing and other deformation methods of ribbon, wire and thick specimens of amorphous alloys. In inhomogeneous plastic deformation, the plastic flow is localised in discrete thin shear bands and the residual volume of the solid remains undeformed [1.1] (Fig. 1.18 *b*).

The characteristic feature of the inhomogeneous plastic deformation by which it greatly differs from the homogeneous deformation is that the inhomogeneous flow results in a decrease (not increase) of the degree of order in the amorphous matrix.

In inhomogeneous plastic deformation, shear steps form on the surface of the samples. These steps correspond to the exit of the shear bands to the surface. They are usually distributed under the angle of 45–55° in relation to the axis of uniaxial tensile loading (compression) or the rolling direction, and are also parallel to the bending axis. The height of the steps above the surface of the specimens reaches 0.1–0.2 μm, and the thickness of the individual shear bands does not exceed 0.05 μm. Consequently, the amorphous alloys are characterised by the very high local plasticity in the region of inhomogeneous deformation.

The structure of the amorphous alloy in the regions, affected by the inhomogeneous plastic flow, is characterised by the changes of the structure because the presence of the shear bands in the amorphous matrix decreases their strength, facilitating further plastic flow in these regions: the flow of the material is easier in the areas where the local shear has already taken place, and not in the areas where the nucleation of a new shear band is required.

One of the methods of investigating the structure in the shear bands is transmission electron microscopy (TEM). The area of the investigated foil, corresponding to the shear step, has a different effective thickness in comparison with the region of the surrounding non-deformed matrix and, consequently, the contrast on the electron microscopic image of the shear band is of the absorption nature (Fig. 1.19).

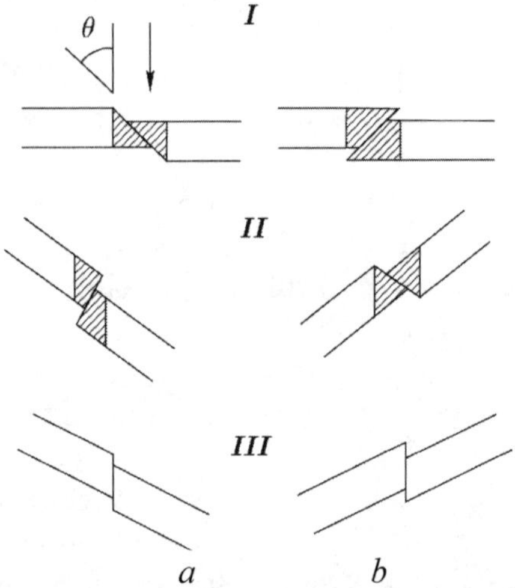

Fig. 1.19. Diagram of the formation of contrast in investigation of the amorphous alloy by the TEM method on the shear bands at different positions of the bands in relation to the incident electron beam in the case of the effect of tensile (a) or compressive (b) stresses: I) $\theta < 180°$; II) $\theta > 180°$, III) $\theta = 180°$.

In cases in which the angle θ between the directions of the incident electron beam and the effective shear, carried out by the band, is smaller than 180° (I in Fig. 1.19 *a* and II in Fig. 1.19 *b*), the shear band is observed in the form of a light band on a dark background (Fig. 1.20 *a*). If the angle θ is greater than 180° (II in Fig. 1.19 *a* and I in Fig. 1.19 *b*), then it is observed in the form of a dark band on a light background (Fig. 1.20 *b*).

The dark field images obtained in the examination of the plastically deformed Fe–Ni–P alloy [1.124] can be used to explain a number of morphological special features of the shear bands: they are 'non-crystallographic', i.e., they easily change the local plane of their orientation and are characterised by the presence of the branching points.

In electron microscopic studies it is important to pay special attention to the problem of the determination of the thickness of the shear bands. In the diagram shown in Fig. 1.19 to simplify examination the shear band is represented as a flat formation of zero thickness.

The shear band with the smoothly changing orientation in transition at certain points through the position corresponding to the angle θ equal to 0 or 180° is observed on the electron microscopic

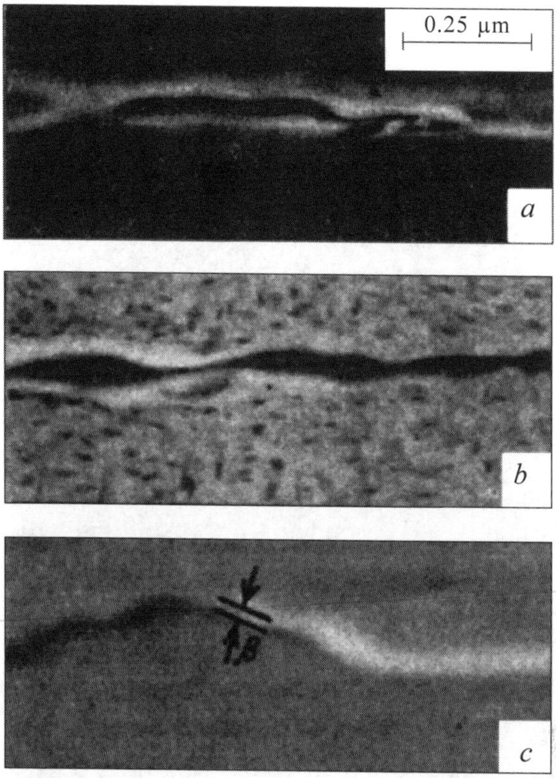

Fig. 1.20. Electron microscopic image of the shear bands, formed during deformation at room temperature in the $Fe_{65}Ni_{17}P_{18}$ alloy: *a*) $\theta < 180°$; *b*) $\theta > 180°$; *c*) $\theta \approx 180°$; β is the measured thickness of the shear band [1.124].

images in the form of a region with light contrast changing to a dark contrast region or vice versa (Fig. 1.20).

In this case, the transition of this type should be accompanied by narrowing of the local region of the contrast extending to the constriction (Fig. 1.21 *a*). The individual areas of the shear band are observed in the form of very narrow regions because at these points the local orientation of the band is close to the position corresponding to $\theta = 0$ or 180°. If the shear band has some thickness (not zero thickness), the variation of the nature of contrast should be accompanied by the formation of a constriction (Fig. 1.21 *b*), with the width of the constriction corresponding in fact to the thickness of the given shear band (or, more accurately, it can be used to determine the maximum values of this parameter). The accuracy of

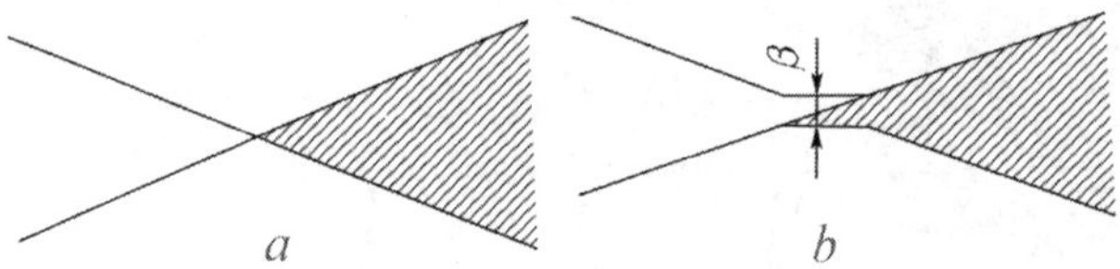

Fig. 1.21. The diagram explaining the determination of the thickness of the shear band β.

measurements on the electron micrographs is not high and requires carrying out a large number of measurements.

In [1.125, 1.126] investigations were carried out to study the relationships governing the development of the shear bands in nano-indentation of the surface and conclusions have been made regarding the nature of formation of the indentation of the indentor:

1. There is a specific stage of deformation of the surface of the amorphous material by the indentor;
2. In the initial stage of plastic deformation there is a range in which the shear bands do not form;
3. When the shear band forms, the contact pressure in the indentation of the indentor is small, but the rate of penetration of the indentor into the amorphous material rapidly increases.

The authors of [1.125, 1.126] assumed that the shear bands support the transfer of elastic energy from the indentation of the indentor which results in the unique short-term decrease of the resistance of the amorphous matrix to the penetration of the indentor.

1. Dislocation mechanisms of plastic deformation

To describe the elementary acts of plastic deformation we should assume undoubtedly that, as in the crystals, the plastic deformation process takes place by the nucleation of interaction and annihilation of defects typical of the amorphous state. Obviously, these defects should enable both the macroscopic and microscopic description of all the special features of the mechanical behaviour of the material. At the same time, they should be characterised strictly at all variants of the non-crystalline structure, and they should be experimentally identified in order to obtain information on the evolution of such defects in all stages of the plastic flow.

The high-intensity plastic flow in the amorphous alloys is observed in a specific temperature range in the form of thick shear bands. This fact enabled some investigators to conclude that to describe plastic deformation it is completely correct to introduce dislocation considerations.

According to the Gilman dislocation model [1.127], the plastic flow takes place by the movement of dislocations with the Burgers vector changing along the dislocation line both in the magnitude and direction. In addition to this, the movement of these dislocations in the amorphous matrix is accompanied by the change of the magnitude and direction of the shear vector at every point of the dislocation line. Consequently, the moving dislocation leaves a high density of the defects of the types of dislocation loops compensating the discrepancy vector of the sliding dislocation. These defects soften the region of the amorphous matrix in the slip plane. The formal reason for the introduction of the dislocation considerations is the observation of distinctive slip steps ('macrodislocations') on the surface of the deformed specimens and also the detection of the shear bands breaking up inside the amorphous matrix. This breaking of the amorphous bands or, more accurately, the regions between the areas where the shear did and did not take place, can be described by the Volterra dislocations. Another reason favouring the dislocation model is the unstable nature of creep in the amorphous alloys. The dislocation model predicts in this case the movement of a group of dislocations forming the shear front. Regardless of its conventional nature, the dislocation model has an evident advantage – it uses efficiently the mathematical facilities of the theory of dislocations for the amorphous alloys.

The dislocation model describes accurately many special features of the mechanical properties of the amorphous alloys, in particular, the experimental values of E/σ_T, and the data for the stress relaxation [1.128].

A number of the plastic deformation models of the amorphous alloys are based on the disclination description of the structure of these materials. As in the case of the purely dislocation considerations, the advantage of the disclination models of the structure is the explanation of the special features of the mechanical behaviour of the amorphous alloys: combination of high strength with the capacity for plastic shape changes, localisation of the plastic flow in the narrow shear bands, absence of any distinctive deformation hardening, the presence of the acoustic emission spectra in the mechanical tests, etc.

It should be stressed that one of the fundamental properties of the disclinations is the capacity for conservative movement accompanied by the emission (absorption) of the dislocations. The non-conservative movement of the disclinations is possible only at high temperatures and is associated with the formation of a large number of point defects. This shows that at low temperatures (below $0.6T_c$) the disclinations are sources of dislocations of almost unlimited power. As shown by the calculations, in movement of a single wedge disclination it is necessary to apply the external stress equal to the Peierls barrier in magnitude.

A number of models of the plastic flow of description of the amorphous alloys is based on the disclination–dislocation models. They are based on the models of the amorphous state, produced by adding a three-dimensional network of disclination dipoles into the ideal cubic [1.129] and hexagonal close-packed (HCP) lattice [1.130]. The model proposed by Morris [1.129] is most complete in describing the mechanical properties of all the disclination–dislocation models of the amorphous alloys.

Thus, the dislocation considerations provide an almost satisfactory description of the nature of the localised plastic flow, observed in the amorphous alloys, and also the nature of buildup of plastic shear in the bands, the ordering in the regions of the plastic flow, the values of the strength in relation to the shear modulus and a number of other special features of the mechanical behaviour of the amorphous alloys. At the same time, computer experiments, simulating the movement of the dislocations in the amorphous structure, indicate, with a small number of exceptions (see chapter 1), the instability of the dislocation defects which break up during their movement into a set of local shear processes not correlated with each other. The latter is completely evident because the dislocation is a crystal defect, and the loss of the translation elements of symmetry should result in the loss of the linearly correlated shear. In addition to this, the dislocations investigated in different previously described models, are 'invisible'. They disappear when the external load is removed so that it is not possible to carry out experimental examination of the defects of this type. At present time, there are no experimental confirmations which would indicate that the dislocations in the amorphous solid are the elementary carriers of plastic shear.

Consequently, the plastic flow process, being a shear process on the macrolevel, is not necessarily reduced to the movement of the individual dislocations but can be described quite efficiently on the basis of the dislocations considerations, if to facilitate examination we introduce the effective dislocation-like defects characterised by a specific (in some cases, fluctuating) Burgers vector and capable of providing the quantitative description of the shear macroscopic processes.

The disclination mechanism of plastic deformation has also been confirmed in computer experiments carried out by the authors in [1.7] where attention was given to the evolution of the structure of the Fe–50% Mn two-component amorphous alloy at high plastic strains. These experiments were carried out under cyclic boundary conditions. The simulation crystal was in the form of a cube consisting of 2000 atoms of manganese and iron, and the atoms in the initial condition were ordered in the FCC lattice.

Analysis of the computer films indicates that the disclination mechanism of plastic flow may take place at high plastic strains. The results confirm the efficiency of using computer experiments for examining the processes of deformation and fracture of the amorphous alloys.

2. 'Atomic' plastic deformation mechanisms

In studies of the structure of amorphous alloys it was observed that the free volume is of considerable importance for understanding the plastic deformation processes. The free volume describes the part of the amorphous matrix with the low atomic coordination in comparison with the other part of the matrix, characterised by the random high-density packing and the same chemical composition. Actually, in the free volume areas the mechanical interaction with the adjacent atoms is relatively weak because the inelastic relaxation becomes possible by means of the local atomic rearrangement without any significant effect on the neighbours. In examining the kinetics of the linear viscoelastic behaviour of glasses of all types at low stresses it was established that these processes of local relaxation are not monoenergetic processes and are characterised by a range of activation energies, depending on the nature of the local coordination in the given section of the free volume.

The Spaepen–Taub model [1.131] is a development of the free volume model with respect to the equilibrium viscous flow. In this model it is assumed that there must be some critical value of the fluctuation of the free volume for carrying out the annihilation of this volume in the local region of the amorphous matrix. The kinetics of the relaxation processes within the framework of this model is determined by the rate of decrease of the size of the free volume. The rate of change of the value of the average free volume per atom is determined by the annihilation of the free volume in each position per unit time and the fraction of the potential annihilation areas. The viscosity is determined in this case as a function of the free volume so that it is possible to carry out experimental verification of the conclusions of the theory. For example, the experiments confirmed the theoretically predicted linear increase of the effective viscosity in relation to the isothermal annealing time [1.38].

Analysis of this model shows that the microflow and relaxation require the component fraction of the atomic size with similar fluctuations of the free volume. The magnitude of the activation barrier for the relaxation (annihilation) of the free volume is smaller than the appropriate value for the iso-configuration flow. Therefore, it was concluded that there are 'jumps' of the atoms where a jump is accompanied by the collapse of the neighbouring atoms around the area released by the 'jumping' atom.

Another important postulate is the assumption according to which the dilation at the stress raisers (microcracks, steps, inclusions, etc) results in a large decrease of the local viscosity. This is very important for examining the conditions of nucleation of the shear bands. The nucleation of a shear band is followed by a gradual softening of the matrix in the shear zone at the rate which increases with increase of the extent to which the process of structural relaxation can restore the initial structure [1.39]. In this model, the softening is caused by the formation of the free volume under the effect of shear as a result of the pressure of the atoms on their neighbours outside the shear band at a relatively high stress level. This process is balanced by the ordering accompanied by the annihilation of the free volume. The processes of the nucleation and annihilation result in the establishment of the stable state in which the free volume and, consequently, viscosity, depend strongly on the level of the applied stresses. This pattern differs from the homogeneous plastic flow in which the viscosity is determined by the thermal prior history or the equilibrium structure of the material. In principle, the model can be used to explain not only the pattern of the inhomogeneous flow but also the slight sensitivity of these processes to the strain rate. In addition, the model can be used to determine the temperature boundary of transition from inhomogeneous to homogeneous deformation.

In the model proposed by Argon [1.132] attention is given to the process of plastic deformation at temperatures lower than T_c, based on two types of thermally activated shear transformations nucleated around the regions of the free volume under the effect of the applied shear stress. In most cases, the size of the regions is approximately 5 atomic diameters. At high temperatures (above 0.6 T_c), the transformation is in fact a diffusion rearrangement which results in the relatively low shear deformation in the approximately spherical region. At low temperatures (below 0.6 T_c) the transformation takes place in a narrow disc-shaped shaped region and resembles the nucleation of the dislocation loop or microcracks. Taking these theory into account, calculations were carried out to determine the possible level of dilation associated with the plastic flow, with the determination of the rate of localisation of the shear. It is important to note that the experiments confirmed both the high rate of localisation of shear deformation and the presence in the shear bands of the displacements which can be described by the formation of submicrocracks in them [1.133].

The model of the relaxation of the free volume has been further developed in [1.134]. In this study, the discrepancy between the theory and experiment, observed in the investigations to study the effect of the stresses and structural relaxation on the flow rate and also the effect of structural changes on the behaviour in the flow (decrease of the rate of homogeneous flow with the development of the processes of structural relaxation), were eliminated. This was achieved by introducing the threshold stress available for the crystals which can be determined from the dependence of the strain rate on stress.

The model proposed by Egami [1.61] can be regarded as most efficient for specific conclusions on the possibility of application to different temperature ranges and intermediate structural states. This model proposed the method of describing the local structure of the amorphous alloys, based on the concept of the *n*-, *p*- and τ-defects. Strictly speaking, in the Egami model, the processes of relaxation and plastic deformation are associated with the annihilation or local movement of the structural defects which have no analogue in the crystals and are determined by the microstresses at the atomic level.

Thus, it can be concluded that the solids contain the 'doubling mechanism' of plastic deformation, associated with the displacement of point defects. This mechanism can operate when, for some reason, the main dislocation mechanism of plastic deformation also operates. These reasons may include, for example, very high values of the Burgers vector of the dislocations (as in the experiments carried out by V.R. Regel' and V.P. Rozhanskii on alumo-yttrium garnets

[1.135]) or generally the absence of the translational symmetry in the distribution of the atoms (as in the case of the amorphous alloys).

V.A. Pozdnyakov analysed theoretically the conditions of development of the shear bands in the amorphous alloys [1.136].

The plastic flow of the amorphous metallic alloys takes place as a result of the formation of localised shear transformations (microregions) [1.132]. The resistance stress results in the formation of the purely shear transformations with the resistance stress τ_s. The local shear results in the generation of the excess free volume in such a manner that the tensor of intrinsic deformation of the local shear transformation contains the dilation component, geometrically associated with the magnitude of shear deformation. As the extent of dilation in the resultant shear microregions increases their energy also increases. In heterogeneous nucleation of the shear regions (with the reduced dilation deformation) the areas with the larger excess free volume produce smaller energy losses. Consequently, the stresses of resistance to shear decrease with increase of the excess free volume in the amorphous metallic alloys.

Introducing the volume fraction of the excess free volume f_V, measured in the units of its limiting value, i.e., the shear resistance stress τ_c = min at $f_V = 1$, for τ_c we can write [1.137]:

$$\tau_c = \tau_s[1-\psi(f_V)], \qquad (1.32)$$

where ψ is the numerical increasing function of f_V, with the modulus smaller than unity, such that at $f_V = 0$ $\psi = 0$, and at $f_V = 0$ $\psi = \psi^*$, where $0 \leq \psi^* < 1$ and $\tau_s [1 - \psi^*] = \tau_s^*$.

At the known dependence of the volume fraction of the generated free volume on the magnitude of plastic shear deformation, we can calculate the function $\tau(\gamma)$ or $\tau(u)$, i.e., the local deformation curve of the material.

In loading of the amorphous metallic alloys, the shear stress increases in the elastic deformation range, at $\tau > \tau_p$ the plastic flow start to take place, and the stress reaches the maximum value, equal to the local flow stress $\tau_m = \tau_p + \eta\gamma(f_{V1})$, where η is the effective viscosity of the material, and then rapidly decreases to the stationary value $\tau_0 = \tau_s^* + \eta\gamma\ (f_{V2})$, corresponding to the minimum shear stress resistance of the plastic flow for the given external conditions. The first terms represent the athermal components, the second terms – thermally activated components of the local flow stress which depend on the plastic strain rate $\gamma[f_V(\gamma)]$.

The local deformation curve of the amorphous metallic alloy can be represented in the simplified form

$$\tau(\gamma) = \mu\gamma \quad \text{at } \gamma < (\tau_m/\mu) \equiv \gamma^*$$
$$\tau(\gamma) = \tau_0 \quad \text{at } \gamma > \tau_m/\mu, \qquad (1.33)$$

where μ is the shear modulus of the amorphous material.

The maximum stresses are obtained at a very small relative displacement u, and at $u > u^*$ $\tau = \tau_0$. The stresses τ_m and τ_0 are in this approach the main microscopic parameters, characterising the plastic behaviour of the amorphous metallic alloys.

We examine the condition of development of the shear band im the amorphous metallic alloy (Fig. 1.22 *a*). If the condition of formation of the local flow is determined by the stress τ_m, and the stress τ_0 is sustained in the resultant band, the shear stress in the end regions of the band should change from τ_m to τ_0.

The component of the displacements in the shear band in x and y directions is equal to $U(x,y)$. The actual thickness of the shear band is $h(x)$ (Fig. 1.22 *b*). As the thickness of the band is small in comparison with its length, we transfer from the two-dimensional field of the components of displacements $U(x,y)$ to the one-dimensional field $u(x)$ averaged out with respect to the half thickness of the band introducing the following displacements [1.137]:

$$u^+(x) = (h/2)^{-1} \int_0^{h/2} U(x,y)\,dy,$$
$$u^-(x) = (h/2)^{-1} \int_{-h/2}^{0} U(x,y)\,dy. \qquad (1.34)$$

In this approach, the shear band is treated as a discontinuity surface on which there is a specific relationship between the relative displacement $u(x) = u^+ - u^-$ and the shear stress $\tau(x)$. As a result we obtained the problem of the section of the body along the plane $y = 0$ in the section $-c < x < c$ and in the shear of the upper part, $y > 0$ – relative to the lower part, $y < 0$ – by the value $u(x) = u^+ - u^-$ (Fig. 1.22 *c*).

The equation expressing the equilibrium condition of the body with such a section [1.138] has the form:

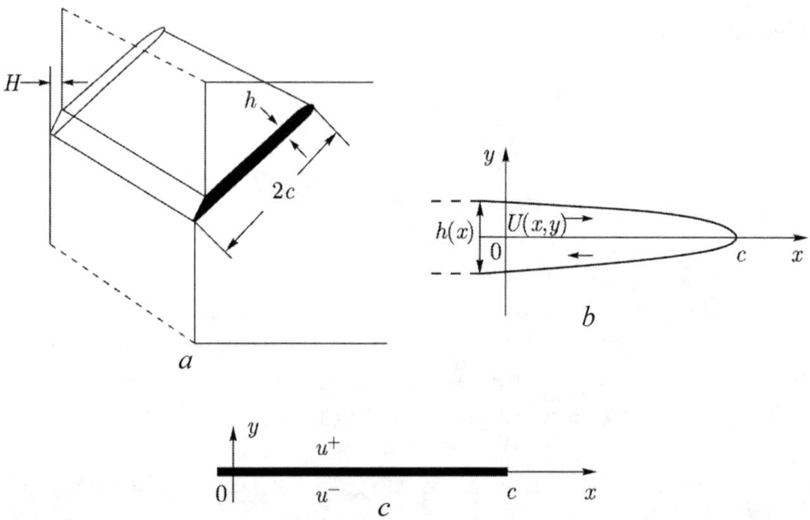

Fig. 1.22. Shear bands: *a*) developed on the surface of the sample of the amorphous alloy; *b*) cross section of the band in the amorphous alloy; *c*) representation of the band in the form of the discontinuity surface of displacements.

$$-A\int_{-x}^{c}\frac{du}{dx^*}\frac{dx^*}{x-x^*}+Y(u)+\tau(x)-\tau_0=0, \qquad (1.35)$$

where $A = \mu/[2\pi(1-\nu)]$. The term $Y(u)$ in (1.35) takes into account the presence of end transition regions. Equation (1.35) can be used to determine the equilibrium distribution of the displacement in the shear band. If we denote $du(x)/dx = D(x)$, then the relationship (1.35) at $Y = 0$ determines the equilibrium distribution of the effective dislocations with density D.

It is assumed that the dilation component of deformation θ in the band is proportional to the shear component: $\theta \sim \gamma$. The corresponding transverse displacements V in the shear band are normal to the plane of the band, i.e. the components of the Burgers vector of the effective dislocations. Assuming that the transverse component of the displacement field is proportional to the longitudinal displacements, $V = qU$, where q = const, taking the transverse displacements into account leads to the addition of the term $q\sigma_n$ to equation (1.35), where σ_n is the component of the external stress normal to the plane of the band. In a general case, the shear band is characterised by the formation of the longitudinal u_e and transverse u_s components in the plane of the displacement band, and also the displacement components u_n normal to the plane of the band and caused by dilation.

The distribution of the stresses in the vicinity of the tip of the shear band with the displacement components u_e, u_s and u_n, assuming that the constant or slightly changing shear stress τ_0 forms inside the band, is identical with the stress field for the buildup of dislocations with similar components of the Burgers vector or the crack of the appropriate modes taking the stress τ_0 into account.

For the shear band we cam introduce the coefficient K_z, identical with the stress concentration factor of the cracks, expressing the special feature of the stress field at the top of the shear band. For the shear bands of the type k with length L under the effect of the external stresses σ_{ij}^a, the stress intensity factor has the following form

$$K_{z0k} = u_k K_z^*,$$
$$K_z^* = \left(\lambda^{-1}\sigma_{ij}^a n_i u_j - \tau_0\right)\left(\pi L/2\right)^{1/2}, \tag{1.36}$$
$$\lambda = u_e^2 + u_n^2 + (1-\nu)u_s^2,$$

here n is the unique vector of the normal to the plane of the shear band. When taking into account a single component of the displacements u_e, under the effect of the homogeneous external shear stress τ_a, the stress intensity factor is

$$K_{z0} = \left(\tau_a - \tau_0\right)\left(\pi L/2\right)^{1/2}. \tag{1.37}$$

Taking into account the end zone $K_z = Kz_0 + \Delta K_z$, and the contribution of the end zone ΔK_z is negative. When the stress intensity factor reaches some critical value determined by the energy dissipation mechanisms during movement of the shear band, the propagation of the shear band becomes unstable.

The condition of propagation of the shear bands in the amorphous metallic materials in the explicit form is determined using the so-called J-integral from the crack mechanics [1.137, 1.139]. We consider the vicinity of the tip of a straight shear band, small in comparison with the characteristic linear dimension of the band. We introduce the Cartesian coordinates x, y with the origin at the tip 0 and with the x axis along the band line (Fig. 1.23). It will be assumed that plane deformation takes place in the plane (x, y). C denotes an arbitrary contour including the point 0 and starting at some point

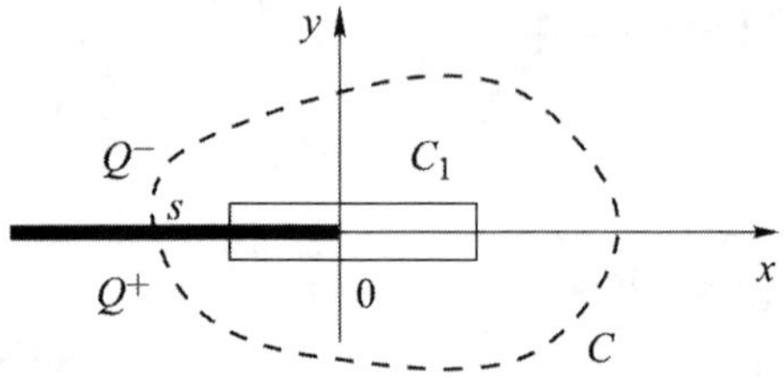

Fig. 1.23. The contour encircling the top of the shear band, for the determination of the J-integral.

Q^- on the lower side of the shear band and ending at some point Q^+ on the upper side of the band.

The components of the unit vector of the external normal to the contour are denoted by n_i, and the components of the displacement vector of the material – by v_i. The displacements v_i inside the band are equal to U, V. The surface forces F_i on the contour C are equal to $F_i = \sigma_{ij} n_j$. Consequently, the J-integral is determined as follows [1.137, 1.139]:

$$J = \int_C \left[\left(W - f_i v_i \right) dy - F_i \frac{\partial v_i}{\partial x} ds \right], \tag{1.38}$$

where f_i are the components of the force acting on the unit volume of the material; ds is an element of the arc of the contour C. The value of the integral does not depend on the integration path and is determined only by the end points Q^-, Q^+. If the contour C is given some special form C^* such that it extends from the point Q^- at the lower boundary of the band to its tip and then returns to Q^+ at the upper boundary, it may be shown that the J-integral has the following form

$$J = \int_C \sigma_{21} \frac{\partial v_1}{\partial x} dx = \int_0^s \tau \frac{\partial u}{\partial x} dx. \tag{1.39}$$

Since τ is an unambiguous function of u, the J-integral (1.39) can be written in the following form

$$J = \int_0^{u^*} \tau(u) du, \tag{1.40}$$

here u^* is the relative displacement in the shear band at the boundary of the end zone with the main part of the band; $\tau = \tau_0$. The value of the relative displacement u^*, corresponding to the deformation γ^*, corresponds to the condition of decrease of the acting shear stress τ in accordance with the curve of local deformation behaviour of the amorphous metallic alloy to the residual value τ_0. Figure 1.24

shows the distribution of the relative displacement and the shear stress along the length of the band. The end zone has the form of the front of the shear band in which the stresses and displacements breach steady-state values. The displacement u in the main part of the band is counted from the steady-state value of displacement in the end region u^*.

The integral in (1.40) will be represented by the sum of two parts: the part corresponding to the steady-state stress τ_0, and the part in which the acting shear stress is higher than τ_0 in the end of each of the band at a low relative displacement:

$$J - \tau_0 u^* = \int (\tau - \tau_0) du. \qquad (1.41).$$

The integral in (1.41) can be expressed by the characteristic value of the displacement u_e in the end part of the band, determined in accordance with the equality

$$G_C \equiv \int (\tau - \tau_0) du = (\tau_m - \tau_0) u_e. \qquad (1.42)$$

The value $J - \tau_0 u^*$ is the energy excess, associated with the increase of the length of the band during its propagation, and the energy dissipated inside the band as a result of overcoming the residual shear resistance stress τ_0. This energy increase for the propagation of the band should be equal to the energy used in the

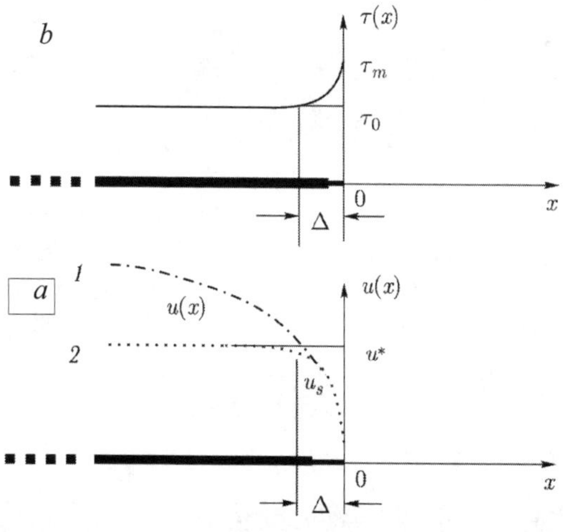

Fig. 1.24. The distribution of relative displacement $u(x)$ (a) and the shear stress τ $\tau(x)$ (b) along the length of the band (Δ is the size of the end zone).

end region when the shear stress in this region is greater than the stress τ_0. The specific plastic deformation work (1.42) is dissipated into heat and is also used for the change of the bulk energy of the material in the band $h^*\Delta\varphi$, where h^* is the mean thickness of the band, and is also used for the formation of the surface energy 2Γ of the band when its length increases by unity. Without separating these components, the energy dissipated in the end zone will be determined by the difference of the stresses $\tau_m - \tau_0$ in accordance with (1.41). When the external stress becomes sufficiently large so that the value $J - \tau_0 u^*$ reaches its critical value $G_C = (\tau_m - \tau_0)\, u_e$, the band becomes active and starts to propagate with a further increase of stress.

Using the field of stresses or displacement of the band without taking the end zone into account, we can calculate directly the J-integral and express it through K_{z0}:

$$J - \tau_0 u^* = (1 - \nu) K_{z0}^2 / (2\mu). \qquad (1.43)$$

Thus, the band propagation criterion can be written in the form of the stress intensity factor for the band K_{z0} (1.37). Assuming that the end region is small in comparison with the length of the band, the propagation criterion has the form which can be expressed through the stress intensity factor of the band:

$$K_{z0} \geq \left[2\mu(\tau_m - \tau_0) u_e / (1 - \nu)\right]^{1/2}, \qquad (1.43a)$$

or through the acting shear stress

$$\tau_p - \tau_0 \geq \left\{4\mu G_C / \left[\pi(1 - \nu) L\right]\right\}^{1/2}. \qquad (1.43b)$$

The relationships (1.43) are the analogues of the Griffith-type energy balance equation. The estimate of the size of the end the region Δ [1.138] is $\Delta \cong 60...100$ nm. For the shear bands with the length $L \cong 10$ μm, comparable with the thickness of ribbon specimens of the amorphous metallic alloys, $\Delta \ll L$.

1.3.4. Fracture

The fracture of the amorphous alloys, like of crystalline materials, may be brittle or ductile. Brittle fracture takes place by cleavage without any features of the macroscopic flow at stresses lower than the yield stress $\sigma_p < \sigma_T$. In uniaxial tensile loading brittle fracture takes place by the fracture of the opposite faces normal to the tensile loading axis. Ductile fracture in the amorphous alloys takes place either after or simultaneously with the process of plastic flow, and the material shows features of macroscopic plasticity to a varying extent. In this case [1.140]: 1) fracture takes place on the planes of the maximum cleavage stresses; 2) fracture is associated always with one (sometimes with two) transition from one plane of the maximum cleavage stresses to another; 3) two characteristic zones form on the fracture surface: almost smooth cleavage areas and areas forming a system of interwoven 'veins'. The thickness of the 'veins' is of the order of 0.1 μm and is usually characterised by the ratio of height to thickness of 2 to 4, with the exception of the triple junction points.

Figure 1.25 shows the typical surfaces of brittle and ductile fracture, produced in uniaxial tensile loading of the $Fe_{82}B_{18}$ amorphous alloy. The majority of the amorphous alloys are characterised by the classic pattern of ductile fracture: the fracture energy decreases with a decrease of the thickness of the ribbon specimens with the change of the crack propagation conditions from the plane stress state to the plane strain state [1.141].

For different amorphous alloys the fracture energy and the stress of the start of plastic flow are linked by the almost linear dependence, and the fracture energy increases when the temperature is reduced below room temperature.

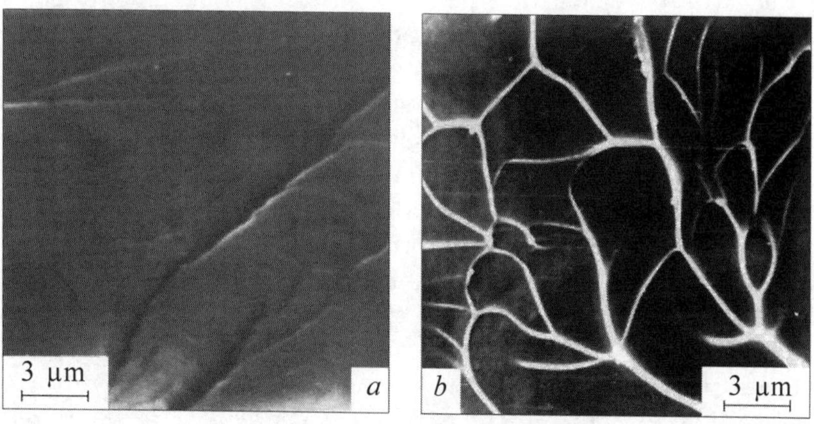

Fig. 1.25. Typical patterns of brittle (*a*) and ductile (*b*) fracture in $Fe_{82}B_{18}$ amorphous alloy. Scanning electron microscopy in the reflected electrons.

As a result of detailed examination of the parameters of the plastic zone at the tip of the ductile crack, and also measurements of the crack opening displacement in dependence on the applied load, the authors of [1.142] observed a direct relationship between the size of the plastic zone, the extent of the crack opening displacement and the magnitude of applied stress. In addition to this, it was concluded that the transition from the mixed to the plane stress state takes place in the early stage of plastic deformation of the crack tip and, consequently, a large part of plastic deformation and fracture takes place in the plane stress state. Fracture takes place when one of the following parameters reaches a critical value: the size of the plastic zone, the displacement of the crack tip or the stress intensity factor.

The process of transition from ductile to brittle fracture was investigated in detail in [1.143]. These experiments were carried out on samples of $Pd_{78}Cu_6Si_{16}$ alloy with a V-shaped notch after annealing at 350°C for different periods of time. In the initial (quenched) condition and after short-term annealing the alloy fractured during uniaxial tensile loading by sliding after the start of the macroscopic flow. After longer annealing, the macroscopic flow was followed by the features of unstable fracture as a result of intermittent crack propagation. This resulted in the final analysis in fracture preceded by the macroscopic flow. A further increase of annealing time resulted in brittle fracture accompanied by the plastic flow on the surface of the specimens. Finally, the longest annealing resulted in brittle fracture below the elasticity limit without any features of the plastic flow even on the specimen surface. Fractographic analysis shows that embrittlement after annealing cannot be attributed only to the increase of the resistance to propagation of the shear bands in the amorphous matrix. The fracture process in fatigue tests starts with the formation of shear bands in the stress concentration areas, and some of these bands become crack nuclei [1.144]. A plastic zone forms around the fatigue cracks. Crack propagation takes place along the shear bands, formed ahead of the cracks, and this fully resembles the processes taking place in the crystals. However, a significant feature is the fact that the fatigue shear bands in the amorphous alloys are not convex or 'pushed in', as in crystals, and have a single sliding step [1.145]. Analysis of the fracture surfaces showed the existence of two crack growth stages: 1) slow growth, accompanied by the formation of a 'terrace-like' surface with fine elements, and 2) rapid growth in which the surface consists of thick striations [1.145].

Stereoscopic experiments show that the so-called 'veins' on the fracture surface typical of the amorphous alloys (Fig. 1.25) have the form of projections on both surfaces. The 'pattern' on the opposite fracture surfaces is similar but not identical. This confirms that the 'veins' formed as a result of local necking during fracture. The 'veins' and smooth fracture surface areas resemble on the whole the pattern of the fracture surface of the crystals, formed at fracture in the immediate vicinity of the ductile–brittle transition temperature.

The fact that the 'veins' are areas of exit of the strongly localised plastic flow to the fracture surface is clearly confirmed on the images obtained simultaneously from the fracture surface and the side surface of the deformed ribbons specimen (Fig. 1.26).

As mentioned previously (section 1.2.2), alloying with surface-active elements strongly influences the strength of the amorphous alloys. It is therefore very interesting to examine fractographic special features of the alloys alloyed with surface-active elements.

Analysis of the fracture surfaces, produced in the tensile test, revealed the presence of qualitative changes, associated with microalloying with the surface-active elements [1.146]. For example, the fracture pattern typical of the amorphous alloys was obtained for the $Fe_{82}B_{18}$ alloy – the combination of cleavage areas with ductile fracture areas (Fig. 1.25). When 0.01 at.% of Nb was added, the overall nature of fracture did not change. However, there was a large (by approximately factor of 3) decrease of the scale of the fragments – both the area of the individual cleavage areas and the geometrical characteristics (height and width) of the 'veins' (Fig. 1.27 *a*).

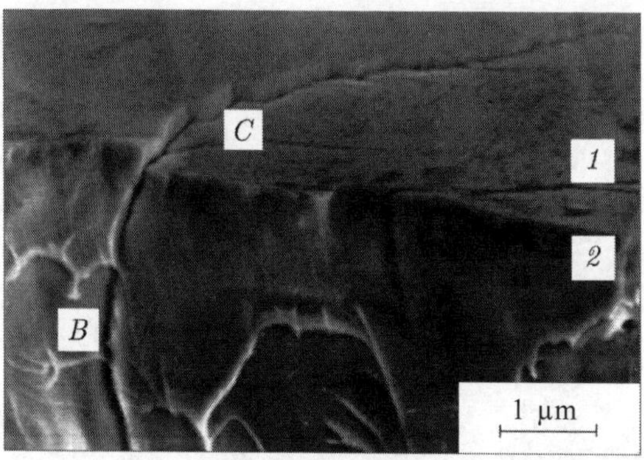

Fig. 1.26. Exit of the shear band to the fracture surface: 1) the side surface; 2) fracture surface; *C* – slip step; *B* – 'vein' (image of two surfaces of the Fe–B alloy was produced by scanning electron microscopy in the reflected electrons).

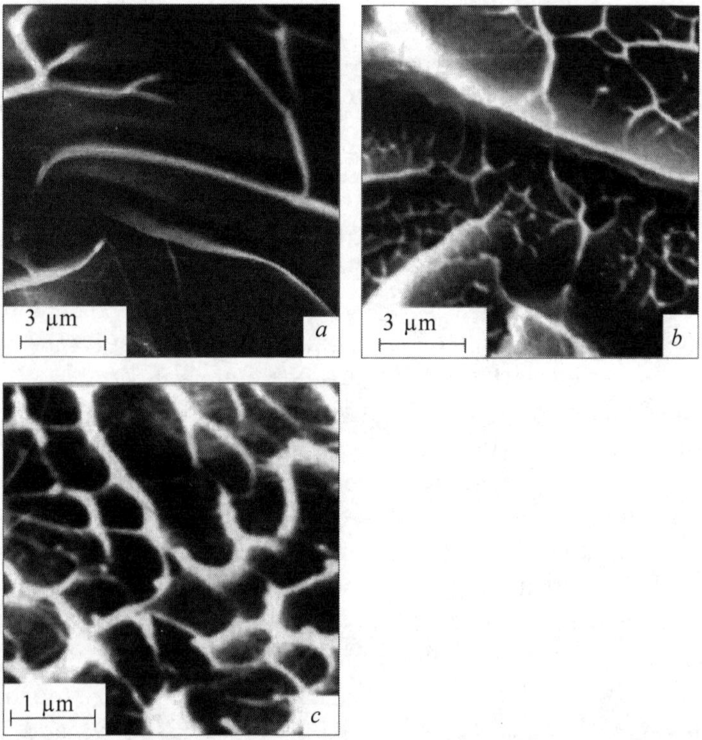

Fig. 1.27. Fractographs of Fe–B–Nb (*a*) and Fe–B–Sb (*b*, *c*) amorphous alloys. Scanning electron microscopy in reflected electrons.

This change of the scale of the fragments resulted in a qualitative difference in comparison with the fracture surface of the basic alloy: the 'veins' in the Fe–B alloys whose thickness gradually decreased together with the height, fractured in the cleavage areas, and in the case of the Fe–B–Nb alloy the 'veins' intersected quite often, forming a relatively symmetric pattern in which the cleavage areas resemble the grains in crystalline alloys, and the 'veins' are their boundaries. The addition of 0.01 at.% of Sb (or Ce) resulted in a qualitative change of the fracture surface. The fracture surface consisted of convex 'veins' almost parallel to each other and the edge of the sample (*X* axis), which greatly differed in the scale and structure from the 'veins' in the Fe–B and Fe–B–Nb alloys. Firstly, the thickness of these 'veins' (transverse dimension) increased by almost one order of magnitude in comparison with Fe–B–Nb. Secondly, the 'veins' were not relatively smooth structureless formations as in the case of the Fe–B and Fe–B–Nb alloys but contained a complicated internal structure, in particular, a system of transverse 'microveins' whose thickness was similar to the cross-section of the 'veins' in the Fe–B–Nb alloys (Fig. 1.27 *b*). The gaps between the 'veins 'were formed in all likelihood by the chains of relatively large grains (500–700 nm). The 'veins' themselves also contained a large number of relatively large pores (200–350 nm) forming chains in some areas. Under high magnification the fracture surface of the Fe–B–Sb and Fe–B–Ce alloys was characterised by distinctive fragmentation and the formation of a continuous network of fine facets with the size of 100 to 500 nm. Inside larger facets there appeared to be a second level of faceting – a continuous field of finer (10–30 nm) facets. The 'borders', restricting the facets, were characterised by a high density of the micropores of different sizes forming often chains (Fig. 1.27 *c*). The fracture surface of the Fe–B and Fe–B–Nb alloys in the initial condition did not contain theses pores in the cleavage areas nor in the structure or contacts of the 'veins'. The change by a factor of three of the details of the fracture surface in the Fe–B–Nb amorphous alloy in comparison with Fe–B is evidently caused by the change of the distribution of the free volume regions in its amorphous matrix [1.58].

Interesting results for the morphological special features of the fracture micropatterns of the amorphous alloys in microindentation on substrates were obtained in [1.93, 1.97]. In the indentation of the amorphous alloys, annealed at temperatures of temper brittleness ($T < T_{br}$) the indentation is surrounded by a deformation zone determined by heterogeneous deformation observed in the shear bands spreading from the indentation (Fig. 1.28 *a*). Heat treatment at $T \geq T_{br}$ resulted in the formation of two temperature ranges, and in transition from one range to another there was a sharp change in the fracture micropatterns of the amorphous alloys as a result of local loading. In the first temperature range in the indentation zone of the metallic alloys there were several radial straight cracks, some of which may join with circular cracks (Fig. 1.28 *b*). Regardless of active embrittlement, shear bands can still form. This range can be regarded as transitional because lower temperatures did not lead to the formation of cracks, and high temperatures did not cause shear bands to form. In the second temperature range, on the 'eve' of crystallisation, indentations in the form of a system of inserted squares of cracks formed preferentially and were mutually perpendicular in the areas of the effect of the edges of the Vickers pyramid (Fig. 1.28 *c*). The formation of these cracks was

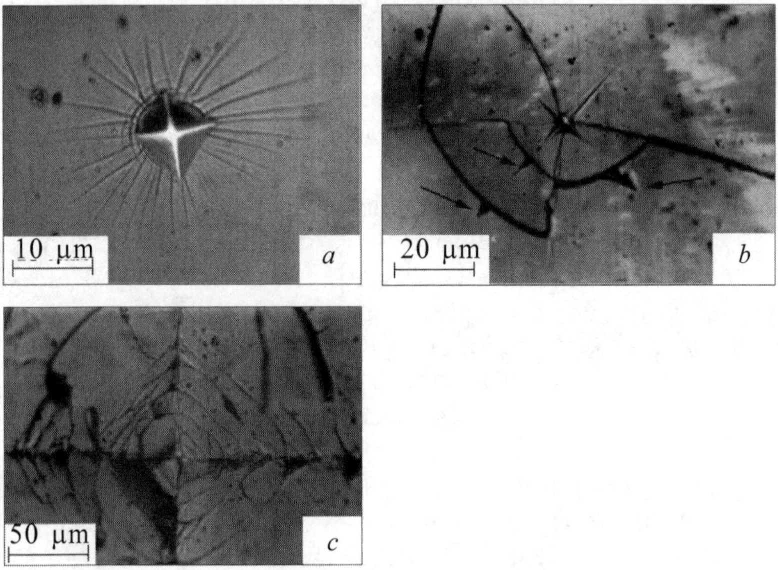

Fig. 1. 28. The characteristic indentations on the surface of the amorphous alloys of the Co–Fe–Cr– Si system in indentation: at $T < T_{br}$ (a), $T_{br} \leq T \leq 718$ K (*b*) (the arrows indicate the marks from the face of the indentor, separated from the main indentation by circular cracks), 718 K $< T \leq 823$ K (*c*) [1.97].

accompanied by the formation of circular cracks with increase of the distance from the indentation zone.

The processes of fracture at the microscopic level are usually carried out by the deformation of the object directly in the column of the scanning or transmission microscope. In [1.117] the authors investigated the process of plastic deformation and subsequent fracture of amorphous alloys based on iron and cobalt in the column of a scanning electron microscope. The results show that in all cases the nucleation of the microcracks was preceded by the process of microplastic deformation. In most cases, the primary crack was nucleated in the region of the widest and most developed shear band or at the intersection of a number of bands. The tip of the crack with a jump-like propagation contained relaxation areas in the form of a developed system or branched shear bands. If the sample was subjected to cyclic deformation in the elastic region prior to deformation, the nature of subsequent fracture greatly changed. Instead of the single main crack, there was an entire system of cracks indicating evidently the increase of the effective surface fracture energy. The kinetics of deformation and fracture at temperatures higher than 250°C was completely controlled by the diffusion mechanism of mass transfer [1.117]. Consequently, the structure was characterised by the formation of a relatively large number of submicropores with the size of 0.1 μm. A small number of studies was carried out to investigate the processes of microfracture by transmission electron microscopy in tensile loading in the column of the microscope [1.147]. The microfracture parameters recorded in these investigations cannot be regarded as characteristic of all amorphous alloys ($Ni_{55}Pd_{35}P_{10}$ amorphous alloy produced by powder metallurgy [1.148], the amorphous film of niobium oxide [1.149]). Only the attachment for tensile loading used in [1.150] ensuring uniaxial tensile loading of a melt-quenched amorphous Fe–P alloy makes it possible to avoid problems in the interpretation of the results which formed as a result of the application of the given tensile mode – bending [1.149].

The formation of the crack in the $Fe_{82}B_{18}$ amorphous alloy was always preceded by the formation of a localised deformation band (Fig. 1.29).

Both the shear band and also the crack corresponding to the band formed in the region of the highest stress concentration. However, note every shear band propagated into the crack. The nucleated crack reproduces accurately almost always the configuration of the shear band preceding its formation. The elastic deformation zone is a powerful 'torch' with a complicated structure in which the characteristic periodic 'loopholes' form in the central part of the

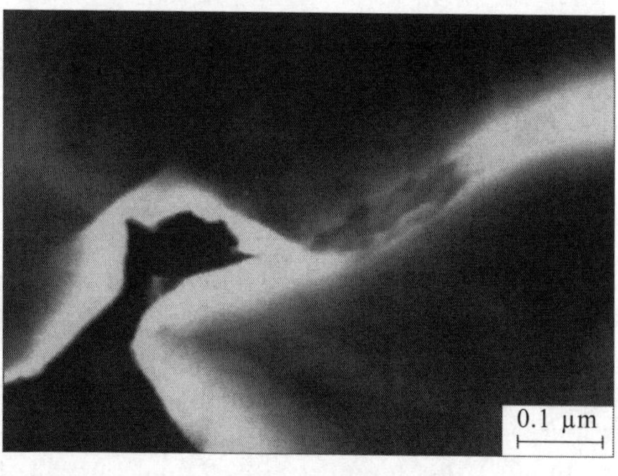

Fig. 1.29. The shear band formed prior to the nucleation of a crack in the $Fe_{82}B_{18}$ alloy, the dark field image produced by transmission electron microscopy.

Fig. 1.30. Characteristic 'loopholes' in the central part of the plastic deformation zone at the crack tip in the $Fe_{82}B_{18}$ amorphous alloy, the dark field image produced by transmission electronmicroscopy.

'torch' (Fig. 1.30). The angle between the general direction of crack growth and the tensile loading axis was approximately 45–60°.

The investigation of the dynamics of crack growth in the Fe–B–Sb (Ce) amorphous alloys always showed the presence of the two main stages:

1. Slow crack growth in the direction of length and width at a constant load. The localised deformation zone remains unchanged. The crack gradually 'eats' the shear zone and is arrested, reaching its edges – steps (Fig. 1.31).

2. The jump-like growth in the direction of length and width with increase of loading. A new localised shear band forms in front of the crack. At the same increase of the load the increase of the length of the crack in this case is an order of magnitude greater than in the Fe–B alloy.

The general direction of crack propagation in the Fe–B –Sb (Ce) alloy forms the angle of 45–60° with the direction of the tensile loading axis. However, in the slow growth stage at a constant load growth stages were detected in the direction normal to the main direction. The angle of approximately 45°C with the direction of the tensile loading axis was retained.

It should also be noted that other characteristic special features of crack growth at a constant load also important (Fig. 1.31):

a. The crack in most cases does not have a distinctive sharp tip. If the crack after increase of the load with jump-like propagation would have a tip and, then in the stage of slow crack growth it would be rapidly 'blunted' as a result of the formation of the frontal area due to the growth in the direction normal to the main crack propagation

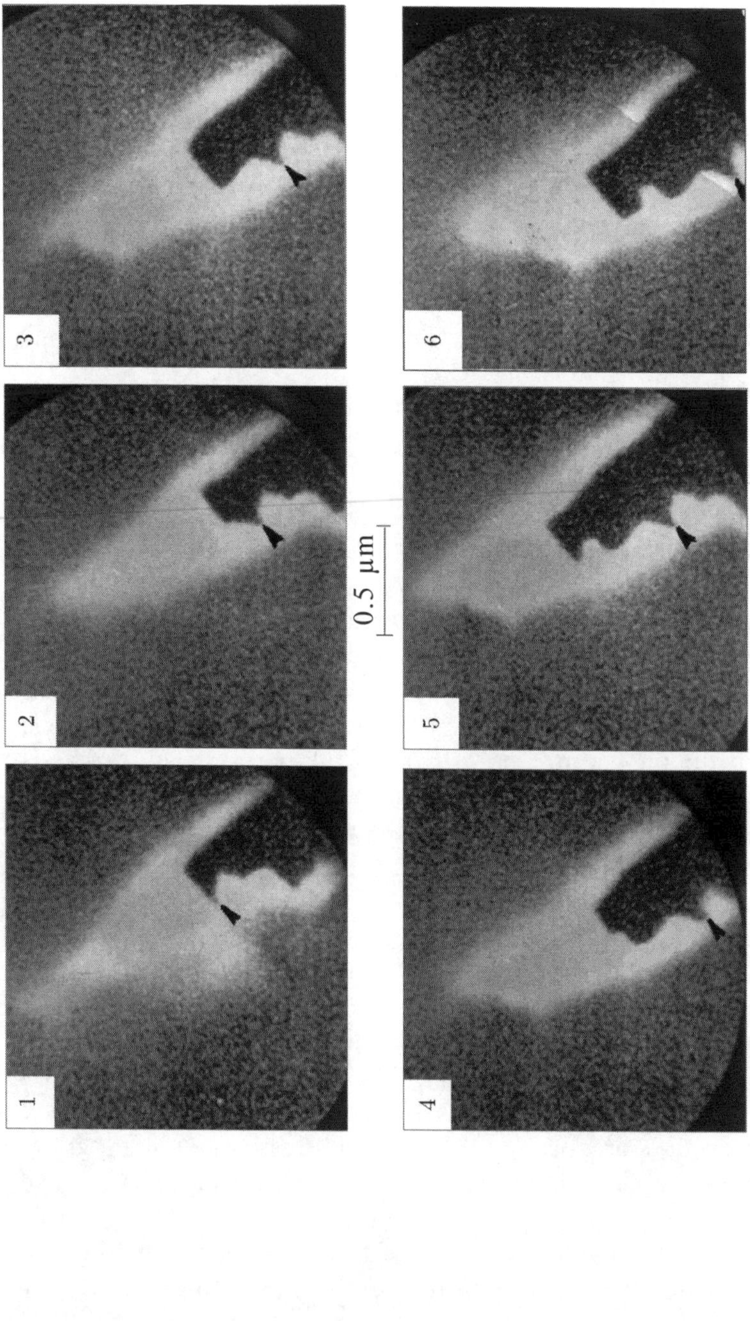

Fig. 1.31. Stages (1–6) of crack propagation at a constant load in the Fe–B–Sb alloy, dark field electron micrographs, photographs taken from the screen of a video tape recorder.

direction. The crack grew having at the tip the frontal region up to several hundreds of angstroms wide.

b. The crack growth was pulsed: slow crack growth with the rate of approximately 2 nm · s^{-1} for a period of 20–40 s alternating with 20–30 s stoppages.

c. Slow crack growth throughout the entire frontal region is periodically replaced by faster (with the rate of 3–4 nm·s^{-1}) growth of a separate region forming approximately 2/5 of the total width of the front;

d. The period of growth in the direction of length as a result of a separate section in the specific stage is replaced by rapid growth in the direction of width (with the rate of 5 nm · s^{-1}) to the restoration of the initial width of the front. This is usually followed by an arrest for 20–30 s during growth.

The resultant crack, as in the Fe–B and Fe–B–Nb alloys, was of the 'zigzag' shape. However, the periodicity of the 'zigzags' formed in the case of the Fe–B–Nb and Fe–B–Ce alloys was far less distinctive. In addition to this, the resultant crack showed the presence of long smooth areas (without 'zigzags'), associated with the catastrophic stages of crack propagation with increasing load. The distance between the 'zigzag' areas was 210–270 nm.

Recently, several theoretical studies have been published examining the role of the formation of pores in the zone of plastic deformation of the crack. In [1.151] the authors proposed a general theory of the nucleation and growth of the pores ahead of the crack tip. The theory does not define the mechanism of nucleation and growth of the pores (depending on the material; this can be diffusion of plastic deformation) and assumes that the stress field ahead of the crack can be determined irrespective of the pore formation process.

The model of the nucleation and growth of the pores in the plastic deformation zone at the crack tip, constructed in [1.152], uses the dislocation theory. It is shown that the nucleation of pores results in the relaxation of stresses and decreases the energy of the deformed region in the vicinity of the crack. The most interesting aspect (with respect to the amorphous alloys) is the one of the consequences of this model: the pores, subjected to shear stresses, not situated in the plane of the crack, can merge with the crack resulting in the cyclic shape of the resultant crack. It should also be noted that this form also results when using the model of formation of the main crack as a result of the interaction with the cavities, initiating the inclusions [1.106].

The largest contribution to the process of plastic deformation is provided by the regions of the free volume with the size of the order of 5 atomic diameters (1–2 nm). In quenching from the melt, the material contains regions of the free volume which must be statistically distributed between the regions with higher density ('anti-free volume') at some (which depends mainly on the parameters of production of the amorphous alloy) value by which the concentration of the regions of the free volume is greater than the concentration of the regions with higher density. The shift of the free volume regions during plastic deformation results regularly in the shift of the regions of higher density. The regions of the free volume in the shear band should be interleaved by the regions with higher density.

Taking these assumptions into account, the crack growth in the amorphous alloys $(Fe_{100-x}Me_x)_{82}B_{18}$ (Me = Nb, Sb, Ce, x = 0; 0.1; 0.01) [1.150, 1.153] can be described as follows. The shear band prior to the crack nucleation already contains a chain of statistically distributed micropores and areas of higher density. During crack growth, the stress field ahead of the crack tip initiates the growth of the micropores at the crack tip (stretching it). The shape and size of the micropore change: the size in the direction of crack growth greatly increases. In this case the micropore becomes the 'emissary' crack. However, in contrast to the crystalline analogue, a region of higher density forms between the main crack and the 'emissary' crack (otherwise the main crack would grow without initiating the growth of the 'emissary' crack). It is possible that the growth of the 'emissary' crack increases slightly the density of this region. This is indicated by the change of the electron microscopic contrast on the section of the transition between the main crack and the growing 'emissary' crack. Finally, the last stage of crack growth – the relatively rapid (in comparison with the growth rate of the 'emissary' crack) merger of the main crack with the pore as a result of the formation of a narrow (an order of magnitude smaller than the transverse dimension of the micropore) connecting region. The most probable model for this stage is the model of the viscous flow based on the assumption of adiabatic heating in the plastic deformation zone [1.147]. The model was proposed to explain all the phenomena of the regions with high density in which the cracks 'rests'. This results in the formation of a higher dislocation concentration, generating suitable conditions for the local adiabatic heating which decreases the level of viscosity of the material to the level required for the viscous flow. It should be noted that the analysis of the model of the free volume, carried out in [1.154], shows that the processes of microflow and relaxation as a result of annihilation of the free volume regions are characterised by similar values of the critical fluctuation of the volume. At the same time, the size of the activation barrier for relaxation in the non-relaxed structural levels of the amorphous alloys is considerably smaller than the activation energy of the iso-configuration flow. This indicates that the atomic configuration in the relaxed positions should be more 'free' (porous). However, in the situation described previously when the entire 'free' atomic configuration is exhausted for the development of the free volume region to the 'emissary' crack it is natural to assume that the activation energy of relaxation increases. This energy increases to such an extent that the viscous flow becomes more favourable from the viewpoint of energy. The activation energy of this flow [1.154] is unchanged for any degree of relaxation of the structure because the structural relaxation may change the number of atoms taking part in the viscous flow process but does not affect the nature of the process.

An important result directly related to the microfracture mechanism of amorphous alloys has been obtained in [1.155]. It was shown that the fracture surface of $Fe_{84}B_{16}$ alloy after inhomogeneous plastic deformation in the column of the electronic Auger spectrometer is characterised by the formation of segregations of interstitial elements in the areas of localisation of deformation (in the 'veins'). The concentration of the atoms of boron, carbon and oxygen in the regions of the 'veins' is considerably higher than the concentration of these elements in any other area of the fracture surface. Correspondingly, the regions, distributed approximately 1 μm from the centre of the 'vein' but outside its limits are depleted in the atoms of these elements. The nature of the observed segregation effects is evidently associated with the local decrease of viscosity in the shear bands and, correspondingly, with the effective diffusion displacement of the atoms–metalloids in the area with the excessive removal from the regions adjacent to the shear bands. Since the fracture of the alloy takes place, as shown previously, at the shear bands, and the 'veins' formed on the fracture surface are areas of the strongest localisation of slip, it is not surprising that the 'veins' draw into themselves the atoms of the elements with the highest mobility. To some sense, the analysis of the local chemical composition of the 'veins can be used to explain the nature of the structural and chemical changes in the shear bands in the process of inhomogeneous plastic deformation.

Unfortunately, at the moment it is not possible to discuss the question why the amorphous alloys show the ductile–brittle transition at low temperatures and other alloys do not. In all likelihood, we should use the classic considerations of A.F. Ioffe according to which at low temperatures the stress of the start of the plastic flow indicating the strong temperature dependence in amorphous alloys (section 1.3) becomes higher than the brittle separation stress. This results in the ductile–brittle transition. In the alloys in which the brittle separation stress is high or the temperature dependence of σ_T is weak because σ_T does not exceed the brittle separation stress to the lowest temperatures, the ductile–brittle transition is not detected. Further investigations should provide the actual physical meaning to this phenomenological model.

In [1.156] investigations were carried out on iron-based alloys to obtain the values of the critical temperature in the range 180–230 K in dependence on the specific chemical composition and, in particular, the nickel content. At the same time, there are alloys which do not undergo the ductile–brittle transition at low temperatures: complete embrittlement of the $Fe_{40}Ni_{38}Mo_4B_{18}$ alloy was not detected up to 77 K [1.158]. In [1.157] it is assumed that the controlling factor in the absence of the ductile–brittle transition temperature in the amorphous alloy is the high nickel content in comparison with the content of the metalloid atoms, and in [1.158] – the presence of disclination elements in the structure of the alloy.

In [1.157] it is reported that it is rational to define two ductile–brittle transition temperatures: T_{br1} – the minimum temperature resulting in the occurrence of macroplastic deformation, and $T_{br2} \approx 0.7\ T_{cryst}$ (T_{cryst} is the crystallisation temperature) resulting in a large increase of the ductility as a result of the change of the plastic deformation mechanism from inhomogeneous to homogeneous. In principle, T_{br2} coincides with the equicohesion temperature T_e (see section 1.1). However, it should be mentioned that if the point T_{br1} is in fact the temperature of the ductile–brittle transition, then the point T_{br2} has a less clear physical meaning because the large increase of the ductility in transition through this temperature is recorded only in the uniaxial tensile test and, for example, this jump is not detected in bending. Nevertheless, the nature of rupture in the ductile homogeneous plastic flow differs from that in the inhomogeneous flow and, therefore, it may be assumed that the fracture mechanism changes at the point $T_{br2} = T_e$.

The amorphous alloys are characterised by very high sensitivity to the embrittling effect of hydrogen and liquid media [1.159, 1.160]. The determination of the effect of hydrogen and deuterium on the structure and properties of the amorphous alloys has been the subject of a number of studies of the authors [1.161–1.165]. The results show a decrease of the elastic properties after hydrogen charging of the amorphous metallic alloys based on cobalt and iron and the subsequent recovery of these properties during holding at 295 K [1.162]. In the Finemet amorphous alloys, hydrogen saturation for ~10 min results in the maximum deformation determined by the loss of the elastic properties [1.163]. The charging time longer than 40 min results in the fracture of the amorphous alloys. The results also show that the nature of deformation of the amorphous alloys in hydrogen charging and subsequent holding in air depend strongly on the natural ageing time (the time passed after preparation). Consequently, it may be assumed that two effects are detected in the hydrogen saturation of the freshly quenched alloys: 1) the reversible loss of shape; 2) the relaxation of quenching stresses and of the free volume. The large variety of the observed deformation effects in the alloys based on iron, nickel and cobalt is explained by the amorphising effect of hydrogen (deuterium) on the matrix. Consequently, the shear modulus decreases dramatically and the topological and compositional short-range orders change [1.163].

1.3.5. Temper brittleness

In temper brittleness on reaching a specific preliminary annealing temperature T_{br} within the limits of stability of the amorphous state ($T_{br} < T_{cryst}$) the amorphous alloys become completely or partially brittle at room temperature. This phenomenon of the loss of plasticity of the amorphous alloys is not only of scientific but also considerable practical interest. In particular, this phenomenon greatly restricts the temperature range of heat treatment of the industrial alloys which, for example, in the case of the magnetically soft amorphous alloys should include the temperature range close to T_{cryst}.

Figure 1.32 shows the typical curves of embrittlement obtained as a result of the mechanical bend test at room temperature for the Fe–B amorphous alloy annealed at different temperatures [1.74]. It may be seen that catastrophic embrittlement occurs in a very narrow

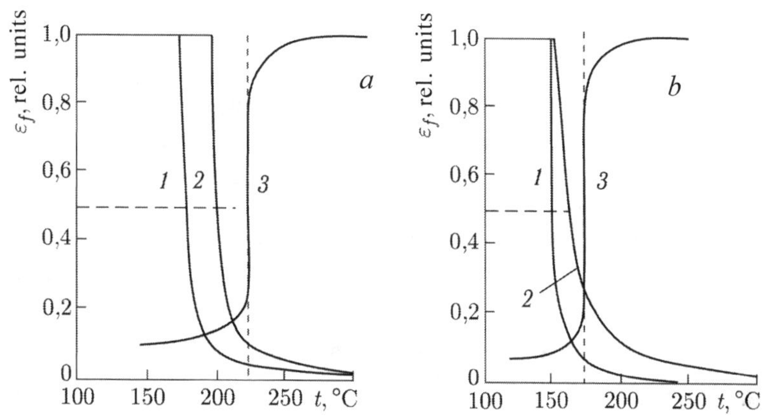

Fig. 1.32. Dependence of the ductility parameter ε_f on the preliminary annealing temperature in vacuum for Fe–B (*a*) and Fe–B–Ce (*b*) alloys: 1) annealing time 6 h; 2) annealing time 1 h; 3 – the $\delta(T)$ dependence defining the equicohesion temperature T_e.

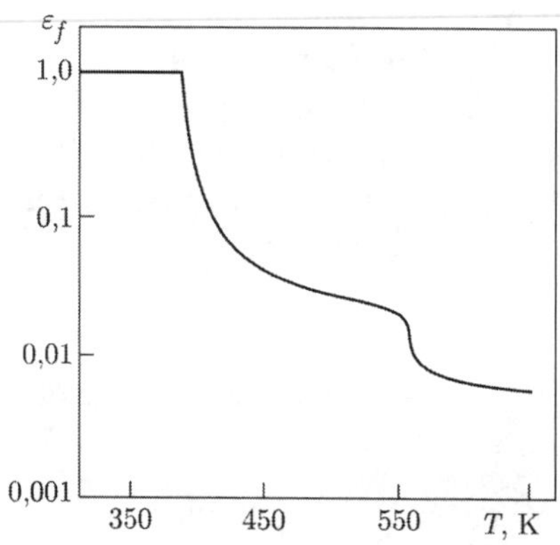

Fig. 1.33. Dependence of preliminary annealing (semi-logarithmic scale) for the $Fe_{40}Ni_{40}P_{14}B_6$ alloy (annealing time 0.25 h).

annealing temperature range where the value of the parameter ε_f (ductility in the bend test described in detail in section 1.2.1) changes from unity to almost 0. Consequently, it is justified to take into account the critical embrittling annealing temperature T_{br}, and the accurate value of this temperature corresponds to halving of the value of ε_f in relation to the initial value.

More accurate measurements show [1.112] that the ductility of the ribbon of the amorphous alloy after such a large jump remains low but differs from zero, and there is a second stage of the jump-like reduction of ductility at a higher preliminary annealing temperature. Figure 1.33 shows the standard curve for the $Fe_{40}Ni_{40}P_{14}B_6$ amorphous alloy where the value ε_f is plotted on the logarithmic scale. Consequently, the two-stage nature of the embrittlement process in annealing can be recorded more accurately. However, it should be borne in mind that the significance of the first low-temperature embrittlement stage in which ε_f decreases from 1.0 to 0.1 is considerably greater and the significance of the second high-temperature stage in which ε_f changes within a few hundredths. Temperature T_{br} will be represented by the annealing temperature resulting in the low-temperature considerably greater embrittlement shown in Fig. 1.32.

Summing up the experimental results of this phenomenon, it is important to mention the following main relationships of the temper brittleness of the amorphous alloys:

1. Each amorphous alloy is characterised by the embrittlement temperature T_{br} which often correlates with the crystallisation temperature in heating the amorphous alloy T_{cryst}. Figure 1.34 shows the concentration dependence of the values of T_{br} and T_{cryst} in the $Fe_{80-x}Ni_xSi_6B_{14}$ alloy [1.74].

The experimental results show that the value T_{br} of the amorphous alloys based on iron and cobalt decreases in alloying with chromium and molybdenum [1.165]. The simultaneous presence in the alloy of two types of metalloid atoms, characterised by the different characteristics of bonding with the iron atoms, also decreases T_{br}. For example, this takes place in the Fe–B–Si, Fe–B–P and Fe–P–C systems [1.666]. The phosphorus-containing amorphous alloys embrittle at a considerably lower temperature than the boron-containing alloys [1.114] and the simultaneous presence in the alloy

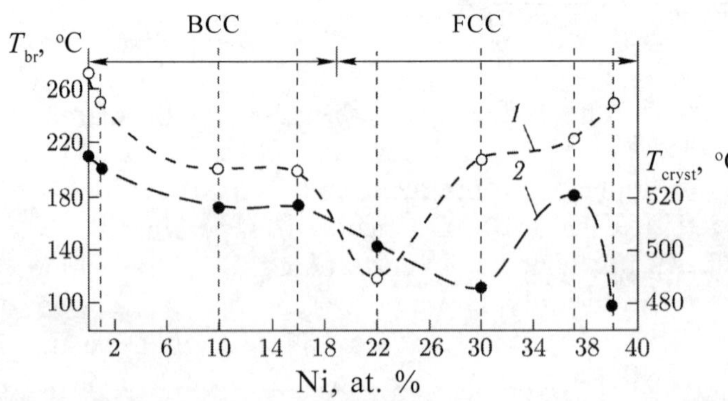

Fig. 1.34. Dependence of the temper brittleness temperature T_{br} (1) and crystallisation temperature T_{cryst} (2) of $Fe_{80-x}Ni_xSi_6B_{14}$ alloys on the nickel content; the arrows indicate the concentration ranges of existence of the BCC and FCC crystal phases which form during crystallisation of the amorphous alloy.

of phosphorus and boron results in an even larger decrease of T_{br} [1.167]. The value of T_{br} slightly increases when boron is replaced by silicon or carbon in the Fe_{80}–$B_{20-x}N_x$ system (N = C, Si, P, Ce), and any replacement of boron by phosphorus or cerium decreases the value T_{br} [1.166]. In the study [1.168] the authors reported the opposite effect of chemical composition on the T_{br} in the Fe–*Me*–B and Ni–*Me*–B alloys (*Me* = Ti, Zr, Hf, Hb, Na). Consequently, it was concluded that the tendency for embrittlement increases with the increase of the binding force between the elements included in the composition of the amorphous alloys. It is important to note that the temper brittleness is typical not only of the amorphous alloys of the metal–metalloid type but also of the metal–metal amorphous alloys [1.169].

2. The temperature T_{br} is the function of the logarithm of the embrittling annealing time and decreases with increase of this time at the rate which depends on the composition of the amorphous alloy.

For example, the T_{br} temperature for the $Fe_{82.5}B_{17.5}$ amorphous alloy is 185°C at the annealing time of 6 h and 215°C at the annealing time of 1 h [1.74]. Undoubtedly, this confirms the thermal activation nature of embrittlement.

3. Brittleness is typical of some amorphous alloys in the initial condition; at the same time, there are alloys with the amorphous structure not showing any features of temper brittleness up to solidification (in most cases, these are amorphous alloys of the metal–metal type).

In reality, this means that the structural states typical of the amorphous alloy after embrittling annealing can form directly after melt quenching. However, this state may not form under any thermal effect which retain the amorphous state. It has been shown that the parameters of production in melt quenching (quenching rate, the superheating temperature of the melt) not only strongly influence the value of T_{br} [1.113] but also in a number of cases determine the possibility of formation of the amorphous alloys in the initial ductile condition [1.170]. In this case, a decrease of quenching rate decreases the T_{br} value of the amorphous alloys [1.171].

4. The value T_{br} greatly decreases in alloying the surface alloys with surfactants. The experimental results show [1.172] that elements such as Te, Se or Sb have a strong embrittling effect on the amorphous alloys based on iron and nickel decreasing the value of T_{br} even when applied in small quantities. In addition, Fig. 1.32 shows the results of determination of T_{br} using the $\varepsilon_f(t_{ann})$ dependence

for the Fe–B alloy [1.74]. T_{br} greatly decreases when adding only 0.01 at.% Ce – approximately by 50°C.

A very interesting problem is the presence or absence of changes in the nature of the local plastic flow at room temperature with increase of the preliminary annealing temperature. The experimental results show [1.74] that after different heat treatment conditions, the density of the shear bands and their average length in the uniaxial tensile test or bending do not change significantly. In particular, there are no changes in the distribution parameters and geometry of the bands after heat treatment in the range of T_{br}. The average local thickness of the shear bands also does not change in the temperature range corresponding to the formation of temper brittleness.

The structural relaxation stage, preceding temper brittleness, also did not cause any changes in the fractographic special features. A characteristic feature is the combination of the regions of almost smooth cleavage with the river-like pattern, formed by a system of veins. The only significant difference in the fractographic features after annealing at temperatures close to T_{br} is the appearance of micropores in the alloys in which they are not detected in the initial condition, and the increase of the density of the micropores in the amorphous alloys in which they were already detected in the initial condition. After annealing, leading to embrittlement, the mechanism of nucleation and growth of the cracks changes qualitatively. Electron microscopic studies show [1.74] that in this case the cracks nucleated at the micropores and the number and size of the latter greatly increased as a result of thermal effects. In front of each crack formed at the pores there is a distinctive region of the local plastic flow (Fig. 1.35)

This again confirms the previously made conclusions according to which the susceptibility to plastic flow in the amorphous matrix is still very high, and the cleavage cracks are in fact quasi-brittle cracks regardless of the macroscopic brittle nature of fracture. The micropores in the amorphous matrix are not only the source of microcracks but also facilitate their further propagation determining in a number of cases the trajectory of propagation of the main crack (Fig. 1.35).

If we are discussing the changes in the structure of the amorphous alloys in the stage in which embrittlement takes place, no changes were observed in this structure by means of direct diffraction methods [1.173]. Changes in the amorphous structure are not recorded by the radial distribution function correlation functions nor by electron

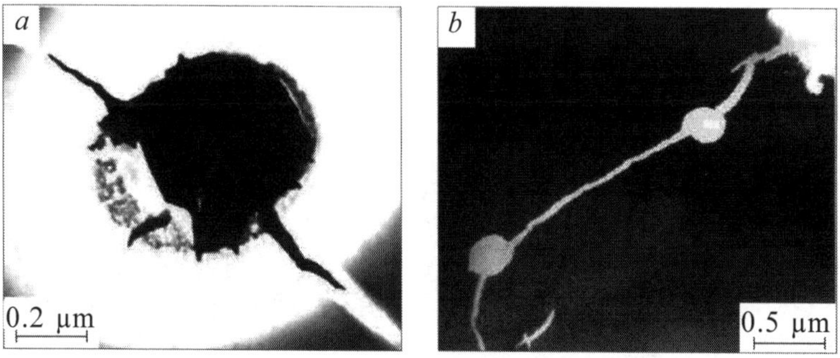

Fig. 1.35. Nucleation (*a*) and propagation (*b*) of cracks at the submicropores in the Fe–B alloy; dark field (*a*) and bright field (*b*) images, produced by the TEM method.

microscopic images in the regime of formation of amplitude or phase contrast. Only the method of small-angle X-ray scattering detects a significant shift in the distribution and average size of the free volume regions [1.174]. As shown by dilatometric experiments, the value T_{br} corresponds with the temperature range of the most active course of the processes of densening of the amorphous matrix as a result of the annihilation of the excess free volume.

In [1.175] it is reported that the brittleness in annealing of the $Fe_{27}Ni_{53}P_{14}B_6$ alloy is observed in the temperature range characterised by the most marked change of electrical resistivity and Curie temperature. Consequently, it may be concluded that the brittleness is associated with the formation of a more stable short-range order structure in the amorphous matrix.

Two groups of models have been proposed to explain the nature of temper brittleness of the amorphous alloys:

1. The 'segregation' model [1.81] which explains the brittleness by the formation of segregations of the atoms–metalloids in certain areas of the amorphous matrix.

2. The 'crystalline' model [1.176] which links the brittleness with the formation in the amorphous matrix of a distinctive short-range order or crystalline phases of a specific type.

Each model is based on a specific set of indirect experimental data but cannot explain a number of experimental factors confirming the alternative model or not fitting in any of these models. For example, the 'segregation' model contradicts the data for the large decrease of the values of T_{br} when adding small quantities of surfactants. In most cases, the conclusions on the realisation of a specific model are 'structureless'. They were proposed not on the basis of details structural investigations of the relationships of plastic flow in fracture in transition through T_{br} and were proposed mostly on the basis of examination of a number of external effects on this characteristic and subsequent, not always justified conclusions according to which the temper brittleness is associated with the structural relaxation processes in the amorphous structure. This is also indicated by the agreement of the temperature ranges of structural relaxation and the occurrence of temper brittleness observed in the studies (on alloys of the Fe–B, Fe–P, Fe–Ni–P, Fe–Ni– P–B systems) and also similar values of the activation energy of structural relaxation and embrittlement (for the $Fe_{40}Ni_{40}P_{14}B_6$ alloy). However, none of the studies propose in this case a specific mechanism according to which to structural relaxation could lead to extensive embrittlement from the viewpoint of the physics of plastic deformation and fracture.

Taking into account the general considerations on the nature of the brittle state, it is important to note two possible processes which may lead to the loss of macroplastic ductility of the amorphous alloys as a result of the thermal effects: 1. Decrease of the susceptibility to the plastic flow, and 2. Easier process of formation and subsequent propagation of the cracks. As already mentioned previously, the susceptibility to plastic shape changes in the amorphous alloys remains almost unchanged after annealing at temperatures higher than T_{br}. In particular, this results from the fact that the value of the parameter δ, characterising the capacity of the amorphous matrix for the plastic flow after such heat treatment is close to the appropriate value for the alloy in the initial condition and does not undergo any significant changes in the range of T_{br} (Fig. 1.36).

To explain the nature of the processes forming the basis of the mechanics of fracture of amorphous alloys and playing a significant role in the appearance of tempered brittleness, an important contribution was provided in the studies of Kimura and Masumoto [1.178, 1.179]. The ductile–brittle transition in annealing is explained by the decrease of the microfracture stress below the maximum value of the yield limit in the longitudinal direction. Consequently, using the fracture criterion proposed by Kimura and Masumoto, it may also be concluded that the temper brittleness is associated with the more favourable conditions for the fracture process and not with the inhibition of the plastic flow process. In this case, the characteristic of the mechanical behaviour of the system such as the microfracture

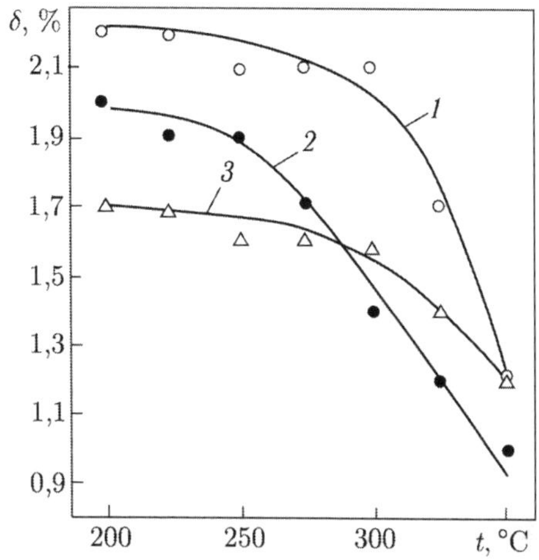

Fig. 1.36. The effect of microalloying with surfactants on the dependence of ductility of the room temperature on the preliminary annealing temperature for six hours of the $Fe_{83}B_{17}$ alloy: 1) no addition; 2) addition of Sb; 3) addition of Nb.

stress becomes the most important physical parameter characterising the brittle fracture susceptibility of the amorphous alloys.

It is also important to note that the microfracture stress is a structure-sensitive characteristic of the material and tends to decrease rapidly under the thermal effects leading to temper brittleness. What are the structural reasons for the large decrease of the microfracture stress in annealing of the amorphous alloys?

All the previously mentioned experimental data and also the above-mentioned thermal activation nature of the temper brittleness can be used to make an unambiguous conclusion on the relaxation nature of this phenomenon. As shown is section 1.1, the structural relaxation of the amorphous alloys is itself a very complicated phenomenon, including several processes which are in close relationship and take place with greatly different rates in different temperature ranges. It is necessary to clarify which structural relaxation processes lead to a change of the fracture mechanism and, correspondingly, to a large decrease of the microfracture stress under specific thermal effects.

A number of experimental factors will now be discussed:

1. T_{br} coincides with the temperature range of the most active occurrence of the processes of densening of the amorphous matrix as a result of the annihilation of the excess free volume (Fig. 1.37)

2. The increase during structural relaxation of the number and dimensions of the micropores recorded by the microscopic examination methods (Fig. 1.35);

3. The appearance in the temper brittleness stage of a high rate of crack formation at the micropores (Fig. 1.38).

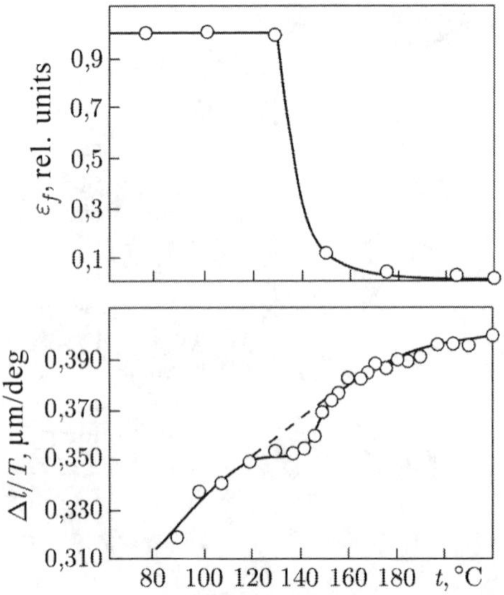

Fig. 1.37. Comparison of the dependences $\Delta l/T$ and $\varepsilon_f(t_{ann})$ for the Fe–Ni–P alloy. [1.74]./

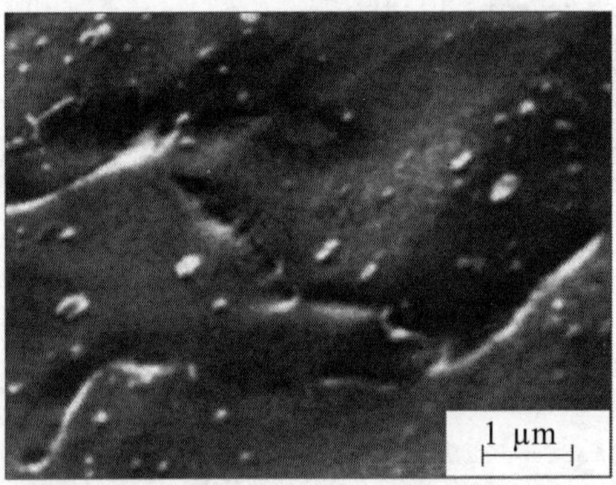

Fig. 1.38. Fracture surface of the Fe–B amorphous alloy annealed at temperatures higher than T_{br} (scanning electron microscopy in reflected electrons).

4. A large decrease of T_{br} when adding to the amorphous alloys atoms of tin, antimony or cerium resulting in the qualitative change of the microfracture mechanism as a result of the effect on the effective surface energy of the cracks nucleated and propagating in the material (see section 1.2.4).

5. The amorphous alloys produced by spraying and characterised by higher porosity have considerably lower values of T_{br} in

comparison with the alloys produced by the melt quenching methods (1.180);

6. The production parameters of the alloys in melt quenching have a significant effect on both the ductility in the initial state and on the value of T_{br} [1.171]. It is important to mention two circumstances: 1) as the quenching rate increases, the temperature of the most extensive compaction of the amorphous alloy increases; 2) the temperature range of intensive compaction is always identical with the value of T_{br}.

Obviously, all these results indicate that the controlling role in the temper brittleness phenomenon is played by the excess free volume and the nature of its evolution under the thermal effect on the amorphous structure. In [1.181] this approach was supplemented by the assumption of the development of zones enriched with the metalloid in the amorphous matrix. The formation of these zones should result, according to the author, in the formation in the amorphous matrix of a region in which the amount of the free volume is considerably smaller than the average value for the entire alloy and, in particular, they cause brittleness in annealing.

Thus, the main mechanism of temper brittleness is the relaxation mechanism which, however, contains the elements of the segregation and crystalline models.

Taking into account the restrictions for the heat treatment of the amorphous alloys caused by temper brittleness, it is also interesting to apply this effect to the amorphous matrix which would reduce the rate of the process of coalescence of the micropores and would consequently increase the temperature threshold of embrittlement.

At the present time, these influences include the thermomechanical [1.182, 1.183], ultrasound [1.183, 1.184] and radiation [1.171] treatments. Thermomechanical treatment is based on the application of a constant load creating uniaxial elastic stresses at temperatures which are considerably lower than the glass transition point of the amorphous alloy but slightly higher than room temperature.

In [1.184] it was attempted to develop a mechanism of the ultrasonic effect which would enable to influence the thermal and time stability of relatively large volumes of the amorphous ribbon. The experimental results for the effect of ultrasound on the temperature T_{br} for the $Fe_{75}Ni_2B_{13}Si_{10}$ alloy are presented in Fig. 1.39. It may be seen that at ultrasound treatment of an amorphous ribbon moving at the given speed ($v_s = 1.3 \cdot 10^{-4}$ m/s) using a flat oscillating end of the waveguide–emitter increases the embrittlement

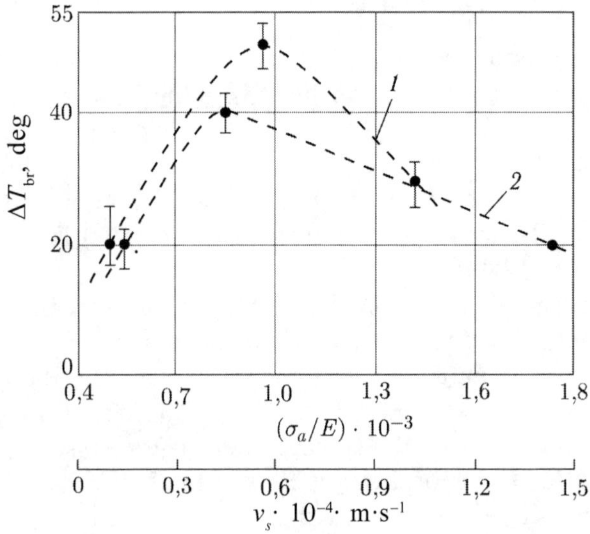

Fig. 1.39. The dependence of the increase of T_{br} on the value of σ_0/E (curve 1) and the speed of travel of the ribbon in the treatment zone v_s (curve 2); 1) $v_s = 1.3 \cdot 10^{-4}$ m/s, 2) $\sigma_0/E = 0.286$ [1.183].

temperature of the alloy in the entire investigated range of the amplitude of alternating stresses σ_0 (curve 1). In this case, the dependence $\Delta T_{br}(\sigma_0)$ has the form of a curve with a maximum of the value of ΔT_{br} at a specific value of σ_0. The identical form is found for the dependence $\Delta T_{br}(v_s)$ at a constant value σ_0 = 45.7 MPa (curve 2) in the conditions of ultrasound treatment of the amorphous ribbon using spheres [1.184].

In [1.171] experiments were carried out with the irradiation by thermal neutrons of the $Fe_{40}Ni_{40}B_{20}$ amorphous alloy in the initial condition and after annealing above T_{br}. The alloy was transferred to the brittle state. Figure 1.40 shows the change of the ductility parameter in bend testing ε_f in dependence on the strength of the neutron flux acting on the specimen (in annealing prior to radiation above T_{br}). Starting with the flux of $2 \cdot 10^{17}$ cm^{-2}, the amorphous alloy is plasticised: the value of ε_f increases to unity. This in fact is equivalent to the increase of T_{br} because annealing at this temperature no longer causes rapid embrittlement. As in the case of the thermomechanical or ultrasound effect, there is the optimum value of the neutron flux increasing the ductility equal to 10^{17}–10^{18} cm^{-2}. When the flux density of 10^{18} cm^{-2} is exceeded, embrittlement again takes place. Approximately at the same value of the neutron

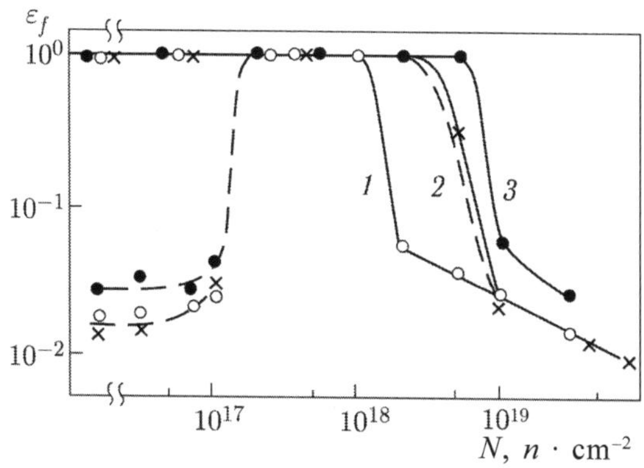

Fig. 1.40. Change of the ductility parameter ε_f in dependence on the dose of preliminary neutron irradiation N of the $Fe_{40}Ni_{40}B_{20}$ alloy. The thickness of the ribbon specimens, μm; 1) 50; 2) 40; 3) 30; the alloy was annealed prior to radiation at temperatures higher than T_{br} [1.171].

flux the initially ductile specimens not subjected to preliminary heat treatment are embrittled.

Therefore, temper brittleness is a structure-sensitive characteristic of the amorphous alloy and can be stimulated or depressed by exerting a corresponding effect on the regions of potential nucleation of quasi-brittle cracks in the amorphous matrix. This effect of stress annealing, ultrasound oscillations or irradiation with high-energy particles on T_{br} is another confirmation of the relaxation nature of the temper brittleness of the amorphous alloys. In addition, this effect confirms the controlling role in embrittlement of micro-discontinuities which are a direct consequence of the existence of the excess free volume in the amorphous alloys.

The thermal–time stability is positively influenced by the deposition of thin crystalline layers on the surface of the ribbon specimens of the amorphous alloys. In [1.185, 1.186] coatings of nickel, titanium and barium titanate were deposited by vacuum spraying in the pulse conditions equipment in the pulsed conditions in which the specimens could not be heated to temperatures higher than 353–373 K. The coating thickness was ~100 nm. The experimental results show that the increase of the thermal and time stability of the mechanical properties of the amorphous alloys is influenced most extensively by the coating of chemical purity nickel increasing the temperature of the start of the ductile–brittle transition at low temperature annealing by a factor of 3–3.5.

1.4. Magnetic properties

Recently, extensive studies have been carried out to investigate the magnetism of amorphous metals and alloys and special attention has been paid to the practical application of the amorphous magnetics. At the present time, the magnetically soft ribbon amorphous ferro- and ferrimagnetics are used. They are the alloys of transition metals with metalloids.

The actual possibility of existence of the amorphous ferromagnetics was reported for the first time by A.I. Gubanov in 1960 [1.187]. The existence of ferromagnetism in the amorphous iron ribbon source described for the first time in 1964 [1.188]. Since then, the explanation of the fundamental reasons for magnetism in the amorphous alloys has been the subject of extensive investigations.

1.4.1. Ferromagnetism and ferrimagnetism of amorphous metals

The properties of the permanent magnets and magnetic cores, produced from crystalline metallic alloys and chemical compounds, are based on the ferromagnetism phenomenon. In particular, it should be noted that the source of magnetism is the existence of the magnetic moment formed as a result of the inherent spin moment of the pulse of the electron. The substances, capable of strong magnetisation which will be referred to as magnetics, can be divided into the so-called ferromagnetics and ferrimagnetics. In the ferromagnetics, all the magnetic moments of the atoms are parallel to each other, in the ferrimagnetics the magnetic moments of the atoms are antiparallel and have different magnitude because the total moment differs from zero. The main reason for the formation of the ferromagnetic state of spontaneous magnetisation in this substances is the internal structure of the atoms.

The ferromagnetism is detected in $3d$-transitional metals (iron, cobalt, nickel), in gadolinium and in some other rare-earth metals, and also in alloys based on them and in intermetallic compounds. The ferrimagnetics are complex oxides, containing ferromagnetic elements. Since the above substances are crystalline, it could be assumed that for the parallel ordering of the magnetic moments we must have the irregular distribution of the atoms. However, in 1947 A. Brenner [1.189] reported the ferromagnetism phenomenon in the Co–P amorphous ribbon, produced by electrolytic deposition. Later,

A.I. Gubanov [1.187] confirmed theoretically that the regularity and symmetry of the atomic configurations are not necessary for the ordering of the magnetic moments. This confirmed that the ferromagnetism may be observed not only in the crystals but also in the liquids and amorphous solids.

Figure 1.41 shows the simple case of the ferromagnetic state – the magnetic atoms are distributed without ordering in space but all the magnetic moments are arranged parallel to each other. It is characteristic that in this case the magnetic polarisation vector has a strictly fixed direction, and the spontaneous magnetisation tends to saturation.

Figure 1.42 shows another case: here the magnetic moments tend to reduce each other and the ferromagnetic state is not saturated.

However, since the angle between the magnetic moments is not equal to 180°, spontaneous magnetisation may take place. A similar magnetic state forms when in addition to the relatively weak volume interaction there is also magnetic anisotropy leading to the formation of disordering in the distribution of the magnetic moments of the atoms. In the amorphous state where there are local differences in the atomic configurations, the magnitude of magnetic anisotropy and its direction should also locally differ. Consequently, the parallelity between the magnetic moments, determined by the volume interaction, maybe partially disrupted and as a result of the competition of the ordering in disordering processes there may be spin configurations similar to those shown in Fig. 1.42. The presence of local magnetic anisotropy has only a flat effect on the magnitude of spontaneous magnetisation, and in this case the Curie temperature decreases [1.190, 1.191].

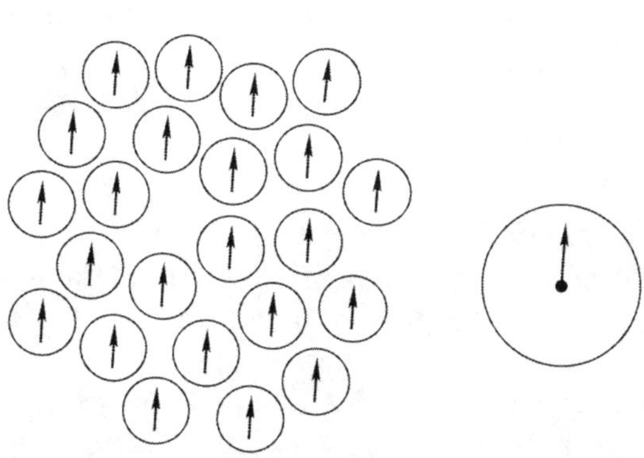

Fig. 1.41. The simple amorphous ferromagnetic.

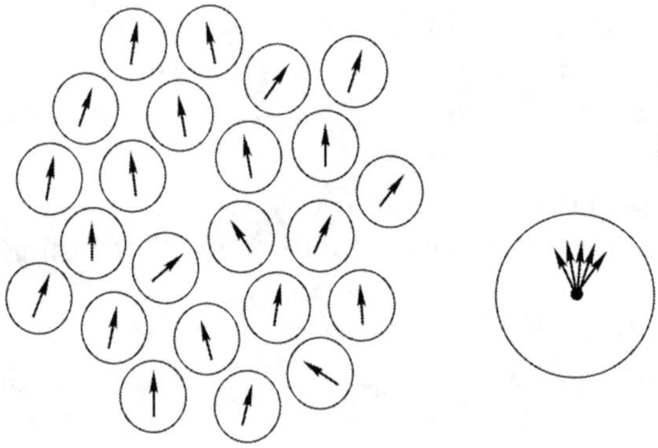

Fig. 1.42. The disorderd amorphous ferromagnetic.

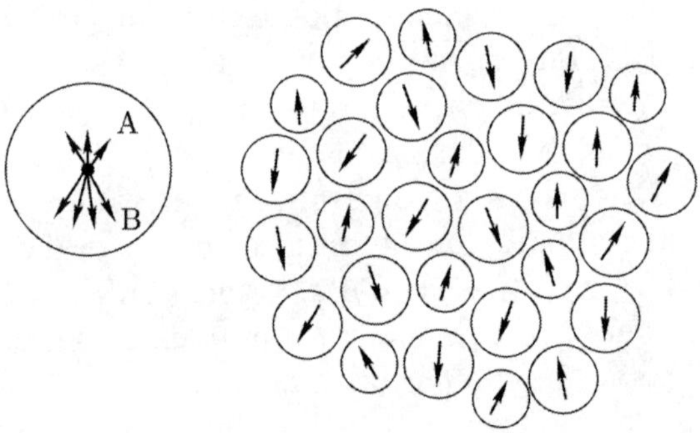

Fig. 1.43. Amorphous ferrimagnetic.

In amorphous metals, there is another type of magnetic disordering observed in various crystalline materials such as oxides (ferrites) and, in particular, it is ferromagnetism. If the AA–BB interaction in the amorphous alloy, containing two types of magnetic atoms, A and B, is positive, and the A–B interaction zone negative, we obtain the state in which the magnetic moments A and magnetic moments B are opposite to each other (Fig. 1.43).

When the magnetic moment *B* is greater than the magnetic moment *A*, or vice versa, spontaneous magnetism appears which is determined as ferrimagnetism. In Fig. 1.43, the magnetic moments *A* and *B* are anti-parallel, but the magnetic moments of the atoms of the same type can be disoriented under small angles and therefore, as in the case shown in Fig. 1.42, the local magnetic anisotropy may have a significant effect. Consequently, depending on the local oscillations of the direction of the magnetic moment, the ferrimagnetics, like the ferromagnetics, may be regarded as disordered ferrimagnetics or super ferrimagnetics [1.192]. As an example of the amorphous ferrimagnetics one can mention the amorphous ribbon alloys based on rare-earth metals (REM) with Fe and Co [1.193, 1.194] which are highly promising materials for application in magnetic recording devices.

The magnetic behaviour of these alloys is characterised by the so-called compensation effect, typical of ferrimagnetics in general [1.195].

In the compensation effect when the sum of the magnetic moments of the atoms of the type *A* and the sum of the magnetic moments of the atoms of the type *B* are equal to each other, spontaneous magnetisation completely disappears. Important factors here are the concentrations of the atoms *A* and *B* (the chemical composition of the alloy) and temperature. The composition of the alloy and temperature at which the compensation effect becomes evident are referred to as the compensation composition and compensation temperature, respectively. Varying the concentration of the atoms in the vicinity of the compensation composition, it is possible to regulate the magnitude of spontaneous magnetisation, including by making it sufficiently small. On the other hand, if there are large differences between the magnetic moments *A* and *B*, strong induced magnetic anisotropy may form. An important consequence in this case is the formation of the bubble domain structure in the amorphous magnetic thin films.

These considerations show that in the amorphous state characterised by the absence of ordering in the distribution of the atoms there may be an ordered magnetic state in which the magnetic moments are more or less parallel. This is the reason for the formation in the amorphous state of the strong spontaneous magnetisation, i.e., ferro- and ferrimagnetism.

The (Fe, Co, Ni)–metalloid amorphous alloys are also characterised by ferromagnetism and are simple ferrimagnetics (Fig. 1.42).

Measuring the magnetisation along the axis in the direction of the length of the amorphous ribbon (the axis of the ribbon) we may detect the phenomenon of magnetic saturation in the hysteresis loop, exactly as in the conventional crystalline ferrimagnetics. This shows that the internal magnetisation in the amorphous metallic ribbons is divided into parts – magnetic domains. It is assumed that the magnetisation of the amorphous metals takes place by displacement of the boundaries of the magnetic domains and rotation of the spontaneous magnetisation vector.

At the start of the magnetisation process the magnitude of magnetisation increases in proportion to the strength of the external magnetic field, but with an increase of the strength of the field the magnetisation tends asymptotically to some maximum value because the asymptotic law of approach to saturation is also satisfied for the amorphous ferromagnetics. In the magnetically soft amorphous metallic ribbons, the magnetisation saturation is obtained at very high values of the strength of the external magnetic field (in many cases, these values equal (8–80) · 10^3 A/m). The magnitude of spontaneous magnetisation decreases with increase of temperature and at the Curie point (T_c) becomes equal to 0. In the development of the magnetic materials it is necessary to use those in which the spontaneous magnetisation M_s in the range from 0 K to room temperature at a high Curie temperature would be relatively high.

The magnetism carriers in the amorphous alloys are the atoms of the transitional metals – iron, cobalt, nickel or chromium, etc, and the atoms stabilising the amorphous state (the metalloids of the type of phosphorus, boron, carbon, silicon, germanium) are non-magnetic. Therefore, the magnetic moment μ is determined only by the magnitude of the magnetic moment of the magnetic atoms of the metal μ_f and by their concentration c in the alloy:

$$\mu = c\mu_f, \tag{1.44}$$

where μ_n is the magnetic moment of the metalloid atom, and its value is equal to 0.

When alloying the amorphous alloys with the transition non-ferromagnetic metals such as manganese, chromium, vanadium, etc, the value of μ_f changes. When alloying iron with manganese, chromium or vanadium in the iron–metalloid amorphous alloys, the value of μ_f decreases almost linearly with the increase of the concentration of the alloying element. The effect of the alloying elements on the value μ_f increases in the sequence Mn, V, Cr, which differs from the crystalline alloys Fe–(Mn, Cr, V) of similar compositions.

It may be concluded that the amorphous metals are characterised by a 'dilution' of the ferromagnetism by the atoms of manganese, chromium and vanadium. In the cobalt-based alloys, μ_f decreases monotonically in alloying with chromium and vanadium, but when manganese is added μ_f initially increases and then starts to decrease only when the manganese concentration is greater than 0.1.

The change of μ_f in the Ni–Mn alloys is similar. Consequently, using this analogy, it is possible to make several assumptions regarding the formation of amorphous ferromagnetism.

The Co–Mn alloys in comparison with other cobalt alloys show relatively strong magnetism. This is important from the viewpoint of practical application of these alloys.

In the Fe–Co and Fe–Ni alloys the differences in the Curie temperature may reach up to 100°C, depending on the type and concentration of the metalloid atoms, but the temperature T_c may be quite high. This reflects the nature of the exchange interactions between the magnetic moments of the atoms which are stronger between the atoms of different source than in the atoms of the same type. The strongest volume interaction is observed between the atoms of iron and cobalt. A short review of the reports for the Curie temperature of the amorphous alloys may be described as follows:

1) the Fe–Co alloys have a high Curie temperature;

2) at a low cobalt content of the Ni–Co alloys the temperature T_c is situated in the range of the climatic temperatures;

3) at a high iron content of the Fe–Co alloy, in addition to the increase of the magnetic moment there is also a decrease of the Curie temperature.

The amorphous metallic materials are characterised by high magnetic permittivity, and to understand this phenomenon it is necessary to consider the main relationships governing the magnetisation processes.

In the amorphous structure shown in Fig. 1.41 all the magnetic moments are parallel to each other, and the direction of the total magnetic moment in this case coincides with the direction of each magnetic moment. This is the ideal case in which neither the local changes of the short-range order, density or chemical composition nor any other deformation are taken into account. There is no magnetic anisotropy in this case. This situation never forms in the crystalline state. In addition, since the disordered amorphous phase is macroscopically homogeneous in the entire volume of the sample, the properties should also be homogeneous. The homogeneity of the structure is indicated mainly by the fact that in the amorphous state there are no defects which would prevent the displacement of the domain boundaries (usually ~10 nm thick), such as grain boundaries, pores, inclusions, etc. It may be expected that because of these features the amorphous ferromagnetics have extremely high magnetic permittivity. In the so-called zero ferromagnetics, having

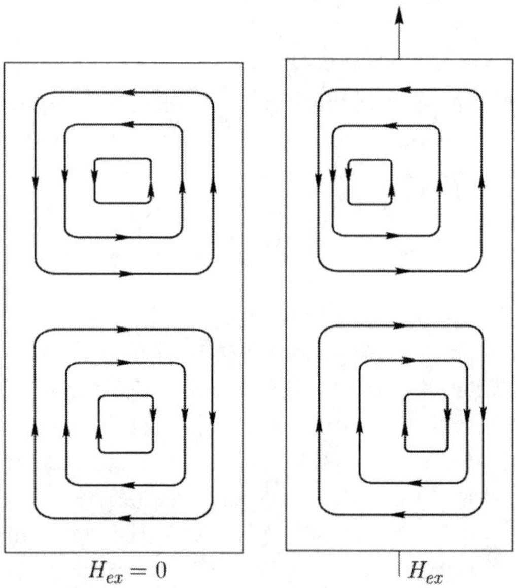

Fig. 1.44. The distribution of magnetisation in the completely isotropic ferromagnetic [1.196] (H_{ex} is magnetic anisotropy).

ideal magnetic anisotropy, the parallelity of magnetic moments is maintained only by the energy of bulk interaction, and the magnetic flux is closed inside the sample as a result of competition with magnetostatic energy. As indicated by the scheme in Fig, 1.44, in this case the direction of rotation of the magnetic moment in some parts of the sample is the same and this may lead to the formation of the so-called circular domain structure.

In crystalline substances in which the direction of the magnetic moment and of the easy magnetisation axis in different domains differ, the situation is completely different. Kittel [1.196] showed that in substances in which the magnetic anisotropy tends to zero, the width of the domain walls greatly increases and this may result in the formation of the domain structure similar to the circular structure. Actually, the circular domain structure is observed in, for example, permalloys but, as indicated by electron microscopic studies, the structure is highly local and has a characteristic fine structure [1.197]. In this case, since the magnetisation process takes place not by the movement of the circular domains and the domain boundaries do not travel to any large distances, the remagnetisation losses are reduced only to the classic losses through eddy currents which are localised in the region of the circular domain. This reduces the total losses and is very important for the materials used as the cores of transformers.

Thus, the amorphous metals as magnetically soft materials are highly attractive. The high magnetic permittivity of the amorphous metals is most evident and, therefore, special attention has been paid to this property. For example, one can refer to the study [1.198] as one of the earliest studies (1967) concerned with AC magnetisation of the $Fe_{80}P_{12.5}C_{7.5}$ alloy. Here, however, the coercive force was ~240 A/m and, therefore, this alloy should not be qualified as a magnetically soft material. Later studies carried out on ribbons of amorphous alloys based on iron, nickel and cobalt, produced by melt quenching, determined the characteristics of the process of static magnetisation: on the $Fe_{80}P_{13}C_7$ alloy [1.199] and then on the (Fe–Ni)–P–B alloys [1.200]. In these materials the coercive force at the saturation induction of 1.0–1.3 T was 0.8–8 A/m and, therefore, these materials are promising as magnetically soft materials. Subsequently, even better characteristics were obtained for the $Co_{70}Fe_5Si_{15}B_{10}$ [1.201–1.203] and CoFePB alloys [1.204, 1.205] characterised at the same time by almost zero magnetostriction. Since then, a large number of investigations have been carried out in this area.

Previously, it was assumed that since the amorphous alloys have the isotropic and magnetically homogeneous structure, they should be easily magnetised. This may be confirmed by the fact that the coercive force does not exceed 8 A/m. However, the amorphous ferromagnetics may show anisotropy in magnetisation, i.e., the domain walls overcome the potential barrier during their movement. This indicates that the amorphous metallic ribbons are not always in the ideal homogeneous magnetic state. The magnetic anisotropy of the amorphous alloys as a consequence of the inhomogeneity of their magnetic state does not fail completely in heat treatment but nevertheless rapidly decreases as a result of the processes of structural relaxation and, consequently, the amorphous alloys become considerably magnetically softer. The possibility of improving the magnetic properties of the amorphous alloys is now a stimulus for the development of new chemical compositions and improvement of the methods of production and the heat treatment conditions. The search for the optimum compositions and the conditions of improving the magnetic properties supports in the final analysis better understanding of the physics of processes of magnetisation of the amorphous ferromagnetics.

1.4.2. Magnetically elastic phenomena in amorphous alloys

Magnetoelastic damping, the ΔE-effect and other phenomena, relating to magnetically soft properties, have been attracting the attention of researchers for many years both from the theoretical and practical viewpoint [1.206]. Recently, the interest in these phenomena has become even stronger as a result of the investigation of amorphous ferromagnetics where they are far more efficient in comparison with the crystals [1.207, 1.208]. For example, the magnitude of the ΔE-effect in some amorphous alloys may reach several hundreds of percent [1.209]. The magnetoelastic properties are determined by the effect of the magnetic order on the elastic characteristics and are associated with the existence of the magnetostriction effect. Although the magnetostriction is a second order effect in the magnetism theory, in the case of the amorphous metallic alloys which are anomalously magnetically soft the effects are in many cases considerably more important than the magnetic and elastic defects of the first order.

The magnetostriction is the most important parameter of the ferromagnetics and is interesting both from the viewpoint of the physics of the magnetic state and practical application. The data on the magnetostriction of the amorphous metallic alloys are essential for solving fundamental problems of the magnetism of disordered structures and provide direct information on the nature of the orbitals of the magnetic electrons.

In spontaneous appearance of the magnetic order below the Curie point the ferromagnetic material changes its form and volume. The application of external mechanical stresses in this case results in the formation (in addition to the purely elastic deformation) of additional magnetostriction deformation, i.e., the total deformation ε_{ij} is equal to:

$$\varepsilon_{ij} = \varepsilon_{ij}^{s} + \varepsilon_{ij}^{m}, \tag{1.45}$$

ε_{ij}^{s} are the deformation components of the ferromagnetic in the saturated magnetic field; ε_{ij}^{m} are the components of magnetostriction deformation.

Experimental investigations of many amorphous iron-based alloys show [1.210] that their magnetic effect, leading to the increase of volume with increase of spontaneous magnetisation M_s, may compensate or even exceed the conventional thermal expansion as a result of the anharmonism of the oscillations of the atoms in a wide temperature range. For example, the thermal expansion coefficient of the $Fe_{83}B_{17}$ amorphous alloy in the temperature range 200–600 K is equal to 0, i.e., this alloy is a typical Invar alloy. Analysis of the experimental results for the Fe–B, Fe–P, Co–B and other amorphous alloys shows [1.211] that the zero coefficient of thermal expansion is determined by the effective number of electrons which is estimated using the charge transfer model as follows

$$n_{\text{eff}} = n_{\text{TM}} + \frac{1}{1+\sum_i x_i} x_i q_i, \qquad (1.46)$$

where n_{TM} is the number of $3d$ + $4s$ electrons for each transition metal, x_i is the concentration of the i-th metalloid, q_i is the number of donor electrons of the i-th metalloid. For B, Si and P, the value q_i is equal to 1, 2 3, respectively. For the above alloys, irrespective of the type of metalloid, the Invar type alloys are characterised by n_{eff} = 8.2. The same value of n_{eff} corresponds to the magnetic instability in the amorphous state and to the maximum value of the magnetic moment for the amorphous iron produced on the cooled substrate.

The change of the shape of the sample (deformation at constant volume) formed as a result of the presence of the uniaxial anisotropy causes that the amorphous ferromagnetic has the linear magnetostriction λ_S.

The experimental values of λ_S for the majority of the investigated amorphous alloys based on Fe–Co and Fe–Ni at 20–25 at.% of the metalloids are in the range $(-10 \div 45) \cdot 10^{-6}$. Applications of electronic and electromagnetic devices require amorphous alloys in the form of thin films and foils or with a high value of magnetostriction or with a negligible magnetostriction (close to 0). For example, for sensors of various types and magnetically elastic transducers we require the materials with a high value of λ_S. The highest value of λ_S, as in the crystalline state, is typical of the amorphous alloys based on rate-earth metals. For example, the amorphous alloys with the composition close to the composition $Ti_{50}Co_{50}$ [1.212] are characterised by giant magnetostriction ($\lambda_S \sim 2 \cdot 10^{-4}$) in considerably smaller saturating fields (H = 3–4 kOe) than those required for the crystalline analogues (H > 100 kOe). It is assumed that the magnetostriction in the Tb–Co amorphous alloys is determined mainly by the process of rotation of the magnetic moments of Tb which form a 'fan' in the zero magnetic field. The angle of exposure of the 'fan' is determined by the competition of local magnetic anisotropy and the effective field of the volume interaction. In the field $H > 0$ the magnetic moments of the Tb ions are oriented along the field and this results in the formation of magnetostriction deformation of the giant type. Further studies of the development of the conditions of production of amorphous alloys with the giant values of λ_S and studies explaining these phenomena are obviously very important.

The devices and systems working at high frequencies require amorphous alloys with the magnetostriction close to zero because the magnetoelastic effects impair the magnetically soft properties. In the initial stage of examinations of the amorphous alloys successes have been achieved with the production of amorphous alloys based on cobalt and iron with zero magnetostriction. Later, different types of the amorphous alloys based on cobalt were produced by spraying. These alloys contained non-magnetic elements Zr, Hf, Ti, Nb, Ta, W, Mo as glass forming elements. Alloys with positive or negative λ_S were produced. They also included the alloys $Co_{85}Nb_{7.5}Ti_{7.5}$ and $Co_{86}Nb_7Zr_{3.5}Mo_{3.5}Cr_{1.5}$ with zero magnetostriction [1.213, 1.214].

ΔE-effect in amorphous alloys

The presence of the magnetoelastic bond results in a change of the elasticity modulus which can be treated as a superposition of the individual contributions [1.215]:

$$E = E_p + E_m, \quad (1.47)$$

where E_p is the 'paramagnetic' elasticity modulus produced by extrapolation of the dependence $E(T)$ from the paramagnetic range; E_m is the 'magnetic' contribution of the elasticity modulus which consists of three components:

$$E_m = \Delta E_a + \Delta E_\omega + \Delta E_\lambda \quad (1.48)$$

here ΔE_a and ΔE_ω are determined by the change of the binding forces in the magnetic ordering and of the volume contribution under the effect of the magnetic field, respectively, and ΔE_λ characterises the change of the domain structure under the effect of external stresses.

Investigations of the ΔE-effects in the ferromagnetics show that the main contribution to the change of the elasticity modulus in the amorphous metallic alloys based on the transition metals iron, cobalt and nickel is associated with the change of the domain structure under the effect of external mechanical stresses.

Figure 1.45 shows the temperature dependence of the Young modulus E in the $Fe_{82}B_{18}$ amorphous alloys after isothermal annealing at $T = 573$ K for two hours [1.216].

The value E in the zero magnetic field is almost independent of temperature and, consequently, at temperatures lower than the Curie point this alloy is a typical Elinvar material. In the field $H = 112$ kA/m at room temperature the value of E changes approximately by 70% (curve 4) and this increase is due to the component ΔE_a and also ΔE_λ (component $\Delta E_\omega \approx 0$). However, the accurate determination of the components $\Delta E_a + \Delta E_\lambda$ is determined by the difference between E_s (measured in the saturation magnetic field) and the elasticity modulus, determined by extrapolation from the paramagnetic state, is difficult because the Curie temperature of the amorphous alloys is close to T_{br}. Consequently, the extrapolated curves are not determined accurately. Therefore, when studying the ΔE-effect in the amorphous alloys, attention is usually given to the component of the elasticity modulus, and the reasons for the formation of this component are closely linked with the existence of magnetostriction deformation ε_m.

Thus, on the basis of a large number of investigations it has been established that the iron-based alloys have excellent magnetically soft properties: low coercive force (0.5–1 A/m) and high saturation magnetisation, exceeding 1.4 T. Even higher characteristics were

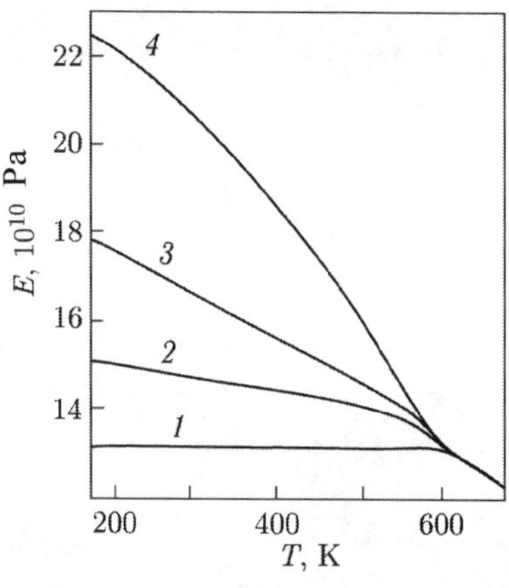

Fig. 1.45. Temperature dependence of the Young modulus of the $Fe_{82}B_{18}$ amorphous alloy in measurement in the magnetic field with a strength H (kA/m): 0 (1), 0.96 (2), 1.36 (3) and 112 (4).

obtained for the $Co_{70}Fe_5Si_{15}B_{10}$ alloys [1.217] and alloys of the Co–Fe–P–B system [1.218] with almost zero magnetoconstriction.

The magnetically soft properties of the amorphous alloys can be improved by relaxation annealing and annealing in a magnetic field. On the whole, the main characteristics of the magnetically soft amorphous alloys based on iron, cobalt and nickel are: high values of remanent induction and small losses in magnetic reversal; high values of magnetic permittivity at high (at a high iron content) or almost zero (at a high cobalt content) magnetostriction. The magnetic properties can also be improved by adding alloying elements, and the values of magnetic permittivity may reach 120 000 [1.3].

The general characteristics of the amorphous alloys include the following:

– the amorphous alloys based on iron or characterised by high values of remanent induction and low magnetic reversal losses;

– the Fe–Co– Ni amorphous alloys with a high iron concentration are characterised by high values of magnetostriction and magnetic permittivity, and a high cobalt content results in the almost zero magnetostriction and high magnetic permittivity;

– the amorphous alloys with a high nickel content have a low Curie temperature and high magnetic permittivity;

The magnetic properties of the amorphous alloys can be improved by low-temperature annealing (which reduces the internal stresses formed after quenching), annealing in a magnetic field (for both magnetically soft and magnetically hard metallic glasses), etc.

References

1.1. Glezer A.M., Molotilov B.V., Structure and mechanical properties of amorphous alloys. - Moscow: Metallurgiya, 1992.

1.2. Amorphous metal alloys: ed. F.E. Lyuborskiy. - Moscow: Metallurgiya, 1987..

1.3. Suzuki K., Fujimori H., Hashimoto K. Amorphous metals, ed. S. Masumoto. - Moscow: Metallurgiya, 1987.

1.4. Khilman Kh., Hiltsinger Kh. R. On the preparation of amorphous ribbons by the method of melt spinning, Fast-quenched metals, ed. M. Cantor. - Moscow: Metallurgiya, 1983. P. 30–34.

1.5. Zolotukhin I.V. Physical properties of amorphous metallic materials. - Moscow: Metallurgiya, 1986.

1.6. Alekhin V.P., Khonik V.A., Structure and physical patterns of deformation of amorphous alloys. - Moscow: Metallurgiya,, 1992.

1.7. Likhachev V.A., Shudegov V.E., Principles of organization of amorphous structures. - St. Petersburg: Publishing house of St. Petersburg University, 1999.

1.8. Glezer A.M., Molotilov B.V., Fiz. Met. Metalloved., 1990. No. 2. Pp. 5–28.

1.9. Glezer A.M., et al., Metallofizika. 1983. Vol. 5, No. 1. Pp. 29–45.

1.10. Zolotukhin I.V., Kalinin Yu.E. Usp. Fiz. Nauk, 1990. Vol. 160, No. 9. P. 75–110.

1.11. Zolotukhin I.V., Barmin Yu.V. Stability and relaxation processes in metal glasses. - Moscow: Metallurgiya, 1991.

1.12. Manokhin A.I., et al., Amorphous alloys. - Moscow, Metallurgiya, 1984. - 160 p.

1.13. Davis H.A. Formation of amorphous alloys. In the book: Amorphous metal alloys. - Moscow, Metallurgiya, 1987. - P. 16–37.

1.14. Kim Y.-W., Lin H.-M., Kelly T. F., Acta Met. 1989. V. 37, No. 1. P. 247–255.

1.15. Gleiter H., Nanostruct. Maters. 1992. V. 1. P. 1–19.

1.16. Andrievsky R.A., Glezer A.M., Fiz. Met. Metalloved., 1999. V. 88, No. 1. Pp. 50–73.

1.17. Andrievsky R.A., Glezer A.M., Fiz. Met. Metalloved., 2000. V. 89, No. 1. Pp. 91-112.

1.18. Handrich H., Kobe S. Amorphous ferro- and ferrimagnetics. - Moscow: Mir, 1982.

1.19. Wagner C.N.J., Ruppersberg H., Atom Energy Rev. 1981. Suppl. 1. P. 101.

1.20. Wagner C.N.J. In: Amorphous Metallic Alloys, Ed. F. Luborsky, Butterworths. - London, 1983. P. 58–73.

1.21. Gaskell P.H., J. Non-Cryst. Solids. 1979. V. 32, No. 1. P. 207–224.

1.22. Bakai A.S., Polycluster amorphous structures and their properties. Part 1. - Moscow: TsNIIatominform, 1984.

1.23. Gaskell P.H., J. Non-Cryst. Solids. 1985. V. 75, No. 2. P. 329–345.

1.24. Likhachev V.A., Khayrov R.Yu. Introduction to the theory of disclinations. - Leningrad State University, 1975.

1.25. Sadoc J.,F. J. Phys. Lett. 1983. V. 44, No. 17. P. 707–715.

1.26. Dobromyslov A.V., et al., Materialovedenie, 2001. No. 10. Pp. 43-46.

1.27. Betechtin VI, et al., Fiz. Tverdogo Tela. 1998. V. 40, No. 1. Pp. 85-89.

1.28. Primak W., Phys. Review. 1955. V. 100, No. 6. P. 1677–1689.

1.29. Primak W., J. Appl. Phys. 1960. V. 81, No. 9. P. 1524–1533.

1.30. Argon A.S., Kuo H.Y., J. Non-Cryst. Sol. 1980. V. 37. P. 241–266.

1.31. Gibbs M.R.J., et al., J. Mater. Sci. 1983. V. 18, No. 1. P. 278–288.

1.32. Kruger P., Kempen L., Neuhauser H., Phys. Stat. Sol. (A). 1992. V. 131. P. 391-402.

1.33. Van den Beukel A., Van der Zwaag S.A., Mulder A.L., Acta Met. 1984. V. 32, No. 11. P. 1895–1902.

1.34. Van den Beukel A., Huizer E. Scr. Met. 1985. V. 19, No. 11. P. 1327–1330.
1.35. Koebrugge G.W., Van der Stel J., Sietsma J., Van den Beukel A., J. Non-Cryst. Sol. 1990. V. 117/118, No. 2. P. 601–604.
1.36. Egami T. Atomic short-range order in amorphous metal alloys. In the book: Amorphous metal alloys. - Moscow, Metallurgiya, 1987. - P. 92–106.
1.37. Spaepen F., Acta Met., V. 25, No. 3. P. 407-415.
1.38. Taub A.I., Spaepen F., Acta Met. 1980. V. 28, No. 10. P. 1781–1788.
1.39. Spaepen F., Taub A.I., Plastic flow and destruction. - In: Amorphous metal alloys. - Moscow, Metallurgiya, 1987. -Pp. 228–256.
1.40. Gibbs M.R.J., Sinning H.-R., J. Mater. Sci. 1985. V. 20, No. 7. P. 2517–2525.
1.41. Kosilov A.T., Khonik V.A., Izv. RAN. Ser. Fiz., 1993. Vol. 57, No. 11. Pp. 192–198.
1.42. Kosilov A.T., Khonik V.A., Mikhailov V. A., J. Non-Cryst. Sol. 1995. V. 192 & 193. P. 420-423.
1.43. Bobrov O.P., Quasistatic and low-frequency mechanical relaxation of metallic glasses. Dissertation, Voronezh, 1996. - 116 p.
1.44. Mikhailov V.A., Creep of metallic glasses under inntense structural relaxation. Dissertation, Voronezh, 1998..
1.45. Van den Beukel A., Huizer E., Mulder A. L., Van der Zwaag S., Acta Met. 1986. V. 34, No. 3. P. 483–492.
1.46. Huizer E., Melissant I-I., Van den Beukel A., Zeitchrift fur Physikalische Chemie Neue Folge. Bd. 1988. V. 157, No. 1. P. 335-339.
1.47. Van den Beukel A., Sielsma J., Phil. Mag. B. 1990. V. 61, No. 4. P. 539–547.
1.48. Koebrugge G.W., Sietsma J., Van den Beukel A., J. Non-Cryst. Sol. 1990. V. 117/118, No. 2. P. 609–612.
1.49. Knuyt G., et al., Modeling Simul. Mater. Sci. Eng. 1993. V. 1. P. 437–448.
1.50. Kasardova A., Ocelik V., Csach K., Miskuf J., Phil. Mag. Lett. 1995. V. 71, No. 5. P. 257–261.
1.51. Chaudhari P., J. Phys. 1980. V. 41, No. 8. P. 267–271.
1.52. Popesku M., Thin Solid Films. 1984. V. 121, No. 2. P. 317–347.
1.53. Shi L.T., Chaudhari P., Phys. Rev. Lett. 1983. V. 51, No. 17. P. 1581–1583.
1.54. Kirsanov V.V., Orlov A.N., Usp. Fiz. Nauk, 1984. V. 142, No. 2. Pp. 219–264.
1.55. Bueche F., J. Chem. Phys. 1959. V. 30, No. 3. P. 748–752.
1.56. Turnbull D., Cohen M.H., J. Chem. Phys. 1970. V. 52, No. 6. P. 3038–3041.
1.57. Glezer A.M., Utevskaya O.L., Fiz. Met. Metalloved., 1984. Vol. 57, No. 6. Pp. 1198-1210.
1.58. Utevskaya O.L., Makarov V.Ya., Glezer A.M., In: Physics of Amorphous Alloys. - Izhevsk: Udmurt State University, 1984. - P. 32–36.
1.59. Brandt E.N., J. Phys. F: Met. Phys. 1984. V. 14, No. 2. P. 2485–2505.
1.60. Steinhardt P. J., Chaudhari P., Phil. Mag. 1981. V.A44, No. 6. P. 1375–1381.
1.61. Egami T., Vitek V., Non-Cryst. Sol. 1984. V. 62, No. 4. P. 499-510.
1.62. Egami T., Vitek V., Srolovitz D., Microscopic model of structural relaxation in amorphous alloys, Proc. Fourth Int. Conf. RQM. (Sendai, Japan). 1981. V. 1. P. 517-522.
1.63. Srolovitz D., et al., J. Phys. F: Metal Phys. 1981. V. 11, No. 12. P. 2209–2219.
1.64. Bakai A.S., Polycluster amorphous structures and their properties. I. Preprint of the Kharkov Institute of Physics and Technology. 84-33. - Moscow: TsNIIatominform, 1984.
1.65. Glezer A.M., Permyakova I. E. Deformatsiya i razrushenie materialov. 2006. No. 3. Pp. 2–11.

1.66. Noskova N.I., et al., Fiz. Met. Metalloved. 1996. V. 82, No. 6. Pp. 116–121.
1.67. Mikhailov V.A., Khonik V.A., Fiz. Tverd. Tela, 1997. Vol. 39, No. 12. Pp. 2186-2190.
1.68. Kosilov A.T., et al., Fiz. Tverd. Tela, 1997. P. 39, No. 11. Pp. 2008–2015.
1.69. Masumoto T., Murata T., J. Mater. Sci. Eng. V.V. 25, No. 1. P. 71–75.
1.70. Davis L.A., Mechanics of metallic glasses, Prep. Second Int. Conf. RQM. Cambr. Univ. - Cambridge, 1975.
1.71. Mulder A.L., Emmens W.C. et al. Influence of annealing and surface conditions on the strength and fatigue of metglass 2826. Int. Conf. Met. Glass: Science and Technology, (Budapest, Hungary). 1980. V. 2. P. 407–413.
1.72. Otselik V., et al., Fiz. Tverd. Tela, 2000. Vol. 42, no. 4. P. 679–682.
1.73. Betekhtin V.I., et al., Fiz. Tverd. Tela, 2000. Vol. 42, no. 8. P. 1420–1424.
1.74. Glezer A.M., et al., Fiz. Met. Metalloved., 1984. Vol. 58, no. 5. P. 991-1000.
1.75. Khonik V.A., Ryabtseva T.N., Scr. Met. Mater. 1994. V. 39, No. 5. P. 567-573.
1.76. Bobrov O,P., et al., Vest. Tamb. Univ., 2003. V. 8, No. 4. P. 525–527.
1.77. Bobrov O.P., et al., Fiz. Tverd. Tela. 1994. V. 36, No. 6. 1703–1709.
1.78. Aronin A.S., et al., Fiz. Tverd. Tela, 1988. Vol. 30, no. 10. P. 3160–3162.
1.79. Kobelev N.P., et al., Fiz. Tverd. Tela, 1999. Vol. 41, no. 4. P. 561–566.
1.80. Kalinin Yu. E. Internal friction and elastic modulus of some amorphous and quasi-amorphous metallic alloys based on lanthanum, copper, nickel. Dissertation, - Voronezh, 1980.
1.81. Pampillo C.A., Polk D.E., J. Mater. Sci. Eng. 1978. V. 33, No. 2. P. 275–280.
1.82. Dzyuba G.A., Fiz. Tverd. Tela, 1991. Vol. 33, No. 11. P. 3393–3399.
1.83. Bobrov O.P., et al., Fiz. Tverd. Tela, 1996. V. 38, No. 10. P. 3059–3065.
1.84. Spivak L.V., Khonik V.A., Zh. Teor. Fiz., 1997. Vol. 67, No. 10. P. 35–46.
1.85. Fursova Yu.V., Khonik V.A., Izz, RAZ, Ser. Fiz., 1998. V. 62, No. 7. P. 1288–1295.
1.86. Bobrov O.P., et al., Fiz. Met. Metalloved., 1996. Vol. 81, No. 3. P. 123–132.
1.87. Kosilov AT, Kuzmischev V.A., Khonik V.A., Fiz. Tverd. Tela, 1992. V. 34, No. 12. P. 3682–3690.
1.88. Zaichenko S.G., Zavod. Lab., 1989. V. 55, No. 5. P. 76–79.
1.89. Kimura H., Ast D. G. in: Proc. Fourth Int. Conf. RQM. (Sendai, Japan). 1981. V. 1. P. 475–478.
1.90. Ushakov I.V., Zavod. Lab. Diagnostika Mater., 2003. V. 69, No. 7. Pp. 43–47.
1.91. Pareja R., J. Mater. Sci. Letter. 1986. V. 5, No. 3. P. 287–289.
1.92. Inoue A., et al., Nanostruct. Mater. 1996. V. 7, No. 3. P. 363–365.
1.93. Fedorov V.A., et al., Izv. RAN. Ser. Fiz. 2005. V. 69, No. 9. P. 1369–1373.
1.94. Grigorovich V.K., Hardness and microhardness of metals. - Moscow: Nauka, 1976.
1.95. Shutin A.M., Korolev L.A., Zavod. Lab. Diagnostika Mater., 1988. Vol. 54, No. 8. P. 81–83.
1.96. Orlova N.A., Fiz. Khim. Obrab. Mater., 1996. No. 1. P. 43–49.
1.97. Permyakova I.Ye., Dissertation, - Tambov, 2004.
1.98. Semin A.P., et al., In: Proceedings of VI Intern. Sympos. "Modern problems of strength" (Staraya Russa, 20–24 October 2003). 2003. V. 1. P. 239–243.
1.99. Glezer A.M., Utevskaya O.L., Development of a technique for measuring the mechanical properties of thin ribbon materials, Compositional precision materials: Subject. Ref. Sat. (Ministry of Ferrous Metallurgy of the USSR), ed. B.V. Molotilov. - Moscow: Metallurgiya, 1983. - P. 78–82.
1.100. Feodorov V.A., et al., Journal of Guangdong Non-Ferrous Metals. 2005. V. 15, No. 2–3. P. 185–187.
1.101. Novikov N.V., et al., Zavod. Lab., 1988. V. 54, No. 7. P. 60–67.

1.102. GOST 25.506-85. Calculations and strength tests. Methods of mechanical metal tests. Determination of the characteristics of crack resistance (fracture toughness) under static loading. - Moscow, Publishing Standards, 1985.
1.103. Davis L.A. Strength, ductility, fracture toughness. In: Metallic glasses. - Moscow: Metallurgiya, 1984. - P. 150173.
1.104. Kimura H., Masumoto T., Scr. Met. 1975. V. 9, No. 3. P. 211–222.
1.105. Fedorov V.A., et al., In the collection: Mechanisms of deformation and destruction of advancedmaterials. XXXV seminar "Actual problems of strength" (September 15-18, 1999, Pskov). 1999. Part 2. P. 471–473.
1.106. Fracture, ed. G. Liebowitz. T. 1. - Moscow: Mir, 1976.
1.107. Glezer A.M., et al., Mechanical behavior of amorphous alloys. - Novokuznetsk: Publishing house of SibGIU, 2006.
1.108. Chen H.S., Krause J. T., Golleman E., J. Non-Cryst. Sol. 1975. V. 18, No. 2. P. 157-172.
1.109. Inoue A., et al., J. Non-Cryst. Sol. 1984. V. 68, No. 1. P. 63–73.
1.110. Inoue A., et al., J. Phys. 1980. V. 41, No. 8. P.C8-831–C8-834.
1.111. Inoue A., et al., Trans. Japan Inst. Metals. 1979. V. 20, No. 10. P. 577–584.
1.112. Kimura H., Masumoto T. Strength, ductility and viscosity - examination within the framework of deformation mechanics and fracture, In: Amorphous metal alloys. - Moscow: Metallurgiya, 1987. Pp. 183–221.
1.113. Latuszkiewicz J., et al., In: Proc. Int. Conf. Metal. Glass, Science and Technology. (Budapest, Hungary). 1980. P. 283–289.
1.114. Stubicar M., J. Mater. Sci. 1979. V. 14, No. 6. P. 1245–1248.
1.115. Hillenbrand H.G., VDI Zeitschrift, 1983. Bd. 125, No. 10. P. 403–410.
1.116. Mil'man Yu.B., et al., Poroshk Metallurigya. 1984. No. 12. P. 69–75.
1.117. Alekhin V.P., et al., MiTOM, 1982. No. 5. P. 33–36.
1.118. Cheremskaya P.G., et al., Pores in a solid body. - Moscow: Energoatomizdat, 1990. .
1.119. Taub A.I., Walter J.L., Mater. Sci. and Eng. 1984. V. 62, No. 66. P. 249–260.
1.120. Homer C., Eberhardt A., Scr. Met. 1980. V. 14, No. 12. P. 1331–1332.
1.121. Zelensky V.A., et al., Fiz. Khom. Obrab. Mater., 1986. No. 2. P. 119–121.
1.122. Khonik V.A., et al., FMM. 1985. Vol. 59, No. 1. P. 204–205.
1.123. Greer A.L., J. Non-Cryst. Sol. 1984. V. 61–62, No. 2. P. 737–739.
1.124. Glezer A.M., et al., DAN SSSR, 1985. Vol. 283, No. 1. 106–109.
1.125. Vereshchagin M.N., et al., Prikl. Mekh. Tekh. Fiz., 2004. V. 45, No. 3. P. 172–175.
1.126. Vereshchagin M.N., et al., FMM. 2002. V. 93, No. 5. P. 101–104.
1.127. Gilman, J.J., J. Appl. Phys. 1973. V. 44, No. 2. P. 675–679.
1.128. Ast D., et al., Scr. Met. 1978. V. 12, No. 1. P. 45–48.
1.129. Morris R.S. J. Appl. Phys. 1979. V. 50, No. 5. P. 3250–3257.
1.130. Zaichenko S.G., Borisov V.T., DAN SSSR. 1982. V. 263, No. 3. P. 622–626.
1.131. Waseda Y., Chen H.S., Sci. Repts. Res. Inst. Tohoku Univ. 1980. V. A28, No. 2. P. 143–155.
1.132. Argon A.S., Acta Met. 1979. V. 27, No. 1. P. 47–58.
1.133. Zielinski P.G., Ast D.G., Phil. Mag. 1983. V. 48A, No. 5. P. 811–824.
1.134. Taub A.I., Luborsky F.E. Creep, stress relaxation and structural change in an amorphous metals, Proc. Fourth Int. Conf. RQM. (Sendai, Japan). 1981. V. 2. P. 1341–1344.
1.135. Akchurin M.Sh., et al., Poverkhnost', fizika, khimiya, mekhanika. 1983. No. 3. Pp. 119–123.
1.136. Pozdnyakov V.A., FMM. 2004. V. 97, No. 1. P. 9–17.
1.137. Cherepanov G.P., Mechanics of brittle fracture. - Moscow: Nauka, 1974.

1.138. Pozdnyakov V.A., FMM, 2002. V. 94, No. 5.
1.139. Raye, J., Fracture, V. 2. - Moscow: Mir, 1975. - P. 204–335.
1.140. Kovneristy Yu.K. Introduction. Physicochemistry of amorphous (glassy) metallic materials. - Moscow: Nauka, 1987. - P. 3–4.
1.141. Chebotnikov V.N., et al., FMM. 1989. V. 68, No. 5. P. 964–968.
1.142. Inoue A., Takeuchi A., Mater. Transact. 2002. V. 43, No. 8. P. 1892–1906.
1.143. Peker A., Johnson W.L., App. Phys. Lett. 1993. V. 63, No. 17. P. 2342–2344.
1.144. Nishiyama N., Inoue A., Mater. Transact. JIM. 1997. V. 38, No. 5. P. 464–472.
1.145. Inoue A., Mater. Sci. and Eng. 2001. V. A304–306. P. 1–10.
1.146. Johnson W.L., MRS Bull. 1999. V. 24. P. 42–56.
1.147. Li Y., Mater. Transact. 2001. V. 42, No. 4. P. 556–561.
1.148. Petrzhik M.I., Molokanov V.V., Izv. RAN. Ser. Fiz., 2001. Vol. 65, No. 10. P. 1384-1389.
1.149. Lu Z.P., et al., Scr. Mater. 2000. V. 42. P. 667–673.
1.150. Lu Z.P., Liu C.T., Acta Mater. 2002. V. 50. P. 3501–3512.
1.151. Nishiyama N., Inoue A., Acta Mater. 1999. V. 47, No. 5. P. 1487–1495.
1.152. Shen T.D., Schwarz R.B., Appl. Phys. Lett. 1999. V. 75, No. 1. P. 49-51.
1.153. Inoue A., et al., Mater. Transact. JIM. 1991. V. 31, No. 5. P. 425–428.
1.154. Taub A.I. Scr. Met. 1983. V. 17, No. 7. P. 873–878.
1.155. Ivashchenko Yu.N., Metallofizika, 1985. Vol. 7, No. 5. 104–106.
1.156. Zhang T., Inoue A., Mater. Sci. and Eng. 2001 V. A304–306. P. 771–774.
1.157. Kim J.-J., Science. 2002. V. 295. P. 654–657.
1.158. Bengus V.Z., Phys. stat. sol. 1984. V. 81a, No. 2. P. K11–K13.
1.159. Flores K.M., et al., Mater. Transact. 2001. V. 42, No. 4. P. 619–622.
1.160. Amiya K., Inoue A., Mater. Transact. JIM. 2000. V. 41, No. 11. P. 1460–1462.
1.161. Zhang Y., et al., Acta Mater. 2003. V. 51 P. 1971–1979.
1.162. Zhang H., et al., Scr. Mater. 2003. V. 49. P. 447–452.
1.163. Inoue A.,et al., J. Non-Cryst. Sol. 2002. V. 304 P. 200–209.
1.164. Inoue A., et al., Mater. Transact. JIM. 2000. V. 41, No. 11. P. 1511–1520.
1.165. Chen H.S., et al., Mater. Sci. and Eng. 1976. V. 26, No. 1. P. 79–82.
1.166. Naka M., et al., J. Phys. 1980. V. 41, No. 8. P. C8-839–C8-842.
1.167. Bresson I., et al., J. Mat. Sci. and Eng. 1988. V. 98, No. 2. P. 495-500.
1.168. Egami T., Mater. Res. Bull. 1978. V. 13, No. 6. P. 557–562.
1.169. Deng D., Argon A.S., in: Proc. Fifth Int. Conf. RQM. Elsevier Sci. Publ. 1985. V. 2. P. 771–774.
1.170. Chen H.S., Proc. Fourth Int. Conf. RQM. (Sendai, Japan). 1981. V. 1. P. 555–558.
1.171. Gerling R., Schimznsky F.P.,in: Proc. Fifth Int. Conf. RQM. Elsevier Sci. Publ. 1985. V. 2. P. 1377–1380.
1.172. Liebermann H. H., Luborsky F.E., Acta Met. 1981. V. 29, No. 6. P. 1413–1418.
1.173. De Zhen Ya. Physics of liquid crystals. - Moscow: Mir, 1977.
1.174. Glezer A.M., Betekhtin V.I., FTT. 1996. V. 38, No. 6. P. 1784–1790.
1.175. Komatsu T., et al., J. Mater. Sci. 1985. V. 20, No. 8. P. 1376–1382.
1.176. Fujita F.E., in: Proc. Fourth Int. Conf. RQM. (Sendai, Japan). 1981. V. 1. P. 301–304.
1.177. Egami T., J. Magn. and Magn. Mater. 1983. V. 31-34, Pt. 3. P. 1571–1574.
1.178. Kimura H., Masumoto T., Acta Metal. 1980. V. 28, No. 7. P. 1663–1675.
1.179. Kimura H., Masumoto T., Acta Met. 1980. V. 28, No. 7. P. 1677–1693.
1.180. Masumoto T., Sci. Rep. Res., Tohoku Univ. 1977. V. A26. P. 246–262.
1.181. Yavari A.R., J. Mater. Res. 1986. V. 1, No. 6. P. 746–751.
1.182. Glezer A.M., FMM ,1988. V. 65, No. 5. P. 1035–1037.
1.183. Aldokhin D.V., et al., Vestn. Tambov. Gosud. Univ. Ser. Est. tekhn. nauki, 2003. T.

V. 8, issue 4, P. 519–521.
1.184. Glezer A.M., Smirnov O.M., FMM, 1992, V. 68, No. 34. P. 1411–1612.
1.185. Aldokhin D.V., Materialovedenie, 2004. No. 7. P. 28–36.
1.186. Glezer A.M., FMM, 1995. V. 80, No. 2. P. 142–152.
1.187. Gubanov A.I., FTT. 1960. V. 2. P. 502.
1.188. Grigson W. B., et al., Nature. 1964. V. 204. P. 173.
1.189. Brenner, A., Riddell G.J., Res. Nat. Bur. Stand. 1947. V. 39. P. 385.
1.190. Harris R., et al., Phys. Rev. Lett. 1973. V. 31. P. 160.
1.191. Cochrane R.W., et al., J. Phys. F, Metal Phys. 1975. V. 35. P. 4–265.
1.192. Bhattacharjee A.K., et al., J. Phys. F, Metal Phys. 1977. V. 7. P. 393.
1.193. Chandhari P., et al., IBM. J. Res. Development, 1973. V. 11. P. 66.
1.194. Orehotsky J., Schröder K., J. App. Phys. 1972. V. 43. P. 2413.
1.195. Mimura Y., et al., J. App. Phys. 1978. V. 49. P. 1208.
1.196. Kittel C., Rev. Mod. Phys. 1949. V. 21. P. 541.
1.197. Nakagawa Y., J. Phys. Soc. Japan. 1971. V. 30, No. 6. P. 1596.
1.198. Duwez P., Trans. ASM. 1967. V. 60. P. 607.
1.199. Fujimori H., et al., Japan J. Appl. Phys. 1974. V. 13. P. 1889.
1.200. Egami T., et al., Appl Phys. Letters. 1975. V. 26. P. 128.
1.201. Kikuchi M., et al., Japan J. Appl. Phys. 1975. V. 14. P. 1077.
1.202. Fujimori H., et al., Sci. Repts. Rep. Instn. Tohoku Univ. 1976. V. A-26. P. 36.
1.203. Fujimori H., et al., Japan J. Appl. Phys. 1976. V. 15. P. 705.
1.204. Sherwood R.C., et al., AIP Conf. Proc. 1975. V. 24. P. 745.
1.205. Chen H.S., et al., Appl Phys. Letters. 1975. V. 26. P. 405.
1.206. Kekalo I.B., Itogi nauki i tekhniki. Ser. Metalloved. term obrab., VINITI, USSR Academy of Sciences, 1973. V. 7. P. 6.
1.207. Zolotukhin I.V., Metallofizika, 1984. Vol. 6, No. 6. P. 58.
1.208. Kobelev N.P., Soifer Ya.M., FTT. 1986, V. 28. P. 425.
1.209. Arai K.I., Tsuya N., Sci. Rep. Ser. A. RITU. 1980. V. 30. Suppl. P. 247.
1.210. Fukamuchi K., Masumoto T., IEEE Trans. Magn. 1979. V. Mag-15. P. 1404.
1.211. Ishio S., Takahashi M., J. Magn. and Magn. Mater. 1985. V. 50. P. 93.
1.212. Nikitin S.A., et al., FTT. 1987 V. 29. P. 1526.
1.213. Kazama N.S., et al., IEEE Trans Magn. 1982. V.Meg-18. P. 1182.
1.214. Sakakima H., ibidem. 1983. V.Mag-19. P. 131.
1.215. Hausch G., Warlimont H., Zs. Metallk. 973. Bd. 64. P. 152.
1.216. Kikuchi M., et al., Phys. Stat. Col. Ser. A. 1978. V. 48. P. 175.
1.217. He J., et al., Journal of Materials Research. 2009. V. 24. P. 1607–1610.
1.218. Fukunaga T., et al., Amorphous Magnetism II, ed. R.A. Levy, R. Hasegawa. - New York and London, Plenum Press, 1977.

2

Nanocrystalline alloys

Nanocrystalline materials are one of the groups of nanomaterials. Under the nanocrystalline materials it is customary to understand such materials in which the size of individual crystals or other structural elements does not exceed 100 nm in at least one measurement [2.1]. It should be borne in mind that an important condition for classifying any material as nanocrystalline is not only the presence in its structure of nanoscale structural elements, but also a significant influence on the properties of the material. The specificity of any dimensional effect (the nature of the dependence of any physicochemical characteristic on the effective size of the structural element) is largely associated with nature of the structural element – a crystal (in the case of nanocrystalline materials). For nanomaterials of various types, the size effect of a certain physicochemical or mechanical characteristic has often its own features.

2.1. Classification of nanocrystalline alloys

The famous specialist in the field of nanomaterials N. Gleiter in his works (for example, [2.2]) used the dimension criterion in one of the classifications of nanomaterials. So he suggests to allocate only three classes of nanomaterials: nanoparticles; nanolayers, films and near-surface structures; bulk nanostructured materials.

In this chapter, special attention is paid to bulk nanostructured (nanocrystalline) materials, their production methods, structure and properties.

A wide interest in nanocrystalline materials arose in the middle of the eighties in connection with the work of N. Gleiter and his colleagues who first drew attention to the increase in the role of interfaces and, especially, of border regions with a decrease in grain size [2.2].

The structural characteristics of the main types of nanocrystalline materials are shown in the diagram (Fig. 2.1) proposed by N. Gleiter.

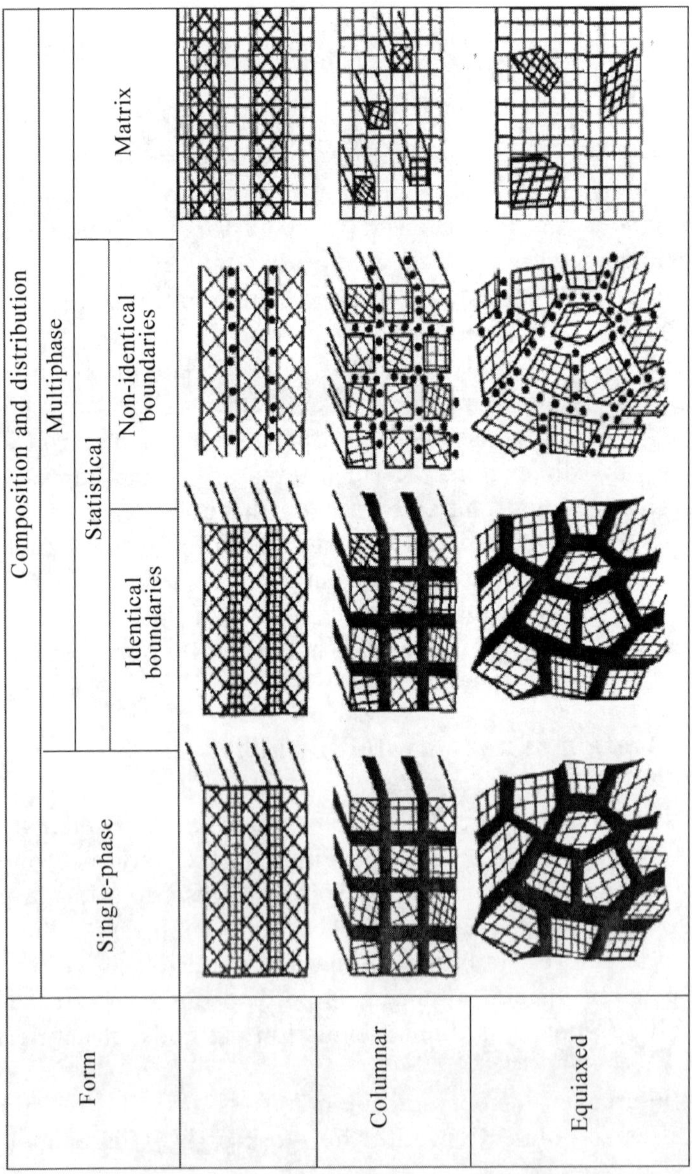

Fig. 2.1. Classification of nanocrystalline materials by composition, distribution and form of structural components.

This classification takes into account the composition, distribution and form of structural components, and also includes structures obtained by a variety of methods [2.2]. As can be seen, there are four types of chemical composition and distribution of structural components (single-phase, statistical multiphase compositions with identical and non-identical interfaces and matrix compositions) and three categories of structure forms (laminated, columnar and containing equiaxial inclusions).

In fact, the variety of structural types may be wider due to mixed variants, the presence of porosity, polymer matrices, etc. The most common are single-phase and multiphase matrix and statistical objects, columnar and multilayer structures.

2.2. Methods for obtaining bulk nanocrystalline materials

To obtain bulk nanocrystalline materials, various methods can be used which are expediently divided into three large groups in accordance with the initial aggregate state from which they are formed: from solid, liquid or gaseous.

The most common methods for obtaining bulk nanostructured materials are the following [2.1]:

– compacting of powders (cold and hot pressing, high-temperature sintering, electrosintering, hot extrusion, etc.);

– controlled crystallisation of the amorphous state;

– severe plastic deformation;

– film technologies (chemical deposition, physical deposition, electrodeposition, sol–gel precipitation, etc.);

– controlled polymerization, copolymerization and self-assembly;

– bionanotechnology;

– intensive irradiation with a flux of high-energy particles

Technologically, methods for obtaining bulk nanocrystalline materials can be subdivided into methods of powder metallurgy followed by consolidation (compaction), methods of severe plastic deformation, methods of controlled nanocrystallisation of an amorphous state, and methods associated with chemical technologies. Let us briefly consider the main of these methods.

1. Consolidation (compacting) of nanopowders

Among the parameters that largely affect the degree of compaction of nanopowders, the following are most important: the average size and shape of nanoparticles, the content of impurities, the state of the surface and the method of compaction. To obtain bulk nanocrystalline materials, the methods of compacting at room temperature are used mainly, followed by sintering at an elevated temperature. For pressing nanopowders, uniaxial pressing is most often used, either static, or dynamic, or vibrational. High-density and homogeneous nanocrystalline materials are processed by uniform (isostatic) pressing: hydrostatic, gas-static or quasi-hydrostatic.

The use of quasi-hydrostatic pressing makes it possible to increase the density of compacts in comparison with uniaxial pressing. At the same level of pressing pressure, a decrease in the size of nanoparticles leads to a decrease in the density of compacts. Therefore, it is not possible to obtain bulk nanocrystalline materials with a density above 70% of the theoretical density by known methods. The method of dynamic magnetic-pulse pressing makes it possible to substantially increase the density of nanocrystalline materials. Unlike static methods, pulsed compression waves cause intense heating of the nanopowder due to rapid energy release during the friction of nanoparticles in the pressing process. If the particles are nanosized, then their warm-up time is noticeably less than the characteristic duration of the pulsed compression waves (1–10 μs). The parameters of this method are easy to control and manage. For example, aluminium nitride nanopowders are pressed by the magnetic-pulse method at a pressure of 2 GPa to a density of 95%. An increase in the uniformity of the density of compacts is also achieved when vibrational ultrasonic compacting is used. The application of the method of severe plastic deformation is also very effective. The compacting of copper nanopowders with an average particle size of 28 nm by shearing under high pressure results in a bulk nanocrystalline material with a grain size of 75 nm and a density of 98%. However, we see that none of the methods considered for consolidating nanopowders allow one to obtain bulk nanocrystalline materials without pores.

Sintering of a nanopowder at a relatively low temperature also does not make it possible to obtain a porous nanocrystalline material. The installation scheme is shown in Fig. 2.2. The use of high temperatures is not possible because of the rapid diffusion growth of nanoparticles with similar sintering. A promising way to produce defect-free bulk nanocrystalline materials is to sinter nanopowders under pressure. To obtain massive, uniformly dense compacts with a homogeneous structure, the methods of hot isostatic pressing (HIP)

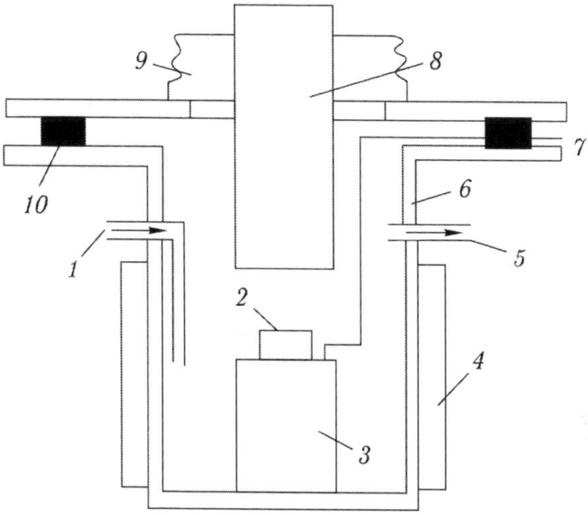

Fig. 2.2. Scheme of installation for sintering under pressure: 1 – input of inert gas; 2 – shaping; 3 – anvil; 4 – heating element; 5 – gas outlet; 6 – working chamber; 7 – thermocouple; 8 – punch; 9 – bellows; 10 – sealing gasket.

Table 2.1. Mechanical properties of compacts

Method	Material	Nanopowder particle size, μm	Material grain size, μm	Yield strength, MPa	Relative elongation, %	Micro-hardness, GPa
HIP	Ni	6	25	440	36	1.0
		0.06	1	545	7	2.6
	Fe	40	55	350	41	0.9
		0.04	1	460	1	2.3
HTGE	Ni	0.06	0.1	700	15	3.0

and high-temperature gas extrusion (HTGE) have proved to be very useful. The latter consists in obtaining a compact by the hydrostatic method at room temperature and its preliminary heat treatment in hydrogen at low temperature and extrusion at elevated temperature. The data on mechanical properties and on the structure of bulk nanocrystalline materials obtained by these methods are presented in Table 2.1.

In order to preserve the nanocrystalline state in the consolidation of nanopowders, in addition to reducing the sintering temperature, it is quite effective to introduce alloying additives that hinder the intensive growth of grains. An example is the production of WC–Co nanocrystalline hard alloys.

The nanopowder composition of tungsten carbide and cobalt is formed during the thermal decomposition of organometallic compounds. To inhibit the growth of grains and reduce the solubility of the tungsten carbide in cobalt, the non-stoichiometric vanadium carbide VC is added to the mixture in an amount of up to 1 wt%. The hard alloy obtained from this nanocrystalline composition is characterized by an optimal combination of high hardness and high strength. It has been established that each WC–Co composite grain having a size of about 75 μm consists of approximately 10^6 WC nanocrystals of less than 50 nm in size distributed in a cobalt matrix. An even finer-grained structure has an alloy containing, in addition to the tungsten carbide, 9.4 wt% Co, 0.8 wt% Cr_3C_2 and 0.4 wt% VC. The nanoalloy is much stronger and much more resistant to destruction compared to the conventional polycrystalline material.

2. Controlled nanocrystallisation of the amorphous state

The preparation of bulk nanocrystalline materials in this case is carried out in two stages: the preparation of amorphous alloys and then their nanocrystallisation by thermal treatment. The first part of the problem can be solved using the method of melt quenching, machining in a ball mill with subsequent consolidation and intensive plastic deformation. As discussed in Chapter 1, to obtain the amorphous state, the melt spinning method is used, consisting in injecting a melt jet under pressure into a rotating drum–cooler and solidifying it with a speed reaching 10^6 K/s in the form of a thin ribbon. The installation scheme for the implementation of the spinning method of the melt is presented in Chapter 1 (Fig. 1.1).The production of nanocrystalline materials controlled by crystallisation has several advantages over other methods of obtaining nanocrystalline alloys (for example, compacting powders). Annealing leads to the formation of a nanocrystalline structure in all alloys that can be obtained in the amorphous state. In addition, the grain size can be easily changed within wide limits by varying the heat treatment parameters. In isothermal annealing, one of the most important factors determining the grain size is the annealing temperature. The annealing time is usually determined by the completion time of the transformation of the amorphous phase into a nanocrystalline phase. An important advantage of the crystallisation method is also the possibility of obtaining non-porous nanocrystalline materials with a homogeneous microstructure.

3. The method of severe plastic deformation

When deforming ordinary materials with very high degrees of plastic deformation at relatively low temperatures (below (0.3–0.4)T_m, where T_m is the melting point), pore-free nanocrystalline states can be formed. This method is based on the principle of the formation of a highly fragmented and disoriented structure, obtained by very large deformations. To date, there are many methods that realize severe plastic deformation (SPD): high-pressure torsion distortion, equal-channel angular pressing, extrusion, mechanomelting, multiple deformation (cold rolling or drawing), shock wave deformation (created by explosion), deformation in the ball mill (attritors), magneto-impulse pressing, external friction.

Currently, three most commonly used methods of creating giant degrees of deformation are actively used: torsion under pressure in a Bridgman chamber (TPBC), equal-channel angular pressing (ECAP), and accumulated rolling.

In the first case, the sample is placed between two anvils, one of which rotates slowly while creating very high hydrostatic stresses (several GPa) (see Chapter 3).

The method of equal-channel angular pressing, realizing the deformation of massive samples by a simple shift, was developed by V.M. Segal. Equal-channel angular pressing (ECAP) is currently the most widely used method of SPD. A specimen of circular or square cross-section is pressed into the matrix through channels that are contiguous at a certain angle. Deformation occurs when the workpiece passes through the zone of their intersection, since the dimensions of the workpiece in the cross section do not change pressing can be carried out repeatedly in order to achieve exceptionally high degrees of deformation. In the process of repeatedly pressing in the workpiece, deformation by shear accumulates, which results in the formation of an ultrafine-grained structure in the material. If necessary, in the case of difficult to deform materials, deformation occurs at elevated temperatures.

In ECAP the direction and number of passes through the channels are very important for structure formation. The following ECAP schemes are most commonly used: the orientation of the workpiece remains unchanged with each pass; after each pass the workpiece rotates about its longitudinal axis by an angle of 90°; after each pass the workpiece rotates about its transverse axis by 180°.

2.3. Structure

As already mentioned above, nanocrystalline (nanoscale) materials are understood to be materials in which the size of individual crystallites or phases constituting their structural basis does not exceed 100 nm in at least one dimension. This limit is quite arbitrary and dictated by considerations of convenience. But at the same time simple estimates show that, starting from these dimensions, the share of border regions with a disordered structure becomes more and more noticeable and amounts to several percent, and with an average size of nanocrystals of 8–10 nm it is 40–50%. On the other hand, the upper limit of the values of the size of nanocrystals must correspond with the characteristic size for a given physical phenomenon (the size of the dislocation loop, the mean free path of electrons, the size of the magnetic domain, etc.). It is quite understandable that the limiting values of the size of nanocrystals for different physical properties and for different metals, solid solutions and compounds will be unequal. Hence the conventionality of the above value of 100 nm.

The structural characteristics of the main types of nanocrystalline materials were considered in section 2.1 of this book and are shown in the diagram (Fig. 2.1). Let us clarify here that there are four types of chemical composition and distribution of structural components (single-phase, statistical multiphase compositions with identical and non-identical interfaces, and matrix compositions) and three categories of structure forms (laminated, columnar and containing equiaxed inclusions).

The dimensions of the structural elements of nanomaterials can be determined by electron microscopy methods in which a magnification of more than ×100 000 is achievable. However, the methods of electron microscopy are characterized by a rather high labour input and require expensive equipment.

Within each of the basic methods of obtaining nanocrystalline materials, the range of structural elements can be very diverse. Thus, for the method of controlled crystallisation from the amorphous state, three types of nanostructural states can be distinguished depending on the melt quenching conditions:

1. Complete realization of crystallisation directly in the process of melt quenching and the formation of single-phase or multiphase both conventional polycrystalline and nanostructures (type I).

2. Crystallisation in the quenching process from the melt does not completely flow, and an amorphous–nanocrystalline structure (type II) is formed.

3. Quenching from the melt leads to the formation of an amorphous state which is transformed into a nanocrystalline state only with subsequent thermal treatment (type III).

For each of the three types listed above, their morphological features and the specificity of thermal stability are characteristic. It is shown that the rate of cooling from the melt affects the transition from the microcrystalline state to the amorphous state, which is accompanied by a non-monotonic change in hardness the maximum of which occurs in type II of amorphous–nanocrystalline structures.

The structure of nanocrystalline materials, formed by the method of severe plastic deformation, has its own peculiarities. The main feature is the presence of a complex system of high-angle and small-angle boundaries that have a complex non-equilibrium structure and are a source of very high elastic stresses. Another source of internal stresses are the triple and quadruple joints of grain boundaries. In the structure there are also disclination inconsistencies.

Dislocations in non-equilibrium grain boundaries and disclinations create long-range stress fields and cause the flow of processes that under ordinary conditions occur at higher temperatures – dynamic recrystallisation, phase transformations, and other phenomena. Strictly speaking, the structure formed after severe deformation in the vast majority of cases can not be attributed to nanocrystalline: grains rarely have a size less than 100 nm. At the same time, high strength properties in combination with sufficient ductility, obtained on aluminium, titanium, iron and their alloys, make it necessary to speak of severe plastic deformation (in particular, about the method of equal-channel angular pressing) as one of the most promising method of increasing the operational characteristics of steels and structural alloys.

For example, Fig. 2.3 shows a photograph of the microstructure of 10G2FT steel (0.1% C–1.12% Mn–0.08% V–0.07% Ti) obtained by transmission electron microscopy after severe plastic deformation by high-pressure torsion at a deformation temperature of up to 500 C. Analysis of the image showed that the average grain size in this case is 85 nm.

At the same time, it is well known that a decrease in the size of crystallites and the appearance of microdistortions lead to broadening of the X-ray diffraction lines. Using various calculation

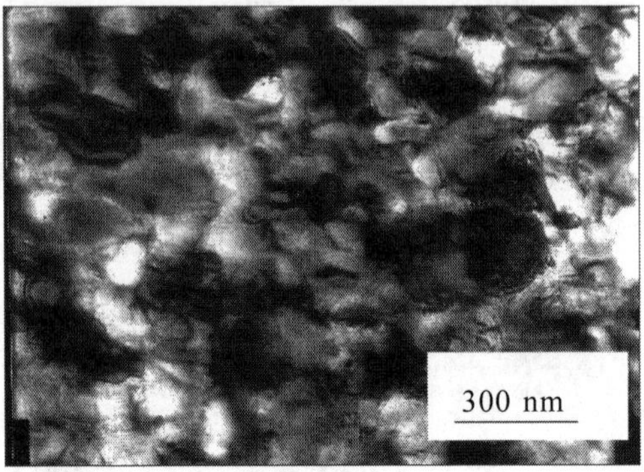

Fig. 2.3. The microstructure of 10G2FT steel after severe plastic deformation by high-pressure torsion.

techniques, by approximating the shape of the x-ray reflections, for example, by a set of Gaussian or Lorentz functions, or a combination of them, one can obtain information on the dimensions of nanocrystallites, the microdeformation of the lattice, and the magnitude of the Debye–Waller factor. An analysis of the neutron data shows that the difference in the determination of the size of nanocrystals by different methods, including microscopic ones, can be quite significant, which, on the one hand, reflects the difference in methodological assumptions, and on the other – quite naturally, taking into account the anisotropy of growth and the non-identical conditions of measurement and averaging. Obtaining reliable information on the size of crystallites in nanocrystalline materials requires, as a rule, the use of 2–3 independent methods with a detailed discussion of the results obtained. Determination of the Debye–Waller factor has shown that practically for all investigated objects (Cu, Pd, Se, etc.), its increase is observed in the transition from ordinary coarse-grained objects to nanocrystalline ones; the background component also increases. All that has been noted can be attributed to the influence of border regions, the proportion of which increases with the decrease in the characteristic size, and the displacements of atoms from the equilibrium positions are larger than those for ordinary coarse-grained objects.

The nature of the interfaces in general and the boundaries of crystallites in particular with reference to nanocrystalline materials continues to be the subject of lively discussions. Thus, for example, the idea of a 'gas-like' structure of the boundaries in nanocrystalline materials was put forward. This position is based on the results of the determination of the coordination number (i.e., the number of the nearest neighbours) from diffuse scattering data for nanocrystalline palladium, on the results of a short-range study using EXAFS spectrometry, Mössbauer spectroscopy, etc. On the other hand, based on the results obtained by the method of high-resolution electron microscopy, it was suggested that there are no special differences in the structure of intercrystallite boundaries in nanocrystalline materials and conventional materials. When studying the structure of the boundaries in nanocrystalline materials obtained by severe plastic deformation, it was concluded that there are non-equilibrium boundaries with long-range stress fields and increased energy in connection with the high density of grain-boundary dislocations. With a decrease in the average grain size in nanocrystalline materials, the volume fraction of not only the border areas increases, but also to a much greater extent – triple joints of grains. This means that the triple joints of grains can have a determining effect on the properties of nanocrystalline materials. The presence of impurities in nanocrystalline metals, alloys and compounds, especially in the case of conventional powder technology, inevitably leaves its imprint on the nature of the interfaces.

In nanocrystalline materials based on silicon carbide and nitride and examined by high-resolution electron microscopy and analytical electron microscopy (energy-dispersive spectroscopy and characteristic energy loss spectroscopy), the presence of amorphous interlayers up to 1–2 nm in thickness (for a silicon nitride matrix) and up to 5–25 nm (for Si_3N_4–SiC interphase boundaries) was recorded. The change in Raman spectra was recorded for nanocrystalline titanium oxide, which is associated with the appearance of vacancies in the oxygen sublattice. The study of other oxide materials in the nanocrystalline state revealed the presence of both usual 'crystalline' (ZrO_2 compacts) and 'amorphous' (NiO films) boundaries. Both these boundaries were observed in the samples of calcium hydroxyapatite. A similar situation occurs for tin oxide films, but the 'crystal' boundaries are predominant. The structure of the boundaries in the materials obtained by the controlled crystallisation of the amorphous state is also manifold. Concentration segregations are characteristic of the dendritic–cellular structure. In nanocrystalline materials obtained by controlled crystallisation from the amorphous phase, submicroscopes, prismatic dislocation loops of a vacancy nature, a wide spectrum of dislocation and disclination structures with different degrees of relaxation processes, as well as twin and semicoherent boundaries, misfit dislocations, etc. were found. A detailed study of the structure of nanocrystalline palladium (grain size 14–15 nm) s howed that according to the data of EXAFS-spectrometry its structure turned out to be identical with the structure of ordinary palladium. The observed decrease in the value of the coordination number could be explained not only by the small grain size, but also by the possible vacancy disordering of the lattice. Thus, the technological background (the method of production) has a decisive influence on the structure of the interfaces in nanocrystalline materials, and it is still difficult to develop a unified concept of the boundary structure since the accumulated experimental information is far from complete. Many studies have been devoted to the interfaces in systems of the metal–oxide type (Cu–Al_2O_3, Nb–Al_2O_3, Nb–TiO_2, etc.). In addition to revealing the fine details of coupling, an important result is the detection of a possible change in the electronic state of the elements at the interfaces. Thus, the existence of Cu^{+1} states for the Cu–Al_2O_3 pair was noted. At the same time, no charge transfer was detected in the Cu–TiO_2 vapour.

The question of linear structural defects in nanocrystalline materials is equally complex and not fully understood. On the one hand, the size of the crystallites in the latter often turns out to be smaller than the known characteristic dimension of the loop of the Frank–Read dislocation source, and the multiplication of dislocations by these sources turns out to be suppressed. In small particles and ultradisperse powders, dislocations, as a rule, are not observed. On the other hand, in nanocrystalline materials, especially those obtained with the use of deformation or representing heterophase compositions, the role of grain boundary dislocations and misfit dislocations is great and their presence and evolution represent the basis of most model representations. Theoretical considerations of the critical stability of dislocations in nanocrystals, taking into account the action of configuration forces (image forces reflecting the effect of interfaces and free surfaces) and friction forces of the lattice make it possible to estimate the characteristic size of a nanocrystal below which the probability of the existence of mobile dislocations within it decreases markedly.

Estimates for prismatic dislocation loops and linear edge dislocations lead to values from 5 to 40 nm, depending on the nature of the nanocrystals.

Experimental studies of dislocations in nanocrystalline materials are not very numerous. They testify to the reasonableness of the above estimates. Twins and, correspondingly, twin boundaries were also observed; the presence of nanopores at the boundaries was confirmed. Irradiation with an electron beam leads to a more ordered distribution of atoms at large-angle boundaries. Japanese researchers, studying by electron microscopy the boundaries in various consolidated nanoobjects (metals, metal–ceramic composites, oxides, fullerites), concluded that there are no special anomalies in the density of atoms at the boundaries. For metals (Pd, Ag), for example, the presence of boundaries characterized by low values of the reciprocal density of nodes ($\Sigma 3$ and $\Sigma 11$), i.e. with a relatively small degree of misorientation, is noted. The presence of boundaries with high values of Σ was observed in a small amount. The presence of $\Sigma 9$ boundaries for consolidated palladium is noted, where small-angle boundaries and the presence of dislocations inside the crystallites, as well as twins and packing defects, have also been observed. The latter were not observed in the study of molybdenum samples in which crystallites situated at an angle of 9° were studied.

High-resolution electron microscopy was used to study grain boundaries in samples obtained by severe plastic deformation. The presence of an increased density of grain-boundary dislocations and significant elastic distortions of the crystal lattice was noted, which made it possible to draw a conclusion about the formation of non-equilibrium boundaries in nanocrystalline materials. Zones of compression and stretching are found in border areas with a width of 6-10 nm.

2.4. Physical and mechanical properties

The hardness, strength, plasticity, elasticity and other mechanical characteristics of solids have been studied very intensively for practically all types of materials, and especially for nanomaterials. Bearing in mind the high structural sensitivity of mechanical properties, the problem of attesting properties with respect to nanomaterials acquires special significance. Earlier, the main features of the structure of nanomaterials were described: a small amount of crystallites and, correspondingly, a large volume fraction of boundaries, border regions and triple joints of grains; a high level of internal stresses, the presence of impurities and other defects both inherent in nanocrystalline materials and introduced in the process of a very complex technology for their production.

The insufficient density of powder nanocrystalline materials with respect to their porosity was at the initial stage of research a source of errors consisting in the fact that nanocrystalline materials are characterized by substantially lower values of the elastic properties. More accurate measurements and taking into account porosity have shown that the modulus of normal elasticity for nanocrystalline materials (Cu, Pd, Fe, Ti) with a grain size of 4–100 nm does not practically differ from the values typical for conventional polycrystalline samples. It is also shown that the decrease in the modulus of normal elasticity for copper subjected to severe plastic deformation is apparently associated with the appearance of a crystallographic deformation texture, rather than with a decrease in grain size. Nevertheless, the problem of the elastic properties of nanocrystalline materials with a very small grain size (<5 nm), when the fraction of atoms at the grain boundaries and in the boundary volumes is very high, continues to be relevant. Similar objects in a certain sense come close to amorphous materials for which elastic moduli are almost 50% lower than for crystalline analogs. In this regard, for nanocrystalline materials with a crystal size <4–5 nm, a significant decrease in the elastic characteristics can be expected. According to theoretical estimates performed by molecular dynamics methods, the decrease in Young's moduli, shear and bulk elasticity for copper is approximately 25, 50 and 8%, respectively.

X-ray and electron microscopic studies show that the orientation of the grains of nanocrystals is usually close to statistical and does not reveal a clear tendency to form a crystallographic texture even after an extrusion operation to reduce the porosity of the nanocrystals. A noticeable texture arises only in the case when nanocrystalline materials are obtained by the method of severe plastic deformation of polycrystals with the usual size of the initial grain. This fact undoubtedly indicates that the role of dislocations in the plastic deformation of nanocrystals is negligible, since with a dislocation plastic flow, as a rule, a clear crystallographic texture is formed. In fact, as shown by electron microscopic studies, only a very small number of dislocations can be observed inside the nanograin. At the same time, very often they are stationary (sedentary) configurations. The low density of dislocations in nanocrystalline materials is associated with the existence of image forces that push out the mobile dislocations from the grains, especially the fine ones. This occurs in a manner similar to how a point charge is pushed out near the free surface of the conductor. In molecular dynamics calculations, it was shown that the appearance of individual dislocations in nanocrystalline materials is possible only with a crystal size

> 9 nm, and the number of dislocations depends on the magnitude of the external pressure used to produce compacted nanocrystalline materials. This effect, of course, negates the role of mobile dislocations in the development of plastic deformation processes, even if it is assumed that new sources of mobile dislocations can be activated in nanocrystals when an external load is applied.

As is known, the grain boundaries are the most important element of the structure of nanocrystalline materials which determine their strength properties [2.5]. From the accumulated experimental data it follows that a decrease in the characteristic dimensions of an object or elements of its structure to less than 1 μm (at least in one of three dimensions) entails a significant change in its mechanical properties [2.6–2.10]. Even stronger dimensional effects arise when the structure parameters are reduced to less than 100 nm, and when they are reduced to less than 10 nm their character changes radically [2.10]. Elucidation of the laws and mechanisms of the nanostructure influence on the parameters of strength and plasticity is described in a large number of papers (for example, [2.11–2.13]).

Nanocrystalline materials have a low plasticity, which is due to the suppression of generation processes and the movement of dislocations due to the small grain size. The study of the phenomenon of superplasticity, which nanomaterials possess, is very important for practical application [2.14].

Due to the difficulties in making nanocrystalline samples for tensile testing, the hardness test was preferred; this test uses small samples of arbitrary shape.

2.4.1. Theoretical strength and theoretical hardness

Strength is the ability of a material to resist plastic deformation and destruction under the influence of external loads. Crystalline solids have the strength the value of which can not exceed the ideal strength. There are theoretical strengths for shearing and tearing off [2.15].

According to [2.16], there is a difference between the terms 'theoretical' and 'ideal' strength. Ideal strength is a stress corresponding to the destruction of an infinite perfect single crystal (an ideal crystal) at zero absolute temperature, the term 'theoretical strength' can be extended to systems with defects [2.17].

The theoretical shear strength was first calculated in 1926 by Ya.I. Frenkel'. Subsequently, the principles laid down in this calculation were refined and supplemented by Mackenzie (1949). In all cases, the strength was calculated on the basis of the interaction of the atoms in the ideal crystal lattice of the metal. Ya.I. Frenkel' calculated the theoretical strength of a crystal using a model that considers the shift of one plane relative to another [2.15, 2.18].

Based on this model, it was found that the maximum shear resistance is:

$$\tau_{theor} = Gb/2\pi h, \tag{2.1}$$

where G is the shear modulus, b is the distance between atoms in the plane, and h is the interplanar spacing.

For metals with a cubic lattice, $\tau_{theor} \approx G/2\pi$. According to Mackenzie [2.15], the ideal shear strength is approximately $G/30$.

The theoretical peel strength was first estimated by Orowan in 1949 for a defect-free solid [2.19]:

$$\tau_{theor} = (E\gamma/a_0)^{1/2}, \tag{2.2}$$

where E is the modulus of elasticity in the direction of tensile loading, γ is the surface energy of one interface formed during detachment, and a_0 is the distance between the planes in the equilibrium state.

Orowan's calculations were further clarified. For example, in [2.20], in order to calculate the ideal peel strength, a model is proposed that represents the general coupling field of the metal atom with the environment in the form of equivalent bonds along three orthogonal axes. The Morse potential was chosen, the parameters of which were found from the Young's modulus of the metal and the coefficient of its thermal expansion. The obtained values of ideal peel strength for 15 metals are in the range $E/6$–$E/10$.

The values of the theoretical peel strength are much higher than the corresponding shear strengths. In the process of sliding, the bonds between atoms across the slip plane periodically resume when the next elementary slip event ends. No new surfaces are formed, except for the steps at the ends of the slip plane, so $\tau_{theor} < \sigma_{theor}$ [2.15].

An analysis of the numerous results of estimating the value of the ideal strength of various materials to date shows that its values are approximately in the range of $E/5$–$E/30$. The maximum achievable strength of the material can be limited by a certain limiting stress σ_{theor}, at which shear or fracture nuclei form in an ideal crystal.

Thus, the theoretical strength of a material is that level of stress upon which a transition occurs from the elastic displacement of individual atoms in its ideal crystal lattice to a rigid mutual displacement of two neighbouring atomic planes. There is a transition from elastic deformation of the material to plastic, carried out by the shearing mechanism.

With respect to loading by indentation, the value of theoretical strength can correspond to the theoretical hardness. In [2.21], S.A. Firstov and T.G. Rogul' introduced the notion of theoretical hardness HV_{IT} – the maximum hardness of a material that can be achieved under the condition that the stress causing a plastic flow in the material under the indenter corresponds to the theoretical shear strength of this material. This characteristic is very important for assessing the extremely hardened state of materials and is:

$$HV_{IT} = \frac{\beta E}{\alpha(1+v)} = \frac{E^*(1-v)\beta}{\alpha}, \tag{2.3}$$

where E^* is the reduced Young modulus equal to $E/(1+\nu)$, ν is the Poisson coefficient, α is the proportionality coefficient between Young's modulus and theoretical shear strength ($\alpha = E/\tau_{theor}$) and β is the proportionality coefficient between HV and σ_y, ($\beta = HV/\sigma_y$). The value of β is in the range 1.5–3.3, and the values of α – in the range of 5–30.

2.4.2. The Hall–Petch relation and its anomaly

The most important element of the structure of nanocrystalline materials, determining their strength properties, are grain boundaries (GB). Influencing them, one can control the physical and mechanical characteristics of materials. As is known [2.22], the dependence of the yield point (hardness) on the grain size in polycrystalline metals and alloys obeys the Hall–Petch relation:

$$\sigma_y(HV) = \sigma_0(HV_0) + k_y \cdot D^{-1/2},$$

where σ_y is the yield point, HV is hardness, σ_0 (HV_0) is the plastic flow stress (hardness) in the grain body; k_y is the proportionality coefficient characterizing the 'transparency' of the GBs, and D is the average grain size (Fig. 2.4).

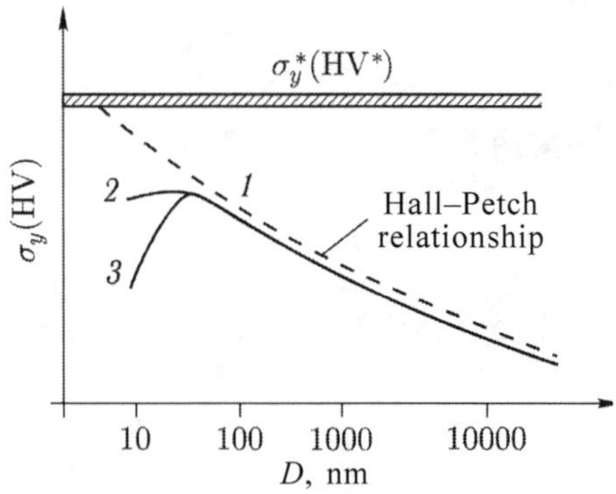

Fig. 2.4. Dependence of yield stress on grain size in polycrystalline material; 1 – the Hall–Petch ratio; 2, 3 – experimental dependences in the region of violation of this ratio; σ_y^* (HV^*) – theoretical (ultimate) strength (hardness).

Reducing grain sizes in crystalline metals and alloys from conventional units of tens of micrometers to tens of nanometers should increase their strength by an order of magnitude, and if we extrapolate the dependence *1* to the maximum achievable grain sizes (see Fig. 2.4), we can achieve theoretical strength σ_y^*, however, in reality the strength is increased 5–6 times. In this case, the brittleness increases and the thermal stability of nanocrystalline materials decreases.

The experimental results obtained on nanocrystals show that they are much stronger than coarse analogues. Nanophase Cu, Pd, and Ag with a grain size of 5 to 60 nm, obtained by compacting ultrafine powders, showed hardness values of 2 to 5 times that of samples with a conventional grain size.

Moreover, the hardness of nanocrystalline copper exceeded the hardness of cold-rolled coarse-grained copper [2.5]. In the same paper it was shown that the hardness of such nanophase materials can increase after low-temperature annealing. As the analysis of a number of experiments demonstrates, a similar tendency is maintained for nanocrystalline materials obtained by other methods (condensation, sputtering, electrodeposition, etc.). Table 2.2 shows the values of the hardness of nanocrystalline materials obtained by compacting. Attention is drawn to the fact that in the case of brittle materials (oxides, nitrides, carbides, intermetallics, etc.), there is no significant increase in hardness as compared to single-compnent alloys.

The best result is achieved for TiN, however, the hardening possibilities are far from being completely used, since in brittle

Table 2.2. The hardness values of some nanocrystalline materials obtained by compacting

Material	Relative density	Grain size, nm	Hardness, GPa
Fe	0.94	15	8
Fe–63 vol.% TiN	0.92	12	13.5
Ni	individual particles	17	5
Ni–64 vol.% TiN	0.97–0.98	10	13
Cu	0.98–0.99	5	2.3–2.5
Ti	individual particles	12	5.7
Ag–76% MgO		2–50	2.5
Nb_3Al	0.97	30	18–22
TiAl	0.99	20	6
ZrO_2–0.25% Y_2O_3	0.985	180	14.2
SiC	0.9–0.95	200–400	22–26
WC–10 vol.% Co	1.0	200	19–20
TiN	0.98–0.99	30–50	28–30
BN	0.95–0.97	25	43–80
Diamond	0.93	20	25

nanocrystalline materials the grain size is usually larger than 20-30 nm. Unfortunately, the difference in hardness values rapidly disappears as the test temperature rises, and even after reaching 300°C they practically coincide. In Ni_3Al and NiAl intermetallics, a strength gain in the production of a grain of about 10 nm in size was also observed.

In a number of works it has been reported that hardness values of more than 50 GPa have been obtained in a composite consisting of Me_nN particles (*Me* = Ti, W, V) smaller than 4 nm in the amorphous Si_3Ni_4 silicon nitride matrix. As the size of the nanoparticles decreases, a noticeable increase in hardness is found, in the limit corresponding to the hardness of the diamond.

A number of important studies were carried out on nanocrystals obtained by crystallisation from an amorphous state. Microhardness measurements of microhardness were carried out both for single-component Se nanocrystals, and for single-phase ($NiZr_2$) and multiphase (Ni–P, Fe–Si–*Me* systems). The general rule is that after the formation of nanocrystals in an amorphous matrix the microhardness always increases. In other words, the hardness of the nanocrystalline state is, with rare exceptions, higher than that of the

Table 2.3 Hardness values (GPa) of some nanocrystalline materials obtained by crystallisation of the amorphous state in comparison with the amorphous and coarse-crystalline state

Material	Hardness, GPa		
	Nanostructured state	Amorphous state	'Coarse-crystalline' state
NiP	10.4 (9 nm)	6.5	11.3 (120 nm)
Se	0.98 (8 nm)	0.41	0.34 (25 nm)
Fe–Si–B	11.8 (25 nm)	7.7	6.2 (1 μm)
Fe–Cu–Si–B	9.8 (30 nm)	7.5	7.5 (250 nm)
Fe–Mo–Si–B	10.0 (45 nm)	–	6.4 (200 nm)
NiZr	6.5 (19 nm)	–	3.8 (100 nm)

corresponding amorphous state. The summarized data are given in Table 2.3.

In more detailed consideration it turns out that the strengthening effect depends on the method used to change the grain size, in particular, there is a tendency to some softening of nanophase materials when measuring the hardness of those samples where the grain size was varied by subsequent annealing. Several studies have shown that when individual samples are annealed to increase the grain size, initial hardening and subsequent softening can occur. Such effects were observed in nanophase copper and palladium, TiAl and NiP. It was suggested that this is due to a decrease in the porosity of compacted nanocrystals, with the transition of grain boundaries from non-equilibrium to equilibrium configurations, with local structural relaxation.

In many experimental works [2.23-2.27] devoted to the study of the mechanical properties of nanocrystalline materials it was found that in the nanometer-sized grain-size range there are significant deviations from the Hall–Petch ratio (curve 1 in Fig. 2.4): in the region $D < 30$–50 nm noticeable deviations from the Hall–Petch relation begin.

The graphs of Figs. 2.5 c, *d* show the data for nanocrystals obtained by annealing the amorphous state of $Fe_{73.5}Cu_1Nb_3Si_{13.5}B_9$ [2.28], $Fe_{81}Si_7B_{12}$ [2.29] and $Fe_5Co_{70}Si_{15}B_{10}$ [2.30]. The results for microhardness and yield stress for different sizes of nanograins (nanophases) D indicate that the Hall–Petch ratio is almost always satisfied if $D \geq 10$–20 nm. At smaller nanophase sizes, the Hall–Petch ratio remains valid only for the $Fe_{73.5}Cu_1Nb_3Si_{13.5}B_9$ alloy.

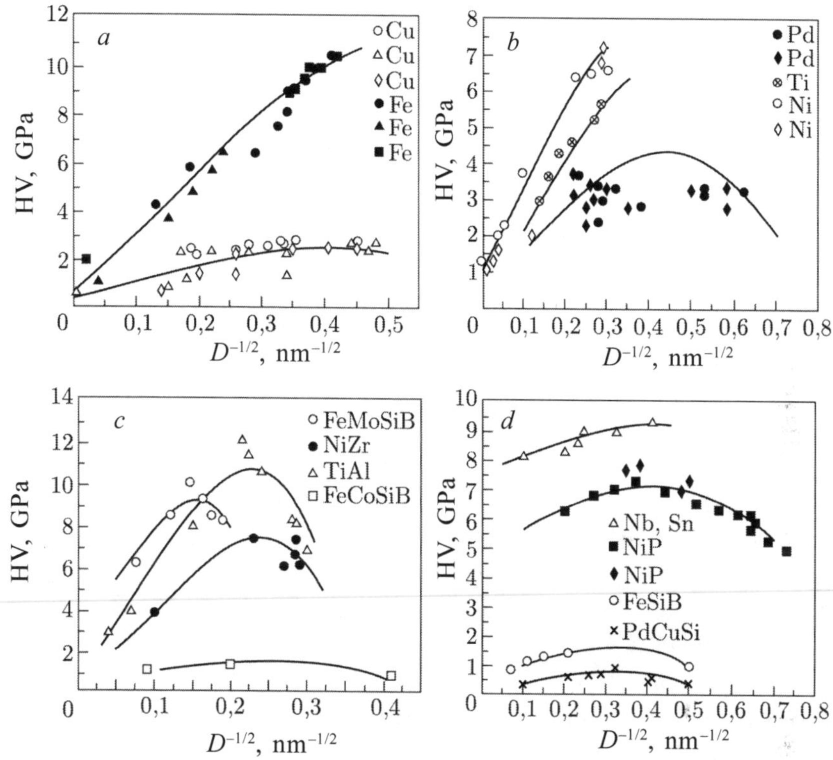

Fig. 2.5. Dependence of hardness on the grain size of nanocrystals for various alloys [2.24].

For Fe–Cu–Si–B nanocrystals, the Hall–Petch ratio has the usual character up to the size of nanocrystals of 25 nm (Fig. 2.6). For the Fe–Mo–Si–B alloy, there is a critical grain size (47 nm), below which the character of the dependence becomes anomalous. A characteristic maximum on the dependence of microhardness on the grain size was also observed for the Ni–W nanocrystals produced by electrodeposition. Anomalous behavior of ceramic materials was also observed as the grain size decreased down to $D_{min} \approx 1$ μm.

Summarizing all the experimental data known to date, it can be stated that the hardness of metals and ceramic materials increases as the grain size passes into the nanophase region. However, the grain size to which hardening occurs depends on a number of factors, and its nature is not entirely clear. Usually, the Hall–Petch ratio is satisfied for a significant part of the investigated nanocrystals only up to a certain critical grain size, and at lower values reverse effects are observed: hardness (strength) decreases with decreasing nanograin

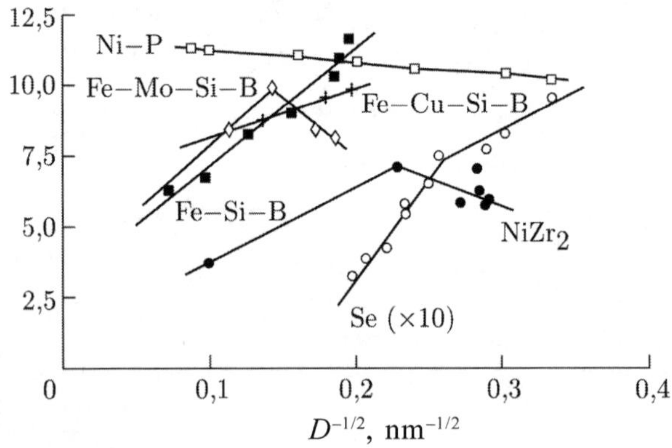

Fig. 2.6. Dependence of hardness on grain size for nanocrystals obtained by crystallisation from the amorphous state.

size. It is characteristic in this connection that the sample subjected to annealing to increase the grain size has a higher hardness value than the sample that had exactly the same grain size immediately after production.

2.4.3. Structural mechanisms of plastic deformation.

The researchers tried to explain the anomaly of the Hall–Petch ratio for nanocrystals [2.31, 2.32]. To describe the mechanical behaviour of nanocrystalline materials, several groups of models have been proposed.

Initially, the Hall–Petch relation was considered from the position of grain boundaries as barriers to the motion of dislocations, and the coefficient k_y as a quantity determining the degree of 'transparency' of grain boundaries for dislocations. In the classical Cottrell theory, the dependence (*1*) in Fig. 2.4 linked the hardness (yield strength) to flat dislocation clusters near the grain boundary, which accumulated shear stresses to activate dislocation sources in the neighbouring grain. In dislocation models, in general, the processes of formation of dislocation clusters in nanograins [2.25] and the kinetics of dislocation structure development are considered [2.33].

In a number of works (for example, [2.34]) it is assumed that below a certain grain size in crystals the formation of flat dislocation clusters is limited, which should lead to softening. As mentioned above, there is a critical grain size D_{cr}, below which flat clusters are not formed. Such an approach can only explain the kink in the values of hardness by decreasing the grain size, but is unable to describe the inverse Hall–Petch dependence, which, as we have seen, is observed in a number of nanocrystalline materials. To explain this phenomenon, it was assumed in the work that annealing nanocrystalline materials leads to a relaxation of the intergranular structure and, accordingly, to a decrease in the excess grain-boundary energy, which causes an anomaly. A similar dislocation model based on the concept of planar clusters was proposed to explain the anomalous strength dependence for multilayer film materials, where the existence of maximum strength at certain thicknesses of individual layers was also observed, and then its decrease with decreasing thickness. In most cases, however, there is a smooth increase in strength (hardness) as the 'wavelength' of the modulation of the composition of the multilayer materials decreases. It should be emphasized that even in the case of two-component multilayer materials, the situation is much more complicated than in 'ordinary' nanocrystalline materials. The propagation of shear through the interphase boundary is influenced by such additional parameters as the degree of coherence of the interface, the processes of mutual diffusion of the components, and also the ratio of the elastic moduli of the components. A different combination of the aforementioned parameters may make it more difficult or easier to operate the dislocation sources as the thickness of the layers decreases and, consequently, determine the specific character of the strength characteristics.

A number of studies have been aimed at explaining the change in the value of the exponent at the D value in the Hall–Petch relation observed in a number of experiments. Thus, data [2.35] show that the change in yield strength or hardness with the grain size is described by a relationship for which the exponent n is not equal to the usual value –1/2, but varies from –1 to –1.4. Each of these values, obtained from theoretical estimates, corresponds to a characteristic mechanism of interaction of the dislocations with the grain boundaries. It is also assumed that there is a definite relationship between the type of crystal lattice and the value of n. In [2.36], when measuring the microhardness of nanocrystalline Se obtained by crystallisation of the amorphous state and practically unaffected by porosity, impurities, and segregation, three clear stages were observed with decreasing grain size from 70 to 8 nm, corresponding to different values of n in the Hall–Petch relation. The maximum value of n is observed in the range of D from 15 to 20 nm.

The second group of models is conditionally called disclination-dislocation models [2.37, 2.38], in which not only the mechanisms of plastic deformation are considered but also disclination–dislocation description of non-equilibrium boundaries, grain joints, a grid of grain boundaries in general, and internal stresses in such materials.

The third group of models represents the grain boundaries as an independent phase, and the yield strength of such two-phase material is expressed by the rule of mixtures or by other means through the mechanical characteristics of its constituents – intragranular and grain-boundary [2.39]. The fourth group includes the models of deformation of nanocomposite materials, the fifth one – models of high-temperature behaviour, the sixth – models of grain-boundary deformation of nanomaterials, the seventh group combines studies on computer modeling of the grain boundary structure and processes of plastic deformation of nanocrystalline materials [2.40].

In [2.41] it is shown for different ranges of copper grain sizes that the coefficient k_y decreases or changes its sign and goes to negative values in the nanoscale region. It is shown that changes in the mechanical properties begin in the grain size range 10–100 nm, which subsequently lead to an anomaly of the Hall–Petch ratio. The pattern of the stages of plastic deformation upon transition to polycrystals with ultrafine grains also changes. The main contribution to deformation mechanisms is made by deformation mechanisms associated with the grain boundaries: slip along the grain boundaries, migration of the grain boundaries and the processes of dynamic recrystallisation.

Scientists have proposed various models to explain the deviation from the Hall–Petch ratio for nanocrystals, but after studies of 1990-2001 it became clear that simple models do not cover the entire complex of problems associated with this phenomenon. Therefore, recently complex models have been developed that partially take into account the combined action of several mechanisms: diffusion along the grain boundaries, dislocation and non-dislocation slip along the grain boundaries, emission of dislocations from the grain boundaries and the absorption of dislocations by grain boundaries, slip in the border region, the reaction of defects at the grain boundaries, accommodative structural processes [2.42, 2.43].

Further development of the idea of the features of dislocations effecting plastic shearing was obtained by considering the dislocation-kinetic model, according to which the appearance of an anomaly is caused by a strong dependence of the annihilation rate of dislocations on the size of the crystallites. It is also assumed that the initial stage of low-temperature (T = 300 K) plastic deformation of polycrystals by generation of dislocations from the grain boundaries is preceded by grain boundary microslipping (GBMS). When the magnitude of the shift at the GBMS reaches a critical value, the stress becomes so large that it generates dislocations at the edges of the GBMS zones.

In the authors' opinion [2.33], the anomalous Hall–Petch dependence and other features of plastic deformation and destruction of nanocrystalline materials are a consequence of the change in the structural mechanism of the plastic flow: as the grain size decreases, the classical dislocation flow is gradually depleted, giving way to grain-boundary microslipping. Taking into account the smallness of grain-boundary deformation, this mechanism was later refined and detailed for the case of bimodal structures [2.44]. Later this model was repeatedly confirmed in both direct and computer experiments [2.44].

The process of plastic deformation of nanocrystals, regardless of the way in which they are obtained, always begins with GBMS, and in itself it is in many respects similar to the process of shear deformation in an amorphous state, since the structure of the grain boundaries can be completely or partially correctly described by means of the amorphous state model. The difficulty of the process of GBMS apparently leads to the brittle behaviour of nanocrystals.

In accordance with modern concepts, any arbitrary grain boundary (GB) can be represented as a limited set of structural elements that coincide with some of Voronoi–Bernal polyhedra proposed for describing the structure of liquids and amorphous bodies [2.45]. It can be assumed that the GBMS under the influence of shear stresses will occur due to the restructuring of the structural elements of the boundaries in the formation of microregions of shear transformations of the GB structure. The stress of resistance to GBMS in a general type GB will be determined by the critical stress of the restructuring of the structural elements of the boundary. Such elementary shear regions are similar to shear transformations that determine the development of heterogeneous plastic deformation in amorphous alloys at low temperatures.

It was shown in [2.4] that as the size of the grain decreases, the share of the grain boundaries gradually increases to a maximum value of 0.45. At very small dimensions, the volume fraction occupied by the grain boundaries decreases, but the volume fraction occupied by the triple junctions increases sharply. Consequently, in the nanometer range of grain sizes (less than 15–20 nm), not only the grain boundaries but also the triple joints of the grain boundaries-

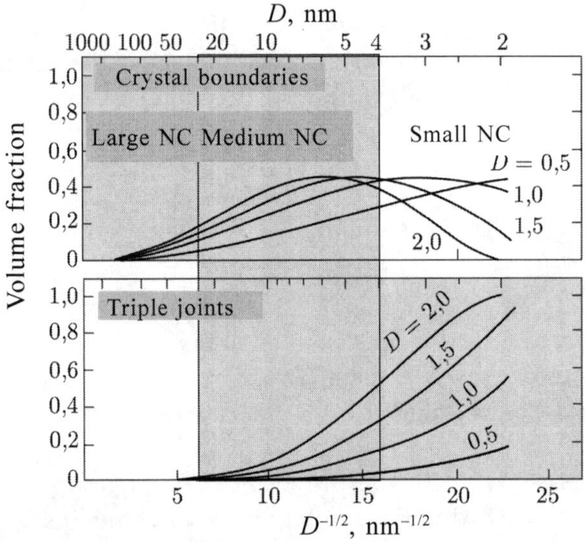

Fig. 2.7. Structural classification of nanocrystalline materials, based on the dominant contribution to the structure of various structural elements and determining the mode of plastic flow; D is the conventional size of the crystalline phase.

the predominant structural elements in this dimensional range of nanocrystals – play a decisive role in various processes (including deformation and fracture processes) .

In [2.46], a structural classification of nanocrystals is suggested from the point of view of their deformation behaviour. According to this classification, there are three dimensional groups: 'large', 'medium' and 'small' nanocrystals. The boundary values of the size of nanocrystals, corresponding to the transition from one type of nanocrystal to another, are very conventional and, perhaps, somewhat blurred. In 'large' nanocrystals, the predominant element of the structure is actually the crystals, in the 'medium' ones – grain boundaries and in the 'small' nanocrystals – triple junctions (Fig. 2.7).

The main idea of this classification is that in the 'large', 'medium' and 'small' nanocrystals, various plastic deformation mechanisms dominate, determined by the predominant element of the structure (proper crystals, grain boundaries or triple grain junctions, respectively). The deformation of 'large' nanocrystals (the grain size is approximately >30 nm) is realized with the help of dislocation mechanisms, therefore, the Hall–Petch relation must be satisfied for them.

When the 'medium' nanocrystals are deformed (the approximate size of the crystallites is 5–30 nm), the grain boundaries play the main role in plastic deformation. Plastic deformation in this case is carried out by means of low-temperature grain-boundary microslipping, which proceeds the easier the smaller the grain size. Both theoretical studies [2.11] and computer simulation [2.47] and structural studies [2.48] have shown the reality of the action of GBMS in these nanocrystals.

In 'small' nanocrystals (the dimensional range of grains is less than 5 nm), the fraction of the volume occupied by the grain body and the grain boundaries decreases, and the triple joints are the dominant element of the structure. Plastic deformation of nanocrystals occurs as a result of plastic rotations of grains which causes the generation of partial disclinations in the joints of grain boundaries. A feature of the mechanical behaviour of 'small' nanocrystals is the existence of the $\sigma_y(D)$ dependence, which corresponds to the inverse Hall–Petch relation, but it may not correspond to the $\sigma_y(D)$ dependence inherent in 'medium' nanocrystals.

The proposed separation of all nanocrystals into three types ('large', 'medium' and 'small') allows a deeper understanding of the cardinal changes in the mechanical behaviour that are the consequence of the size effect. Two important observations need to be made.

• In the transition from one type of nanocrystal to another, or if there is a distribution in size in the polycrystalline ensemble encompassing nanocrystals of various types, it is possible to realize mixed deformation mechanisms that are obviously more complex.

• The proposed structural classification from the standpoint of deformation behaviour can be attributed not to all nanomaterials, but only to nanocrystals. In other nanostructured materials the processes of plastic flow obey other, yet completely unexplored laws. For example, this consideration does not explain the plastic deformation of materials that have undergone severe deformation, since this group of nanomaterials refers not to nanocrystalline materials but to nanofragmented materials.

Summarizing all of the above, we come to the important conclusion that the process of plastic deformation of nanocrystals, regardless of the way in which they are obtained, always begins with grain-boundary microslipping, and the process of microslipping itself is carried out in a similar manner to the process of shear deformation in the amorphous state, since the structure of the grain boundaries is described in full or in part by means of the amorphous state model. The difficulty of the microslip process leads to the brittle behaviour of nanomaterials. This conclusion was fully confirmed in the experiments with computer simulation of the plastic flow of nanocrystals. First, it was established that the dependence of the deforming stress and the yield stress on the size of the nanocrystals obeys the inverse Hall–Petch dependence and, secondly, that plastic deformation is realized along the grain boundaries in the form of a large number of small shifts when only a small number of atoms moves relative to each other. The same pattern of atomic displacement is observed in the computer simulation of the processes of plastic deformation in amorphous metallic materials. Thus, the anomalous Hall–Petch dependence is also possible in the absence of porosity and is associated with slipping along grain boundaries, even in the absence of thermally activated processes.

2.4.4. Destruction

The study of the features of the propagation of cracks in nanocrystalline materials is important in connection with the possibility of increasing the fracture toughness (fracture toughness) of K_{Ic} of brittle nanocrystalline materials. The available information on this subject is still contradictory. On the one hand, a number of results indicate that the increase in K_{Ic} is not observed in nanocrystalline materials, and the coarse-grained oxide, carbide, nitride and other single-phase brittle materials have better values for this parameter $D \approx 1$–10 μm). On the other hand, in the early studies of nanocrystalline materials it was noted that it was possible to increase the ductility of brittle materials at room temperatures due to the transition to the nanocrystalline state. More or less reliable experimental data on the increased characteristics of K_{Ic} for brittle nanocrystalline materials are available only for multiphase objects (ZrO_2–Al_2O_3, TiB_2–TiN, Si_3N_4–SiC, etc.). Apparently, taking into account the preferential intergranular fracture, combining the sizes of the phase components in the nanocomposites, it is possible to obtain a significant gain in the fracture toughness characteristics by increasing the fracture energy (increasing the length of the trajectory of the intercrystalline crack).

As is known, the phenomenon of superplasticity is very useful for optimizing the pressure treatment modes, especially with regard to brittle materials, but at the same time it can have catastrophic consequences when operating heat-resistant structures. Naturally, the study of creep and superplasticity of nanocrystalline materials is in the centre of attention. In the experimental study of these phenomena, the main difficulties are in the preparation of well-qualified samples (especially in the case of tensile tests) and in preventing or strictly taking into account their recrystallisation, therefore, variations in the grain size, stresses, etc., are usually small in experiments. In connection with this, it is also popular to study high-temperature deformation characteristics in indentation. In themselves, the achieved plasticity indicators are quite impressive: tens and hundreds of percent at moderately high temperatures (650–725°C for Ni_3Al, 700°C for TiO_2, 1150–1250°C for ZrO_2). Due to the transition to the nanocrystalline state, the superplasticity manifestation temperature in comparison with conventional fine-grained materials was reduced by about 300–400°C, however, the stress level remains approximately an order of magnitude higher than that of industrial superplastic metallic materials.

The most important element in the structure of nanocrystalline materials, which largely determines their macroscopic properties, is, of course, grain boundaries. It is known that they can have a significant effect on the process of destruction of polycrystals. At the boundaries which are places of stress concentration and reduced strength, the processes of nucleation and propagation of cracks can be facilitated. In nanocrystalline materials with a very high density of boundaries and their joints, the influence of the boundaries on the development of cracks should be much more significant than in traditional materials. Depending on the method of obtaining nanocrystalline materials, such as nanopowder compacting, mechanical alloying, nanocrystallisation of amorphous alloys, or severe plastic deformation, a grain structure with various degrees of structural non-equilibrium, a spectrum of misorientations, defectiveness and chemical composition of the boundaries can form. The structure of nanocrystalline materials can be characterized by the presence of uncompensated joints of grain boundaries.

Fractographic studies of brittle fracture surfaces of nanocrystalline materials have revealed the dominant role of intergranular fracture mechanism. The indentation method was used to measure the fracture toughness of the nanocrystalline FeMoSiB alloy obtained from the amorphous state with the grain size of the α-Fe phase from 11 to 35 nm and it was established that the dominant fracture mechanism is the intercrystalline propagation of the crack. It was found that the average value of the fracture pits on the fracture surface for samples with the grain size D = 11, 25 and 35 nm, respectively, is 0.5, 2 and 5 μm. The crack resistance for samples with a grain size increasing from 11 to 35 nm increases from 2.7 to 4.6 MPa · $m^{1/2}$. The change in the fracture resistance with the grain size variation does not depend on the plastic deformation of the nanocrystalline materials.

A study of the features of the development of cracks in nanocrystalline materials is important in connection with the search for opportunities to increase the fracture toughness (crack resistance) of brittle materials with a dispersed structure. There are experimental data on the increased fracture toughness of multiphase brittle materials in the nanostructured state. On the other hand, a number of data indicate that for nanocrystalline materials an increase in ductility is not achieved. Only the first theoretical papers begin to appear in which dimensional effects have been studied.

If we introduce the concept of 'the ideal nanocrystalline material' (a homogeneous single-phase polycrystal without dislocations with a grain size of the order of 10 nm and with grain boundaries corresponding to coarse-grained materials), analysis of the effect of the grain size on the mechanisms and conditions for the fracture of the 'ideal nanocrystalline material' allows one to single out dimensional strength effects. For nanocrystalline materials with $D < 10$ nm, the volume fraction of triple joints becomes comparable with the volume fractions of grain boundaries and intragranular material. In this case, their contribution to the fracture energy must be taken into account. In intercrystalline fracture, the contribution of the linear tension of the crack surface to the fracture energy becomes significant for nanocrystalline materials. An estimate of the increase in the specific fracture energy for nanocrystalline materials with a grain size of 10–20 nm due to taking into account triple junctions and surface tension of the crack yields 20%.

With such methods for producing nanocrystalline materials, such as nanopowder compacting, mechanical alloying or severe plastic deformation, a grain structure with a high degree of structural non-equilibrium can be formed. The structure of nanocrystalline materials can be characterized by the presence of uncompensated joints of grain boundaries. An analysis of the influence of the relaxation of the fields of uncompensated triple junctions and grain boundary defects on the crack propagation conditions shows that the advancing crack tip will lead to the relaxation of the stress fields of the joint disclinations located on the front line and at some small distance from it. There is a dual effect of the change of the Griffith criterion due to internal stresses – the effective energy of fracture decreases, but an additional contribution to the stress comes from the loss of stability.

An important mechanism for increasing the fracture toughness of materials is the formation of structures that facilitate the realization of the formation of bridges at the mouth of the crack. This is of particular importance in the case of ceramic nanocomposites. Since nanostructural components of composite materials can not be deformed plastically, it can be concluded that the source of increase in fracture toughness could be internal stresses that increase friction when drawing out the nanograins of a different phase when bridges are formed. Analysis of the fracture mechanisms of nanocrystalline materials shows that the most significant source of increasing their viscosity is grain-boundary plastic deformation. If the stress of resistance to grain-boundary microslipping in a nanocrystalline material obtained by a particular technology is less than the formation stress and/or the propagation stress of a crack, such materials can undergo a plastic flow prior to failure and exhibit significant viscosity.

2.4.5. Magnetic properties

One of the possible effects associated with the magnetic properties of very small metal particles (nanoparticles) was predicted in the early 1960s by the well-known Japanese theorist R. Kubo [2.49]. The idea he proposed is well-known and simple. Imagine clusters composed of a small number of atoms with an odd number of electrons (for example, they may be atoms of alkali metals). It is clear that, containing an odd number of such atoms (with an odd number of electrons), the clusters will have the Curie paramagnetism. On the other hand, it seems equally obvious that clusters with an even number of such atoms should always be described by the usual Pauli paramagnetism for metals, with the exception of the region of very low temperatures, where an additional spin pairing occurs and the magnetic susceptibility is lower than the Pauli energy. At low temperatures, this phenomenon should lead to a complex behaviour of mixed ensembles composed of clusters with an even and odd number of atoms. Moreover, the effect is not limited to only unusual magnetic properties, since in the described systems at low temperatures there should be a decrease in the specific heat relative to the usual value.

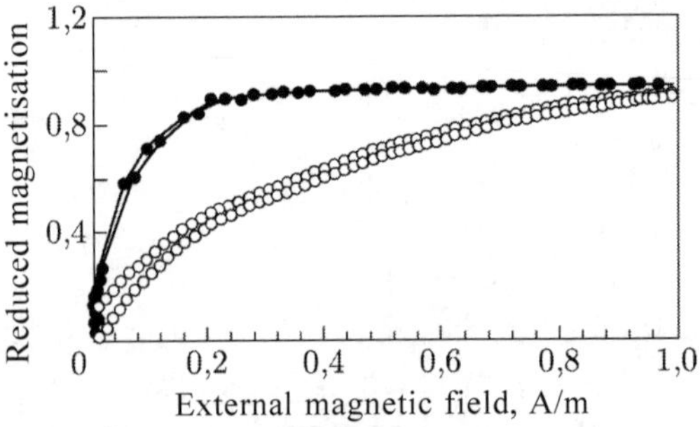

Fig. 2.8. Relative magnetization (with respect to the saturated state) of nanocrystals 20 nm in diameter (dark dots) and 73 nm (light circles) at room temperature [2.51].

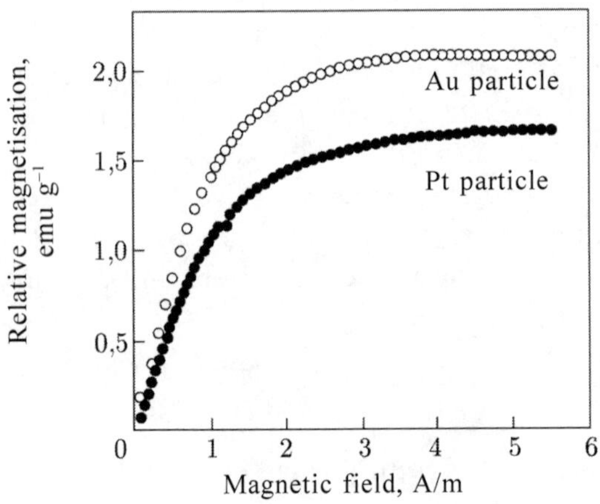

Fig. 2.9. Magnetization of gold and platinum nanoparticles with an average diameter of 2.5 nm at 1.8 K [2.52].

It should be noted that the saturation magnetization of nanocrystalline samples of nickel or chromium is only slightly (only a few percent) different from the values for bulk polycrystalline samples. In this case, the difference in the form of the hysteresis loop of the corresponding materials was recorded [2.50]. With decreasing crystal grain sizes, the magnetic hysteresis loops for such samples become steeper and narrower, so that the saturation magnetization is reached at lower values of the external field, as illustrated in Fig. 2.8. This observation confirms the assumption that the reorientation of spins is more easily realized in small-sized crystals.

An exceptionally strong superparamagnetic behaviour at temperatures below 4 K was observed for gold and platinum microclusters (for elements that are normally non-magnetic) with an average diameter of about 2.5 nm and a fairly narrow size distribution [2.52]. The saturation magnetization in these systems (see Fig. 2.9) is about 20 μ_B and 30 μ_B for the Pt and Au particles, respectively, which indicates a large number of unpaired spins, apparently located near the surface.

From the data presented it follows that (at least) the Au particles can be considered metallic under these conditions. Anomalous magnetization increasing with decreasing particle size and reaching about 6.3 μ_B (for a particle with a diameter of 2.5 nm) was also observed in experiments with monodisperse platinum particles [2.53].

The data obtained are in good agreement with theoretical calculations of the spin density for neutral or negatively charged clusters of the PdN type. For neutral clusters with $2 \leq N \leq 7$, the theory predicts the existence of triplet ground states, and for a cluster with $N = 13$, the presence of a spin nonet. More calculations predict that the spin density should be localized just near the surface of the clusters [2.54].

At present, nanocomposites with particle sizes of 10–100 nm have increasingly become used in the production of magnetically hard materials. As in ordinary ferromagnets, in the absence of an external field, the direction of magnetization in such nanocomposites coincides with the axis of easy magnetization of the material, determined by the crystal lattice of the matrix and the anisotropy of the shape of individual particles. At very small particle sizes, their rotation is so easy that the energy advantage of the easy magnetization axis becomes meaningless, and even small thermal fluctuations can change the position of the particle, after which any set of isolated microparticles loses its ferromagnetic properties and the substance becomes a superparamagnet. It is the temperature of disappearance of the 'width' of the hysteresis loop above which the material becomes superparamagnetic, which is called the blocking temperature. In this case, the smallest crystals of matter (about 1 nm in size) are converted into an exclusively magnetically soft material whose magnetic permeability approaches 10^5 [2.55].

The dependence of the coercive force of nanocrystalline materials on the dimensions and temperature was studied by Herzer [2.56].

The main result is that the dependence of the coercive force H_c and the magnetic susceptibility μ on the diameter of the crystal grains D have a completely different character, depending on the diameter of the crystal grains D and the length (distance) of exchange interaction L_{ex}.

In particular, for $D < L_{ex}$ these dependences have the form:

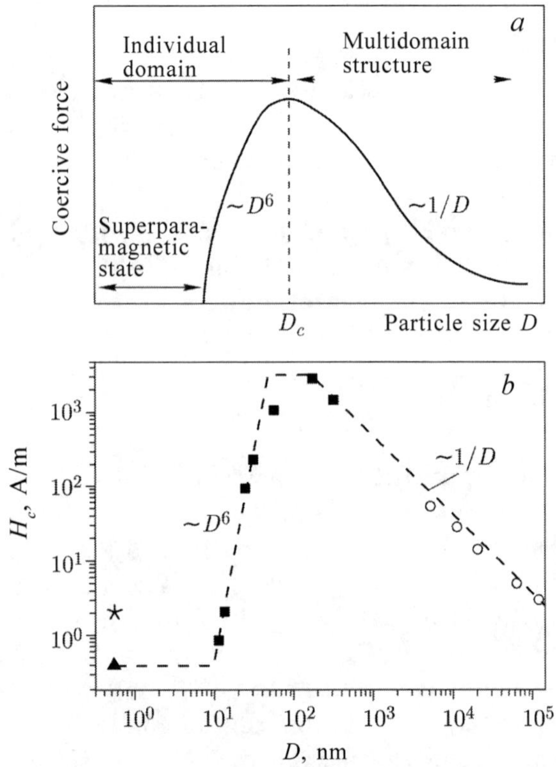

Fig. 2.10. Dependence of the coercive force H_c of nanocrystals on their size D: *a* – diagram of the general pattern of change in the coercive force; *b* – value of H_c for some materials and nanocrystalline alloys at room temperature ($Fe–Co_{0-1}Nb_3(SiB)_{22.5}$ (squares), 50% Fe–Ni alloy (circles), amorphous Co (triangles), amorphous Fe (asterisk)) [2.56].

$$H_c \sim D^6 \text{ and } \mu \sim H_c^{-1} \sim D^{-6}, \tag{2.4}$$

and at $D > L_{ex}$:

$$H_c \sim D^{-1} \text{ and } \mu \sim D. \tag{2.5}$$

The obtained dependences were confirmed by studies of the characteristics of various soft magnetic alloys [2.56]. Figure 2.10 shows some of their data obtained in [2.56] which confirm that the magnetic characteristics in which we are interested can vary in the nanoclusters by up to four orders of magnitude. The maximum of the coercive force H_c is reached at the maximum nanoparticle size D_c corresponding to the size of the individual domain.

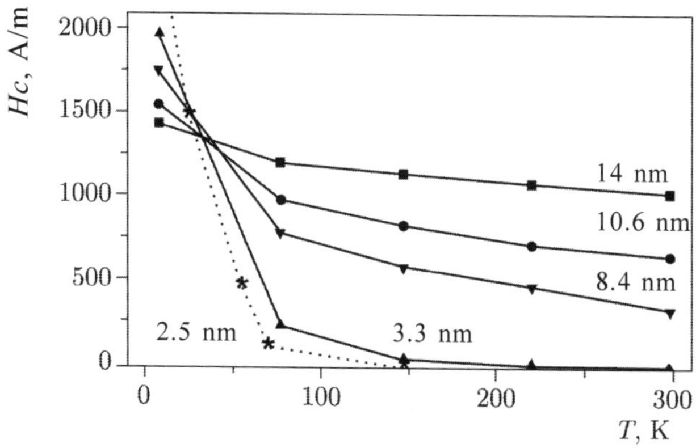

Fig. 2.11. Dependence of the coercive force H_c on temperature for Fe nanoparticles of different radii encased in ferrous oxide [2.57].

In addition, a second critical value for the diameter of nanoparticles at which the anisotropy energy becomes so small that the coercive force (and consequently the hysteresis effect) disappears altogether, so that such a nanoparticle can move in an external magnetic field without additional restrictions. It is this behaviour that corresponds to the described superparamagnetic characteristics of matter.

For larger nanoparticles, the temperature dependences of the coercive force are more moderate, but very sharp changes in the magnetic properties are observed for particles with diameters of only a few nanometers, with the intersection of the characteristics occurring at a temperature of about 30 K, as shown in Fig. 2.11 [2.57].

The authors of [2.58] investigated systems of iron oxide nanoparticles with particle sizes in the range of 2–8 nm by a combined technique including X-ray diffraction, transmission electron microscopy, and magnetic measurements, which allows one to evaluate the parameters by independent methods. The volume of the particles was estimated from the magnetization curves (Fig. 2.12) in accordance with the Langevin function for superparamagnetic particles having a log–normal size distribution. The value obtained is in good agreement with the volume estimates obtained by the Scherrer formula and from the TEM data. Saturation of larger particles is carried out with greater ease, and the saturation magnetization M_s decreases with decreasing particle size.

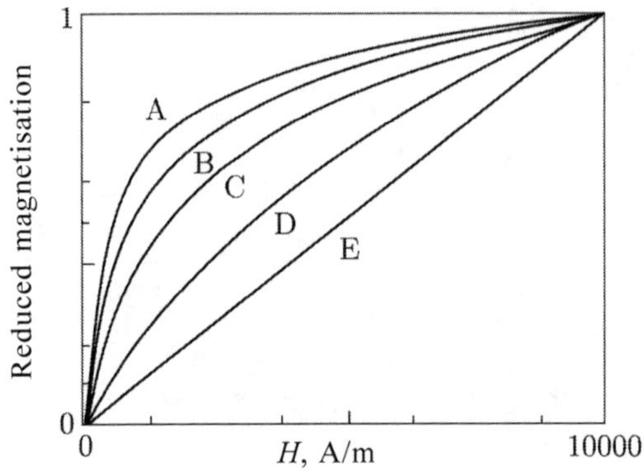

Fig. 2.12. Dependence of the reduced magnetization at room temperature for iron nanoparticles with an average size of 7.3 nm (A), 5.6 nm (B), 3.6 nm (C), 2.4 nm (D) and 1.9 nm (E) [2.58].

Interesting results were obtained in the study of the magnetic characteristics of the bimetallic FePt system [2.59–2.61]. Although platinum in bulk does not belong to magnetic materials at all, its presence in a bimetallic material leads to an appreciable increase (up to 1.56 times) of the saturation magnetization of bulk samples of the FePt alloy and to an even higher (up to 4 times) increase for nanoparticles of the same type alloy with a diameter of 2.5 nm [2.59]. The system studied is a mixture of magnetically hard FePt particles with high coercive force H_c and magnetically soft particles Fe_3Pt with a very small value of H_c and high magnetization. Using the self-organization of microparticles from the magnetically soft and magnetically hard phases, the authors of [2.62] managed to obtain a nanocomposite with a high energy product, which indicates the presence of an exchange interaction between these particles. The high value of the energy product (associated with the area of the hysteresis loop) is one of the most important parameters by which the quality of the materials used in the production of permanent magnets is evaluated.

References

2.1. Golovin Yu.I., Introduction to nanotechnology. - Moscow, Mashinostroenie, 2007.
2.2. Gleiter N., Acta Materialia. 2000. V. 48. P. 223–315.
2.3. Pokropivny V.V., Skorokhod V.V., Mater. Sci. Eng. C, 2007. V. 27. P. 990–993.
2.4. Andrievsky R.A., Fundamentals of nanostructured materials science. Opportunities and problems. - Moscow: Binom. Laboratory of Knowledge, 2012.
2.5. Andrievsky R.A., Glezer A.M., Usp. Fiz. Nauk, 2009. V. 179, No. 4. P. 337–358.
2.6. Andrievsky R.A., Ragulya A.V., Nanostructured materials, textbook. - M .: Publishing centre Akademiya, 2005.
2.7. Malygin G.A., FTT. 2007. V. 49, No. 6. P. 961–982.
2.8. Shuh C.A., et al., Acta Materialia. 2007. V. 55, No. 12. P. 4067–4109.
2.9. Golovin Yu.I., FTT. 2008. V. 50, No. 12. P. 2113–2142.
2.10. Koneva N.A., et al., Izv. RAN. Ser. Fiz. 2006. V. 70, No. 4. P. 582–585.
2.11. Roduner E., Size effects in nanomaterials. ed R A. Andrievsky. - Moscow: Tekhnosfera, 2010.
2.12. Noskova N.I., Mulyukov R.R., Submicrocrystalline and nanocrystalline metals and alloys. - Ekaterinburg: UrO RAN, 2003.
2.13. Malygin G.A., FTT. 2010. V. 52. P. 48–55.
2.14. Shpeizman V.V., et al., FTT. 2003. V. 45, No. 11. P. 2008–2013.
2.15. Kelly A. High-strength materials (trans. from English). - Moscow: Mir, 1976.
2.16. Cerny M. Theoretical Strength and Stability of Crystal from First Principles: Habilitation thesis. - Brno. Univ. Tech., 2008.
2.17. Firstov S.A., Rogul' T.G., Deform. Razrush. Mater., 2011. No. 5. P. 1–7.
2.18. Goldstein M.I., et al., Metal physics of high-strength alloys. - Moscow: Metallurgiya, 1986.
2.19. Orowan E., Rep. Progr. Phys. 1949. V. 12. P. 185–232.
2.20. Slutsker A.I., FTT. 2004. V. 46, No. 6. P. 1606–1613.
2.21. Firstov S.A., Rogul' T.G., Reports of the National Academy of Science of Ukraine. 2007. No. 4. P. 110–114.
2.22. Shtremel' M.A., Strength of alloys. Part 2. - Moscow: MISiS, 1997.
2.23. Andrievsky R.A., Glezer A.M., FMM. 2000. V. 89, No. 1. P. 91–112.
2.24. Golovin Yu.I., Introduction to nanotechnology. - Moscow: Mashinostroenie-1, 2003..
2.25. Nieh T.G., Wadsworth J., Scr. Metall. Mater. 1991. V. 25. No. 4. P. 955–958.
2.26. Shen T.D., et al., Acta Materialia. 2007. V. 55. No. 15. P. 5007–5013.
2.27. Noskova N.I., Deform. Razrush. 2009. No. 4. P. 17–24.
2.28. Glaser A.A., FMM. 1992. No. 8. P. 96–100.
2.29. Noskova N.I., FMM. 1992. No. 2. P. 83–88.
2.30. Noskova N.I., et al., Nanostruct. Materials. 1995. V. 6, No. 5–8. P. 969–972.
2.31. Scattergood R.O., Koch C.C., Scr. Metall. Mater. 1992. V. 27, No. 9. P. 1195–1200.
2.32. Carlton C.E., Ferreira P. J., Acta Materialia. 2007. V. 55, No. 11. P. 3749–3756.
2.33. Glezer A.M., Deform. Razrush. Materialov, 2006. No. 2. P. 10–14.
2.34. Nazarov A.A., Scr. Met. 1996. V. 34. No. 5. P. 697–701.
2.35. Christman T., Scr. Met. Mater. 1993. V. 28. P. 1495–1500.
2.36. Lu K., et al., J. Mater. Res. 1997. V. 12. P. 923–930.
2.37. Palumbo G., et al., Scr. Metall. Mater. 1993. V. 24, No. 12. P. 2347–2350.
2.38. Zaichenko S.G., Glezer A.M., FTT. 1997. P. 39, No. 11. P. 2023–2028.
2.39. Glezer A.M., Pozdnyakov V.A., Nanostructured Materials. 1995. V. 6. P. 767–770.
2.40. Pozdnyakov V.A., FMM. 2003. V. 96, No. 1. P. 114–128.

2.41. Kozlov E.V., Fiz. Mezomekh., 2004. V. 7, No. 4. P. 93–113.
2.42. Kozlov E.V., ibid, 2006. V. 9, No. 3. P. 77–88.
2.43. Kozlov E.V., et al., Izv. RAS. Ser. Fiz., 2009. TV 73, No. 9. P. 1295–1301.
2.44. Pozdnyakov V.A., Izv. RAS. Ser. Fiz., 2007. V. 71. P. 1751–1763.
2.45. Orlov A.N., et al., Grain boundaries of grains in metals. - Moscow: Metallurgiya, 1980.
2.46. Glezer A.M., Deform. Razrush. Mater., 2010. No. 2. P. 1–8.
2.47. Schiotz J., et al., Nature, 1998. V. 391/5, No. 2. P. 561–563.
2.48. Cai B., et al., Mater. Sci. Eng. A. 2000. V. 286. P. 188–192.
2.49. Kugo R., J. Phys. Soc. Jpn. 1962. V. 17. P. 975.
2.50. Przenioslo R., et al., Phys. Rev. B. 2001. V. 63. P. 54408.
2.51. Billas I.M. et al., Science. 1994. V. 265. P. 1682.
2.52. Nakae Y., et al., Physica B. 2000. V. 284. P. 1758.
2.53. Yamamoto Y., et al., Physica B. 2003. V. 329–333. P. 1183.
2.54. Moseler M., et al., Phys. Rev. Lett. 2001. V. 86. P. 2545.
2.55. Larionova J., et al., Angew. Chem. 2000. V. 112. P. 1667.
2.56. Herzer G., IEEE Trans. Magn. 1990. V. 26. P. 1397.
2.57. Gangopadhyay S., et al., Phys. Rev. B. 1992. V. 45. P. 9778.
2.58. Lopez-Perez J.A., et al., J. Phys. Chem. B. 1997. V. 101. P. 8045.
2.59. Schmauke T., et al., J. Mol. Catal, A Chem. 2003. V. 194. P. 211.
2.60. Stahl B., et al., Phys. Rev. B. 2003. V. 67. P. 014422.
2.61. Zheng H., et al., Nature. 2002. V. 420. P. 395.
2.62. Stampanoni M., et al., Phys. Rev. Lett. 1987. V. 59. P. 2483.

3
Amorphous–nanocrystalline alloys

The amorphous state of metals and alloys is a metastable state, so there are thermodynamic incentives for its transition to a crystalline state. The process of transition of the amorphous phase to the crystalline phase is associated with the overcoming of the energy barrier. The activation energy of this process depends on the nature of the crystallizing phase. Very often, the transition from an amorphous state to a stable crystalline state passes through a number of metastable states. The process of transition from the amorphous to the nanocrystalline state can be regarded as a disorder–order transition.

3.1. Methods for the preparation of amorphous–crystalline materials

Various methods are used to form the amorphous crystal structure: quenching from the melt; thermal or deformation effect on the solid amorphous state; pulse photon and laser processing; plasma treatment; ion implantation and a number of others. Each has its advantages and disadvantages.

In the manufacture of coatings for friction units and tool products, as well as in the technology of other functional nanomaterials, film amorphous–nanocrystalline composites have become widely used due to their high wear resistance, various tribological and physico—mechanical properties. Let us briefly consider each of the methods of obtaining materials with an amorphous–crystalline structure.

3.1.1. Melt quenching

Melt amorphisation requires that the melt be cooled at a sufficiently high rate in order to prevent the crystallisation process from proceeding as a result of which the disordered configuration of atoms 'freezes' [1.1]. When cooled for a sufficiently long period of time so that the thermodynamically equilibrium state of the liquid becomes possible, the melt crystallizes at the solidification temperature T_m. However, at a high cooling rate, the liquid does not crystallize even when it is supercooled below T_m. A liquid in this state is called supercooled. Further, if the cooling rate is maintained sufficiently large, the liquid does not become a crystal, the liquid structure is maintained to fairly low temperatures, but eventually the liquid solidifies.

The supercooled liquid solidifies at a temperature called the glass transition temperature T_g.

As was discussed in detail in Chapter 1 (section 1.1), there are several ways to produce alloys with an amorphous structure, but the most common is the spinning method, in which the melt is fed under pressure to a rapidly spinning disc [1.4]. The result is a ribbon with a thickness of 20 to 100 μm the structure of which depends on the alloy composition and the cooling rate. The cooling rate reaches 10^6 K/s, as a result the alloy does not have time to crystallize, and the obtained material has an amorphous structure. The installation scheme is shown in Fig. 3.1.

In all quenching installations from the liquid state, the general principle is that the melt fed into the crucible quickly solidifies

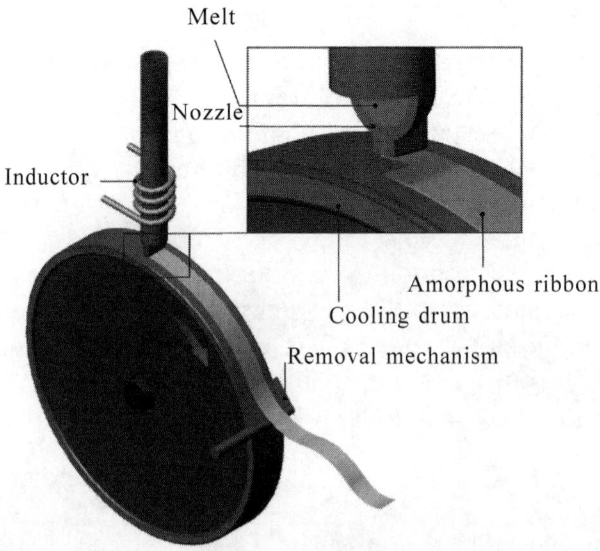

Fig. 3.1. Scheme of installation for melt spinning.

using an inductor, spreading a thin layer over the surface of the rotating refrigerator drum and turning into a final product in the form of a thin ribbon. When the composition of the alloy is constant, the cooling rate depends on the thickness of the melt and the characteristics of the cooler. For various alloys, the cooling rate also depends on the properties of the melt itself (thermal conductivity, heat capacity, viscosity, density). In addition, an important factor is the heat transfer coefficient between the melt and the cooler in contact with it.

When crystallisation of amorphous systems is realized under conditions of melt quenching at a rate close to critical, materials with an amorphous–crystalline structure are obtained. This speed corresponds to a certain cooling rate, above which the system after quenching from the melt is in the amorphous state, and below – in the crystalline state. When cooled at a critical rate, crystallisation proceeds under conditions of severe heat deficiency: the temperature at the front of the growing crystal decreases sharply. This leads to the fact that at a certain stage the growth of the crystals is suspended, and the remaining melt that has not been converted becomes solidified to form an amorphous state. This picture contrasts sharply with that which occurs when the amorphous state is heated, when the crystallisation process takes place under conditions of constant heat supply from the outside, and also in conditions of additional local heat release associated with the crystallisation process.

As a result of this cooling, the amorphizable melt, and then the solidified amorphous matrix contain particles of the crystalline phase evenly distributed in the bulk.

Melt quenching which directly leads to nanocrystallisation, significantly increases the ductility of those alloys that tend to brittle fracture in the ordinary polycrystalline state. In some cases, a clear viscous–brittle transition is observed. So, for example, judging by the nature of the fracture, the FeCo alloy has brittle fracture behaviour in the ordinary state and ductile behaviour – after quenching from the melt. The reasons for the increase in plasticity are: small grain size, the presence of a developed fragmented substructure, a lower degree of long-range order, and, finally, the presence of easily moving dislocations in the structure. It is characteristic that along with the increase in ductility during quenching, the strength of the melt essentially increases. This is due, first of all, to a decrease in grain size in accordance with the Hall–Petch ratio. In addition, this is due to the presence of a high bulk density of vacancy-type defects and/or particles of the second phase of nanocrystalline dimensions. Experiments have shown that the ductility of alloys in the transition amorphous–nanocrystalline state, although considerably lower than in the amorphous state, is substantially higher than in the brittle crystalline state. Consequently, the transition state has not only very high strength but also a sufficient margin of ductility. A study of the pattern of slip bands showed that the deformation process in the transition state is close to that observed in amorphous alloys. For example, the measurement of the height of slip steps performed by scanning electron microscopy showed in both cases a very high degree of localisation of plastic shear (the step height was 0.3–0.4 μm), which corresponds to a local degree of plastic deformation in shear bands of several hundreds of percent. Electron microscopic images in shear bands showed no signs of the existence of dislocations. At the same time, it is established that the shear bands are located in the amorphous matrix and, as it were, bypass the nanocrystalline precipitates. Thus, we can conclude that the process of plastic flow in nanocrystals occurs along amorphous interlayers that have survived during nanocrystallisation, and is analogous to a certain extent to the process of grain-boundary sliding.

3.1.2. Controlled crystallisation

One of the most common methods of obtaining amorphous–crystalline structures is the thermal effect on the solid-phase amorphous state, obtained, in turn, by quenching from the melt. At a certain stage of heat treatment a structure is formed consisting of two structural components: amorphous and crystalline. The nature of the structure in this case depends to a certain extent on the rate of quenching from the melt and subsequent heating and also the temperature and annealing atmosphere.

The crystallisation process takes place under conditions of constant heat input (annealing) and under additional local heat release associated with the crystallisation process. As a result, a two-phase mixture of amorphous and crystalline phases is formed [3.1].

Nanocrystallisation of a pre-amorphized material is carried out under strictly controlled thermal treatment conditions at atmospheric or elevated pressure. In most amorphous alloys, the rate of crystallisation is very high and crystals exceeding the nanometer range grow in very short annealing time intervals.

The main principle of the crystallisation method is the control of the kinetics of crystallisation by optimizing the parameters of heat treatment (temperature and annealing time). In the formation of a microstructure in the crystallisation process, it is possible to obtain grains of smaller size with an increase in the nucleation rate of primary crystals and with a decrease in the rate of subsequent growth of the crystals. Therefore, crystallisation of initial amorphous materials can be successfully used to produce materials with an amorphous–crystalline structure in various alloy systems. To control the process of nucleation and growth of crystals, elements are introduced into the composition, on the one hand, facilitating their nucleation and, on the other hand, inhibiting their growth. Such elements for amorphous Fe–Si–B alloys are copper and niobium in an amount of 1–3 at.%. The annealing of such an amorphous alloy at a temperature of about 800 K leads to the release of Fe–Si BCC nanocrystals in the amorphous matrix with a size of 10–15 nm. In a similar way, a nanocrystalline state was obtained in other alloys (for example, Fe–Co–Zr–B).

The production of amorphous–nanocrystalline materials by controlled crystallisation has a number of advantages over other methods of obtaining alloys with an amorphous–nanocrystalline structure [3.2]. Annealing leads to the formation of a nanocrystalline structure in all alloys that can be obtained in the amorphous state. In addition, the grain size can easily be changed over a wide range by varying the parameters of heat treatment – temperature and annealing time [3.3]. An important advantage of the crystallisation method is also the possibility of obtaining non-porous nanocrystalline materials with a homogeneous microstructure.

3.1.3. Deformation effect

Extreme effects have a significant influence on the structure and properties of solids [3.4]. The principle of deformation processing of materials for obtaining in them an ultrafine-grained structure consists in carrying out large plastic deformations of the material. In recent years, interest in this method of controlling the structure of metallic materials has increased significantly, since it makes it possible to significantly improve their physico-mechanical properties [3.5]. To a large extent, this is due to the formation of nanostructured states of various types and, in particular, to nanocrystallisation processes in the processing of crystalline and amorphous alloys of different phase and chemical composition.

It is important to separate the methods of large plastic deformations that allow forming the nanostructured state in the bulk and on the surface of the material. Below, consider each of them separately.

3.1.3.1. Volumetric strain. Terminology

According of the pioneers in the study of ultrahigh plastic deformations V. Segal and R.Z. Valiev [3.6, 3.7], similar plastic deformation at which the value of true plasticity *e* has values above 1 and can reach values of 7–8 called severe plastic deformation (SPD). This term, however, does not seem entirely successful. In fact, under intensive processes in nature we usually understand processes occurring at high speed [3.8]. In the case of very large deformations, as shown by the estimates, the strain rate is in the range 10^{-1}–10^{1} s^{-1} (i.e., in the transition region between static and dynamic deformation rates) corresponding to the speed realized, for example, in conventional rolling. In this regard, this deformation can not be called intensive. The term *severe plastic deformation* [3.9] used in the foreign scientific literature seems to be more successful, since it can be translated into Russian as 'strict', 'tough', 'deep', and, most likely, 'strong' plastic deformation [3.10].

In [3.11] another physically more rigorous Russian term was introduced instead of 'intense' plastic deformation. The new term goes back to the general philosophical conception of our ideas about the surrounding matter. As is well known [3.8], natural science considers three scale levels of the material world (Fig. 3.2): the microworld, where the scale of individual atoms and molecules is realized (Fig. 3.2 *a*), the macroworld is the scale of the human perception of the world: meter, kilogram, second (Fig. 3.2 *b*) and megaworld – astronomical scale (Fig. 3.2 *c*). There is a direct analogy between the above-described scale levels of organization of matter and levels of plastic deformation. In fact, the process of microplastic deformation observed before the macroscopic yield point is reached is well known, and the process of macroplastic deformation, realized at stresses above the yield point [3.12].

Thus, continuing this analogy, one should call a very large plastic deformation as megaplastic deformation (MPD), which corresponds to the general logic of development of any material phenomenon (Fig. 3.3).

If the boundary between microplastic deformation and macroplastic deformation is defined quite clearly – the degree of deformation

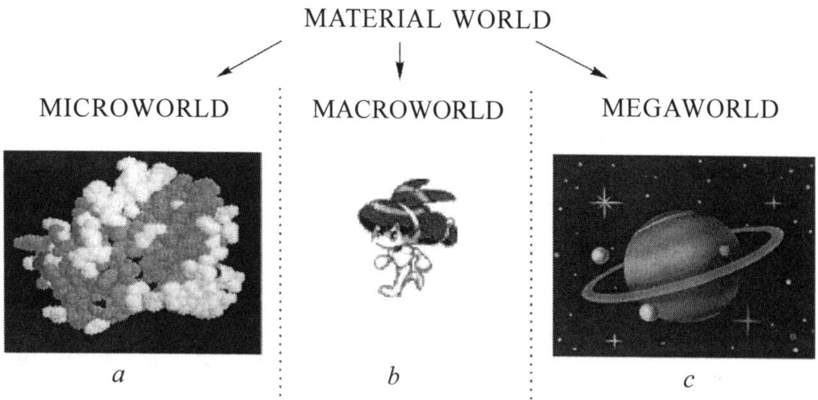

Fig. 3.2. Scale levels of the material world.

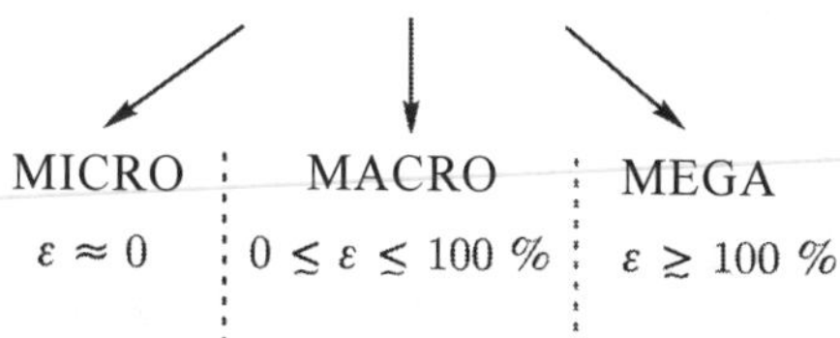

Fig. 3.3. Scale levels of plastic deformation.

corresponding to the macroscopic yield point (relative strain $\varepsilon = 0.05$ or 0.2%), then the boundary between macroplastic deformation and megaplastic deformation (MPD) remains uncertain. Conditionally, we will consider the relative deformation $\varepsilon \approx 100\%$ or the true deformation $e \approx 1$ as the boundary region. Later we will present a more rigorous, physically justified value of the plastic deformation corresponding to the transition in the MPD area.

What is known about megaplastic deformation. The main advantage of methods based on the deformation effect is that they allow to produce non-porous samples of various dimensions with a homogeneous ultrafine-grained structure.

To date, there are many methods realizing megaplastic deformation: deformation by torsion under high pressure, equal-channel angular pressing, extrusion, mechanomelting, repeated deformation (cold rolling, accumulated rolling, or drawing), deformation by a shock wave (created by an explosion), deformation in a ball mill (attritors), magnetoimpulse pressing, external friction.

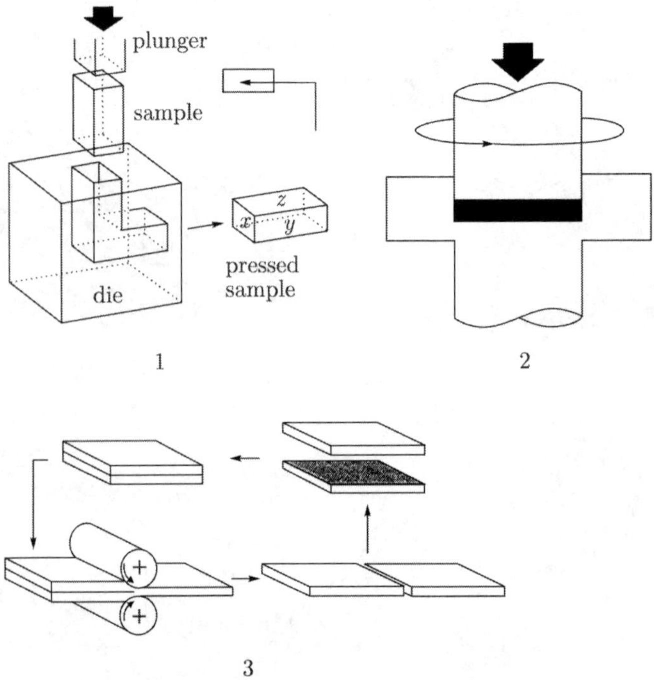

Fig. 3.4. Schemes of the most common methods for creating ultrahigh deformations: 1 – equal-channel angular pressing; 2 – torsion under pressure in the Bridgman chamber; 3 – accumulated rolling.

At present, the most common MPD methods are deformation by igh-pressure torsion and equal-channel angular pressing.

Figure 3.4 schematically shows the three most commonly used methods for creating giant degrees of deformation: equal-channel angular pressing (ECAP) (1), torsion under pressure in a Bridgman chamber (TPBC) (2), and accumulated rolling (3).

In the first case, the sample is deformed according to the shearing scheme (Fig. 3.4, *1*) – it is pushed through two channels of different size located at a certain angle to each other (up to 90°), there is the possibility of repeated deformation using different routes (Fig. 3.5).

In the second case, the sample is placed between the strikers and is compressed under an applied pressure of several GPa. When the upper striker rotates under the effect of surface friction force the specimen is deformed by shear. When deformed by torsion under high pressure the resultant sample obtained is disk-shaped. The diagram of deformation in the Bridgman chamber is shown in Fig. 3.4, *2*. The geometric shape of the samples is such that the bulk of the material

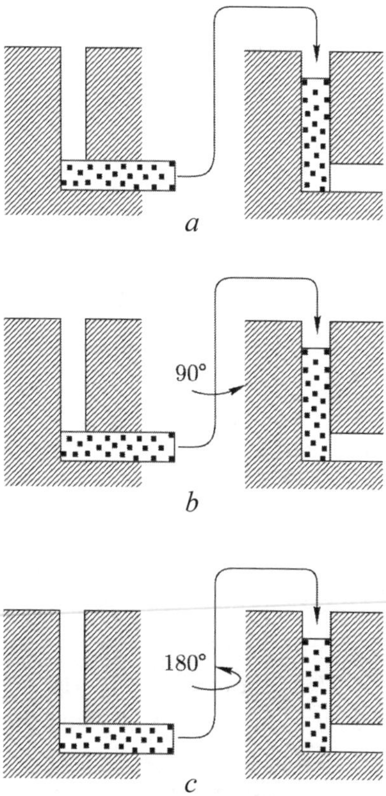

Fig. 3.5. Variants of equal-channel angular pressing: *a* – unchanged orientation of the workpiece; *b* - rotation of the workpiece by 90°; *c* – rotating the workpiece by 180°.

deforms under hydrostatic compression conditions under the action of applied pressure and pressure from the outer layers of the sample. As a result, the deformed sample does not collapse, despite the high degree of plastic deformation.

The plastic strains realized in this case are so high that the usual values of the relative degrees of deformation lose their meaning, and one should proceed to the true logarithmic strains *e*. Their values are defined as follows.

For the TPBC [3.7]:

$$(e = \ln\left(1+\left(\frac{\varphi \cdot r}{h}\right)^2\right)^{0.5} + \ln\left(\frac{h_0}{h}\right), \tag{3.1}$$

where r and h are, respectively, the radius and height of the sample in the form of a disk processed in the Bridgman chamber, and φ is the rotation angle of the movable anvil. The number of complete turns of the movable anvil N corresponds to the deformation at which $\varphi = 2\pi N$.

For ECAP [3.6]:

$$e = 2 \cdot n \cdot \mathrm{ctg}\left(\frac{\phi}{2}\right) \Big/ \sqrt{3}, \qquad (3.2)$$

where n is the number of passes and φ is the angle of rotation of the channels.

In recent years, a new effective method for creating megaplastic deformation has been developed – screw extrusion [3.13].

Formed at such giant deformations, the structural conditions are very unusual and difficult to predict. Unfortunately, the vast majority of authors who study the effect of ultrahigh plastic deformations are limited to studying final structures and the corresponding properties of materials without analyzing those physical processes that occur directly under giant degrees of plastic flow.

3.1.3.2. Surface treatment

Many methods have also been developed for the hardening surface treatment with megaplastic deformation that ensure the formation of a nanocrystalline structure of near-surface layers (Fig. 3.6).

Let us consider in somewhat more detail the method of surface treatment by large plastic deformations – surface hardening ultrasonic treatment (UST). The nature and modes of this deformation effect are critical (ultrasonic action is performed by a tool that produces about 6–10 thousand strokes per square millimeter), which implies a sharp improvement in mechanical, functional and tribological properties as a result of surface nanostructuring [3.14].

The peculiarity of materials after such treatment is the formation in their volume of a gradient structure – the presence of a nanostructure in a thin surface layer while maintaining a coarse-grained structure inside the volume of the material being processed.

UST of materials is based on the use of the energy of mechanical vibrations of the working tool – indenter. The oscillations are performed with an ultrasonic frequency (20 kHz) and an amplitude of oscillations of 0.5–50 μm.

Technological equipment for ultrasonic treatment has a constant circuit regardless of the physical and mechanical properties of the material being processed: power source, process control equipment, mechanical oscillatory system and pressure drive. The installation scheme for the UST is shown in Fig. 3.7.

The operation principle of the UST is as follows. High-frequency

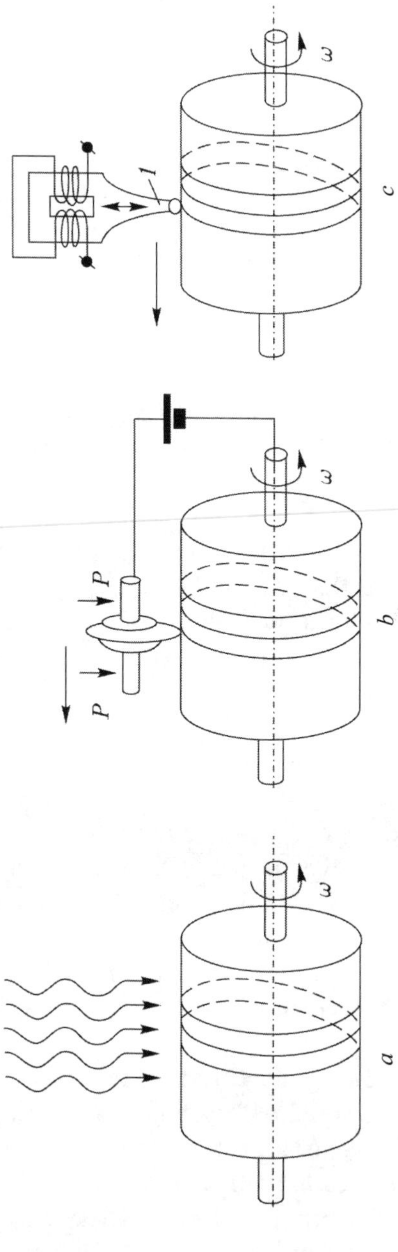

Fig. 3.6. Schemes of processing megaplastic deformation of near-surface layers: *a* – strengthening by different types of irradiation and plasma action; *b* – rolling with current transmission; *c* – rolling with ultrasonic action (*1* – oscillating waveguide); ω is the angular velocity of rotation of the workpiece.

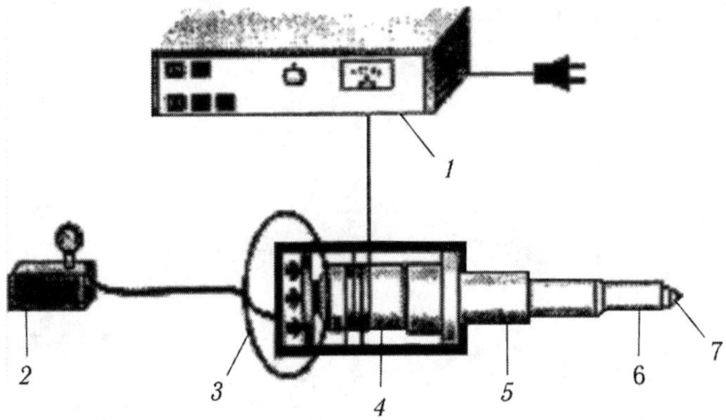

Fig. 3.7. The scheme of installation for UST: 1 – ultrasonic generator; 2 –air compressor for providing a static force of pressing the tool to the surface to be treated; 3 – mechanism for adjusting the force that provides the pneumatic system; 4 – piezo- or magnetostrictive transducer; 5 – booster; 6 – waveguide; 7 – working tool (indenter).

electric current, passing through the winding, creates an alternating magnetic field under the influence of which the transducer oscillates. To convert electric energy into the energy of mechanical oscillations of ultrasonic frequency the well-known physical phenomena of magnetostriction or piezoelectric effect are used. But the obtained magnitudes of the transformations are small. In order to increase them and make them suitable for useful work, first, the entire system is adjusted to resonance, and secondly, a special concentrator-waveguide is attached to the transducer, which converts small amplitudes of oscillations over a larger area into larger amplitudes over a smaller area. A working tool (indenter) is attached to the end of the waveguide. The indenter, together with the entire oscillatory system, is pressed against the surface of the material to be worked. The surface of the processed material is plastically deformed and hardened.

The indenter (its working part and concentrator) is the most important element of the oscillatory system. The working part of the indenter wears out during operation under the influence of dynamic and thermal pulses which leads both to a deterioration in the quality of the surface layer of the processed material and to a decrease in the processing capacity. The question of changing the working part of the indenter is solved on the basis of the requirements for the stability of the physico-mechanical and geometric state of the treated surface layer.

Studies show that the state of the surface of the treated material is strongly influenced by a static force – a force that presses the ultrasound tool to the component and provides acoustic contact; the amplitude of the displacement and the oscillation frequency of the tool, which determine the minimum rate of deformation of the treated surface layer and the intensity of the ultrasonic wave; the dimensions and shape of the working tool.

3.1.4. Compacting of powders

The problem of obtaining finely dispersed powders of metals, alloys and compounds and ultrafine-grained materials from them, intended for various fields of technology, has long been discussed in the literature.

The difference between the properties of small particles and the properties of a massive material has been known for a long time and is used in various fields of engineering. Examples are high-performance catalysts from finely dispersed powders or ceramics with nanometer-sized grains; radio-absorbing ceramic materials used in aviation, in the matrix of which fine-dispersed metal particles are randomly distributed; widely used aerosols. Suspensions of metallic nanoparticles (usually iron or its alloys) with a size of 30 nm to 1–2 μm are used as additives to engine oils to restore worn parts of automobiles and other engines directly in the process. Nanoparticles are widely used in the manufacture of modern microelectronic devices [3.15].

X-ray and ultraviolet optics use special mirrors with multilayer coatings of alternating thin layers of elements with large and low density, for example, tungsten and carbon or molybdenum and carbon; A pair of such layers has a thickness of the order of 1 nm, where the layers must be smooth at the atomic level. Other optical devices with nanosized elements intended for use primarily in X-ray microscopy are Fresnel zone plates with the smallest band width of about 100 nm and diffraction gratings with a period of less than 100 nm.

Ceramic nanomaterials are widely used for the manufacture of parts that operate under conditions of elevated temperatures, non-uniform thermal loads and corrosive environments. The superplasticity of ceramic nanomaterials makes it possible to obtain products of complex configuration used in aerospace engineering with high dimensional accuracy. Nanoceramics based on hydroxyapatite due to its biocompatibility and high strength are used in orthopedics for the manufacture of artificial joints and in dentistry. Nanocrystalline ferromagnetic alloys of the Fe–Cu–M–Si–B systems (M is the transition metal of groups IV–VI) find application as excellent transformer soft magnetic materials with very low coercive force and high magnetic permeability.

The methods for producing isolated nanocrystalline particles, nanoclusters and nanopowders are very diverse and well developed (in particular, this applies to the most well-known methods – gas-phase evaporation and condensation, precipitation from colloidal solutions, plasma-chemical synthesis, various thermal decomposition variants).

For the production of nanoparticles or nanopowders, the methods of spraying a melt jet with liquid or gas are most simple and efficient (Fig. 3.8).

Low-active or inert gases are used as dispersing media: nitrogen, argon, etc. or liquids: water, alcohols, acetone, etc. These methods usually produce powders of metals and alloys with particle sizes of about 100 nm. If it is necessary to obtain particles with dimensions of a few tens of nanometers, a double spray method is used in which the melt is first saturated with a high pressure soluble gas and then sprayed and dispersed by an insoluble gas. Rapid cooling of droplets results in an explosive release of the dissolved gas and their destruction into smaller particles.

Another frequently used technique is evaporation–condensation of the material. Rapid heating and evaporation can be provided by a plasma jet, a laser beam, an electric arc, an electrical explosion

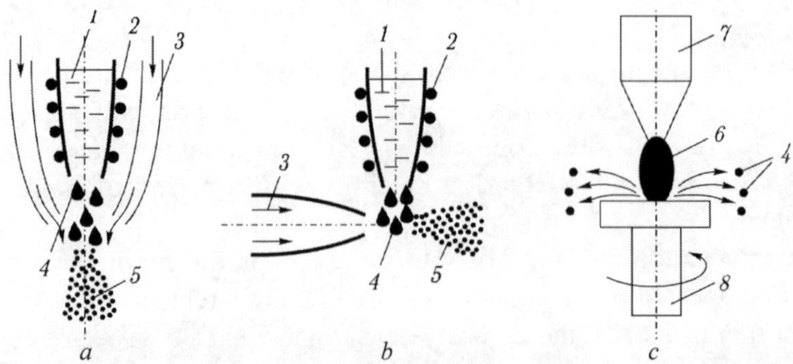

Fig. 3.8. Schemes for spraying a liquid melt: a – coaxial flow of an inert gas; *b* - perpendicular flow; *c* - in an electric arc on a rotating electrode. *1* – melt; *2* – heater; *3* – inert gas; *4* – melt drops; *5* – dispersed material; *6* – electric arc; *7* – fixed electrode; *8* – rotating electrode.

of the conductor. Cooling and condensation of the vapour to form nanoparticles can occur in a vacuum, in an inert gas environment, and also on a solid or liquid substrate. Depending on the specific implementation and regimes, powders of various metals and alloys with particle sizes from 10 to 100 nm can be obtained.

The methods of mechanical grinding of solids are also used widely. They are carried out in mills of various types: ball, planetary, jet, vortex, vibratory, disintegrators, attritors (Fig. 3.9).

These methods can produce powders of metals with particle sizes of tens of nm, their oxides – with sizes of several nm, disperse polymers, components of ceramics, etc. A variation of mechanical methods can be considered the processing of raw materials by detonation waves. This method makes it possible to obtain nanopowders of Al, Ti and other solid materials, including diamond particles smaller than 10 nm.

Compact nanocrystalline materials have been produced only in the last 10–15 years (pioneering work [3.16–3.18] on the compaction of nanopowders refer to 1981–1986). At present, the compaction of nanopowders is one of the most common methods for obtaining compact nanomaterials.

A great fundamental interest in compact nanocrystalline materials is due primarily to their convenience for study. This is what caused the wide popularity and popularity of the technique for obtaining compact nanocrystalline materials (Fig. 3.10), proposed by the authors [3.19, 3.20]. The technology described in these works uses the evaporation and condensation method to obtain nanocrystalline particles deposited on the cold surface of a rotating cylinder; Evaporation and condensation are carried out in a rarefied inert gas atmosphere, usually helium (He); at the same gas pressure, the transition from helium to xenon, i.e. from a less dense inert gas to a more dense gas, is accompanied by an increase in the particle size several times. Particles of the surface condensate usually have facets. Under identical conditions of evaporation and condensation, metals with a higher melting point form smaller particles. The precipitated condensate is removed from the surface of the cylinder by a special scraper and collected in a collector.

After evacuation of the inert gas in vacuum, a preliminary (at a pressure of ~1 GPa) and a final (under a pressure of up to 10 GPa) compression of the nanocrystalline powder is carried out. As a result, cylindrical plates with a diameter of 5–15 mm and a thickness of 0.2–3.0 mm with a density reaching 70–90% of the theoretical

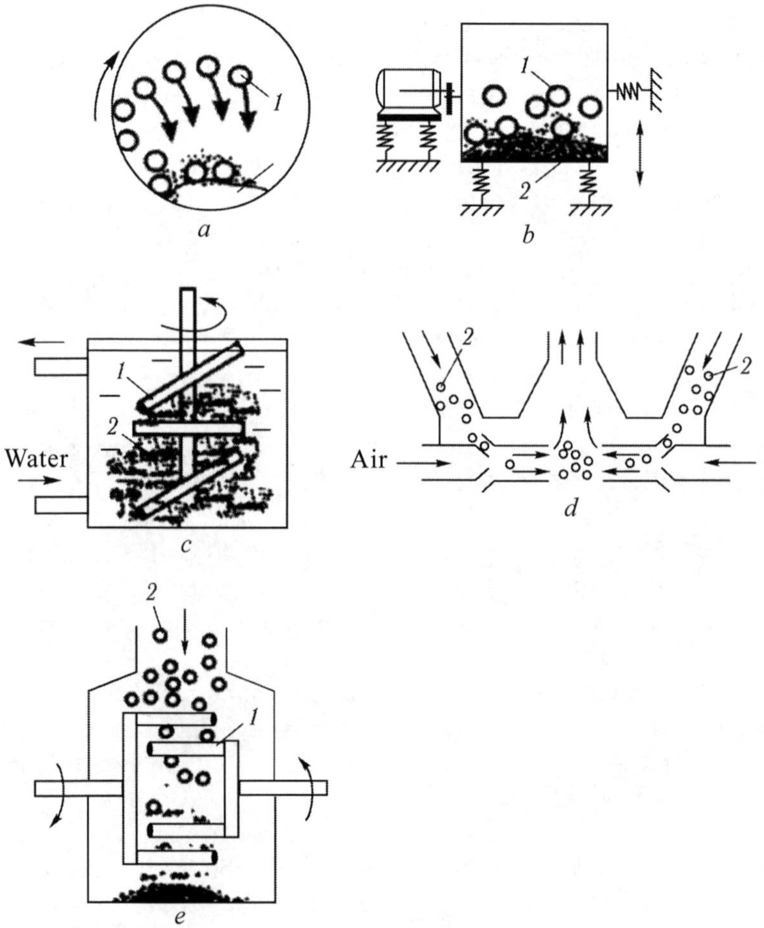

Fig. 3.9. Schemes of mills for fine mechanical grinding of raw materials (*1* – grinding balls or fingers, *2* – raw materials): *a* – rotating ball mill, grinding by falling balls; *b* – vibrating mill, grinding the product with pulsating balls; *c* – attritor, abrading product with rotating fingers; *d* – jet mill, grinding the product 'in the counter beams'; *d* – disintegrator, crushing the product by fingers rotating towards each other.

density of the corresponding material are obtained (up to 97% for nanocrystalline metals and up to 85% for nanoceramics [3.15]).

The compact nanocrystalline materials obtained in this way, depending on the conditions of evaporation and condensation, consist of particles with an average size *d* from 1–2 to 80–100 nm.

Due to the exclusion of contact with the environment during the production of a nanopowder and its subsequent pressing, it is possible to avoid contamination of compact nanocrystalline samples, which

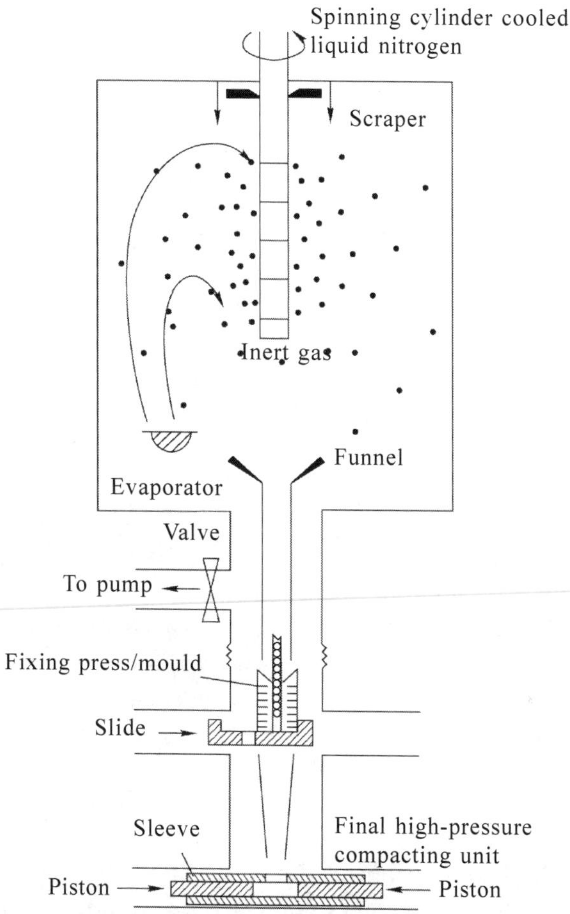

Fig. 3.10. Diagram of the chamber for producing compact nanocrystalline materials [3.20]: the substance, evaporated or sprayed from one of several sources is condensed in the atmosphere of the rarefied inert gas and deposited on the cold surface of a spinning cylinder; the condensate is removed by scratching, collected and pressed in vacuum (after pumping away the inert gas).

is very important in the study of the nanocrystalline state of metals and alloys. The apparatus described in [3.16–3.21] can also be used to produce nanocrystalline compounds – oxides and nitrides; in this case, the evaporation of the metal is carried out in an oxygen- or nitrogen-containing atmosphere.

The porosity of nanoceramics obtained by compacting powders is related to triple joints of crystallites. The decrease in the size of the powders is accompanied by a marked decrease in their compactability when pressed using the same pressure [3.22]. The lowering and more even distribution of the porosity is achieved by pressing at such an elevated temperature which does not yet lead to intensive recrystallisation. Thus, the conventional sintering of a highly dispersed zirconium oxide powder with a particle size of 40–60 nm at 1370 K for 10 s allows to achieve a relative density of 72% with an average grain size in the sintered sample of 120 nm; hot pressing at the same temperature and pressure of 1.6 GPa allows to obtain a sintered material with a relative density of 87% and an average grain size of 130 nm [3.23]. Reducing the sintering temperature to 1320 K and increasing the sintering time to 5 h allowed to obtain compact zirconium oxide (ZrO_2) with a relative density of more than 99% and an average grain size of 85 nm [3.24]. The authors [3.25] produced compact samples with a density of 98% of theoretical density by hot pressing of a powder of titanium nitride ($d \approx 80$ nm) at 1470 K and a pressing pressure of 4 GPa, but (according to diffraction data) after hot pressing due to intensive recrystallisation the average grain size was not less than 0.3 μm. The study [3.26] showed that the densest samples (with a relative density of 98%) of titanium nitride are obtained by sintering samples compacted from the smaller nanopowders ($d \approx 8$–25 nm) with a minimal dispersion of grain sizes.

On the whole, to produce compact nanocrystalline materials, especially ceramic materials, it is promising to use pressing with subsequent high-temperature sintering of the nanopowders. When using this method, it is necessary to avoid growth of the grains in the sintering stage of pressed specimens. This is possible at a high density of the pressings (no less than 0.7 of the X-ray density) when the sintering rate is sufficiently high and at a relatively low-temperature $T < 0.5\ T_m$ (T_m is the melting point). The production of such high density pressings is a difficult task because the nanocrystalline powders have low pressing capacity and the transitional static pressing methods do not result in sufficiently high porosity. The physical reason for low pressing capacity of the powders is the presence of the interparticle adhesion forces whose relative magnitude rapidly increases with a decrease of the particle size.

To retain the small grain size in the compact nanomaterials, in addition to decreasing the sintering temperature another effective measure is to add alloying additions preventing rapid grain growth. A suitable example is the production of nanocrystalline WC–Co hard alloys [3.27–3.31]. The nanocrystalline powder composition produced from the tungsten and cobalt carbides was produced by thermal dissociation of organic metal precursors followed by carbothermal reduction in the suspended layer resulting in retaining high dispersion. To reduce the grain growth rate and the solubility of the tungsten carbide in cobalt, the non-stoichiometric vanadium carbide is added in the amount of up to 1% to the mixture. The hard alloy produced on the basis of this nanocrystalline composition is characterised by the optimum combination of high hardness and high-strength [3.28–3.30]. In [3.31] it is shown that each WC–Co nanocomposite particle with the size of ~75 μm consists of several millions of the WC nanocrystalline grains smaller than 50 nm and distributed in the cobalt matrix. Sintering of the nanocomposite mixture of the tungsten carbide with 6.8 wt.% of Co and 1 wt.% VC produced alloys in which the size of 60% of WC grains was smaller than 250 nm and 20% – smaller than 170 nm. An even finer structure was observed for the alloy containing, in addition to the tungsten carbide, 9.4 wt.% Co, 0.8 wt.% Cr_2C_3 and 0.4 wt.% VC. Even after sintering at a relatively high temperature of 1670 K 60% of the tungsten carbide grains in this alloy were smaller than 140 nm and 20% smaller than 80 nm. Comparison of the nanoalloy and the conventional polycrystalline alloy showed that the nanoalloy has a considerably higher strength and much higher resistance to fracture.

The compacting of the nanocrystalline powders is carried out efficiently by the magnetic pulse method proposed by the authors of [3.32, 3.33]. In contrast to the stationary pressing methods, the pulsed compression methods are accompanied by rapid heating of the powder as a result of the generation of a large amount of energy in friction of the particles during packing. If the size of the particles is sufficiently small ($d < 0.3$ μm), the heating time of the particles by the diffusion of heat from the surface is considerably shorter than the characteristic duration of the pulsed compression waves (1–10 μs). Under certain conditions, by selecting the parameters of the compression wave it is possible to carry out the hot pressing of the ultrafine powders as a result of the high surface energy of the powder. The method of magnetic pulsed pressing produces pulsed compression waves with the amplitude of up to 5 GPa and the duration of several microseconds. This method, based on the concentration of the force effect of the magnetic field of the powerful pulsed currents, makes it possible to control relatively simply the parameters of the compression wave, is ecologically clean and considerably safer than the dynamic methods using explosive substances. The aluminium nitride (AlN) powders produced by electric explosion is pressed by the magnetic pulse method under the pressure of 2 GPa to a density of 95% theoretical density, and Al_2O_3 – to 86%. The magnetic pulse pressing method is used for producing components of different shape, and in the majority of cases these components do not require any additional machining. In particular, in operation with the superconducting oxide ceramics [3.33], components with the density higher than 95% theoretical density were produced.

The application of pulsed pressures results in a high density of the pressings in comparison with static pressing. This indicates that the interparticle forces in rapid movement of the powder medium are efficiently overcome.

The magnetic pulse method is used for pressing the nanocrystalline powders of Al_2O_3 [3.34, 3.35] and TiN [3.36]. The results obtained in [3.36] showed that the increase of the pressing temperature to ~900 K is more effective than the increase of pressure in cold pressing. At a pulse pressure of 4.1 GPa and a temperature of 870 K it was possible to produce compact specimens of the nanocrystalline titanium nitride with the grain size of ~80 nm and the density of approximately 83% of theoretical density. A decrease of pressing temperature to 720 K was accompanied by a decrease of density to 81%.

3.1.5. Pulsed treatment

The method of activation of physical and chemical processes by irradiation with the electrons, ions and light is used widely for the modification of subsurface layers of the materials. The application of the pulsed energy sources for annealing offers a number of advantages in comparison with the method of traditional pulsed annealing. The most widely methods are pulsed photon treatment and pulsed laser treatment. The main differences of these methods are in the different degrees of monochromaticity of the light, different density of the energy flux and different pulse times.

3.1.5.1. Photon treatment

From the viewpoint of the short duration of the heat treatment of the materials, the method of pulsed photon treatment (PPT) is very interesting. In this method, the duration of the effect of the light flux from a high power radiation source is several microseconds [3.37].

The method can be used to localise energy in the subsurface layer and induce the nucleation of crystalline phases at a lower thermal load on the sample. In pulsed photon treatment, the rapid introduction of the energy should accelerate the process of formation in the amorphous alloy of nuclei of crystalline phases of different composition, i.e., all the activation barriers of nucleation are overcome. The increase of the rate of nucleation of the crystalline phases results in high density of the phases at smaller dimensions thus increasing the strength whilst retaining the ductility of the alloy.

Pulsed photon treatment by radiation of powerful xenon lamps (radiation range 0.2...1.2 μm) results in higher heating rate (approximately 1000 K/s) in comparison with high-speed thermal annealing [3.38], and the presence in the radiation spectrum of the xenon lamps of the shortwave region (radiation spectrum from 200 nm) results in the activation of solid-phase processes.

The pulsed photon treatment effect is manifested in the acceleration of the processes of diffusion, synthesis of thin films of compounds, recrystallisation, in a decrease of the temperature thresholds of phase formation, in an increase of the dispersion of the synthesised structures; in the formation of metastable phases; in nanocrystallisation of the amorphous metallic alloys (a number of amorphous alloys based on iron, aluminium), increasing the microhardness whilst retaining the ductility. At small radiation doses of the amorphous alloys the modulus rapidly increases. This increase is associated with the formation of clusters – the nuclei of nanocrystalline phases [3.39].

The pulsed light radiation, used for annealing, affects the relatively large surface of the material and this is one of the reasons why it is used in practice. Initially, the method was proposed for cleaning the surface of the substrate prior to vacuum spraying [3.40, 3.41]. However, later it was found that a secondary consequence of this cleaning may be the changes in the structure of the substrate material [3.42].

3.1.5.2. Laser treatment

Laser treatment is one of the methods of high-speed annealing of the surface of materials. The currently available methods of pulsed laser treatment of the materials are used for local thermal heating of the solid material controlled o the basis of time and temperature distribution. The main factor resulting in the interest of researchers in the methods of laser annealing is the rate of transformation of the amorphous layer to the monocrystalline structure [3.43].

The application of laser radiation results in the possibility of carrying out unique heat treatment which cannot be achieved by other methods. It is possible to ensure the concentration of the light energy in small volumes and over short periods of time. The effect of pulsed laser energy on the material results in high heating and cooling rates of the material. The effect of short laser radiation pulses on thick specimens with efficient heat removal from the surface layers is accompanied by high heating and cooling rates so that quenching, surface amorphisation, and other methods can be carried out [3.44].

Until recently, insufficient attention has been paid to the methods of laser treatment of amorphous–nanocrystalline and amorphous metallic alloys. These materials are promising required for various applications. Many amorphous alloys have satisfactory bend ductility, high mechanical strength, wear resistance and toughness. Some amorphous alloys are characterised by high resistance to corrosion and radiation fracture and high magnetic permittivity. In some cases, the amorphous and amorphous–nanocrystalline metallic alloys require additional heat treatment. The application of the advanced methods of laser treatment is capable of not only increasing the efficiency of annealing but also providing the treatment conditions which cannot be achieved by other methods so that the material with new properties can be produced [3.43, 3.46]. However, the high localization of this method in irradiation of the amorphous metallic materials is in fact a shortcoming.

3.1.6. Production of thin films

The method of production of solid thin films of inorganic materials can be divided into the following groups: physical, chemical and physico-chemical [3.45].

The methods in the first group are based on the processes of transfer (transport) of the substance from the source to the substrate surface not accompanied by chemical reactions. The methods in this group used most extensively in thin-film technologies are based on two main processes: formation of the vapour phase of the substance of the source and physical deposition (condensation) from the vapour on the substrate surface. The vapour phase is produced by evaporation (thermal, laser) or sputtering (ion-plasma). The group of the physical methods also includes: deposition of powders from a suspension followed by melting together the layer, diffusion, ion doping, etc.

The methods of the second group are based on the chemical reactions of different types. The large number of the methods in this group can be divided into the following subgroups: chemical deposition from the gas (vapour) phase, chemical interaction of the substrate with the gas medium (for example, oxidation), deposition from the solution (melts), electrochemical deposition, solid-phase reactions.

The physical–chemical methods are based on the combination of the physical processes of formation of the vapour phase from the components of the given material (evaporation or sputtering) and the processes of chemical interaction of the material in the vapour or solid-phase with other components of the material: cathode sputtering in the active gas (reactive cathode sputtering), thermal evaporation and condensation in high vacuum followed by heat treatment of the condensed phase in the appropriate active gas medium, thermal evaporation and condensation in the active gas medium.

The application of the corresponding method is determined by the considerations of the optimum method of obtaining the required parameters of the films and, correspondingly, the component: economic justification, fulfilment of the ecological requirements. In the technology of production of microelectronic devices the main methods are chemical deposition from the gas phase, ion-plasma sputtering, thermal (electron beam) evaporation, solid-phase reactions, ion doping. Many of these methods are also used after appropriate changes in the technology, for the production of materials with poly- or even single-crystal structure [3.47].

Improvement of the apparatus for the transitional methods resulted in the creation of independent methods, such as molecular beam epitaxy (MBE), ionised cluster beam deposition (ICBD), metal-organic chemical vapour deposition (MOCVD).

In a general case, the entire variety of the methods of producing thin films in the amorphous state can be conveniently divided into three large groups according to the initial aggregate state: production from the gas, liquid or solid phase state.

There are many methods of producing thin films and foils with the amorphous or nanocrystalline structure from the gaseous state by condensation of the atoms on the substrate [3.48].

Films are produced by gas phase deposition in different chemical and physical variants: magnetron deposition, electron beam deposition, laser synthesis, plasma-activated processes, and others. Deposition on the substrate may be carried out from the vapours, plasma or a colloidal solution. In deposition from the vapours the metal evaporates in vacuum, in the oxygen- or nitrogen-containing atmosphere, and the metal vapours or the vapours of the resultant compounds (oxides, nitrides) condense on the substrate. The size of the crystals in the film can be regulated by changing the rate of evaporation and substrate temperature. In most cases, this method is used to produce nanocrystalline films of metals [3.49, 3.50].

Each method has its own advantages and shortcomings, depending on the sputtering material and application.

Some of the sputtering methods, used frequently, will now be described briefly.

In *thermal spraying* the thin film is produced as a result of heating, evaporation deposition of the substance on the substrate in a closed chamber with the gas pressure in the chamber lower than 10^{-4} torr [3.51]. The currently available systems for vacuum spraying use a working chamber formed by a hood made of stainless steel situated on a support plate (Fig. 3.11).

The vacuum-tight joint between the base of the hood and the support plate is produced using a rubber gasket. The working chamber contains a technological jig: the substrate holder, the evaporator of the sprayed substance, a screen, controlled with an electromagnetic or electric drive for interrupting the flow of the sprayed substance, electric power supply terminals. The industrial systems can be used

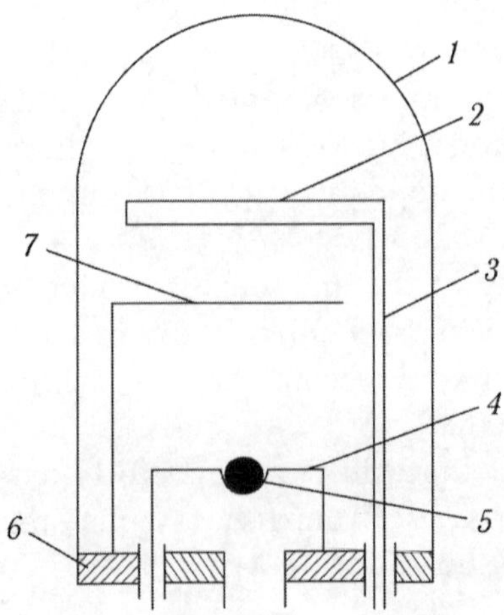

Fig. 3.11. Equipment for vacuum sputtering: 1) hood of the vacuum chamber; 2) the substrate; 3) the substrate holder; 4) the evaporator; 5) the evaporated substance; 6) the support plate; 7) the screen.

for multiple spraying without opening the working chamber. This is carried out using the carousels for the substrates and evaporators capable of moving in vacuum in relation to each other. The presence of carousels makes it possible, evaporating the substance from different evaporators, to produce multicomponent and multilayer films. Evaporation is carried out in most cases from the liquid phase, less frequently from the solid phase (sublimation). The rarefaction in the chamber dictates the requirements on the purity of the produced coatings, the chemical resistance of the evaporated material, the distance to the substrate which should not exceed the free path of the evaporated atoms. The thermal evaporation method has several varieties which differ by the method of heating the evaporated material: resistance heating, explosion (discrete) evaporation, laser heating, induction heating, electron beam heating.

The methods of *cathode sputtering* are based on the application of the energy of the positive ions, formed in a glow discharge and bombarding the cathode from the sputtered material [3.48]. Cathode sputtering may be used to produce films of refractory metals, different alloys and mixtures without disrupting the ratio (in percent) of the components.

In *DC cathode sputtering* the cathode is the evaporated material and the substrate is placed on the earthed anode. A high voltage is maintained between the electrodes and produces the glow discharge. A rarefaction to 10^{-3}–10^{-4} Pa is produced in advance in the working volume, and subsequently the inert gas (usually argon) is supplied into the chamber to a pressure of 10^{-2}–1 Pa. To ensure the passage of current between the electrodes, it is necessary to generate constant electron emission from the cathode which forms under the effect of the high voltage between the electrodes. If the applied voltage is higher than the ionisation potential of the investigated gas, the collision of the electrons with the molecules of the gas results in ionisation of the gas. As a result of the effect of high voltage, the positive ions are accelerated by the electric field and bombard the cathode and this results in the transfer of energy of the ions to the atoms of the cathode material and the material is sputtered. This method is suitable mostly for sputtering of metals and alloys.

In *AC sputtering* the high-frequency AC (usually 13.6 MHz) is used instead of the constant voltage. In this case, the gas discharge is localised in the space between the rods, sputtered alternately only during one half period when they receive the negative voltage and are used as the cathode of the discharge. This method decreases the contamination of the substrate with the residual gases, lowers the charge accumulated on the substrate and is more suitable for the sputtering of dielectric layers.

The efficiency of the diode cathodes sputtering systems decreases at pressures lower than 10^{-1} Pa because of a decrease of the concentration of the ions of the working gas, whereas to obtain the films not filled with the gas it is necessary to reduce the pressure in the working chamber. From this viewpoint, it is preferred to use the methods of ion plasma sputtering in which the discharge is artificially maintained by using either the thermal emission cathode or a high-frequency field.

The *methods of ion plasma sputtering* can be used to produce materials in which the chemical and phase deposition, microstructure and, consequently, the characteristics will greatly differ from those produced by the transitional methods.

At present, the majority of the plasma technologies are based on the most extensively examined electric discharges with direct and alternating current of industrial frequency, high-frequency and superhigh frequency discharges, used traditionally for the generation of plasma [2.6].

In this method, sputtering is carried out by bombarding the target with the ions of the low-pressure plasma gas discharge formed between the thermal cathode and the independent anode [3.52]. The distinguishing feature of ion plasma sputtering is the higher vacuum in comparison with cathodes sputtering (~0.67 Pa) producing cleaner films. The electrical circuits of the discharge and sputtering are not connected in this case. There are different modifications of this method: the triode circuit with a constant potential, the triode circuit with an isolated source, the magnetron circuit, ion-beam sputtering.

In the triode circuit with *the constant potential* on the target the target made of the sputtered material is under the constant negative potential in relation to the plasma potential. The sputtered atoms are deposited on the substrate situated parallel to the target.

In the circuit with the *isolated plasma source* the plasma is generated in an auxiliary ionisation chamber from which the narrow ion beam, formed by the strong magnetic fields, diffuses into the main distribution chamber with the target positioned in it having the potential sufficient for sputtering of the target material.

In deposition from plasma the electrical discharge is maintained using an inert gas. The continuity and thickness of the film and the dimensions of the crystals in the film can be regulated by changing the gas pressure and discharge parameters. The sources of metallic ions in deposition from the plasma are metallic cathodes ensuring a high degree of ionisation (from 30 to 100%); the kinetic energy of the ions equals from 10 to 200 eV and the deposition rate is up to 3 μm/min.

A variety of deposition from plasma is *magnetron sputtering* in which cathodes are produced not only from metals and alloys but also different compounds and the substrate temperature is reduced by 100–200 K or less. This widens the possibilities of producing amorphous and nanocrystalline films. In magnetron sputtering the high deposition rate is achieved by increasing the ion current density by localisation of the plasma at the sputtered surface of the target using a strong transverse magnetic field (Fig. 3.12) [2.6].

The lines of force f the magnetic field close between the poles of the magnetic system. The surface of the target located between the areas of entry and exit of the lines of force of the magnetic field is extensively sputtered and has the form of a closed track with the geometry of the track determined by the form of the poles of the magnetic system. This method is suitable for sputtering all solids, with the exception of magnetic ones.

However, the degree of ionisation, the kinetic energy of the ions and the deposition rate in magnetic sputtering are lower than when using the electric arc discharge plasma.

Magnetron sputtering is highly universal and can be used not only for metallic and also non-metallic targets (to produce appropriate films). In magnetron sputtering the substrate temperature is not high (100–200°C) which offers extra possibilities for producing nanostructured films with a small grain size and amorphous films [2.6].

Of obvious interest are the pulsed plasma sources. When they are used for plasma generation at the power of electric energy sources of tens of kilowatts it is possible to obtain, in the plasma pulse lasting 10^{-4}–10^{-5} s, peak powers from tens to thousands of megawatt and heat the plasma to (4–5) · 10^4 K followed by auto-quenching of the plasma with the rates of 10^7–10^8 K/s [3.52].

3.1.7. Ion implantation

Implantation of the ions into the surface of the metallic materials may lead to the formation of the amorphous state under favourable conditions. The implantation process includes the ionisation of the

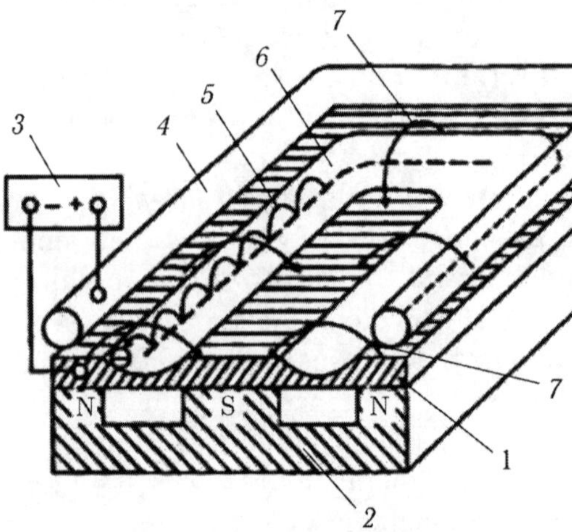

Fig. 3.12. The magnetron sputtering system: *1*) the cathode–target; *2*) the magnetic system; *3*) the power source; *4*) the anode; *5*) the trajectory of movement of the electron; *6*) the sputtering zone; *7*) the line of force of the strength of the magnetic field.

implanted atoms and their acceleration to high energies in the electric field [3.55]. The accelerated ions of the impurity bombard the initial material, penetrate into this material to a depth of several tenyhs of a micrometre and are arrested. The diagram of ion implantation equipment is shown in Fig. 3.13. The equipment consists of the ion source *1*, the magnetic mass analyser *2* separating the undesirable ions from the beam, the ion accelerator *4*, the electric lenses and deflecting plates *3* for focusing the beam, the vacuum chamber *5*, connected with the vacuum system *6*. In most cases, to produce the amorphous structure, the ions of tungsten, boron, phosphorus and also other metalloid are implanted in the subsurface layer of the metallic materials. Since at room temperature these elements are in the solid state, the ion sources are represented by the molecules of their gaseous compounds. The boron ions are produced using BF_3 or BCl_3, and the phosphorus ions – PH_3 or PF_3. Usually, the atoms are ionised by the bombardment with the arc discharge electrons or the electrons of the cold cathode. This type of bombardment results in the formation of ions of several types. In the mass analyser the ion beam is deflected by the magnetic field. Due to the mass difference they are deflected in different ways and, therefore, the output circuit of the magnetic field lets through only the ions of the required kind.

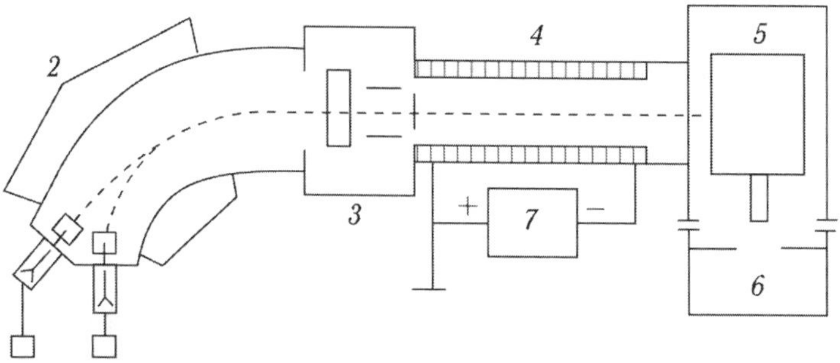

Fig. 3.13. Diagram of the ion beam accelerator: *1*) the ion source; *2*) the magnetic mass analyser, separating the undesirable ions from the beam; *3*) electric lenses and deflecting plates for beam focusing; *4*) the accelerator tube; *5*) the vacuum chamber in which the specimens to be processed are placed; *6*) the vacuum system; *7*) the high voltage power source.

Depending on the applications, different modifications of the ion implantation equipment are produced. In the systems with the high beam current, the beam can be fixed and scanning can be carried out by rotating the target with simultaneous reciprocal movement. In other systems, the beam carries out the scanning movement and the target rotates.

For the majority of systems the accelerating voltage is in the range from 10 to 200 kV. The ion beam is usually focused on a target to a diameter of ~1 cm. The beam current which can be measured accurately is in the range from 10 μA to 2 mA. The number of ions per unit area is referred to as the unit dose Φ. The ion dose is linked with the beam current i, its cross-sectional area S and the irradiation time t by the equation

$$\Phi = \frac{It}{qS}, \tag{3.3}$$

here q is the ion charge. If the beam current is equal to 1 mA for 1.6 ms, and the cross-sectional area of the beam is 1 cm^2, the resultant irradiation dose at the ion charge of $1.6 \cdot 10^{-19}$ C is $\Phi = 10^{13}$ ion/cm^2. At the known ion dose and the cross-sectional area of the beam it is possible to determine the appropriate geometry and the scanning speed [3.56].

To produce thin metallic layers with the amorphous structure the irradiation dose is usually $\Phi > 10^{16}$ ion/cm^2. The advantages of the ion implantation method are the stability of the process, easy control of the parameters, and the composition of the amorphised surfaces can be varied over a wide range and it is possible to produce layers on relatively large areas. The main shortcoming of the method is the cumbersome, complicated and expensive equipment, and also the small thickness of the layer, equalling several hundreds of nanometres.

3.2. The structure of alloys

The amorphous state of the metals and alloys is metastable and, therefore, in the application of amorphous alloys in practice the most important problem is the problem of their thermal and time stability. The investigations of the crystallisation processes of the amorphous alloys are essential for determining the relationships and links between the set of the mechanical properties and the changes of the structure taking place during transition from the amorphous to crystalline state [1.1].

The amorphous–nanocrystalline state may demonstrate a qualitatively new level of the mechanical properties: the properties differ from both completely crystallised and from amorphous materials [3.57–3.59]. For example, partial crystallisation of the alloys with the amorphous structure increases the elastic moduli of the material. When the requirements on the material include not only high strength and toughness but also relatively high values of the Young and shear moduli, it is sufficient to use the material in the amorphous–nanocrystalline state [1.107]. The majority of the magnetically soft materials are amorphous–nanocrystalline alloys in which the formation of a small number of crystalline particles smaller than 10–20 nm results in a low coercive force and high magnetic permittivity [3.60].

However, the formation of the crystalline phase in the amorphous material may also have undesirable consequences, for example, the loss of plasticity, structural inhomogeneity, the formation of local stresses, defects, etc. Many of these phenomena are determined by the conditions resulting in the formation of the crystalline phase because this determines the morphology, phase composition and the amount of the structural components in the amorphous–nanocrystalline state [1.1].

3.2.1. Special features of the transition from the amorphous to crystalline state under thermal effects

Crystallisation thermodynamics

The conventional crystallisation of the amorphous metallic alloys is regarded as the solid-phase transformation governed by the classic thermodynamics of crystallisation of the supercooled liquid. The changes of the enthalpy ΔH, entropy ΔS, and the free energy ΔG in crystallisation of amorphous metallic alloys [3.61] are equal to:

$$\Delta H^{c-a}(T) = -\Delta H_m + \int_T^{T_m} \left(C_p^a - C_p^c\right) dT,$$

$$\Delta S^{c-a}(T) = -\Delta H_m / T_m + \int_T^{T_m} \left[\left(C_p^a - C_p^c\right)/T\right] dT, \qquad (3.4)$$

$$\Delta G^{c-a} = \Delta H^{c-a}(T) - T\Delta S^{c-a}(T),$$

here ΔH_m is the difference of the enthalpies of the amorphous and crystalline states at the melting point T_m, C_p^a and C_p^c are the specific volume (or molar) heat capacities of the amorphous and crystalline phase, respectively. Identical (3.4) thermodynamic parameters $\Delta H^{n-a}(T)$, $\Delta S^{n-a}(T)$ ΔG^{n-a} can be determined [3.61, 3.62] for the transformation from the amorphous to nanocrystalline state:

$$\Delta H^{n-a}(T) = -\Delta H_m + \int_T^{T_m} \left(C_p^a - C_p^c\right) dT,$$

$$\Delta S^{n-a}(T) = -\Delta S_0^{n-a} + \int_T^{T_m} \left[\left(C_p^n - C_p^a\right)/T\right] dT, \qquad (3.5)$$

$$\Delta G^{n-a} = \Delta H^{n-a} / (T) - T\Delta S^{n-a}(T).$$

Here $\Delta H_m(T_k)$ is the difference of the enthalpies of the amorphous and nanocrystalline states at the crystallisation temperature T_k. The value of the enthalpy difference can be measured by the calorimetric method, ΔS_0^{n-a} is the difference of the entropies of the nanocrystalline and amorphous state at 0 K, which can be estimated from the concentration of the free volume in two states [3.63], C_p^n is the specific heat of the nanocrystalline state.

The processes of crystallisation and nanocrystallisation have considerable thermodynamic differences. In the crystallisation process, the change of the entropy is negative $\Delta S < 0$. In this case, ΔS decreases with increasing temperature and ΔG increases. With increasing temperature ΔS and ΔH increases and ΔG decreases for the given grain size.

Crystallisation mechanisms

The processes of crystallisation, taking place as a result of thermal effects on the amorphous state, have been studied most extensively. There are four crystallisation mechanisms [3.62].

1. Polymorphous crystallisation in which the products of transformation have the same composition as the amorphous matrix. The form of the crystals is determined by the anisotropy of the growth rate in different crystallographic directions. The amorphous alloys usually correspond to the eutectic compositions characterised by the large difference in the composition of the crystalline phases and the amorphous matrix. This type of crystallisation is encountered relatively rarely and especially in the formation of metastable crystalline phases which then change to equilibrium phases with the extensive concentrational redistribution.

2. Eutectic crystallisation in which the amorphous matrix crystallises with the simultaneous formation of two phases which are in a close structural relationship. In this case, the components at the crystallisation front are redistributed, although in a number of cases this redistribution is not sufficiently extensive. Although the transformation takes place in the solid state, the resultant colonies are termed the eutectics and not eutectoids because the amorphous matrix is a supercooled liquid [1.1]. Like the polymorphous crystallisation, the eutectic crystallisation is a continuous process. The general composition of the crystal in the amorphous phase is the same and the matrix remains without changes until it is absorbed by the interface. In contrast to polymorphous crystallisation, eutectic crystallisation requires diffusion essential for the distribution of the dissolved component between the two growing faces. This diffusion may take place either in the amorphous phase ahead of the growing crystal or at the amorphous phase – crystal interface.

3. Primary crystallisation in which the crystals with the composition different from the composition of the amorphous phase form in the initial stage. The growth rate of the crystals is controlled by the diffusion of one or several components in the initial structure. After enrichment of the amorphous matrix with these components, polymorphous or eutectic transformation takes place in the matrix with the formation of some crystalline phases. This crystallisation mechanism is the initial stage of crystallisation of many amorphous alloys and is typical of the alloys of the hypoeutectic or hypereutectic composition and also of the alloys whose composition slightly

differs from the composition of alloys produced by polymorphous crystallisation.

4. Crystallisation of the delaminated amorphous matrix which takes place in two stages. If the amorphous matrix is susceptible to delamination already prior to crystallisation, the resultant amorphous 'phases' crystallises independently and at different temperatures.

The morphology of the crystalline particles, precipitated during the examined crystallisation mechanisms, greatly differs – spherical, plate shaped, barrel-like and dendritic [3.1].

The crystallisation of the amorphous metallic alloys is a complicated physical–chemical process of formation (nucleation and growth) of the multiphase state, and the general rate of transformation reflects the time and temperature dependences of both processes of formation of the crystals.

Regardless of the dependence on the specific crystallisation mechanism, it is possible to produce in the optimum treatment conditions the amorphous–crystalline mixture and analyse its mechanical properties in dependence on the volume fraction, the type and morphology of crystalline phases distributed in the amorphous matrix.

An important conclusion is the one according to which the first phase, formed in the amorphous matrix during heating, is the phase which is the first phase to form during slow cooling of the melt. This in fact indicates that the amorphous matrix contains a large number of nuclei of the crystalline phase inherited from the liquid. It is fully possible that this nucleus is in fact a cluster embedded in some manner in the structure of the amorphous alloy or an even internal element of such a structure.

Crystallisation kinetics

The crystallisation kinetics of the amorphous alloys is a result of the effect of thermodynamic factors and kinetic parameters [3.64]. The kinetics of a given process depends on the set of the parameters, namely on the crystallisation mechanism, the number of 'frozen-in' crystallisation centres, the activation energy of diffusion, the driving force – the difference of the free energies of the amorphous and probable crystalline phase. In addition to this, the nature of crystallisation is influenced by the surface quality, the effect of the external factors (radiation, pressure, deformation).

In the case of isothermal holding the fraction of the crystallised material can be represented by the classic Johnson–Mehl–Avrami equation [3.65–3.67]:

$$x(\tau) = 1 - \exp[-b\tau^n], \tag{3.6}$$

Here $x(\tau)$ is the volume fraction of the crystalline phase formed during time τ; b is the rate constant; n is the exponent, its value can be used to assess the process mechanism (n varies from 1.5 to 4).

If $n = 1\pm0.5$, the grain growth is controlled by diffusion, there is no nucleation, at $n = 2.5\pm0.5$ nucleation takes place at a constantly increasing nucleation rate, the value $n = 3.5$–4 indicates that the crystallisation rate is controlled by the growth rate of the crystals at a constant rate of formation of the crystallisation centres, or corresponds to the eutectic breakdown [3.68].

The dependence of the rate constant b on temperature is described by the Arrhenius equation

$$b = b_0 \exp\left(-\frac{\Delta E}{RT}\right), \tag{3.7}$$

here b_0 is the pre-exponential multiplier; ΔE is the activation energy; R is the gas constant [3.62].

The value of the crystallisation temperature T_{crys} is not strictly fixed – the transformation may take place, although at a lower rate, during isothermal holding at a temperature several degrees lower than the value assumed to be T_{crys}. The crystallisation temperature T_{crys} is usually the temperature at which transformation takes place at a high rate (10^{-3}–10^{-1} of the volume of the specimen per minute).

Activation energy E_a and the latent heat of transformation Q play the controlling role in the crystallisation kinetics of the amorphous alloys. The relationship is unambiguous: as the purity of the specimen increases, the values of T_{crys} and E_a decrease and Q increases [3.69, 3.70]. The stability of the amorphous alloys increases with a decrease of the thermodynamic stimulus and increase of the efficiency of the kinetic barrier for their breakdown (lower heat and higher activation energy of crystallisation). The activation energy of the crystallisation process, as shown by the experiments, changes in a very wide range from 40 to 400 kJ/mole [1.1].

At a constant heating rate, the crystallisation kinetics is described by the Kissinger equation:

$$\left(\frac{\partial x}{\partial \tau}\right)_T = k(1-x), \tag{3.8}$$

where x is the fraction of the material solidified during time τ at temperature T, k is the rate constant determined by the following equation

$$k = A\exp\left(-\frac{\Delta E}{RT}\right), \tag{3.9}$$

where ΔE is the activation energy, R is the gas constant [3.71].

If the temperature changes with time, the reaction rate is described by the following equation

$$\frac{dx}{d\tau} = \left(\frac{\partial x}{\partial \tau}\right)_T + \left(\frac{\partial x}{\partial \tau}\right)_T \frac{dT}{d\tau}. \tag{3.10}$$

Since the number of nuclei at fixed time is also constant, and the position of the components in the system is constant, then $\left(\frac{\partial x}{\partial T}\right)_\tau = 0$ and combining (3.9) and (3.10) we obtain the equation which can be used at any temperature

$$\frac{dx}{d\tau} = A(1-x)\exp\left[-\frac{\Delta E}{RT}\right]. \tag{3.11}$$

Differentiating (3.11) with respect to time we have

$$\frac{d}{d\tau}\left(\frac{dx}{d\tau}\right) = \left[\left(\frac{\Delta E}{RT^2}\right)\left(\frac{dT}{d\tau}\right) - A\exp\left(-\frac{\Delta E}{RT}\right)\right]\left(\frac{dx}{d\tau}\right). \tag{3.12}$$

At the very beginning of the crystallisation process, the rate of the process is close to 0, and at the maximum point of the exothermic peaks on the DTA curves the rate is maximum and, consequently, at these temperatures $dx/d\tau = 0$ and

$$A\exp\left(-\frac{\Delta E}{RT_m}\right) = \frac{\Delta E}{RT_m^2}\left(\frac{dT}{d\tau}\right), \tag{3.13}$$

here T_m is the temperature corresponding either to the maximum of the peak or the start of the transformation. Denoting the heating rate by $dT/d\tau = \beta$, we obtain

$$-\left(\frac{\Delta E}{R}\right)\left(\frac{1}{T_m}\right)=\ln\left(\frac{\beta}{T_m}\right)+\text{const.} \tag{3.14}$$

From the slope of the straight-line in the coordinates $\ln\left(\frac{\beta}{T_m^2}\right)=f\left(\frac{1}{T_m}\right)$ it is possible to determine the activation energy of the crystallisation process ΔE on the pre-exponential multiplier A.

It is well-known that the amorphous metals and alloys usually crystallise in heating to a specific temperature T_k. The value of T_k is 13–23 K for the amorphous films of pure metals with a thickness of $D > 20$ nm and may reach several hundreds of degrees for the amorphous alloys. For the majority of metallic glasses T_k is in the range $(0.4–0.65)T_m$ for the conventional heating rates used in DSC investigations equal to 10–100 K/s (here T_m is the equilibrium melting point of the crystalline alloy). The absolute values of the nanocrystallisation range known at the present time are 400–1500 K (for the amorphous alloys based on refractory metals).

The crystallisation of the amorphous metals and alloys in heating takes place by the formation of crystalline nuclei in the amorphous matrix followed by their growth. The kinetics of nucleation of the crystalline phases has been the subject of many discussions. There are disputes regarding the relative role of the intrinsic initiation in the amorphous matrix and the athermal growth of the pre-precipitates or 'frozen-in nuclei'. Additional difficulties of interpretation are associated with the measurements of the nucleation rate. The nucleation rate should be determined on the basis of the measurement of the growth rate and general transformation kinetics. In the first case, a large error may be made if the effect of the small cross section of the this foil is not assessed reliably. In addition, in both cases, the results are not correct in the early stage of crystallisation when the number and size is of the crystals are small.

The nucleation of the crystalline phases in the amorphous matrix takes place, regardless of the dependence on the specific crystallisation mechanism, usually by a homogeneous mechanism, and the nucleation rate is approximately constant with time at the given temperature. However, homogeneous nucleation is only one of the possibilities which is often imposed by heterogeneous nucleation, athermal nucleation and even nucleation determined by the 'frozen-in' crystallisation centres [3.72].

The types and morphology of the nanocrystallisation products are determined by the crystallisation mechanisms closely linked with the chemical composition and thermodynamic characteristics of the resultant crystalline phases [3.64]. The grain size of the nanocrystalline structure, formed in crystallisation of the amorphous alloy, is strongly affected by the heat treatment conditions and the chemical composition of the metallic glass. One of the most important factors which determine the grain size in isothermal annealing is the annealing temperature. The annealing time is usually determined by the time of completion of the transformation of the amorphous phase to nanocrystalline phase.

Surface crystallisation

Surface crystallisation is a special form of the crystallisation of the amorphous alloys produced by melt quenching. In this case, the nanocrystalline phase forms, as is well-known, mostly in the surface layers of the rapidly quenched amorphous ribbon. This process is characterised by different states of the surfaces (internal, adjacent to the disc, and the external that is in contact with the atmosphere). The phenomena of accelerated crystallisation of the surface may be detected on both sides of the ribbon but there are cases of preferential crystallisation of one of the sides [3.73, 3.74]. The reasons for this phenomenon have not as yet been completely determined. At the same time, the surface crystallisation play a significant role in the vertical application of the amorphous alloys. For example, the crystallisation of the surface layers is reflected in their catalytic activity and corrosion resistance [3.74, 3.75] and in the properties of magnetically soft materials [3.76, 3.77].

The investigations show that the crystallisation of both surfaces of the amorphous alloys based on iron and cobalt starts earlier than in the volume [3.78, 3.79]. However, the activation energy of this process on the external surface is the same as in the volume, and on the internal surface it is considerably lower (this is especially evident at moderate temperatures). Taking into account the data obtained by small-angle X-ray scattering and Auger electron spectroscopy [3.78, 3.79], the authors concluded that the accelerated crystallisation of the external surface of the amorphous alloys is determined to a large extent by the presence in the surface layer of the excess free volume, the high concentration of the submicropores, whereas the acceleration of crystallisation of the internal side of the alloy is determined by the change of the chemical composition of these layers.

At the present time, the number of the experimental data for the effect of the alloying elements on the grain size of the nanocrystalline phase, formed during crystallisation of the metallic glasses, is small. In particular, it has been established that the additions of C and Si in the iron-based amorphous alloys increase the diffusibility of the metalloids and, consequently, increase the growth rate of the primary crystallisation products [3.80]. These additions may also decrease the concentration of nuclei and cause at the same time the formation of the structure with larger grains. The additions of Cu or Au to the iron-based glasses increase the nucleation rate of α-Fe crystals by several orders of magnitude. The addition of the elements which slow down diffusion, such as Nb, Zr of Mo, decreases the growth rate of the crystal and increases the dispersion of the structure [3.81]. The microadditions of Cr, Co, Ni or Pd have no significant effect on the primary crystallisation of the metallic iron-based glasses [3.82].

The experimental results show that the minimum possible size of the nanocrystals for the polymorphous and eutectic nanocrystallisation is several nanometres, and in primary crystallisation it is considerably greater (15–30 nm). For the specific crystallisation mechanism the minimum grain size does not depend on the number of elements in the alloy. Different values of the size of the nanocrystals for different crystallisation mechanisms indicate that the limiting size is determined by the nucleation mechanism and the structure of the interphase and also by the structure of the amorphous matrix.

Crystallisation and glass transition temperature

A number of studies have been published [3.83–3.88] in which it was established that the crystallisation of the alloys takes place by different mechanisms at temperatures higher or lower than the glass transition temperature.

At temperatures higher than the glass transition temperature the alloy is in the supercooled liquid state, and at temperatures lower than the glass transition temperature it is in the amorphous state. It is also well-known that the properties of the alloys greatly change in transition through the glass transition temperature.

The glass transition temperature Tg of the $Ni_{70}Mo_{10}P_{20}$ alloy is lower than the crystallisation temperature T_{crys} (T_g = 703 K and T_{crys} = 730 K at the heating rate of 0.33 K/s), and in [3.83] it was shown that annealing prior to the start of crystallisation of the specimens of this amorphous alloy at temperatures both lower and higher than the glass transition temperature results in the delamination of the initial amorphous phase in two regions with different chemical composition and/or the short-range order (different amorphous phases).

The results of X-ray diffraction and electron microscopic analysis show that annealing below the glass transition temperature results in the simultaneous formation of two crystalline phases: metastable FCC phase and the stable phase Ni_3P. The crystal precipitates have the form of dendrites (Fig. 3.14) and the amorphous interlayer is retained between the dendrite arms. Inside the dendrites there are clearly visible plate-shaped precipitates. Phase analysis shows that these precipitates correspond to the equilibrium Ni_3P phase.

The crystallisation of this alloy at temperatures higher than the glass transition temperature results in the formation in the amorphous matrix of the crystals of the metastable FCC phase and a small number of nickel nanocrystals. The resultant crystal precipitates of the metastable FCC phase are dendritic, the size of the crystals of this phase is 300–600 nm. The amorphous phase is retained between the dendrite arms, and the transverse size of the dendrite arms is 10–20 nm (Fig. 3.15) [3.84].

The metastable FCC phase is the metastable nickel phosphide with the composition close to the composition of the equilibrium phases $(Ni (Mo))_3P$ [3.84].

Annealing of the $Ni_{70}Mo_{10}B_{20}$ alloy below the glass transition temperature prior to the start of crystallisation does not result in changes of the initial amorphous structure. The alloy crystallises by the eutectic mechanism resulting in the formation of the FCC phase Ni(Mo) and the Ni_3B phases. The FCC Ni(Mo) phase amounts to 19.4%, and the Ni_3B phase to 80.6% of the total amount of the crystalline part [3.85].

The transformations at temperatures higher than the glass transition temperature can be divided into the transformations in the amorphous phase, taking place prior to the start of crystallisation, and the intrinsic crystallisation processes. Heating of the $Ni_{70}Mo_{10}B_{20}$ amorphous alloy is accompanied by the increase of the half width of the diffusion maximum, and it was concluded that areas enriched and depleted in molybdenum appear and, therefore, the amorphous phase is inhomogeneous prior to the start of crystallisation. A further increase of temperature or annealing time results in the crystallisation of the alloy. The crystallised samples contains three crystalline phases – the first phase, similar to pure nickel; the second phase is a solid solution of molybdenum in nickel with the Mo content of approximately 15 at.%, and the third phase is the phase of the Ni_3B

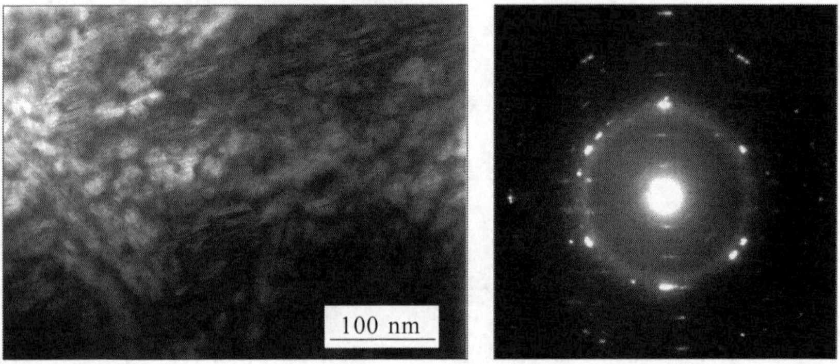

Fig. 3.14. Electron microscopic images and the diffraction pattern of a sample of $Ni_{70}Mo_{10}P_{20}$ amorphous alloy after annealing for 500 hours at 670 K.

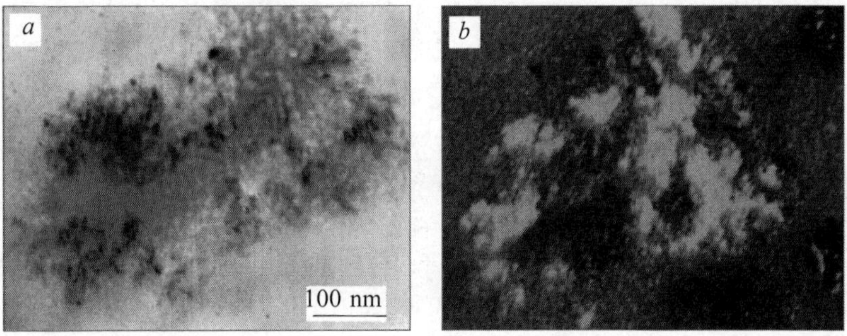

Fig. 3.15. Microstructure of the $Ni_{70}Mo_{10}P_{20}$ alloy annealed at T = 723 K for 60 s: a) bright field image; b) dark field image (TEM).

type with the orthorhombic lattice. Examination by high-resolution electron microscopy shows that the nanocrystals of phase I have the size of 2–5 nm and are defect free (Fig. 3.16).

The nanocrystals of phase 2 (Fig. 3.17) are considerably larger (20–50 nm) and contain a large number of defects [3.84, 3.85].

On the basis of the experimental results the authors of [3.85] analyse the differences in the phase compositions and crystallisation mechanisms of the $Ni_{70}Mo_{10}B_{20}$ amorphous alloy at temperatures higher and lower than the glass transition temperature. It was shown that the observed differences are determined by the delamination of the amorphous phase in two regions enriched and depleted in molybdenum, the FCC crystals of the solid solution of molybdenum in nickel form and contain approximately 15 at.% Mo, and in the regions enriched with boron, crystallisation results in the formation

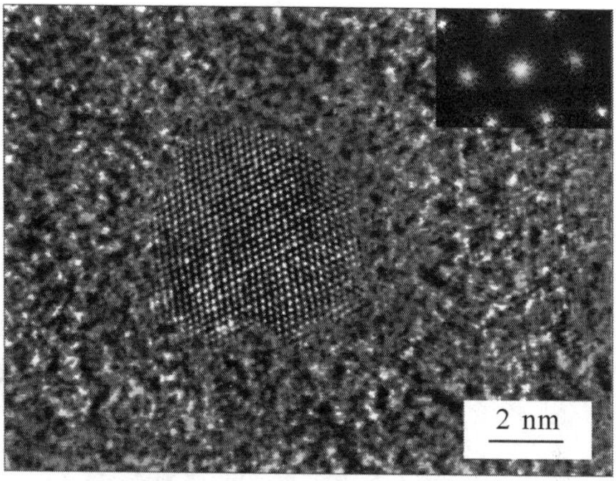

Fig. 3.16. High-resolution electron microscopic image of a nickel nanocrystal as a result of the Fourier transform of this image.

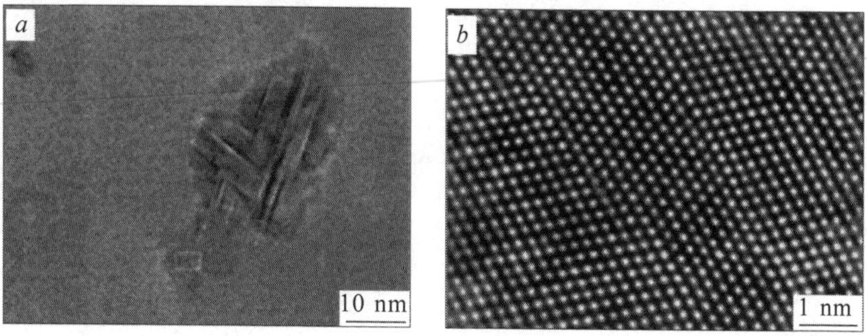

Fig. 3.17. High-resolution electron microscopic image of the nanocrystal of the solid solution of Mo in Ni (*a*), magnified image (*b*) of the area indicated in the frame in (*a*).

of borides, and in the regions depleted in molybdenum and boron – the FCC phase, similar to pure nickel [3.85].

It has been established [3.86, 3.87] than the formation of the nanocrystalline phase in the $Ni_{70}Mo_{10}P_{20}$ and $Ni_{70}Mo_{10}B_{20}$ eutectic alloys is observed only as a result of the preliminary delamination of the initial amorphous phase in two regions with different chemical composition and/or the short-range order (amorphous phase) and takes place by the primary mechanism.

The authors of [3.88, 3.89] investigated the crystallisation of the metallic glasses based on Al–RE–Ni–Co system (RE – the rare-earth metal). The results show that the aluminium-based metallic glasses contain different mechanisms of crystallisation above and below the glass transition temperature. Annealing of the metallic

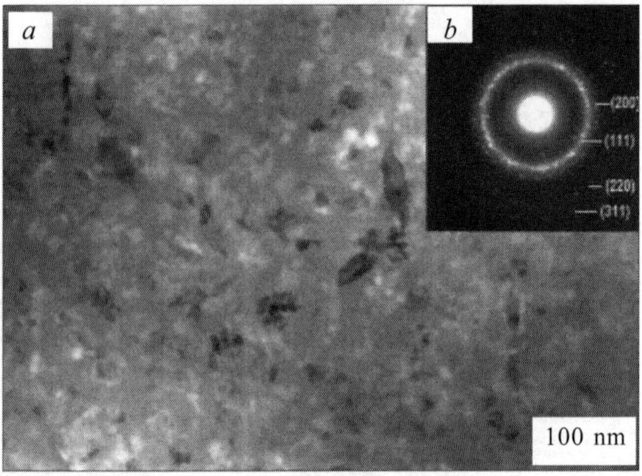

Fig. 3.18. Structure of the $Al_{85}Y_8Ni_5Co_2$ alloy after heat treatment using an isothermal calorie meter at 533 K for 5.3 ks, TEM, bright field (*a*); the diffraction pattern on which the thin rings are formed by α-Al, and the spots by the intermetallic phase (*b*) [3.88, 3.89].

glasses $Al_{85}Re_8Ni_5Co_2$ and $Al_{85}Y_4Nd_5Co_2$ above the glass transition temperature is accompanied by the formation of the primary nanosized particles of α-Al in the amorphous matrix. In the same alloys in isothermal annealing below the glass transition temperature examination showed the primary intermetallic phases or intermetallic phases together with the α-Al particles. The size of these intermetallic phases is approximately 50 nm, they are metastable and have a multicomponent composition (Fig. 3.18).

The study [3.90] describes the effect of annealing on the microstructure of bulk metallic glasses. On the example of the $Fe_{36}Co_{36}B_{19.2}Si_{4.8}Nb_4$ alloy it was shown that annealing below the glass transition temperature results in the formation of atomic clusters with a pseudo-tenfold symmetry and a close bond with the $Fe_{23}B_6$ phase. Annealing at relatively high temperatures leads to the formation of Fe_2B and FeB stable phases and the Fe(Co) solid solution.

The structure in early crystallisation stages

The thermal stability of the nanocrystals is strongly affected by the processes of normal and anomalous grain growth, phase transformations, relaxation, diffusion and other factors [3.91].

Since the transition from the amorphous to nanocrystalline state is the phase transition of the first kind, nanocrystallisation is usually accompanied by the formation of two-phase structures which are interpreted as the amorphous–nanocrystalline structures. The striking feature of the alloys with the amorphous–nanocrystalline structure (ANS) is primarily the fact that the structural (phase) components of such system greatly differ because they have the maximum (crystal) and minimum (amorphous state) degree of atomic ordering.

For correct analysis of the effects, associated with the crystallisation of the amorphous state, it is convenient to separate several stages of transition from the amorphous to crystalline state (Fig. 3.19).

If the completely amorphous (Fig. 3.19 *a*) and completely crystalline (Fig. 3.19 *d*) states are excluded, there are in principle two transition amorphous–crystalline states: the amorphous matrix with the uniformly distributed particles of the crystalline phase with the volume fraction $V_v \leq 0.5$ (Fig. 3.19 *b*) and the incomplete

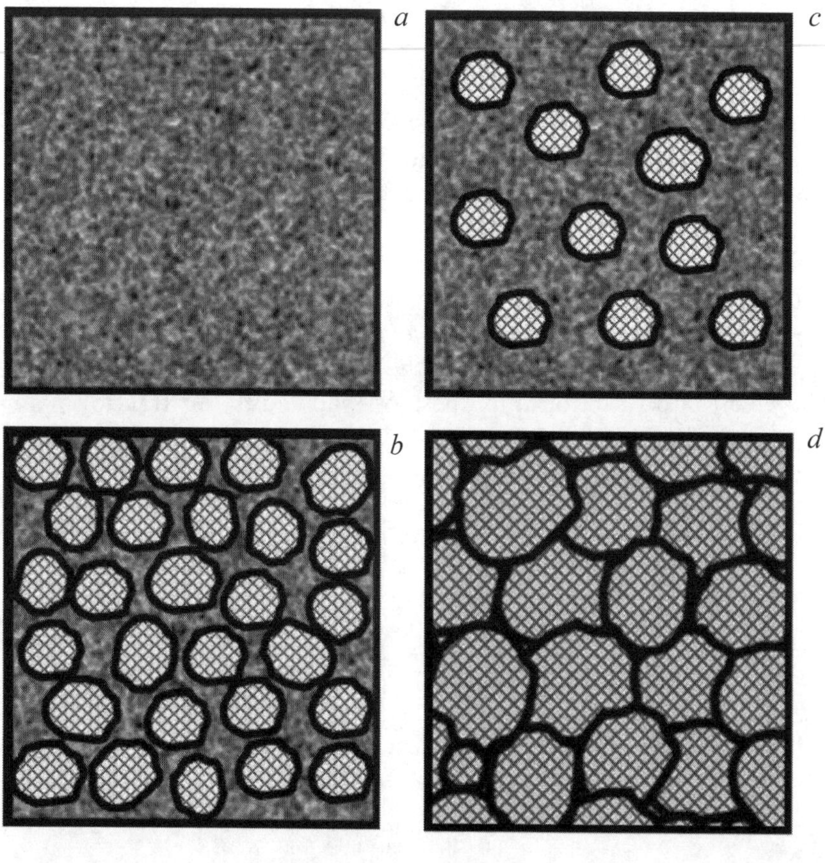

Fig. 3.19. Crystallisation of the amorphous state.

crystalline state with the amorphous interlayers, for which $V_v \geq 0.5$ (Fig. 3.19 *c*).

In this section, we summarise the results of investigations of the alloys with the amorphous–crystalline structure obtained in the initial stages of controlled crystallisation. The results are presented of the investigation of the structure and the quantitative analysis of its main parameters (the type of crystal lattice, composition, the mean size and shape, bulk density and bulk fraction of the crystalline phase) in dependence on the temperature–time regimes of heat treatment of the amorphous alloys.

In [3.92] the authors investigated crystallisation of $Fe_{70}Cr_{15}B_{15}$ alloy which is a modelling alloy having the unique set of the stress, magnetic and corrosion-resisting properties and forms a basis for a number of industrial alloys superior to the crystalline analogues. Using differential scanning calorimetry (DSC) the crystallisation temperature T_{crys} of the investigated alloy was determined. Figure 3.20 shows the DSC curve obtained in heating the specimen at a rate of 20°/min.

As indicated by Fig. 3.20, the crystallisation of the $Fe_{70}Cr_{15}B_{15}$ alloy takes place in a single stage. The most active precipitation of the crystal phase takes place in the amorphous alloy at a temperature of 526°C.

Electron microscopic studies after annealing the initial amorphous alloy at $T < 450$°C showed the typical amorphous structure without any features of crystallisation (Fig. 3.21). The absorption contrast of the 'salt–pepper' type, typical of the amorphous state, was observed. The electron diffraction micropatterns showed the diffraction pattern

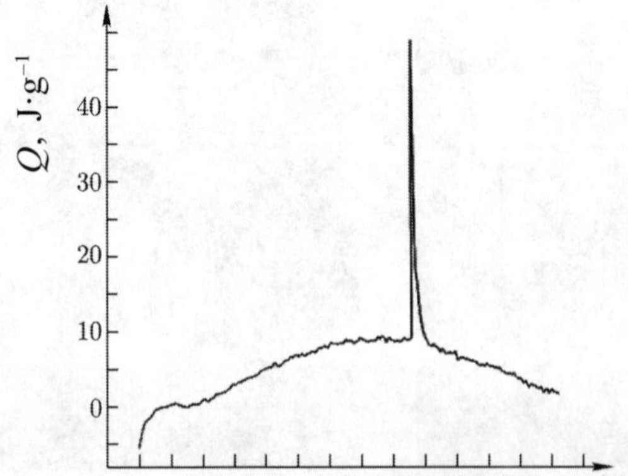

Fig. 3.20. The DSC curve of the $Fe_{70}Cr_{15}B_{15}$ amorphous alloy.

Fig. 3.21. The structure of the $Fe_{70}Cr_{15}B_{15}$ alloy after annealing for one hour at 400°C, TEM (bright field).

characterised by the high-intensity strongly eroded first ring (halo) and difficult to see rings (halo) of the higher orders.

When the annealing temperature is increased to 450–470°C, depending on the annealing time the structure shows the process of eutectic crystallisation in which the amorphous matrix crystallises in the form of eutectic colonies with the simultaneous formation of two phases with a close structural relationship (α-Fe–Cr and Fe_3B) (Fig. 3.24). In this case, the stepped redistribution of the components at the crystallisation front takes place. Although the transformation takes place in the solid body, the resultant crystalline colonies are referred to as eutectics (not eutectoids) since the amorphous phase is a supercooled liquid.

The results of X-ray diffraction analysis of the partially crystallised $Fe_{70}Cr_{15}B_{15}$ alloy shows that in addition to the amorphous state, the structure actually contains two crystalline phases: the BCC solid solution of chromium in α-Fe (α-(Fe–Cr)) and the Fe_3B boride (tetragonal close-packed lattice).

Figure 3.22 shows the X-ray diffraction diagram of the alloy after annealing at 450°C, 2 h.

The spacing of the crystal lattice of the α-(Fe–Cr) phase equals 0.287 nm, and the spacings of the lattice of the tetragonal phase Fe_3B is α = 0.862 nm and c = 0.428 nm. The effective size of the α-(Fe–Cr) crystals, determined from the broadening of the X-ray lines, is 28 nm, and the size of the Fe_3B phase is 44 nm. The volume ratio of the Fe_3B and α-(Fe–Cr) phases in the eutectic is approximately 3:1 and changes only slightly in dependence on the heat treatment

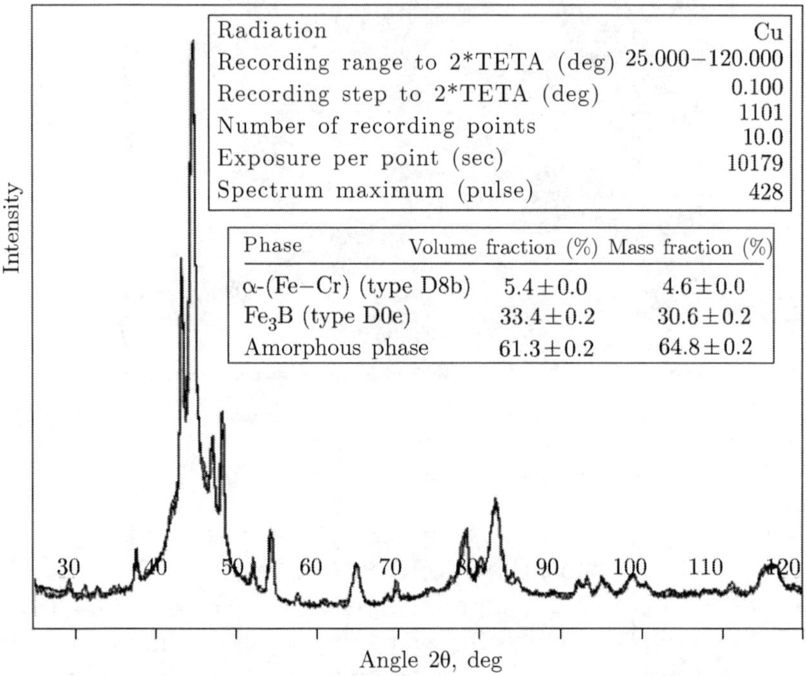

Fig. 3.22. The X-ray diffraction pattern of the $Fe_{70}Cr_{15}B_{15}$ alloy after annealing at 450°C, 2 h.

conditions. The structural components of the eutectic have the form of very fine (no more than 10 nm in thickness) alternating place situated in a strict orientation relationship determined by the TEM method:

$$\langle 100\rangle_{\text{boride}} \| \langle 100\rangle_{\text{Fe–Cr}},$$
$$\langle 100\rangle_{\text{boride}} \| \langle 121\rangle_{\text{Fe–Cr}}.$$

Since the solubility of boron in α-Fe is very small, the oscillations of the boron concentration between the structural components of the eutectic exceed 30 at.%.

Figure 3.23 shows the electron microscopic images indicating the evolution of the structure of the given alloy during annealing. As an example the photographs of the α-Fe–Cr eutectic colonies and the Fe_3B boride distributed in the amorphous phase after annealing for one hour at 460°C in the form of barrel-shaped crystals with the average size of approximately 1 μm. The form of the eutectic colonies does not change during growth up to contact of the individual colonies with each other with increasing temperature and/or annealing time.

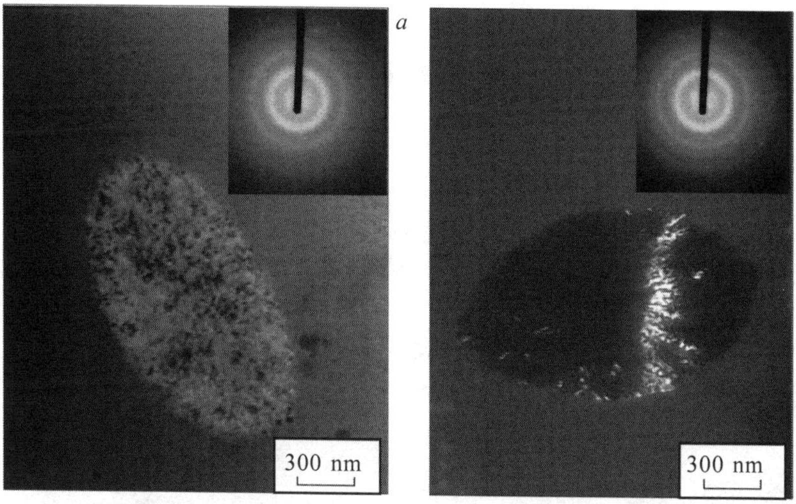

Fig. 3.23. Structure of the $Fe_{70}Cr_{15}B_{15}$ alloy after annealing for one hour at 460°C. TEM (*a* – bright field, *b* – dark field).

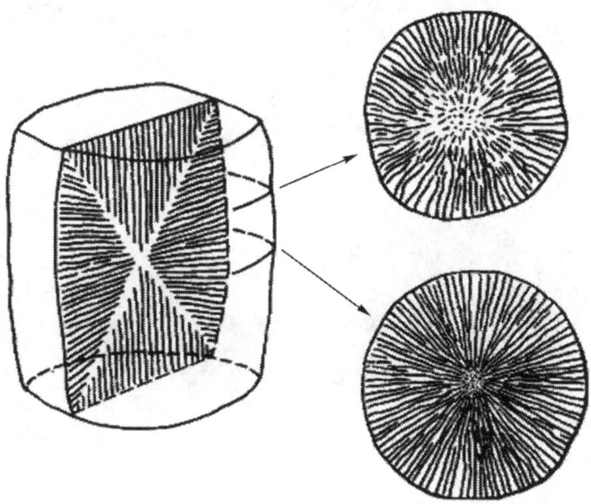

Fig. 3.24. Diagram of spatial distribution of the phases in the eutectic formed in crystallisation of the $Fe_{70}Cr_{15}B_{15}$ alloy; the bars correspond to the Fe_3B phase, the matrix to the α-(Fe–Cr) phase.

The stereometric analysis results were used to reproduce the three-dimensional form of 'kegs' in the distribution of the phases inside them (Fig. 3.24). Evidently, the general form of the eutectic crystals is determined by the growth morphology of the boride tetragonal phase because the axis of the 'keg' was always parallel to the tetragonality axis with the boride phase. The identical pattern of

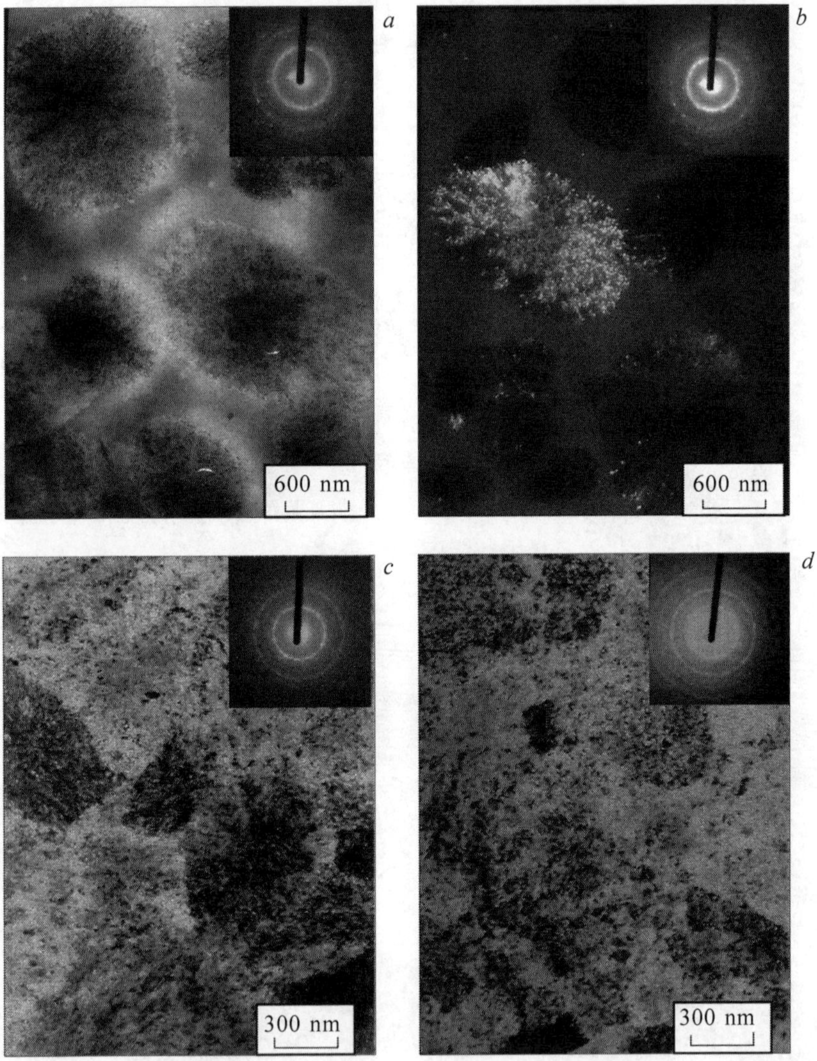

Fig. 3.25. Structure of the $Fe_{70}Cr_{15}B_{15}$ alloy after annealing at 470°C (*a*, *b*), 480°C (*c*) and 510°C (*d*), 1 h. TEM (*a, c, d* – the bright field, *b* – the dark field).

the eutectic crystallisation was found previously in the $Fe_{40}Ni_{40}P_{14}B_6$ amorphous alloy [2.51], with the only difference being that the volume ratio of the metallic and boride phases was opposite.

Figure 3.25 shows later stages of crystallisation of the investigated alloy in annealing for one hour at 470°C (Fig. 3.25 *a*, *b*) and also before (Fig. 3.25 *c*) and after (Fig. 3.25 *d*) completion of crystallisation.

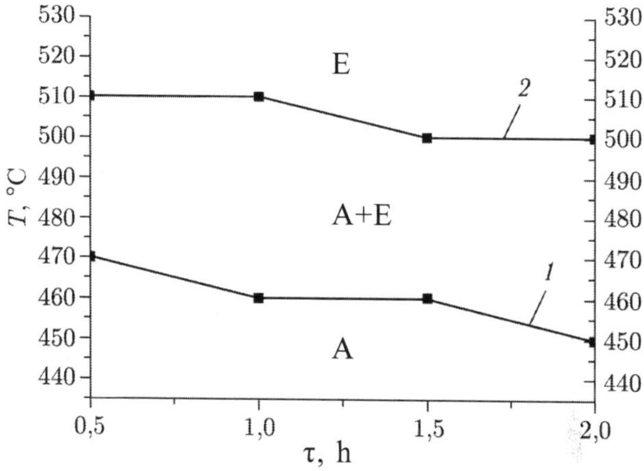

Fig. 3.26. Temperature–time diagram of the $Fe_{70}Cr_{15}B_{15}$ amorphous alloy: A – amorphous state, E – the eutectic, *1* – the line of the start of crystallisation, *2* – the line of the completion of primary crystallisation.

The authors of [3.92] constructed the temperature–time diagram (TTD) for the efficient selection of the conditions of controlled annealing of the amorphous state. Using the diagram, it is possible to predict reliably the structural state of the amorphous alloys corresponding to a specific temperature range and annealing time (Fig. 3.26). Initially, the values of the point of transition from one structural state to another were determined 'approximately' by analysis of the states with the temperature step of 30–40°C. Subsequently, the detailed analysis of the electron diffraction micropatterns and also bright and dark field images with the temperature step reduced to 10°C determined these values more accurately. The error in the determination of the boundary values of the annealing temperature in this case was not greater than 5°C. The TTD can also be used for reliable protection of the structural state of the amorphous alloy corresponding to specific temperature and annealing time ranges.

Attention should be given to the fact that T_{crys}, determined by the DSC method, is in this case 526°C, and the first signs of crystallisation, determined by TEM, were recorded after annealing at a lower temperature, 450°C. These facts do not contradict each other because the DSC method records the temperature of the most intensive occurrence of the crystallisation processes in the conditions of continuous heating of the amorphous state.

It may be noted that the temperature range between the start and completion of crystallisation (between the lines *1* and *2* in Fig. 3.26) is only 30–40°C. In addition, at a longer annealing time the crystallisation processes start at lower temperatures.

The structural parameters and crystalline phase (average size D, bulk density N_v and bulk fraction V_v), precipitated in the amorphous matrix during annealing, were determined by bright field electron microscopic images.

The calculation of the linear parameters of the grains and nanocrystalline particles was carried out by the secant method, the error in the determination of the average size was not greater than 5%. The structural parameters, such as N_v and V_v of the crystalline particles were calculated using the following equations [3.93]:

$$N_v = \frac{N_A}{t + \frac{\pi \cdot d}{2}}, \tag{3.15}$$

where N_v is the bulk density of the crystalline particles, μm^{-3}, N_A is the surface density, μm^{-2}, i.e., the number of particles on the investigated area, t is the foil thickness, μm, d is the average particle size on the plane, μm:

$$V_v = \frac{\pi \cdot D^3 \cdot N_v}{6}, \tag{3.16}$$

where V_v is the volume fraction of the crystalline particles, D is the actual average size of the particles (in the volume), μm.

Using the TEM and the equations (3.50) and 3.16) which were used for calculating the bulk density N_v^e and the bulk fraction V_v^e of the crystalline phase, experiments were carried out to determine the structural parameters (D^e, N_v^e , V_v^e) of the eutectic in the temperature range 450–500°C of the existence of the two-phase state assuming the unchanged form of the 'kegs' during their growth.

Figures 3.27 and 3.28 shows the dependences of D^e, N_v^e and V_v^e on the annealing temperature (at a fixed annealing time of 1 h) and on the annealing time (at a fixed temperature of 480°C). The dependences obtained for other heat treatment parameters are similar.

It may be seen that the variation of D^e in contrast to the change of N_v^e is non-monotonic. With increase of the annealing temperature the amount of the eutectic (N_v^e) smoothly decreases, but in the annealing temperature range 460–470°C the average size D^e increases abruptly (almost by a factor of 1.5). It may be assumed that this effect is associated with the jump-like variation of the conditions of diffusion redistribution of the components between the initial and growing phases. In all likelihood, this is caused by the barrier effect associated with the boron atoms which are almost completely insoluble in the α-Fe–Cr phase. The jump of V_v^e in Fig. 3.27 is caused evidently by the fact that V_v^e is a function of both D^e and N_v^e (expression (3.16)) [3.92].

The amorphous alloys of the Fe–Ni system ($Fe_{58}Ni_{25}B_{17}$, $Fe_{50}Ni_{33}B_{17}$) are modelling alloys of the $Fe_{83-x}Ni_xB_{17}$ three-component system in which the nanocrystalline Fe–Ni phase, formed in annealing, is characterised by the different type of the crystal lattice: BCC or FCC, respectively.

The authors of [3.94] carried out detailed investigation of the initial stages of the crystallisation of the $Fe_{58}Ni_{25}B_{17}$ and $Fe_{50}Ni_{33}B_{17}$ alloys so that the results can be used to determine the effect of the type of crystal lattice of the nanocrystals on the mechanical behaviour of the materials with the amorphous–nanocrystalline structure in different crystallisation stages.

When heating the $Fe_{58}Ni_{25}B_{17}$ amorphous alloy the DSC curve shows the exothermic peaks corresponding to primary crystallisation, at 419°C (Fig. 3.29).

According to the TEM data in annealing the amorphous alloy at $T < 370$°C the samples contain the amorphous structure identical with the structure in Fig. 3.21 for the $Fe_{70}Cr_{15}B_{15}$ alloy.

Depending on the annealing time, at the annealing temperatures of 370–380°C the crystal particles appear in the amorphous matrix. The alloy crystallises by the primary crystallisation mechanism.

The results obtained by X-ray diffraction analysis show that the lattice spacing of the primary BCC crystals is $a = 0.287$ nm, and their structure is a BCC solid solution of Ni in α-Fe at a ratio of the components of 1:4.

Figure 3.30 shows the structure of the alloy after annealing at $T = 380$°C for 1 h, investigated by the TEM method. The electron diffraction pattern shows, in addition to the circular reflections from the amorphous phase, a small amount of point reflections, clearly indicating the presence of the crystalline phase. The primary nanocrystals in the alloy are cubic with the faces oriented on the crystallographic planes of the type {100}.

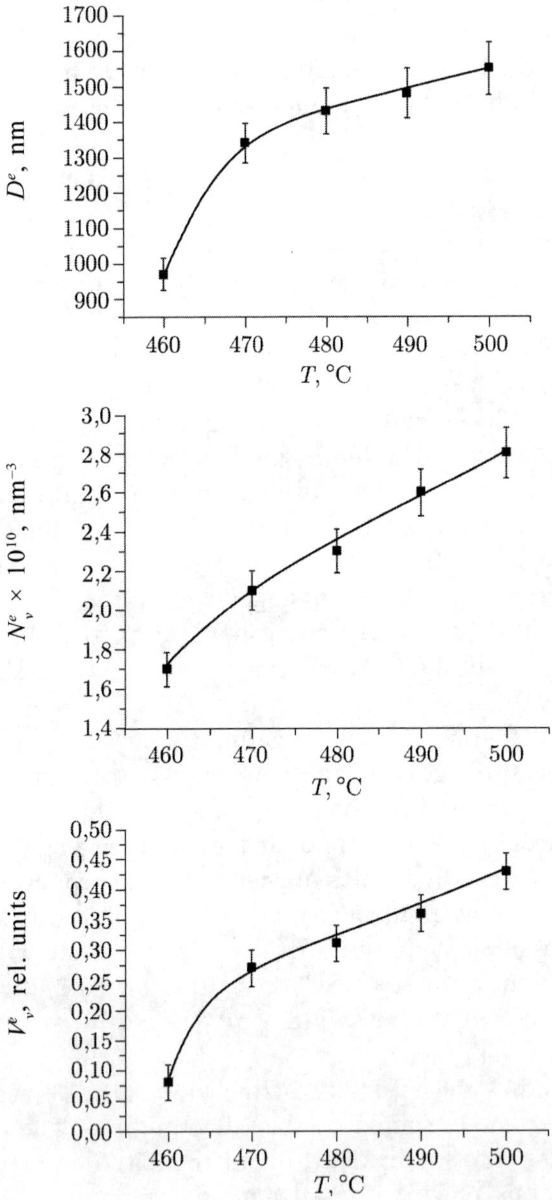

Fig. 3.27. Dependence of the structural parameters on T at $\tau = 1$ h in the $Fe_{70}Cr_{15}B_{15}$ alloy.

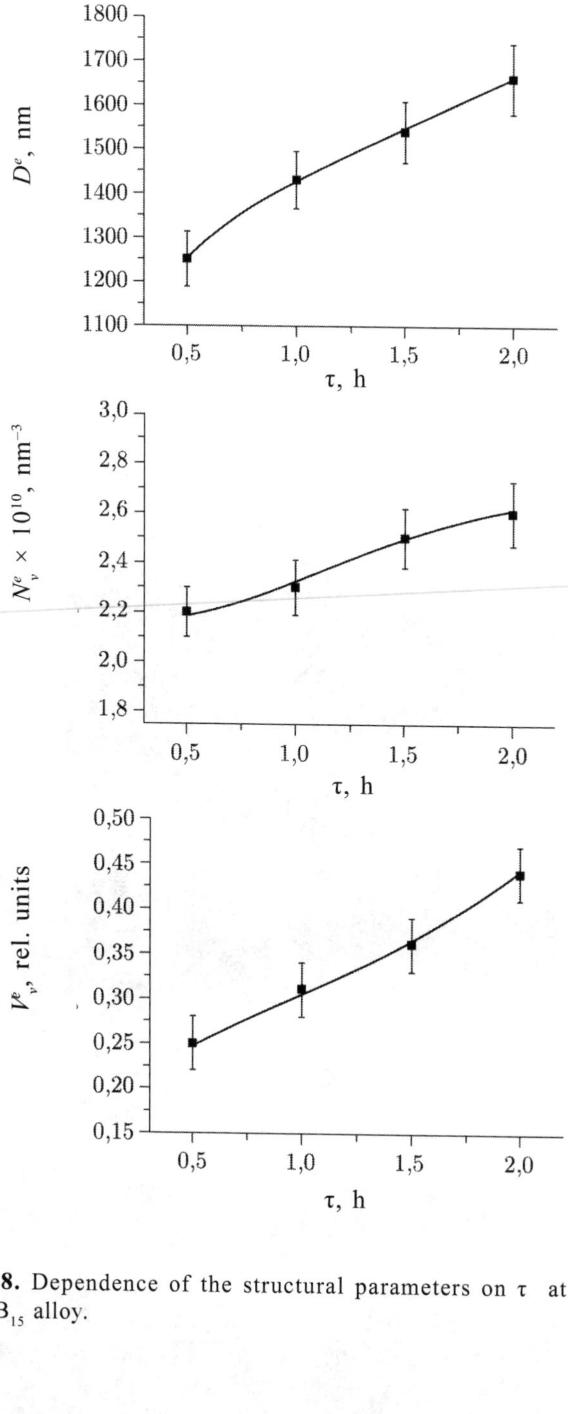

Fig. 3.28. Dependence of the structural parameters on τ at T = 480°C in the $Fe_{70}Cr_{15}B_{15}$ alloy.

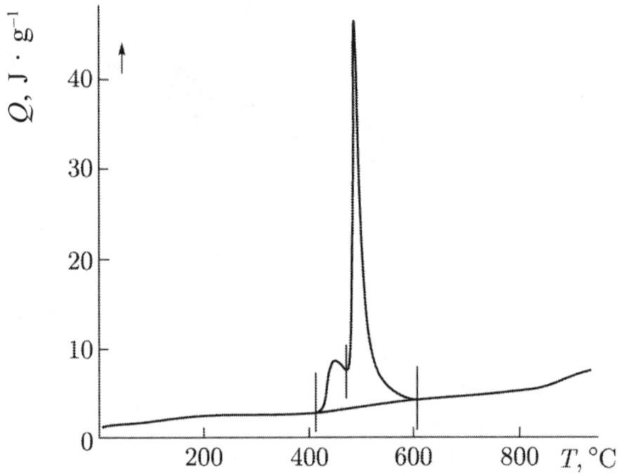

Fig. 3.29. DSC curve of the $Fe_{58}Ni_{25}B_{17}$ alloy.

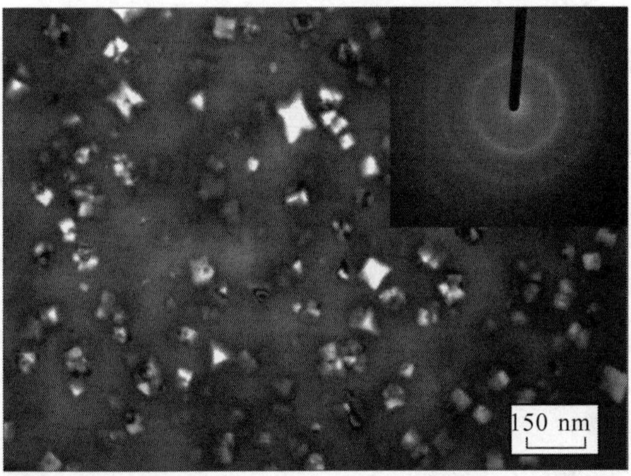

Fig. 3.30. The structure of the $Fe_{58}Ni_{25}B_{17}$ alloy after annealing at T = 380°C for 1 h. TEM (the dark field in the reflection of the first ring).

The temperature–time diagram (TTD) for the given alloy is shown in Fig. 3.31 (it was constructed using the same procedure as for the $Fe_{70}Cr_{15}B_{15}$) [3.92].

Analysis of the TTD shows that in annealing of the amorphous state nanocrystallisation starts at temperatures of 370–380°C depending on annealing time, and the process of crystallisation with the formation of borides starts at 390–410°C. The temperature range of existence of the two-phase structure is narrow, 20–30°C.

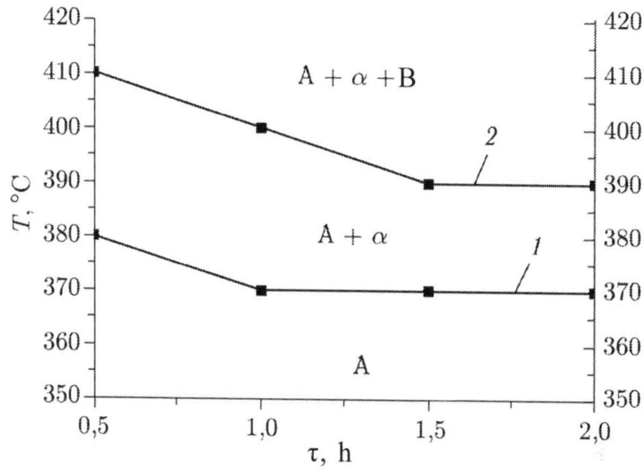

Fig. 3.31. The temperature–time diagram of the $Fe_{58}Ni_{25}B_{17}$ alloy: A – the amorphous state, α – BCC phase, B – the boride (1 – the line of the start of crystallisation, 2 – the line of completion of primary crystallisation).

The TEM and the equations (3.15), (3.16) were used to determine the structural parameters (D, N_v, V_v) of the crystalline phase in the temperature range of co-existence of the amorphous phase and a single crystalline phase.

The results of the investigations carried out on the given alloy were used to determine the dependences of the structural parameters of the crystalline phase on annealing temperature at a fixed annealing time and on annealing time at a fixed temperature. The results are presented in Figs. 3.32 and 3.33. It is important to note the normal growth of all structural parameters in the selected annealing conditions in the range of existence of the two-phase amorphous–nanocrystalline structure [3.94].

To confirm the accuracy of the calculations of the structural parameters of the nanocrystalline phase in this alloy by the TEM method, the authors of [3.94] determined additionally the volume fraction of the crystalline phase by X-ray diffraction analysis.

X-ray diffraction analysis was used to measure the structural parameters (D, V_b) on the specimens after annealing in the conditions 370°C, 1 h and 390°C, 1 h. In annealing of the $Fe_{58}Ni_{25}B_{17}$ alloy at 370°C, 1 h the X-ray diffraction diagram in Fig. 3.34 shows the peak only from the amorphous phase and there are no signs of the nanocrystalline phase. The investigations of the same state by TEM show that the volume fraction of the crystalline phase is 0.04 [3.94]. In this case, it should be taken into account that the X-ray studies

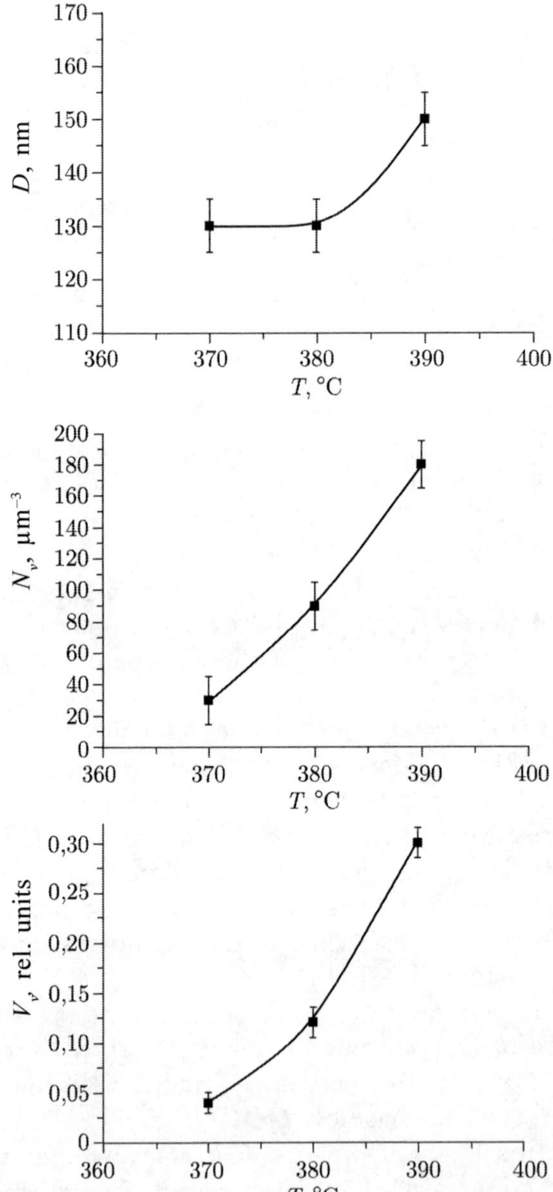

Fig. 3.32. Dependence of the structural parameters on T at $\tau = 1$ h in the $Fe_{58}Ni_{25}B_{17}$ alloy.

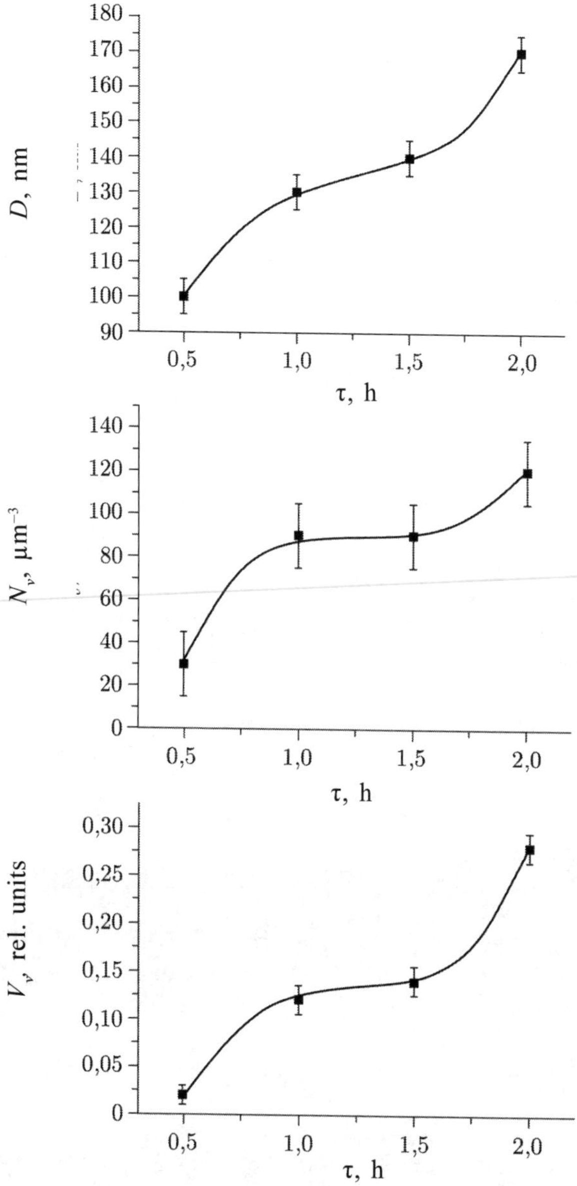

Fig. 3.33. Dependence of the structural parameters on τ at T = 380°C in the $Fe58Ni25B_{17}$ alloy.

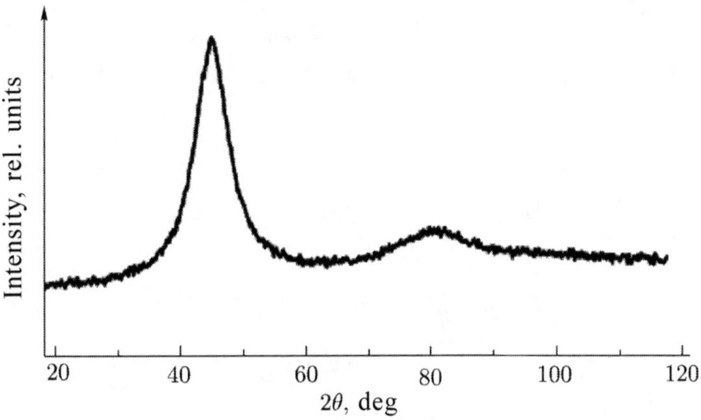

Fig. 3.34. The X-ray diffraction diagram of the $Fe_{58}Ni_{25}B_{17}$ alloy after annealing for 1 h at 370°C.

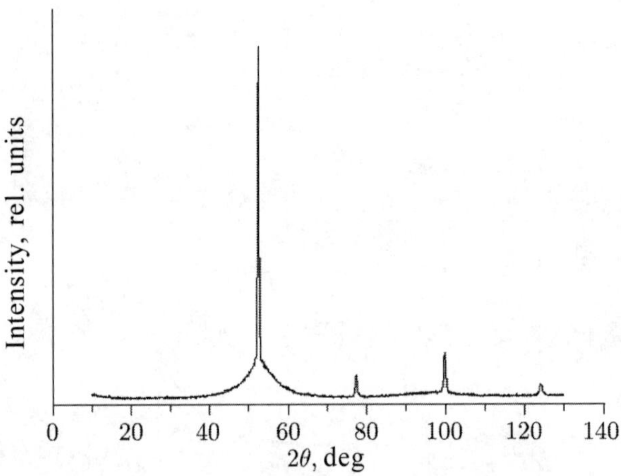

Fig. 3.35. The X-ray diffraction diagram of the $Fe_{58}Ni_{25}B_{17}$ alloy after annealing for 1 h at 390°C.

are characterised by the sensitivity of detection of the crystalline phase with the volume fraction of approximately 0.05.

Figure 3.35 shows the X-ray diffraction diagram obtained in studying the $Fe_{58}Ni_{25}B_{17}$ crystallised alloy after heat treatment at 390°C, 1 h.

On the X-ray diffraction diagram the peaks from the amorphous and nanocrystalline α-phase are clearly visible. The volume fraction of nanocrystals was determined from the relative intensity of the

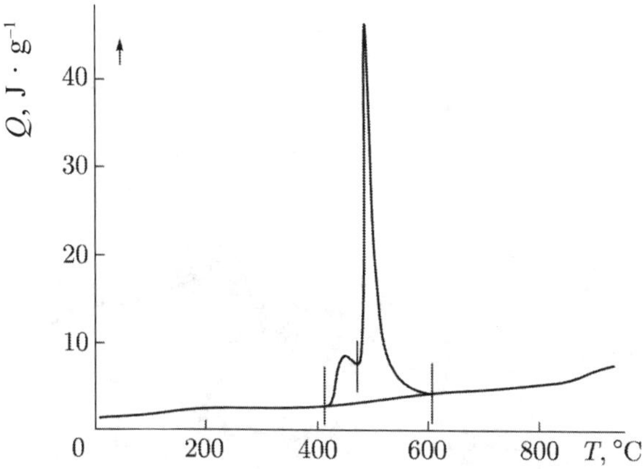

Fig. 3.36. DSC curve of the $Fe_{50}Ni_{33}B_{17}$ amorphous alloy.

X-ray lines, it was 0.2 (according to the TEM data, 0.3 [3.94]). The mean size of the nanocrystals was determined from the degree of broadening of the diffraction maxima, which was 143±8 nm (according to the TEM data D = 150 nm). Thus, the results of calculating the structural parameters of the nanocrystalline phase from the TEM data and X-ray diffraction studies are in sufficient agreement with each other.

When the $Fe_{50}Ni_{33}B_{17}$ alloy was heated by the DSC method, it was found that the primary crystallisation temperature of this alloy was 410°C (Fig. 3.36).

According to the TEM data at an annealing temperature of an amorphous alloy below 360°C an amorphous structure is formed. In the temperature range 370–380°C, signs of crystallisation are observed depending on the annealing time. The alloy crystallizes according by the mechanism of primary crystallisation with the formation of equiaxed nanocrystals. Analysis of the electron diffraction patterns showed that the nanocrystalline particles have an FCC lattice (Fig. 3.37).

A typical X-ray diffraction pattern for the amorphous–crystalline state is shown in Fig. 3.38 where the peaks from the amorphous and crystalline γ-phases are clearly visible. According to X-ray data, the lattice period of FCC crystals was a = 0.357 nm, and the evolving phase is a γ-Fe-based phase.

The temperature–time diagram for a given alloy, constructed on the basis of a TEM using a procedure similar to that described above for the $Fe_{70}Cr_{15}B_{15}$ alloy, is shown in Fig. 3.39.

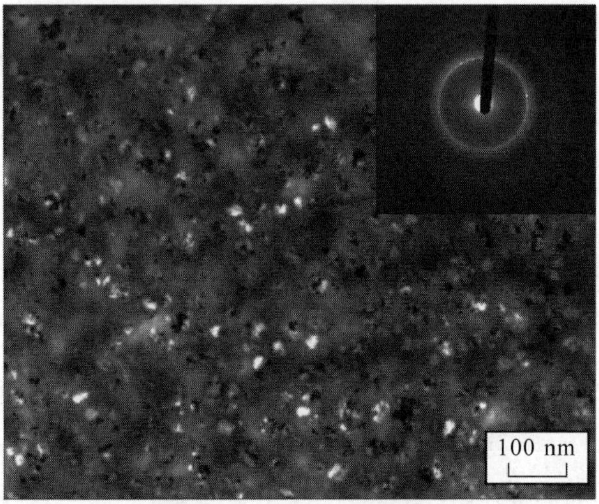

Fig. 3.37. The structure of the $Fe_{50}Ni_{33}B_{17}$ alloy after annealing at a temperature of 370°C for 1 h. TEM (light field).

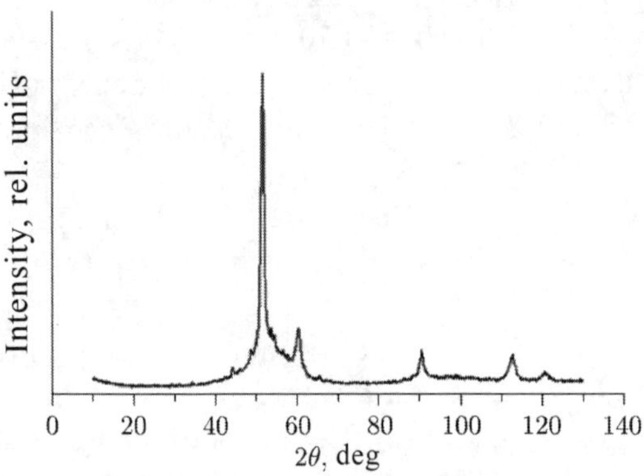

Fig. 3.38. X-ray diagram of $Fe_{50}Ni_{33}B_{17}$ amorphous alloy after annealing for 0.5 h at 400°C.

Analyzing this diagram, it can be noted that the temperature range for the studies is small, but it is somewhat larger than for the $Fe_{58}Ni_{25}B_{17}$ alloy (Fig. 3.31), and is 40–50°C (Fig. 3.39). The temperature of the onset of primary crystallisation (line 1) does not depend on the annealing time, i.e. the crystallisation process with the formation of γ-nanocrystals at any annealing time begins at the same temperature of 360°C. The termination temperature of primary crystallisation (line 2) decreases with the annealing time: at an annealing time of 0.5 h it is 410°C, and for an annealing time of 1 h or more it is 400°C.

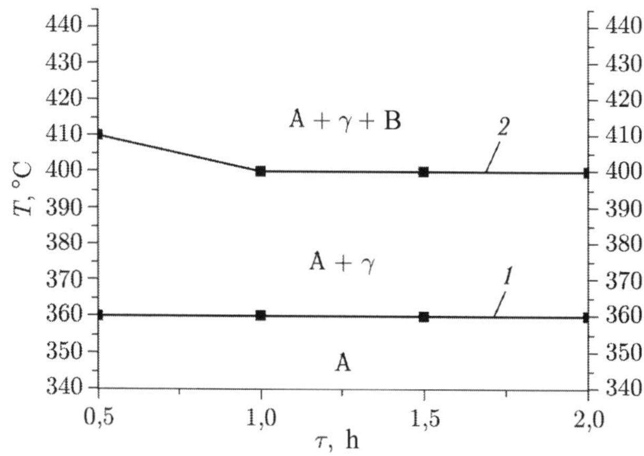

Fig. 3.39. Temperature–time diagram of amorphous alloy $Fe_{50}Ni_{33}B_{17}$: A is the amorphous state, γ is the FCC phase, B is a boride (*1* is the crystallisation start line, and *2* is the end of primary crystallisation line).

Figures 3.40 and 3.41 show the dependence of the structural parameters of the crystalline phase on the annealing temperature (for a fixed annealing time $\tau = 0.5$ h) and on the annealing time τ (at a fixed temperature $T = 370$°C).

As TEM studies show, in all the stages of crystallisation studied, the D values of nanoparticles were constant within the error range and amounted to about 20 nm (Fig. 3.40). X-ray data confirmed this fact. Therefore, the growth of V_v of nanocrystals was due only to an increase in N_v. Taking this into account, the dependences of N_v and V_v on T and τ were of a similar nature.

In the range 360–370°C an incubation period of crystallisation is detected, and only from a temperature of 380°C there is intensive growth of the parameter V_v (Fig. 3.40). The graphs of the dependences N_v and V_v on annealing time τ at a constant temperature were of a similar nature [3.94].

Based on a TEM study of the structure after annealing, it was established [3.94] that nanocrystalline γ-phase particles (FCC) in the $Fe_{50}Ni_{33}B_{17}$ alloy have a shape close to equiaxed and have a size of 20 nm for any annealing parameters; the α-phase nanoparticles (BCC) in the $Fe_{58}Ni_{25}B_{17}$ alloy exhibit a clear faceting in {100} planes and grow from 100 to 170 nm during annealing.

The amorphous alloy $Ni_{44}Fe_{29}Co_{15}B_{10}Si_2$ belongs to industrial functional (magnetic–acoustic) materials and is used in practice for the production of high-performance sensor devices and information protection devices.

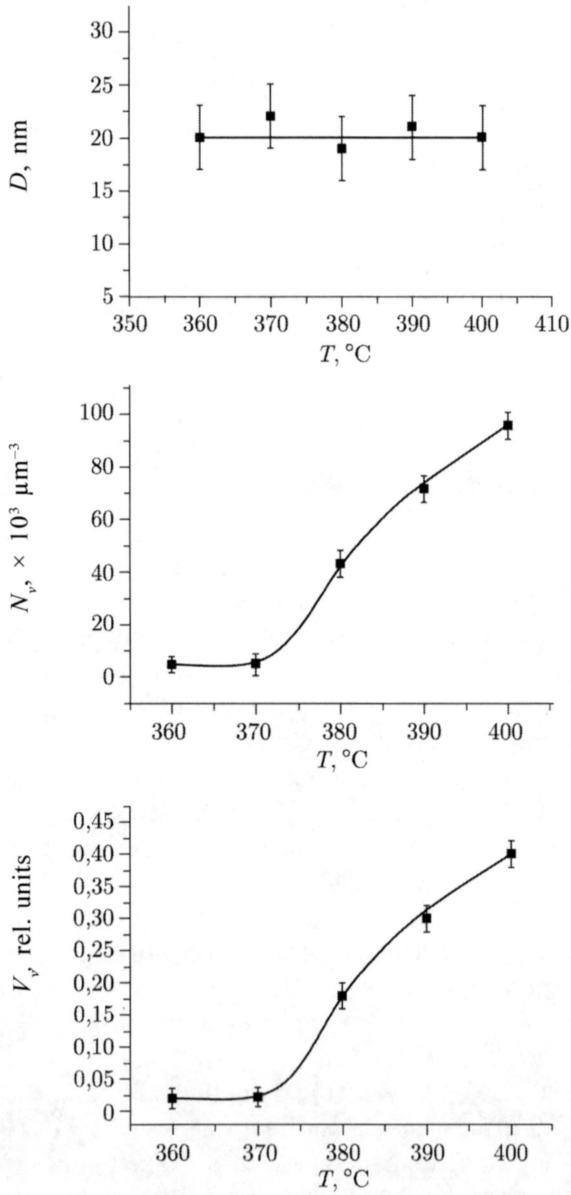

Fig. 3.40. Dependence of the structural parameters on the annealing temperature for an annealing time of 0.5 h in $Fe_{50}Ni_{33}B_{17}$ alloy.

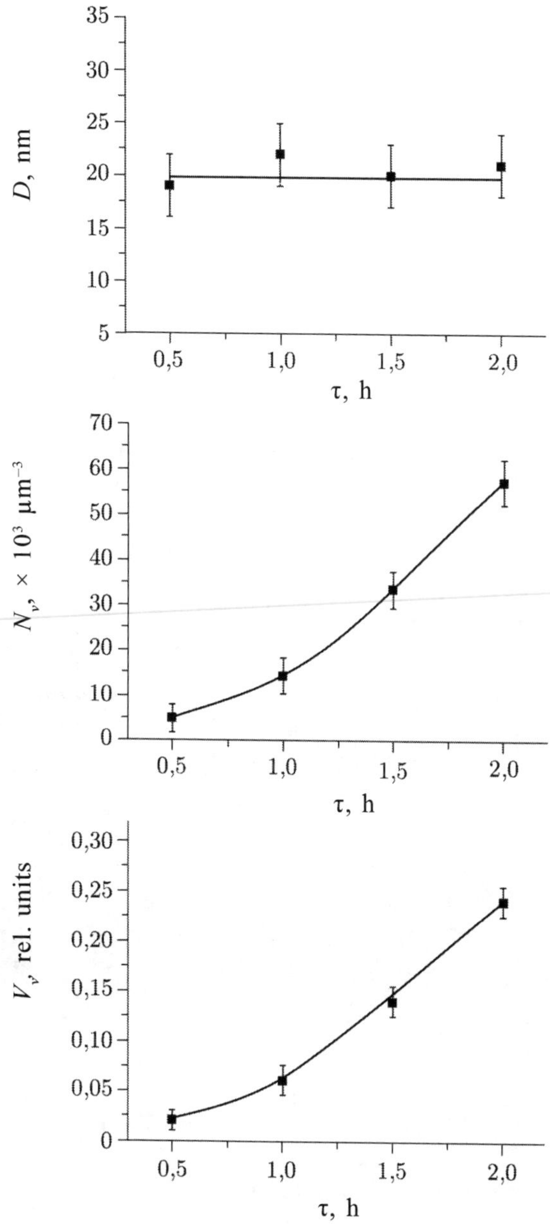

Fig. 3.41. Dependence of structural parameters on annealing time at an annealing temperature of 370°C in the $Fe_{50}Ni_{33}B_{17}$ alloy.

When this alloy is heated the DSC curve (Fig. 3.42) shows two exothermic peaks, therefore, the crystallisation of the amorphous $Ni_{44}Fe_{29}Co_{15}B_{10}Si_2$ alloy proceeds in two stages: the active precipitation of the first crystalline phase corresponds to 409°C.

The alloy structure at annealing below 330°C is still amorphous according to the TEM data. However, an increase in the annealing temperature to 340–350°C, depending on the holding time time, leads to the crystallisation of an amorphous alloy with the formation of equiaxed particles. Analysis of the electron diffraction patterns showed that the crystalline particles have an FCC lattice.

On the X-ray diffraction pattern of the alloy after annealing at 410°C for 1 h peaks from the amorphous and nanocrystalline γ-phase are observed for 1 h (Fig. 3.43). According to the X-ray data, the lattice period of FCC crystals was a = 0.357 nm, and the precipitated phase is based on γ-Ni–Fe–Co phase [3.94].

An electron microscopic image of the structure of this alloy is shown in Fig. 3.44.

The temperature–time diagram (TTD) for this alloy is shown in Fig. 3.45 [3.92].

It can be noted that the temperature range of the coexistence of the amorphous phase and one crystalline phase is quite wide and amounts to 110–120°C. The temperatures corresponding to lines *1* and *2* on the TTD vary slightly with the annealing time: line *1* is in the range 340–350°C, and line *2* is in the range 450–460°C.

Based on the calculations of the structural parameters of primary nanocrystals, the TEM data show the dependences of D, N_v and V_v

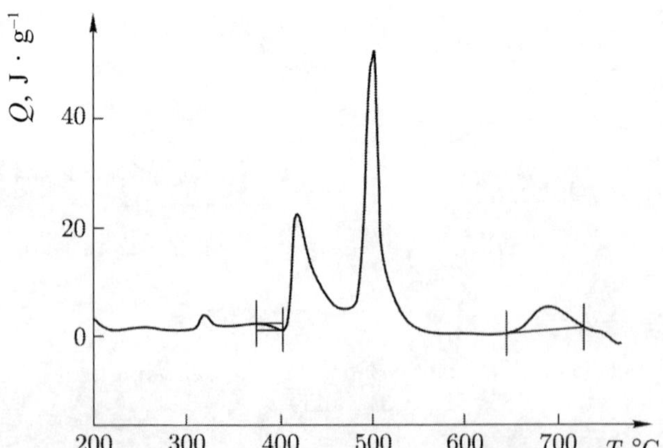

Fig. 3.42. The DSC curve of amorphous $Ni_{44}Fe_{29}Co_{15}B_{10}Si_2$ alloy.

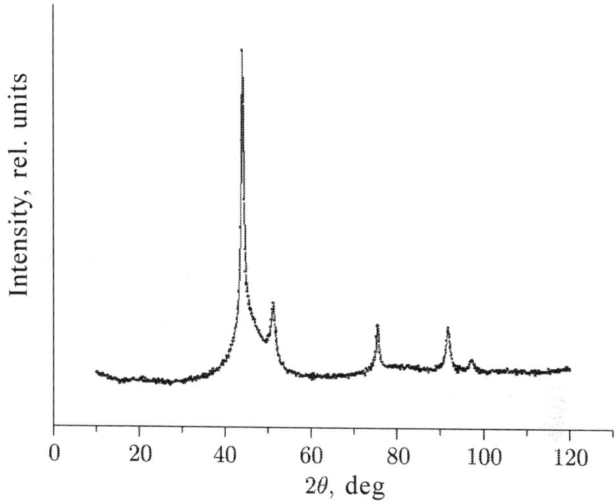

Fig. 3.43. The X-ray diffraction pattern of the $Ni_{44}Fe_{29}Co_{15}B_{10}Si_2$ alloy after annealing at 410°C for 1 h.

on T (for fixed τ = 1 h) and on τ (for fixed T = 360°C) (Fig. 3.46 and 3.47).

In this alloy increasing annealing temperature resulted in the growth of growth of N_v and, correspondingly, V_v of nanocrystals, while the value of D within the measurement error practically did not change and was 20 nm, as in the $Fe_{50}Ni_{33}B_{17}$ alloy. Due to the constancy of the value of D, the dependences for N_v and V_v on T and τ were of a similar nature. It is important to note that the increase in the values of N_v and, respectively, V_v in different temperature ranges occurs with different intensities.

At temperatures of 350–380°C and 420–460°C, the growth rate of N_v and V_v is higher than in the intermediate range 380–420°C (Fig. 3.47). At other heat treatment parameters the dependences were similar.

It has long been known that metallic materials with an amorphous-nanocrystalline structure are of great importance which for a number of properties can exceed both amorphous and nanocrystalline alloys. Thus, alloys of the Fe–Si–B system with small additions of Cu and Nb ($Fe_{73.5}Si_{13.5}B_9Nb_3Cu_1$), termed Finemet [3.95–3.97], possess in the amorphous–nanocrystalline state very high soft magnetic properties, exceeding the properties of the corresponding amorphous materials.

The Finemet alloys were first reported in [3.57] and later in [3.98]. It was shown that in the $Fe_{76.5-x}Si_{13.5}B_9M_xCu_1$ alloys it was possible to reduce the size of the formed crystals mainly due to

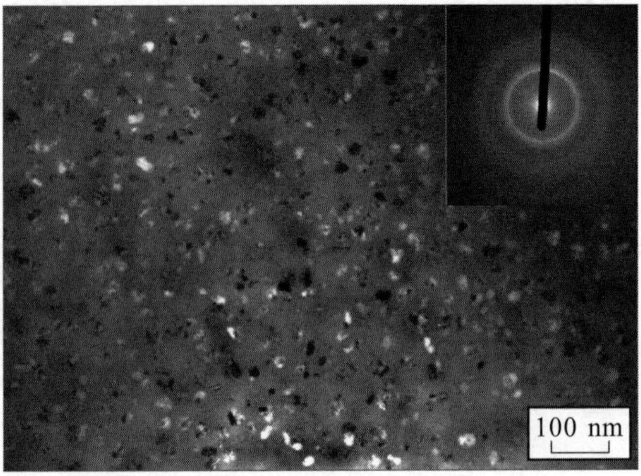

Fig. 3.44. The structure of $Ni_{44}Fe_{29}Co_{15}B_{10}Si_2$ alloy after annealing at 360°C for 1 h. TEM (light field).

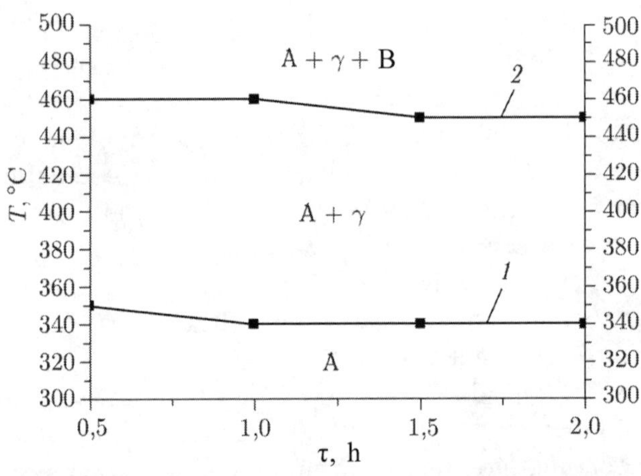

Fig. 3.45. Temperature–time diagram of the amorphous alloy $Ni_{44}Fe_{29}Co_{15}B_{10}Si_2$: A – amorphous state, γ – FCC phase, B – boride (1 – crystallisation start line, 2 – termination line of primary crystallisation).

the introduction of niobium (Nb = Ta > W > V > Cr) and to obtain unique magnetic properties. Optimal magnetic characteristics were obtained for composition $Fe_{73.5}Si_{13.5}B_9Nb_3Cu_1$ [3.96, 3.99].

Many authors (for example, [3.100]) established by the DSC method that the crystallisation of the $Fe_{73.5}Si_{13.5}B_9Nb_3Cu_1$ alloy proceeds in two stages: at 510°C the first crystalline phase is released in the amorphous alloy, and at 689°C the second one (Fig. 3.48).

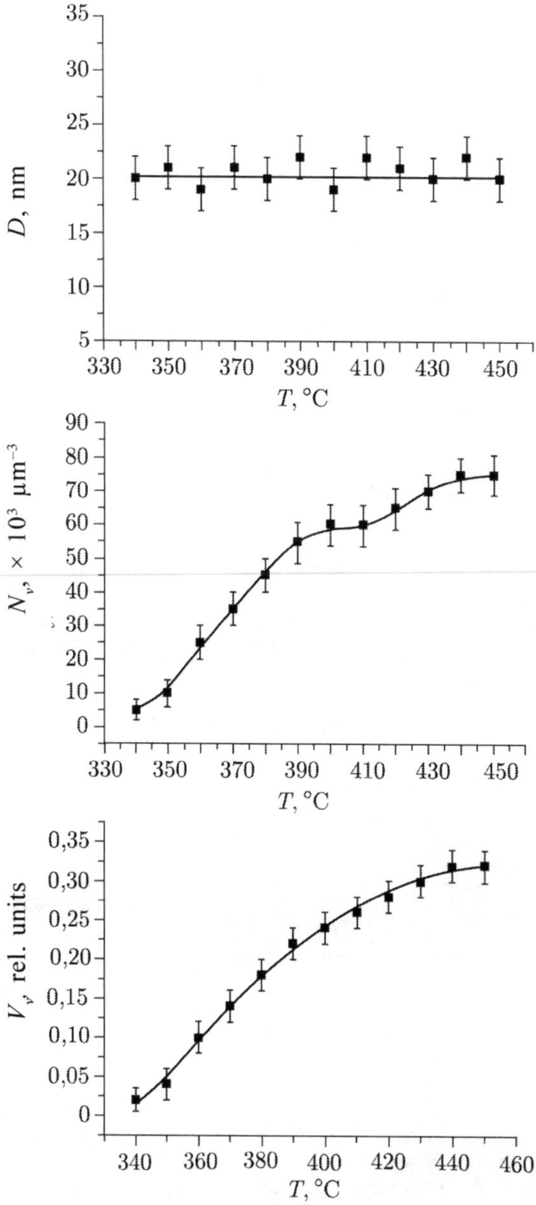

Fig. 3.46. Dependence of the structural parameters on the annealing temperature for an annealing time of 1 h in the $Ni_{44}Fe_{29}Co_{15}B_{10}Si_2$ alloy.

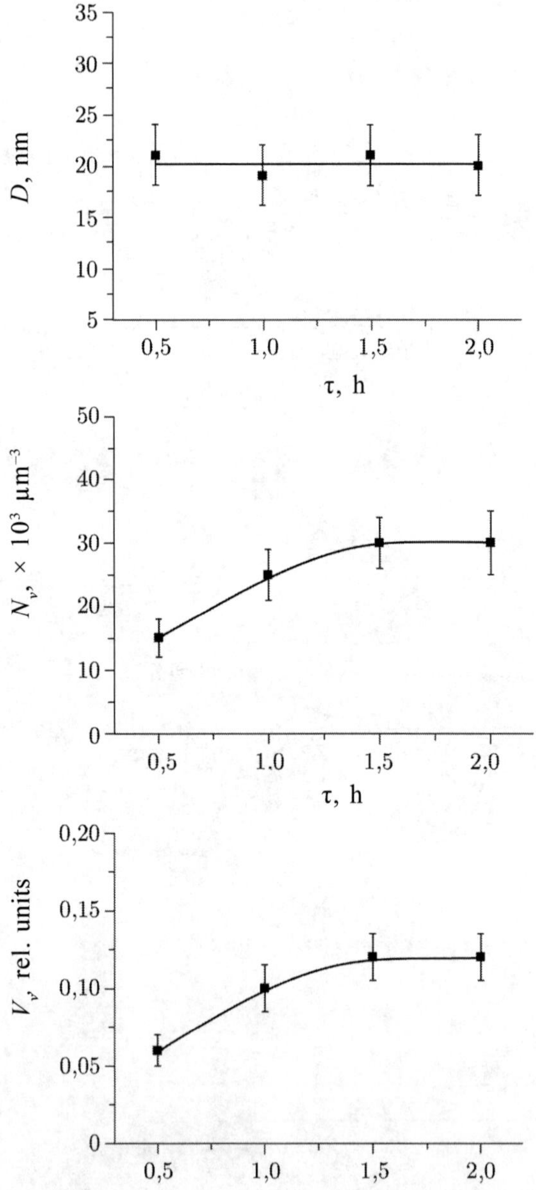

Fig. 3.47. Dependence of structural parameters on the annealing time at an annealing temperature of 360°C in the $Ni_{44}Fe_{29}Co_{15}B_{10}Si_2$ alloy.

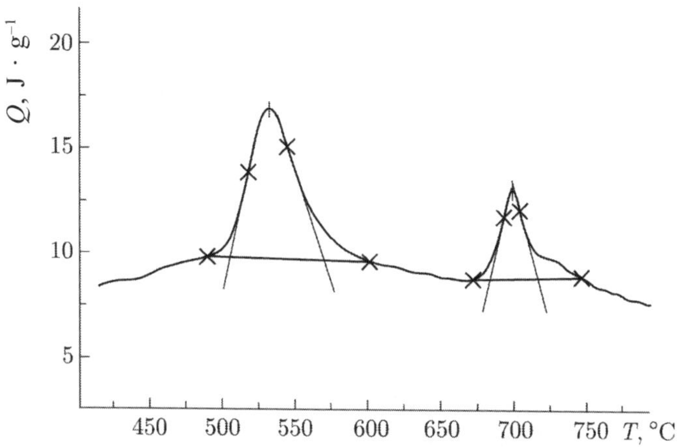

Fig. 3.48. The DSC curve of amorphous $Fe_{73.5}Si_{13.5}B_9Nb_3Cu_1$ alloy.

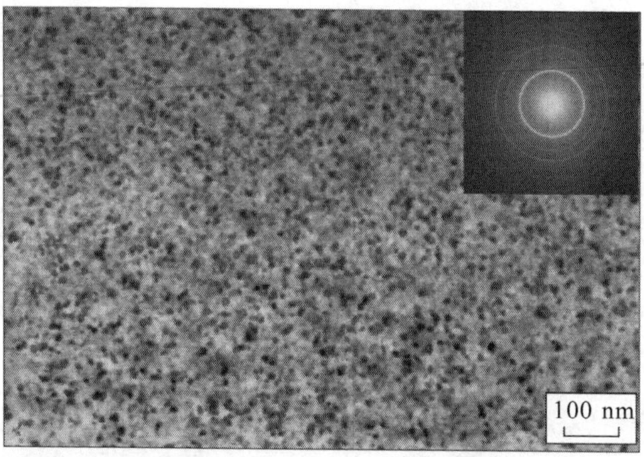

Fig. 3.49. The structure of the $Fe_{73.5}Si_{13.5}B_9Nb_3Cu_1$ alloy after annealing at a temperature of 500°C for 1.5 h. TEM (bright field).

At annealing temperatures $T < 460$°C, the TEM method revealed an amorphous structure [3.101]. When the annealing temperature rises to 460–480°C, depending on the annealing time, primary crystallisation occurs, with the precipitation of equiaxed nanocrystalline particles in the amorphous matrix. Analysis of the electron diffraction patterns shows that these particles have a BCC lattice.

Figure 3.49 is an electron microscopic image of the structure of an amorphous alloy after annealing at a temperature of 500°C for 1.5 h [3.101].

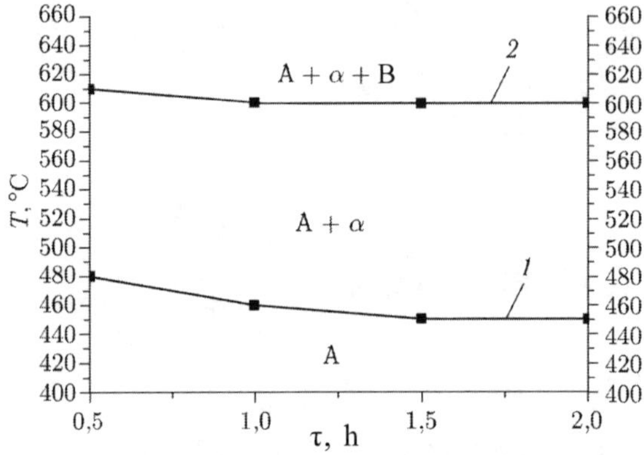

Fig. 3.50. Temperature–time diagram of the $Fe_{73.5}Si_{13.5}B_9Nb_3Cu_1$ alloy: A – amorphous state, B – boride, α – BCC phase (*1* – crystallisation start line, *2* – termination line of primary crystallisation).

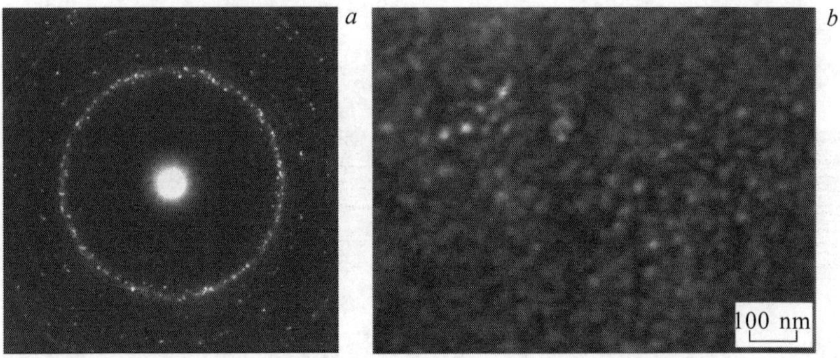

Fig. 3.51. Microdiffraction pattern (*a*) and structure (*b*) of the alloy after annealing at 560°C, 1.5 h. TEM.

The TTD for this alloy is shown in Fig.3.50 (the procedure is similar to the $Fe_{70}Cr_{15}B_{15}$ alloy) [3.92].

The primary crystallisation range is fairly wide and is 130–140°C.

Based on the X-ray diffraction studies, it was established [3.101] that the lattice period of nanocrystalline particles is a = 0.284 nm, which is lower than the lattice period of pure iron (a = 0.287 nm). According to Vegard's law, the decrease in the lattice period of particles indicates the enrichment of the BCC phase with silicon to 16–18% Si. Using the TEM method, it is established that in the diffraction patterns there are superstructural reflections corresponding to a phase ordered by the $D0_3$ type (Fig. 3.51 *a*). Figure 3.51 *b* shows the dark-field image under the action of superstructural reflexes.

The results of the X-ray diffraction and TEM studies showed crystallisation is accompanied by the precipitation of the nanocrystalline particles of the α-BCC Fe(16–18)% Si phase in the matrix with the particles ordered by the $D0_3$ type. This fully confirms the data available in the literature, for example, [3.102].

The kinetics of structural and phase transformations in such alloys is considered in sufficient detail in the annealing temperature range 480°C ÷ 900–950°C [3.102]. The main attention in these papers was given to the range T_{ann} = 480–550°C. It is believed that there are no noticeable structural changes at temperatures below 480°C in the amorphous matrix [3.103]. However, in [3.102], using small-angle X-ray scattering, structural-phase changes were also observed in the low-temperature region.

The known literature data on the kinetics of the transition from the amorphous to the nanocrystalline state using the Johnson–Mehl-Avrami equation differ significantly in studies by different authors both from the values of the effective crystallisation energy E^* and from values of the exponent n, which characterizes the features of transformation by the mechanism of nucleation and growth [3.104, 3.105].

In [3.100], X-ray diffraction studies (including the method of small-angle X-ray scattering) were carried out to study the kinetics of the formation of a nanocrystalline structure in the $Fe_{73.5}Si_{13.5}B_9Cu_1Nb_3$ alloy under isochronous and isothermal (430 and 550°C) annealing. The data on small-angle X-ray scattering show that the transition from the amorphous to the nanocrystalline state under these heat treatment conditions is characterized by the same non-monotonic dependence of the intensity of small-angle scattering on annealing temperature (time). This non-monotonicity indicates that at the initial stages of the phase transformation the contributions of the processes defining the transformation are different. There is an increase in the number of scattering centres of electron density – nanoscale phase formations arising in the amorphous matrix. Although the relaxation of the free volume characteristic of amorphous alloys can contribute to the intensity of small-angle scattering (in particular, with the formation of micropores [1.1]), nevertheless, the obtained dependences can definitely be associated with different ratios of contributions to the transformation of the nucleation and growth stages of the new phase.

According to numerous electron microscopic data, the number and size of nanocrystals in the Finemet alloy increase with the annealing temperature (time), which is consistent with the X-ray data. As follows from the data in [3.106], the values of the Avrami index are significantly lower than the theoretical value $n = 2.5$. In explaining such discrepancies, it is assumed that the presence of copper in the alloy causes an increase in the number of initial samples of frozen-in crystallisation centres formed during the preparation process, followed by their athermal transformation upon heating [3.106].

Thermally activated nucleation can also be superimposed on this process [3.107]. In addition, factors such as the diffusion of Si and B, hampered by the presence of niobium, may contribute to the impeded growth of nanoparticles, which imposes a limitation on the growth of α-Fe (Si) crystals, the presence of local heterogeneities in the chemical composition across the ribbon thickness, which are caused by large temperature gradients during quenching of the melt etc. [3.107, 3.108].

It can be assumed that at low-temperature annealing (430°C) when the diffusion processes are still rather difficult (diffusion coefficient $K_{diff} \approx 7 \cdot 10^{-22}$ m²/s), the main contribution to the transition from the amorphous to the nanocrystalline state is made by the athermal transformation at 'frozen centres'. The increase in the small-angle scattering at this stage is logically related to the increase in the realised 'growth points', and its subsequent decline with a simultaneous increase in the size of nanocrystals from 3.8 to 4.1 nm – with the process of coalescence.

The experimental results indicate the dominance of the growth factor over nucleation during the transition of the Finemet alloy to the nanocrystalline state. The results of calorimetric studies are described more adequately by kinetic processes controlled by the growth of nanoparticles [3.109], rather than by processes involving the stage of nucleation and growth.

The features of the nanocrystallisation processes of the Finemet alloy are attributed by many authors to their original microheterogeneous (nanocluster) structure. A large amount of experimental material has been accumulated concerning the presence of regions with near (topological and/or compositional) order for amorphous alloys, as for melts, and the polycluster model of the structure [3.104] of amorphous alloys adequately describes their structure-sensitive properties. On the basis of experimental data obtained with the help of modern high-resolution techniques [3.105], it is believed that the structure of the alloy consists of Fe–B-enriched regions, as well as clusters containing predominantly Fe and Si atoms [3.100]. Such microheterogeneity of the initial alloy causes the formation of a nanocrystalline magnetically soft α-phase upon heating.

The point of view described in [3.98, 3.106] is very popular, according to which the structure of the Fe–Si–B–Cu–Nb system melt is most adequately described by a microinhomogeneous model, and the structural components in the melt are clusters of iron, silicon and intermetallic iron compounds with niobium and boron. The main component (iron with alloying elements) is characterized by short-range order in the BCC type by the set of structural parameters. The microheterogeneous (nanocluster) state formed during amorphisation of the melt leads to the fact that the process of crystallisation of the amorphous alloy $Fe_{73.5}Si_{13.5}B_9Cu_1Nb_3$ occurs practically without nucleation by athermal transformation of 'frozen-in' crystallisation centres into an amorphous matrix and formation of an amorphous–nanocrystalline structural state at the initial stage of the crystallisation process [3.102].

The generally accepted ideas about the role of alloying elements in the alloys of the Fe–Si–B–Cu–Nb system are as follows. Copper, reducing the activation energy for the formation of the α-Fe (Si) phase, contributes to its earlier and rapid release from the amorphous matrix, which ultimately leads to a reduction in the dimensions of the nanocrystals. Niobium, increasing the activation energy of the formation of borides, shifts their precipitation into a higher-temperature region [3.107], and also niobium inhibits the growth of α-Fe(Si) nanocrystalline particles [3.108].

A structural model for the formation of nanocrystals in Finemet alloys was proposed in [3.109]. During quenching, an amorphous state with copper nanoclusters is formed from the melt. In the early stages of annealing at thee optimum temperature, new copper clusters are formed along with the growth of the already existing ones after quenching. When nanocrystals of α-FeSi are formed, copper nanoclusters with an FCC lattice are places of heterogeneous nucleation. The concept of heterogeneous nucleation of α-FeSi nanocrystals on copper clusters is based on minimizing the migration energy of boundaries [3.110]. Finally, at the growth stage of α-FeSi the copper clusters no longer play the role of potential nucleation sites, growing to a size of about 5 nm.

A great importance in understanding the nature of structural transformations in Finemet-type alloys was played by the work [3.111] performed with the use of atomic probe microscopy. It was found that all five components of the alloy (Fe, Si, B, Cu, Nb) were uniformly distributed in the original amorphous matrix. After nanocrystallisation, a two-phase state was detected: an α-Fe–Si solid solution in the form of a nanocrystalline phase and enriched in B and Nb, and a depleted Si amorphous phase. Secondary crystallisation included the amorphous phase located between the nanoparticles. Since Cu atoms are insoluble in Fe, the removal of Cu from the amorphous matrix should occur at the earliest stages of crystallisation and inhibit the nucleation of α-Fe–Si nanocrystals. In the process of nanocrystallisation at 550°C the following sequence of phenomena [3.111] was observed:

1) the original matrix is completely amorphous;

2) in the initial stages of crystallisation, Cu atoms are collected into clusters of several nanometers in size;

3) α-FeSi nuclei are formed at the sites of the existing Cu clusters;

4) α-FeSi nanocrystals grow, displacing the Nb and B atoms in the surrounding amorphous matrix.

After nanocrystallisation, three phases must exist to create the optimal structure: BCC-solid solution (~20 at.% Si) of Fe–Si, the remaining amorphous phase with ~5 at.% Si and 10–15 at.% Nb and B, and the third phase with ~60% Cu, <5 at.% Si, B and Nb and ~30% Fe. These observations suggest that the formation of Cu clusters stimulates chemical aggregation and the nucleation of the Fe–Si phase. Nb and B, which have a small solubility in Fe, are displaced from the BCC of the Fe–Si nanocrystals. The amorphous matrix enriched with Nb and B surrounds nanocrystals. The structural model of retardation of growing nanocrystals was proposed in [3.112] for alloys with the effect of stabilizing the dimensions of nanocrystals, the realization of which is based on the fulfillment of several conditions for the properties of the components that make up the alloy.

The authors of [3.101] carried out quantitative calculations of the structural parameters (average size D, bulk density N_v and volume fraction V_v) of nanocrystalline phases for the structural states formed at various parameters of the controlled annealing of the Finemet alloy based on the experimental data obtained by the TEM and X-ray diffraction methods [3.101] .

Figures 3.52 and 3.53 shows the dependences of D, N_v, and V_v on the annealing regimes for the given alloy, obtained by the authors of [3.101]. The average size D of the α-phase crystals grows in the nanometer range in the annealing temperature range

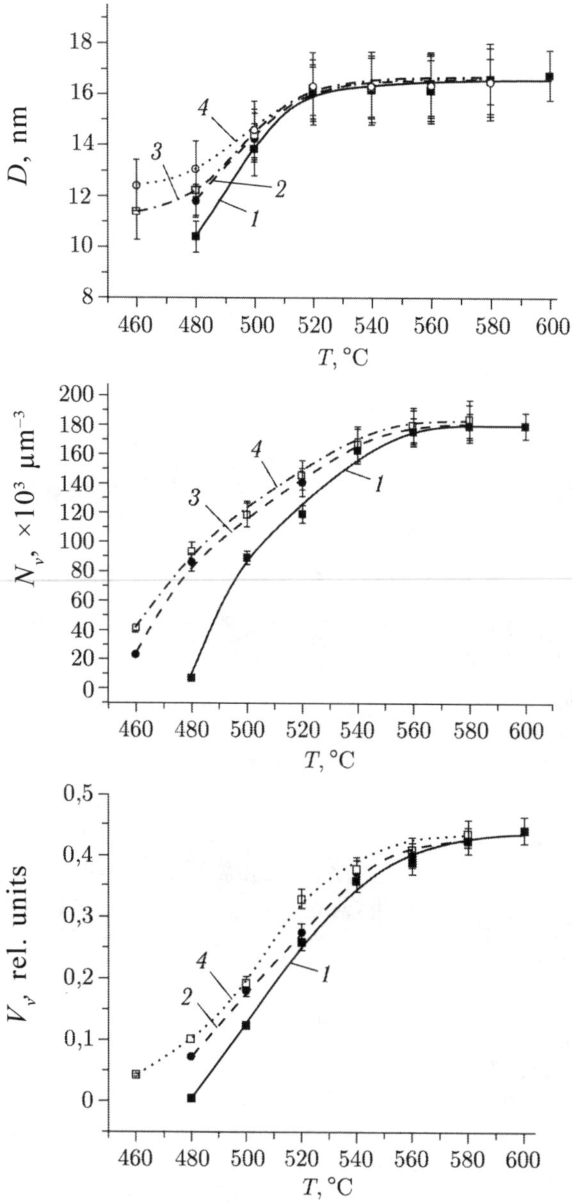

Fig. 3.52. Dependence of structural parameters on annealing temperature T at different annealing times in $Fe_{73.5}Si_{13.5}B_9Nb_3Cu_1$ alloy: *1* – 0.5, *2* – 1, *3* – 1.5; *4* – 2 h [3.101].

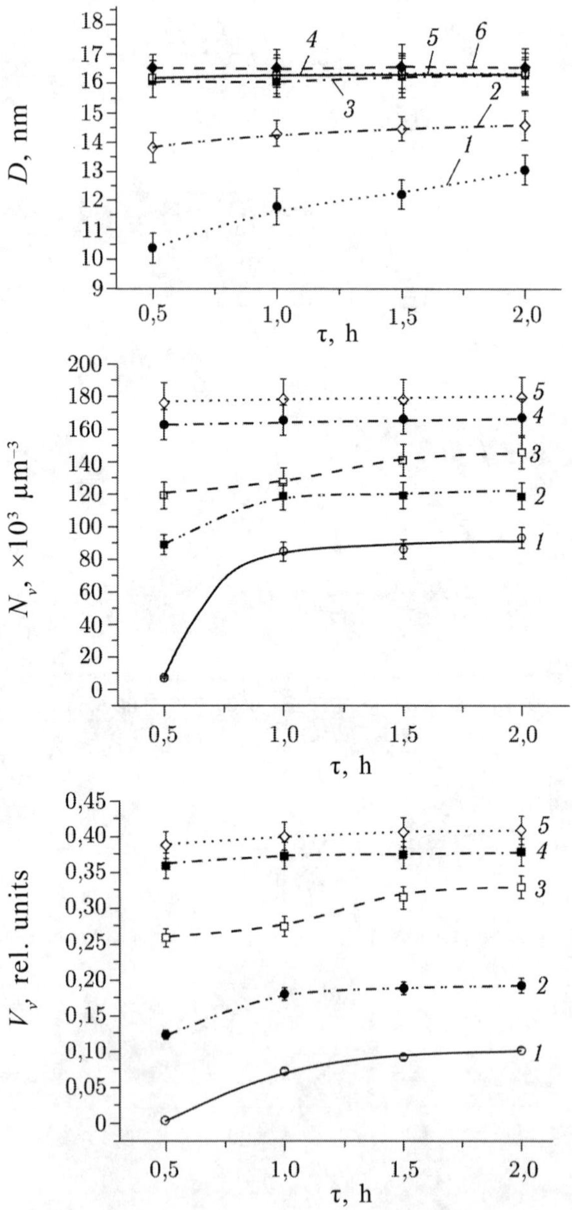

Fig. 3.53. Dependence of structural parameters on annealing time τ at various temperatures in the alloy $Fe_{73.5}Si_{13.5}B_9Nb_3Cu_1$: *1* – 480, *2* – 500, *3* – 520, *4* – 540, *5* – 560, *6* – 580°C [3.101].

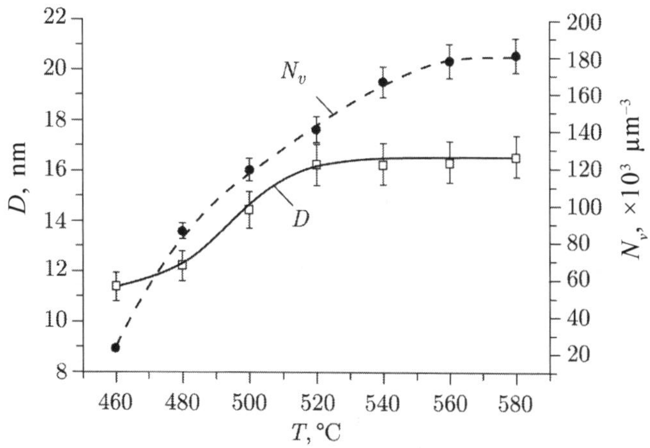

Fig. 3.54. Dependence of the parameters of nanoparticles D and N_v on annealing temperature T (τ = 1.5 h) in the $Fe_{73.5}Si_{13.5}B_9Nb_3Cu_1$ alloy.

460–520°C and then remains practically unchanged at higher annealing temperatures, reaching 17 nm (Fig. 3.52). The dependence of D on τ at various temperatures is shown in Fig. 3.53. The limiting size D of the α-phase nanocrystals is reached at any annealing time above 520°C and increases smoothly with increasing holding time at lower temperatures, but does not reach the limiting value of 17 nm.

The bulk density N_v of nanoparticles increases smoothly with increasing T and the rate of this increase increases with increasing annealing time, but in any case at an annealing temperature of 560°C and higher it reaches a limiting value of $180 \cdot 10^3$ μm^{-3} and does not increase (Fig. 3.52) . In other words, as follows from Fig. 3.54, when annealing below 520°C both D and N_v increase simultaneously in the amorphous matrix of the alloy; in the annealing temperature range 520–560°C the particle size D does not change any more, and the value of N_v continues to grow, and finally, at the annealing temperature of 560°C and above, both these parameters of the nanoparticles are constant, i.e., the two-phase amorphous-nanocrystalline structure stabilizes.

The dependences in Fig. 3.53 show that the kinetics of the determination of the bulk density of nanoparticles in a given alloy strongly depends on the annealing temperature, and at temperatures of 520°C and lower, for a selected annealing time, the limiting value $N_v = 180 \cdot 10^3$ μm^{-3} is not achieved at all.

The effect of stabilizing the dimensions of the crystalline phase
The effect of stabilizing the dimensions of nanocrystals (the Finemet effect), that is, the formation of a very high bulk density of nanocrystals during crystallisation during annealing in th amorphous matrix which in subsequent stages practically do not increase their dimensions, is inherent not only to Fe–Si–B–Nb–Cu alloys, but also a number of other amorphous alloys.

For example, magnetically soft nanocrystalline Fe–M–B–C alloys (M = Zr, Nb, Hf) [2.73] (named Nanoperm) were patented; their composition was selected in such a way as to achieve nanoscale BCC α-Fe particles located in an amorphous matrix. In the Thermoperm alloys [3.113–3.115] nanocrystalline phases α and α′ are formed with the BCC lattice and the superstructure of type *B*2 (FeCo), respectively, with significantly improved high-temperature magnetic properties in comparison with the first two alloys.

In the $Fe_{50}Ni_{33}B_{17}$ and $Ni_{44}Fe_{29}Co_{15}B_{10}Si_2$ alloys growth of the nanocrystalline γ-phase (D = 20 nm) is inhibited [3.94]. Thus, the anomalous stabilization of nanocrystalline phases in the early stages of the formation of the amorphous–nanocrystalline structure is not a characteristic feature of the Finemet alloy and is inherent in a significant number of amorphous alloys.

In [3.116], the Finemet effect was also observed in the study of the thermal stability of the structure in the hypoeutectic Ni–Mo–B alloys as a function of the content of the metalloid (boron) and the refractory component (molybdenum) (Fig. 3.55). The authors found that crystallisation begins with the precipitation of crystals of a solid solution of boron and molybdenum in nickel from an amorphous matrix. As the annealing time increases, the size of the crystalline particles first increases insignificantly, and then practically does not

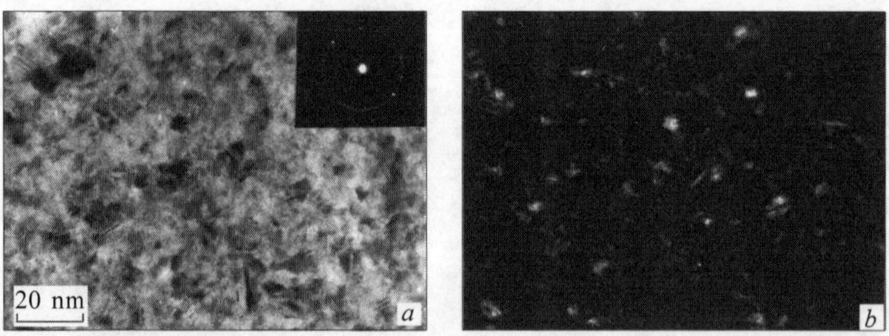

Fig. 3.55. The microstructure of the $(Ni_{70}Mo_{30})_{90}B_{10}$ alloy after annealing at 600°C for 144 h (*a* – bright field, *b* – darkfield image (TEM)).

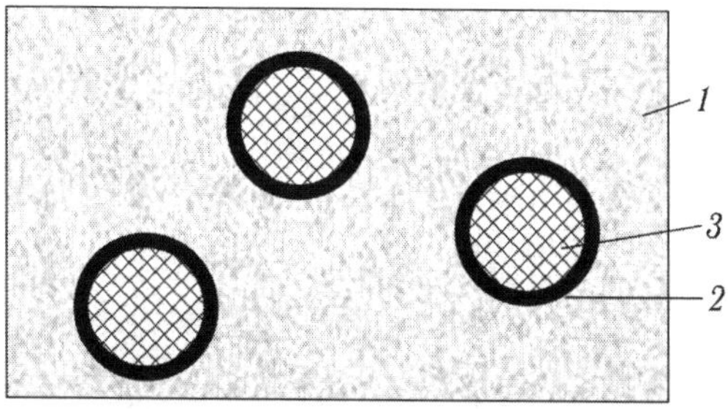

Fig. 3.56. Scheme showing retardation of growth of α-Fe nanoparticles: *1* – amorphous matrix, *2* – layer of amorphous phase enriched with boron, *3* – nanocrystal.

change. In all investigated alloys of this system, the average size of the crystalline particles does not exceed 28 nm.

Several structural models have been put forward to explain this effect. It seems to us that the most physically correct is the structural model 'Crystallization & Stop', proposed in [3.112]. In accordance with this, the retardation of nanocrystals growing in an amorphous matrix is possible if the following three conditions are met.

1. The presence in the amorphous alloy of at least one active alloying element, which increases the crystallisation temperature of the amorphous matrix.

2. It is necessary that the active alloying element dissolves poorly in the lattice of the nanocrystals formed.

3. In the amorphous matrix, conditions must exist for the nucleation of a large number of nanocrystals.

All three conditions were realized by the authors of [3.112] in the massive amorphous Fe–B–Y–Nb–Cu alloy. During crystallisation, α-Fe nanocrystals of 11–13 nm in size were isolated and around them 'barrier' regions enriched with Y and B atoms were formed in the amorphous matrix, which, on the one hand, did not dissolve in nanocrystals and, on the other hand, increased the thermal stability of the amorphous state near the inhibited nanocrystals (Fig. 3.56).

The same situation is formed in the $Fe_{50}Ni_{33}B_{17}$ and $Ni_{44}Fe_{29}Co_{15}B_{10}Si_2$ alloys and the alloys of the Ni–Mo–B system, where boron plays the role of a component that poorly dissolves in the crystal lattice of the growing nanoparticles, and in the Ni–Mo–B system alloys there is additionally present a refractory component (molybdenum). Moreover, an increased concentration of boron atoms in the regions bordering on growing nanocrystals increases the stability of the amorphous Fe–Ni–B matrix of Ni–Fe–Co–Si–B

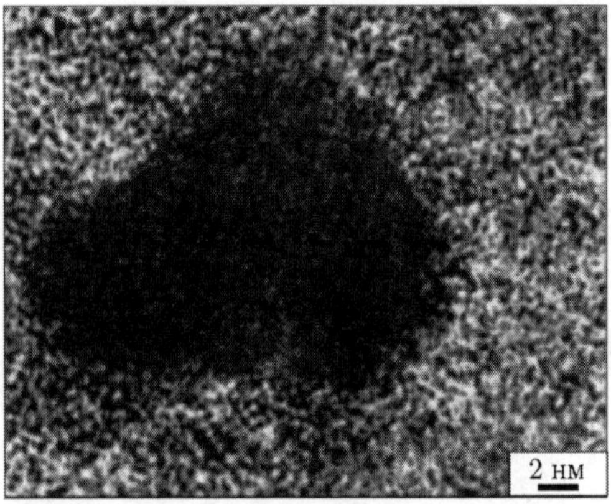

Fig. 3.57. High-resolution electron microscopic image of an aluminum nanocrystal.

and Ni–Mo–B, thus creating a barrier effect. During annealing, the chemical composition of the amorphous matrix changes: increasing the concentration of boron and molybdenum (Ni–Mo–B system) in alloys leads to an increase in the crystallisation temperature of the amorphous matrix.

Structure of nanocrystals in an amorphous matrix

The nanocrystalline structure obtained by the controlled crystallisation of the amorphous phase, as already described above, is in most cases biphased and consists of nanocrystals of one phase and the surrounding amorphous matrix. When analyzing amorphous–nanocrystalline materials, questions arise about their plastic deformation, the possibility of nucleation and propagation of dislocations in them. This requires a study of the structure of nanocrystals.

When studying amorphous–nanocrystalline materials, it was noted that in alloys of different composition, nanocrystals of the same type can be perfect or contain a considerable number of defects.

It was established in work [3.117] that in nanocrystalline aluminum alloys, the nanocrystals emitted from the amorphous matrix are pure aluminum and are perfect, do not contain defects both at the initial stages of crystallisation and at the final stages (Fig. 3.57). The size of nanocrystals does not exceed 50 nm.

In Ni–Mo–B-based alloys in the early stages of nucleation and growth, FCC nanocrystals have a regular shape and are defect-free at sizes less than 5 nm (Fig. 3.58).

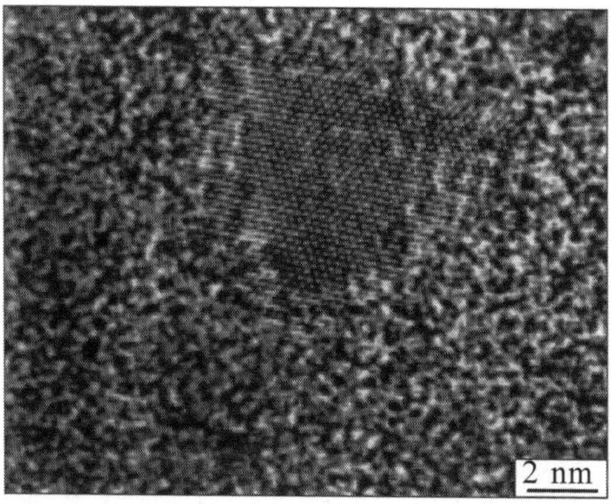

Fig. 3.58. High-resolution image of nanocrystals in an Ni-Mo-B-based alloy at the initial stage of crystallisation.

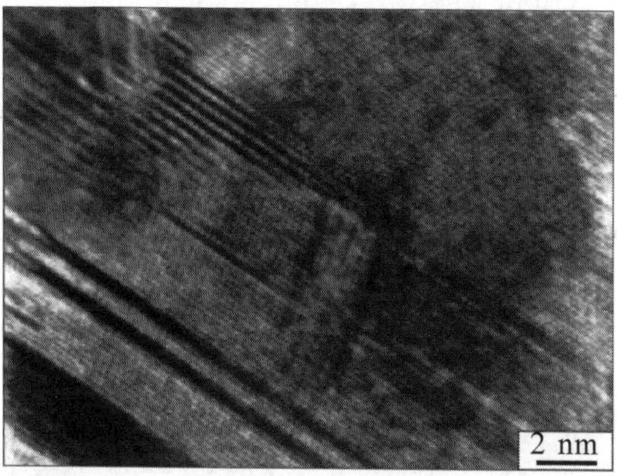

Fig. 3.59. High-resolution image of nanocrystals in the Ni–Mo–B-based alloy after the first stage of crystallisation.

As the nanocrystals grow, the microstructure changes – a considerable number of microtwins and stacking faults are observed, nanocrystals contain numerous twins, defects, etc. (Fig. 3.59).

It was noted in [3.118] that there is a certain critical size of a nanocrystal above which crystals necessarily contain defects. However, the authors of [3.117] established that nanocrystals of the same size in different systems can be both defect-free and contain a considerable number of defects, i.e., the size factor is not a universal quantity. There are other reasons that determine the defectiveness of the structure, for example, the composition of nanocrystals plays an important role in this. On the one hand, a change in composition leads to a reduction in the energy of the defects in the package and facilitates their formation. On the other hand, a certain influence can also be exerted by the uneven distribution of the doping component inside the nanocrystal. Such a non-uniform distribution can lead to a local change in the elastic characteristics and facilitating the nucleation of defects in such places.

3.2.2. Features of the transition of an amorphous state to a crystalline state under deformation effects

Summing up a large number of experimental studies of the structure of materials subjected to megaplastic deformation (MPD), we can state that a complex combination of defect structures containing small-angle and high-angle grain boundaries in different percentages is observed, as well as defective structures within grains of various degrees of perfection. The three-dimensional statistical evaluation of such a grain structure popular in literature gives at best a correlation between high-angle and small-angle boundaries in the structure of the material and little information about the nature of those physical processes that occur directly under MPD. In addition, this information, unfortunately, is quite contradictory and ambiguous, since in various experiments the authors observe, as a rule, different structural states in the same materials under similar deformation conditions. In this case, the true nanostructural state ($d < 0.1$ μm) is not always formed. In steels and alloys MPD is often accompanied by phase transitions (separation and dissolution of phases, martensitic transformation, amorphisation) [3.119, 3.120]. Amorphisation, as a rule, occurs in intermetallics or multicomponent systems.

The most coherent concept of large plastic deformations was proposed by V.V. Rybin [3.121]. On the basis of the concept of

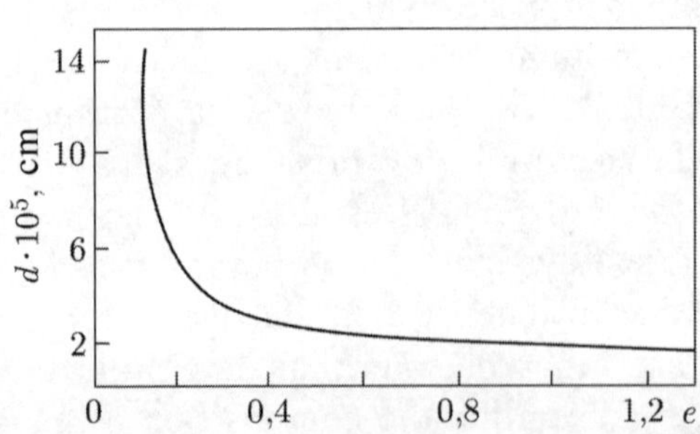

Fig. 3.60. Dependence of the size of grains (fragments) on the magnitude of plastic deformation (V.V. Rybin).

the dominant role of the disclination mode in the implementation of large plastic deformations and the fragmentation processes associated with it, it was possible to correctly describe phenomena occurring at significant degrees of deformation close to $e = 1$. In accordance with the disclination concept, the size of the fragments – main structural elements – gradually decreases with increasing deformation, reaching a constant minimum value of 0.2 μm (Fig. 3.60). This essentially means that the transition to the nanostructured state and the size of fragments (grains) of less than 100 nm (0.1 μm) are hardly possible under the action of the disclination mode. Analysis of numerous experimental data, especially for BCC crystals, led Rybin to the idea of a limiting (critical) fragmented structure, the further evolution of which becomes impossible within the disclination mode [3.121]. Fracture areas are formed along the boundaries of fragments which divide the areas as a rule, free from dislocations. The critical fragmented structure is, in the author's opinion, the ultimate product of plastic deformation, it is incapable of resisting the increasing influence of external and internal stresses and must lead to destruction. It should be noted that the above discussion related essentially to the early stages of MPD ($e \leq 2$) for uniaxial tension or rolling with a relatively small contribution of compressive stresses.

The works of S.A. Firstov et al. [3.122] draw attention to the fact that the transition to a highly deformed state is accompanied by abrupt changes in the structure of the material that occur at a certain critical value of e_c. For the case of technically pure iron $e_c \approx 1$. In addition, the mechanical behaviour of the material changes: the strain hardening instead of the parabolic law obeys the linear law for large deformations.

A detailed systematization of defective structures arising in various materials with increasing degree of plastic deformation was carried out by E.V. Kozlov and N.A. Koneva [3.123].

They showed that when approaching the MPD region, depending on the nature of the material, a successive change of some structural states to other (cellular, strip, fragmented structures, etc.) occurs, similar to structural phase transitions. In this case, internal stresses and conditions for the manifestation of anomalies in the mechanical behaviour of crystals change. Here one should mention the paper [3.124] in which it is shown that with the increase in the degree of MPD, a very large number of excess point defects (mainly vacancies)

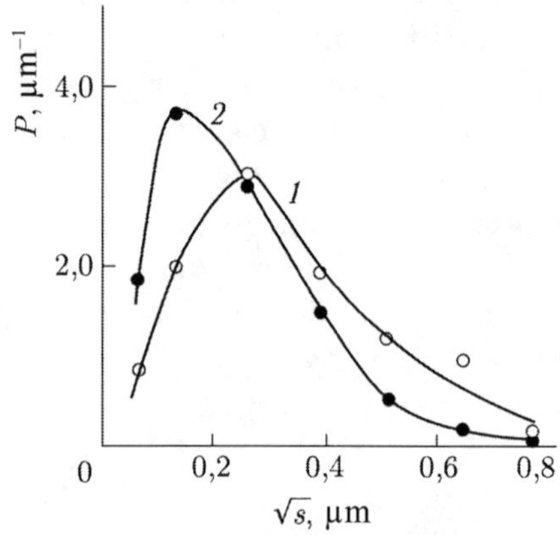

Fig. 3.61. Density of probability P of the formation of fragments with the size S fo smooth (*1*) and notched specimens directly in the crack mouth (*2*): *1* – e = 1.6; *2* – e = 3.7 (V.A. Likhachev et al) [3.126].

is formed in the structure, which can stimulate diffusion phase transitions during deformation.

Extremely widespread was the hypothesis that MPD results in the formation of 'special' non-equilibrium grain boundaries [3.125]. These boundaries, in the opinion of many authors, are responsible for the abnormal phenomena of sliding, diffusion, interaction with lattice defects and, as a result, can be responsible for high values of strength and plasticity.

One should particularly highlight the results of the unjustly forgotten and rarely quoted work of V.A. Likhachev and co-workers [3.126], in which a copper wire was subjected to MPD (e = 1.6 and 3.7). The authors observed a cyclic structure change with increasing deformation: a fragmented structure (d = 0.2 μm) ⇒ a recrystallized structure ⇒ a fragmented structure (d = 0.1 μm), where d is the average fragment size. It is interesting to note that the fragmented structure of the 'second generation' is two times more dispersed than the structure of the 'first generation' (Fig. 3.61). In principle, this implies the achievement of a nanostructural state in the third and more distant cycles.

In fact, to date, there have been only several serious attempts to describe the phenomena occurring under large deformations that are only 'at the threshold' of the MPD. The authors of the above-mentioned studies analyzed the patterns of plastic deformation at $e \leq 1.5$–2.0, and the researchers who conducted experiments in the field of MPD ($e \geq 1.0$–1.5) confined themselves to a descriptive

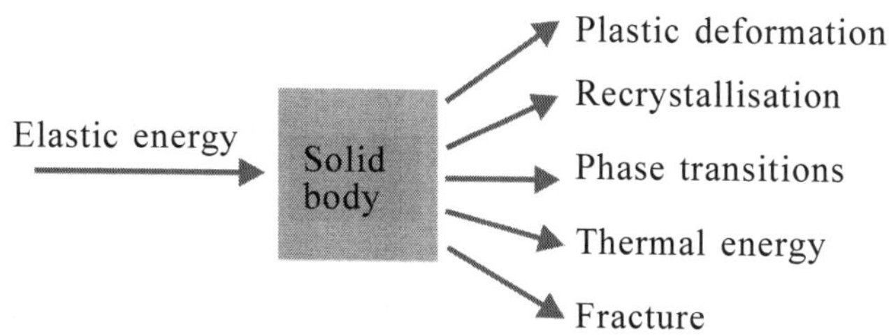

Fig. 3.62. Possible channels for the dissipation of elastic energy when subjected to mechanical action on a solid body.

analysis of the final structures without discussing the mechanism by which the giant plastic deformations are realized. To a certain extent, only the work [3.126] can be considered an exception.

Energy principles of mechanical action on a solid body

A full-fledged theory of the MPD should, from our point of view, be able to unequivocally answer the following questions:

- What structural and phase transformations occur in the MPD process?
- What are the prerequisites for the implementation of the MPD for this or that scenario?
- What are the conditions for the formation of a true nanostructured state at MPD with the crystallite size less than 100 nm, separated by high-angle boundaries or a different phase [3.127]?
- What are the distinctive structural features of the MPD process? What is the difference between MPD and 'normal' plastic deformation?
- What determines the boundary value of the magnitude of the deformation from which we can say that we are in the area of the MPD?

Let us first consider the energy aspects of the behaviour of a solid under load (Fig. 3.62) [3.128]. In mechanical action on a solid body of finite dimensions certain amount of elastic energy is 'pumped' into. The obvious 'dissipation channel' of this energy is plastic deformation. When it is exhausted, another channel can be realized – mechanical destruction. However, at considerable values of elastic energy, plastic deformation can in principle initiate additional 'dissipation channels': dynamic recrystallisation, phase transformations and release of thermal energy.

In the case of MPD, when the component of the compressive stresses is large, the formation and growth of splitting cracks is partially or completely suppressed and, consequently, the realization of the fracture process is substantially complicated. In other words, using equal-channel angular pressing (ECAP), torsion under pressure in a Bridgman chamber (TPBC) or similar loading schemes, we force the solid body to deform without breaking. If the concept of V.V. Rybin [3.121] is correct, then the plastic deformation is effective up to a certain limit, corresponding to the formation of the critical defect structure, and then other physical processes should be the main dissipation channels: dynamic recrystallisation, phase transformations and/or heat generation. Three possible scenarios for the development of further events were proposed in [3.129] (Fig. 3.63). In the case where the processes of dislocation (disclination) rearrangements are facilitated in the material (for example, in pure metals), low-temperature dynamic recrystallisation (the upper branch in Fig. 3.63) occurs after plastic deformation. Local regions of the structure are 'cleared' of defects, and in the new recrystallized grains the process of plastic flow starts again with the help of dislocation and disclination modes. In this case, dynamic recrystallisation acts as a powerful additional channel for the dissipation of elastic energy.

In the case when the mobility of the plastic deformation carriers is relatively low (for example, in solid solutions or intermetallics), a powerful additional channel for the dissipation of elastic energy is the phase transition (the lower branch in Fig. 3.63). Most often this transition is 'crystal ⇒ amorphous state'. As a result, the plastic flow is localized in the amorphous matrix without the effects of strain hardening and the accumulation of large internal stresses. Apparently, there is an intermediate case (the middle branch in Fig. 3.63), when disclinational rearrangements can serve as an additional dissipation channel, which leads to stabilization of the fragmented structure observed in some experiments as the MPD develops.

Obviously, the transition from one scenario of structural changes to another also depends on the parameter (T_{MPD}/T_m), where T_{MPD} is the MPD temperature taking into account the possible effect of heat release and T_m is the melting point.

Low-temperature dynamic recrystallisation

Proposing the first scenario of structural changes in MPD, we thereby *a priori* ascertained that the process of recrystallisation during the MPD can be realized even at room temperature. Since

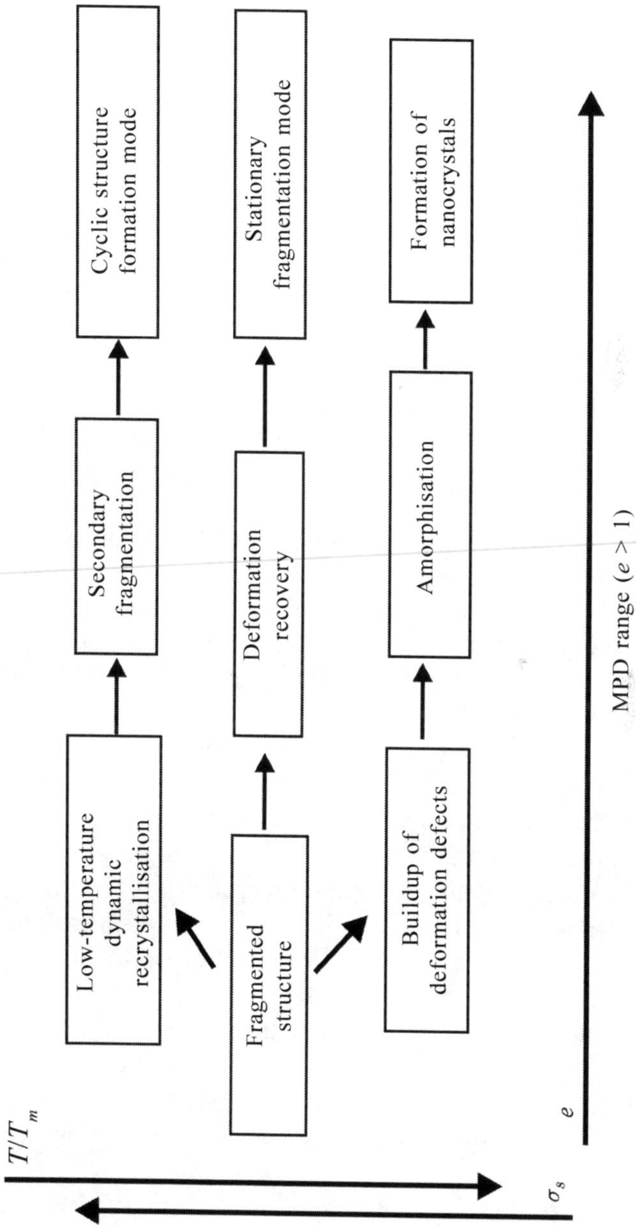

Fig. 3.63. The main scenarios for the development of structural processes in MPD.

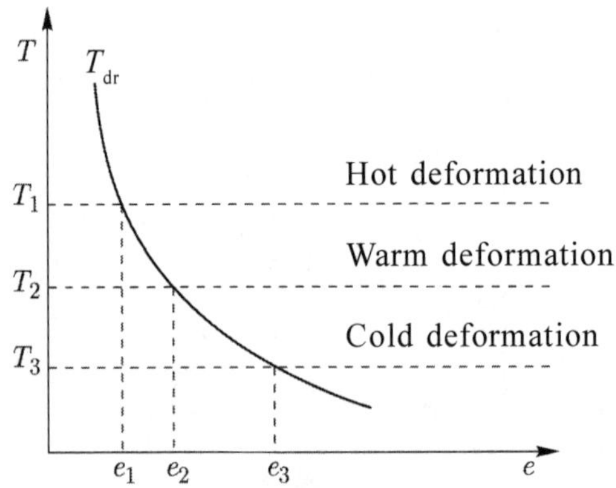

Fig. 3.64. Dependence of the degree of deformation e for the onset of dynamic recrystallisation from temperature; T_{dr} is the temperature of the onset of dynamic recrystallisation.

the process of recrystallisation (including the dynamic one) is purely diffusion [3.130], we thereby assume that the diffusion and self-diffusion processes of the substitutional atoms, necessary for the formation of recrystallisation nuclei and their subsequent growth, can be successfully realized in iron, nickel, and aluminum, titanium and other metals, as well as alloys based on them, for which MPD experiments were carried out. At first glance such a statement looks incorrect. However, a number of considerations confirming the correctness of this hypothesis will be presented.

1. It is known [3.130] that the temperature of the onset of dynamic recrystallisation T_{dr} depends on the degree of deformation at the specified temperature. The higher the degree of deformation, the lower the deformation temperature at which recrystallisation begins (Fig. 3.64). In the region of cold deformation, this dependence is also realized, and there are no physical limitations on extending it to the region of room temperature and temperatures close to it.

In this case, very high degrees of deformation must correspond to dynamic recrystallisation and namely such deformations (MPD) are considered here.

2. Many researchers, considering the possibility of certain diffusion processes under MPD conditions, do not take into account the significant influence of internal stresses on the diffusion fluxes. At the same time, the term 'diffusion under stress' has long been well known [3.131]. Since the processes of plastic deformation are characterized by inhomogeneity, an appreciable role is played

by the gradients of elastic stress, especially manifested in MPD. The resulting gradient of the chemical potential in accordance with the known second Onsager postulate [3.132] should lead to the appearance of diffusion fluxes. It is entirely possible that the diffusion processes are further accelerated by the enormous supersaturation of the material subjected to the MPD by point defects [3.124]. An additional contribution to the acceleration of diffusion can also be associated with the dynamic capture of atoms by ensembles of individually and collectively moving dislocations and disclinations.

3. There are many examples where the acting stresses shift the region of realization of physical processes to the region of lower temperatures. Grain-boundary slip is known to be a process controlled by diffusion and under normal conditions it is observed at high temperatures. Nevertheless, experiments on computer simulation have shown the possibility of slipping along grain boundaries at room temperature under the conditions of large deforming stresses [3.133, 3.134].

As shown in the review [3.135], after MPD of pure copper, the values of the activation energy of a number of diffusion processes are substantially lower than in the usual material. For grain-boundary diffusion 0.64–0.69 eV/at, for Coble creep 0.72 eV/at, for the process of grain growth 0.7 eV/at. The general trend towards a noticeable decrease in temperature at which diffusion processes can be realized under conditions of very large plastic deformations is clearly manifested, as will be shown below, also in MPD of amorphous alloys.

4. Figure 3.65 shows electron-microscopic images of the structure of pure iron processed at room temperature by the TPBC method (4 full turns, $e = 5.6$).

Against the backdrop of a matrix with a high defect density, small regions (100–200 nm) that are completely free from dislocations and appear to be, apparently, embryos of recrystallisation. Obtaining such images is certainly a rare even, since the MPD process was stopped as soon as the recrystallisation embryos only appeared and did not have time to grow and/or 'acquire' dislocations as a result of the ongoing MPD process. The investigations carried out have shown [3.136] that the transition from a pure metal to a solid solution based on it makes the dynamic recrystallisation difficult. The martensitic transformation initiated by deformation is affected in a similar way.

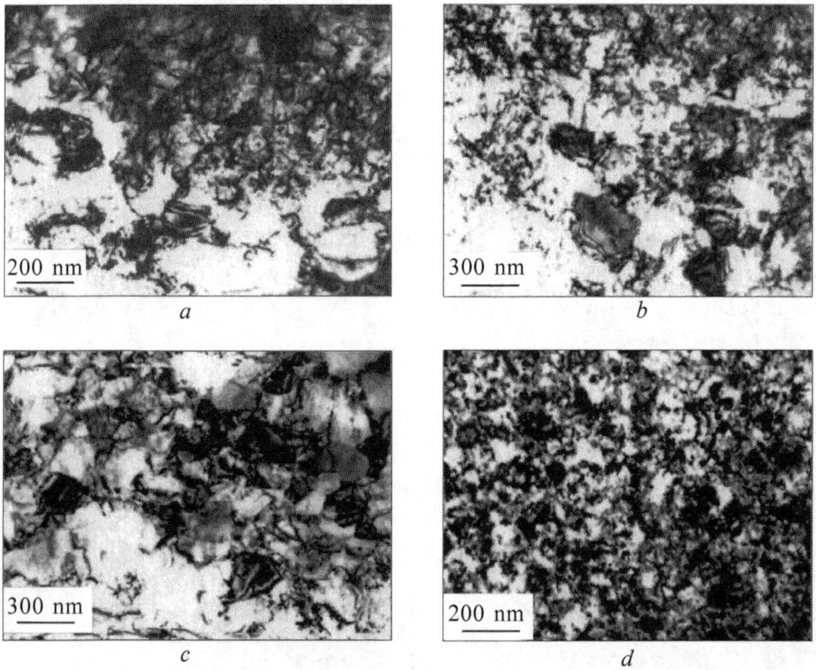

Fig. 3.65. Electron micrographs of early stages of dynamic recrystallisation in pure iron (*a–c*) and nanoparticles of strain martensite in the Fe–32% Ni alloys (*d*) after four complete rotation at room temperature by the TPBC method. TEM, bright field.

It also leads to a noticeable dispersion of the structure, transforming it into a nanocrystalline state (Fig. 3.65 *c*).

Mechanisms for the formation of a recrystallisation nucleus and its subsequent growth in the MPD process may coincide with those mechanisms that are already known for 'ordinary' dynamic recrystallisation at high temperatures, but may be substantially different and possible only at low temperatures under MPD conditions.

5. The paper [3.126] convincingly demonstrates the phenomenon of dynamic recrystallisation in the process of MPD of pure copper at room temperature. Although the authors of the work created the conditions for the MPD in a very original way (local deformation in the zone of a growing crack), this does not in the least reduce the significance of their result, especially since they discovered the phenomenon of secondary fragmentation.

Thus, it can be considered completely established and theoretically justified that in the MPD or pure metals (Fe, Al, Cu, etc.) and solid solutions on their basis, the process of dynamic recrystallisation is observed, which is a powerful additional channel for dissipating elastic energy introduced into a solid body in the MPD process.

The principle of cyclicity in MPD

The classical notions of plastic deformation are based on the fact that an increase in the degree of deformation leads to accumulation of dislocation defects. The higher the degree of plastic deformation, the more defects the deformable crystal must contain. The first exception to this rule arose when large plastic deformations were made with the active participation of disclination modes: the fragments had fine boundaries and were almost completely free from dislocations. However, when moving to the MPD area, as we have seen, there are cardinal structural rearrangements due to additional channels of dissipation of elastic energy. The abrupt change in the structure and properties during the transition to MPD was reported by the authors of the paper [3.122]. If we consider a specific microvolume of the deformed sample, then, following dynamic recrystallisation or amorphisation, the process of plastic deformation begins 'from scratch' in the newly formed recrystallized grain or in the region of the amorphous phase. Further, in the microvolume under consideration under the action of deforming stresses, defect accumulation takes place again, and the process is repeated. A similar cyclicity for MPD was directly observed by the authors of [3.126].

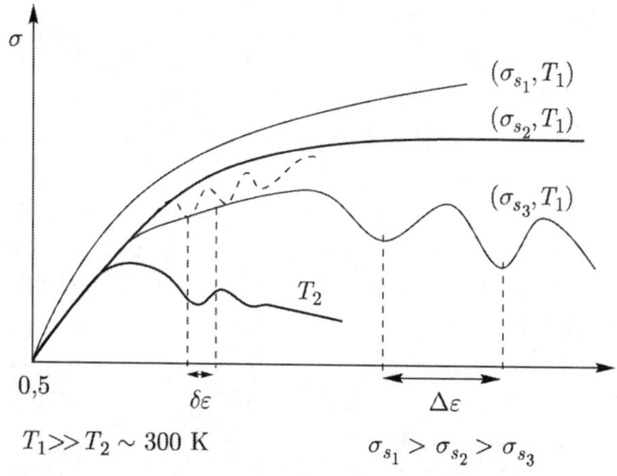

Fig. 3.66. Possible types of strain curves in the MPD range for materials with different Peierls barriers $\sigma_{s1} > \sigma_{s2} > \sigma_{s3}$ and at temperatures $T_2 \gg T_1 = 300$ K.

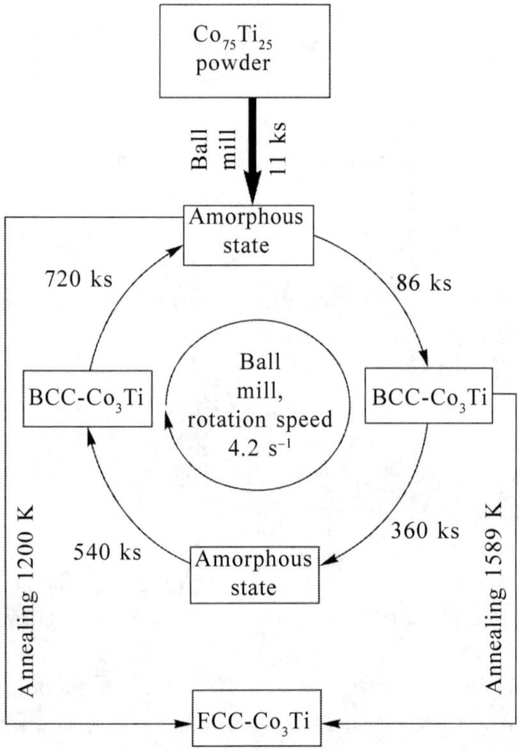

Fig. 3.67. The cyclical nature of phase transitions is the amorphous state ⇒ BCC crystals during mechanoactivation of Co_3Ti alloy.

Figure 3.66 shows the calculated curves of plastic flow, taking into account the existence of a dissipation channel, obtained for materials with different degrees of mobility of dislocations and at different temperatures [3.137]. It is seen that at low values of the Peierls barrier σ_s and in the presence of an effective dissipation channel the curve of plastic flow has a cyclic character with a 'wavelength' $\Delta\varepsilon$. The fact that we practically never record such flow curves in the experiment does not contradict this consideration. The process of plastic flow at all stages of its development proceeds, as is well known, extremely heterogeneously, and the various regions of the deformed crystal are at various stages of their evolution.

The cyclicity of the 'crystal–amorphous state' transition was most effective in the case of mechanoactivation processes very close to MPD [3.138] (Fig. 3.67).

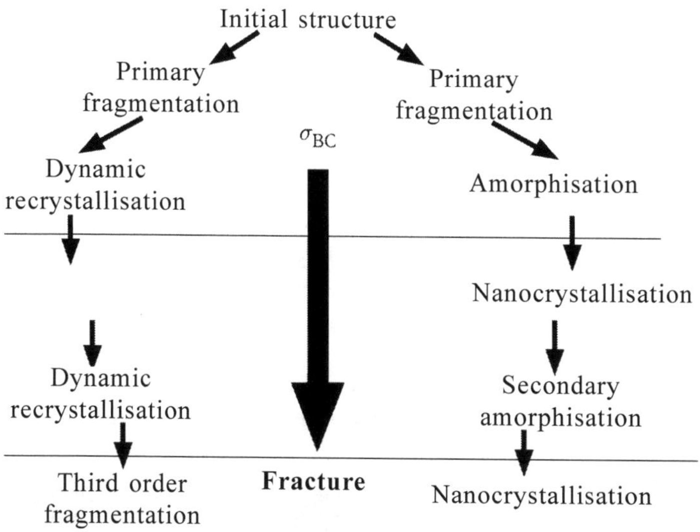

Fig. 3.68. General scheme of possible structural transformations in the MPD process.

When the $Co_{75}Ti_{25}$ intermetallide powder was processed in a ball mill with a processing time of up to 720 ks, radiographic examination revealed the cyclic phase transitions BCC–$Co_{75}Ti_{25}$ ⇒ amorphous state ⇒ BCC–$Co_{75}Ti_{25}$ ⇒ amorphous state ⇒ BCC–$Co_{75}Ti_{25}$ ⇒ amorphous state, etc.

Figure 3.68 shows a general scheme of the flow of structural processes that demonstrates the principle of cyclicity at MPD [3.137]. The process of destruction is compensated by the stresses of all-round compression and is derived from the analysis. Two branches of structural transformations at MPD correspond to the flow of either dynamic recrystallisation or amorphisation of alloys. The scheme is obviously simplistic and does not take into account a number of additional conditions that can complicate the overall picture. But the principle of cyclicity is, in our opinion, fundamental in considering the MPD.

We can now give exhaustive answers to all the questions posed at the beginning of this section [3.139].

- In the process of MPD, additional channels of elastic energy dissipation must be effectively implemented, in addition to plastic deformation. Structural changes in MPD are characterized by a certain cyclicity.
- The specific route of structural rearrangements for MPD is determined by a number of factors: temperature, the Peierls barrier of dislocations and their ability to diffusive

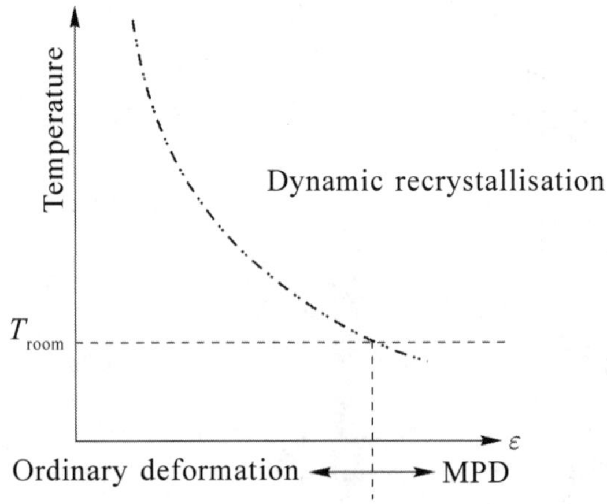

Fig. 3.69. A diagram explaining the position of the boundary region of deformations that separates the macroscopic and megascopic deformations.

rearrangements, free energies of crystalline and amorphous states.

- The flow of MPD does not at all guarantee the formation of a nanocrystalline state with a crystallite size of less than 100 nm, separated by high-angle or interphase boundaries. For example, in pure metals with high dislocation mobility, this is practically excluded. An important factor in the formation of nanostructures in MPD is the occurrence of phase transformations of the martensitic and diffusion type, as well as the transition to the amorphous state. By stimulating phase transformations by varying the temperature and chemical composition of the materials, we are able to obtain nanostructures of various types.
- A distinctive feature of the MPD is the existence of additional effective channels for the dissipation of elastic energy. Such channels, in our opinion, are four (if we exclude the processes of mechanical destruction): dynamic recrystallisation, disclinational rearrangements, phase transformations (including transition to an amorphous state), and the release of latent heat of deformation origin. Under ordinary (macroplastic) deformations, elastic energy accumulates, and only at the MPD stage are powerful dissipative processes included.
- It is possible to define very precisely the boundary deformation region where the macroscopic deformation becomes megaplastic. Figure 3.69 shows a slightly modernized

dependence of the temperature of the onset of dynamic recrystallisation on the strain at a given temperature, shown in Fig. 3.63 [3.140]. Suppose that we carry out a deformation at room temperature (in principle, it can be any one satisfying the relation $T_d/T_m < 0.4$). At deformation degrees below the boundary value ε_b dynamic recrystallisation does not occur, and we are in the region of macrodeformation. For values $\varepsilon > \varepsilon_b$, the process of plastic deformation begins to include dynamic recrystallisation and we pass to the MPD region.

Thus, the boundary of the MPD realization in the case of the action of one of the powerful dissipation channels is clearly defined. In the case of another channel (amorphisation), the appearance of the amorphous phase in the structure of microregions may serve as a sign of the transition to the MPD region. If both the aforementioned dissipation channels act simultaneously (a relatively rare case), then the deformatyion boundary value corresponds to the smaller of them.

Let us now try to give a rigorous definition of the MPD.

Megaplastic (severe) deformation is a process of plastic flow at a temperature $T_d < 0.4\ T_m$, satisfying the following two conditions.

1. In the stress state diagram of a deformable solid, there is an essential component of compressive stresses, preventing mechanical failure.

2. The magnitude of the plastic deformation is so great that the plastic flow is accompanied by cyclic processes of dynamic recrystallisation and/or amorphisation of the structure, which proceed at the same temperatures, taking into account the effects of the release of latent heat.

In concluding this section, a few brief remarks [3.139]:

1. Within the framework of the MPD model under consideration, there is no need to involve concepts of 'special' strongly non-equilibrium grain boundaries. Although, of course, the boundaries formed during dynamic recrystallisation are far from perfect, but should have the same properties as any other grain boundaries.

2. The deformation behaviour of a material under MPD conditions is inherently very close to the behaviour of the material in superplasticity. A similar analogy can prove to be productive for clarifying the nature of superplasticity.

3. MPD is a phenomenon that occurs only in the late stages of deformation; for its implementation, any stress state scheme (for example, conventional rolling) can be used provided that high hydrostatic stresses are created.

The phenomenon of nanocrystallisation

An additional channel of dissipation in intermetallics in other materials with low mobility of dislocations may be amorphisation. The most typical example is titanium nickelide, for which the transition to the amorphous state is observed after the TPBC [3.119] and after cold rolling [3.141]. The transition to an amorphous state at MPD is most pronounced in those alloys that are prone to amorphisation during superfast quenching from a melt. Apparently, a crystal containing a very high concentration of linear and point defects is thermodynamically unstable to transition to the amorphous state, especially if the difference between the free energies of the crystalline and amorphous states is small.

What will happen if the amorphous state obtained, for example, by quenching from a melt or some other way, is deformed under MPD conditions? Based on the foregoing, the amorphous state must remain amorphous. However, as it turned out [3.142–3.144], deformation with the help of the TPBC leads to nanocrystallisation: nanocrystals with a size of about 10–20 nm are homogeneously or heterogeneously located in the amorphous matrix.

1. Amorphous alloys of the metal–metalloid type

The appearance at room temperature of nanocrystals up to 20 nm in size, uniformly distributed throughout the volume of the amorphous matrix, is difficult to explain within the framework of the classical concepts of the thermally activated nature of crystallisation processes. In work [3.145] an attempt was made to analyze in detail the features of the structure and properties under the influence of MPD on a series of amorphous alloys such as metal–metalloid. The $Ni_{44}Fe_{29}Co_{15}Si_2B_{10}$ alloy, obtained by melt quenching, was mainly studied. But in addition, in [3.146], $Fe_{74}Si_{13}B_9Nb_3Cu_1$ (Finemet), $Fe5_{7.5}Ni_{25}B_{17.5}$, $Fe_{49.5}Ni_{33}B_{17.5}$ and $Fe_{70}Cr_{15}B_{15}$ were studied. Figure 3.70 shows the X-ray diffraction patterns of the amorphous Ni–Fe–Co–Si–B alloy in the initial state (*a*), after $N = 4$ at 293 and 77 K (*b*) and after $N = 8$ at the same deformation temperatures (*c*) (N is the number of complete turns in the Bridgman chamber). It can be seen that after the MPD crystallisation processes began in the alloy, expressed more noticeably after deformation at room temperature. A special computer program [3.146] allowed to determine the volume fraction and the size of the crystalline phase in the case of the X-ray patterns shown in Fig. 3.70. For example, for $N = 4$ and $T = 293$ K, the fraction of the crystalline α-phase and the average

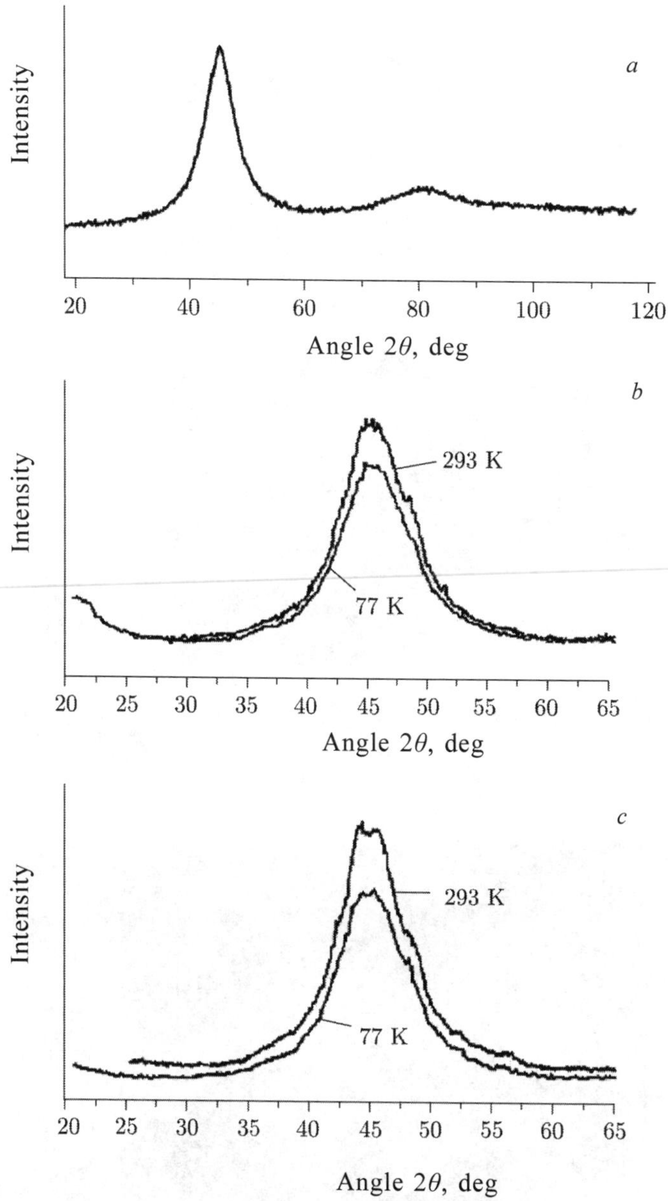

Fig. 3.70. X-ray diffraction patterns of the amorphous alloy $Ni_{44}Fe_{29}Co_{15}Si_2B_{10}$ in the initial state after quenching from the melt (*a*), after $N = 4$ at 293 and 77 K (*b*) and after $N = 8$ at the same deformation temperatures (*c*).

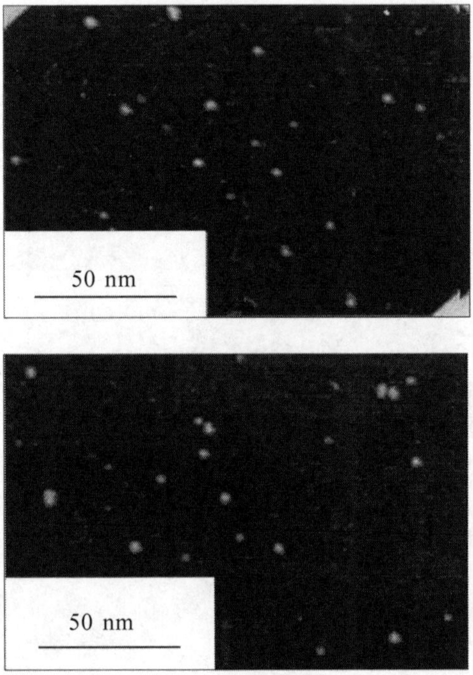

Fig. 3.71. The dark-field images of the $Ni_{44}Fe_{29}Co_{15}Si_2B_{10}$ amorphous alloy after $N = 4$ at 293 K (*a*) and after $N = 8$ at 77 K (*b*).

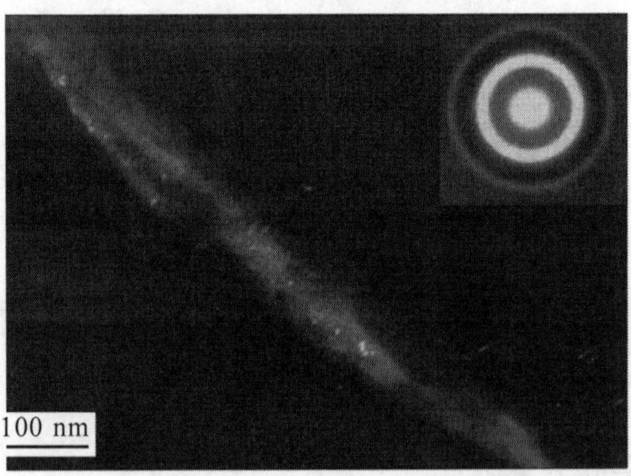

Fig. 3.72. Electron microscopic image of a crystallized local shear band in amorphous $Ni_{44}Fe_{29}Co_{15}Si_2B_{10}$ alloy after MPD ($N = 0.5$; $T = 293$ K).

crystallite size D are 8% and 3 nm, respectively. It is interesting to note that the values of these parameters almost completely correspond to those observed after $N = 8$, but at $T = 77$ K ($\alpha = 6\%$ and $D = 2$ nm). Electron microscopic observations (Fig. 3.71) confirmed these results qualitatively and quantitatively.

In the region of a sharp drop in HV (see section 3.3.4), inhomogeneous plastic deformation is observed with the formation of coarse shear bands, which is inherent in all amorphous alloys at temperatures well below the transition point to the crystalline state [1.107]. The local shear bands can be observed by transmission electron microscopy on deformed samples due to the fact that crystallisation effects are observed on them (Fig. 3.72).

Otherwise, the contrast on the electron microscopic image can only be of an absorbing nature, and it is necessary to first prepare a sample for electron microscopic studies, and then to deform it [1.107]. Theoretical estimates show that the local temperature increase in the shear bands can reach 500°C [1.1]. In this case, the local temperature in the plastic shear zone may exceed the crystallisation temperature of the amorphous alloy (in our case, $T_{cr} = 410$°C) and lead to the formation of a primary FCC phase in the shear bands.

In the later stages of MPD ($N \geq 1.0$), the picture of the deformation changes radically. Shear bands are not observed. Instead, we observe nanoparticles of a crystalline phase up to 10 nm in size, homogeneously distributed throughout the sample (Fig. 3.71). It can be stated that the process of plastic deformation of the amorphous alloy ceased to be highly localized, is inhomogeneous and, most likely, transformed into a 'quasi-homogeneous' alloy. Such a plastic flow pattern is inherent in amorphous alloys at very high temperatures close to the glass transition point, under the conditions of a sharp decrease in the dynamic viscosity of metallic glass [1.107].

In this case, to achieve such a 'softened' state at room temperature, and even more so at 77 K, is hardly possible. We are apparently confronted with the manifestation of a fundamentally new structural mechanism of plastic deformation of amorphous alloys, which manifests itself only under MPD conditions.

One of the possible explanations for this development is the following (Fig. 3.73). As the shear band propagates in the amorphous matrix during the MPD, its temperature rises constantly, with the temperature at its front always being maximum. There comes such a phase of the distribution of the band when the local temperature

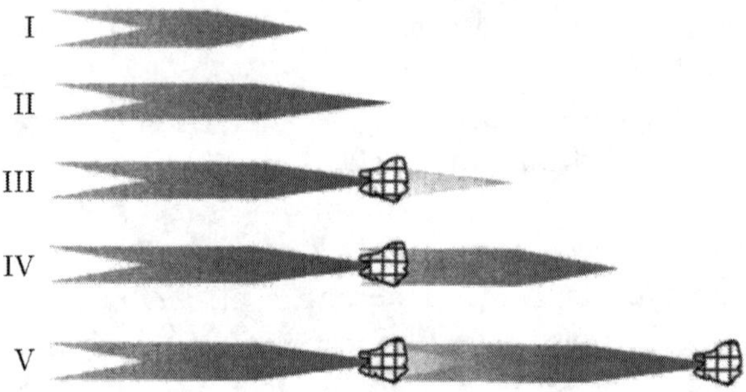

Fig. 3.73. The mechanism of 'self-blocking' of the shear band propagating in an amorphous matrix.

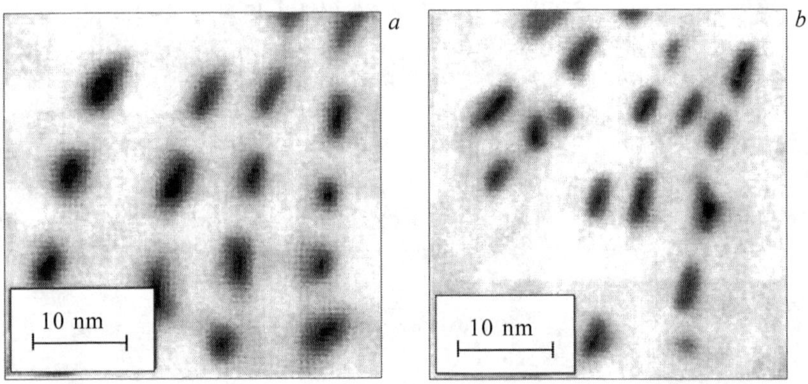

Fig. 3.74. Chains of equidistant nanocrystals arising at MPD of an amorphous Fe–Ni–B alloy. Transmission electron microscopy.

on the front reaches the crystallisation temperature, (II in Fig. 3.73), and a nanocrystal appears at the front of the growing band, sharply inhibiting the plastic flow zone since the resulting crystal has nanoscale dimensions and is incapable of the dislocation plastic flow. There are two options for further developments. Firstly, under the effect of the shear band the nanocrystal accumulates a high level of residual stresses so that a new shear band forms in the amorphous matrix by the mechanism of elastic accommodation (III in Fig. 3.73). In this case, the plastic flow process takes place by the relay mechanism generating nanocrystals in the shear band equidistantly distributed along the trajectory of movement of the shear band in the amorphous matrix.

This is confirmed by the electron micrograph in Fig. 3.74, where the chains of equidistantly located nanocrystals actually appear as a result of MPD. Secondly, the branching of shear bands, which are

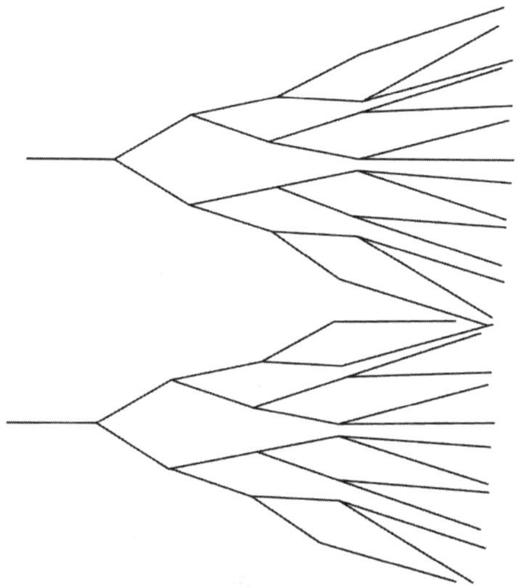

Fig. 3.75. The mechanism of multiplication of shear bands interacting with frontal nanocrystals.

inhibited due to frontal formation of nanocrystals, is possible. This process, somewhat reminiscent of the multiplication of dislocations on non-interruptible particles, is schematically shown in Fig. 3.75. As a result of this 'self-retardation' of the shear bands on the frontal nanocrystals, there is a delocalization of the inhomogeneous plastic flow at the later stages of the MPD. The observed effect of the transition to homogeneous nanocrystallisation on shear bands in fact means that the plastic flow is characterized by a high bulk density of shear bands and, as a consequence, a homogeneous character of the nanocrystal precipitation in 'thinner' shear bands.

Thus, nanocrystallisation is a consequence of local heat generation due to plastic deformation processes. The thermally activated nature of nanocrystallisation and, consequently, its appearance as a result of a local temperature increase is also evidenced by the fact that the structural state and, correspondingly, the microhardness value after $N = 4$ at room temperature exactly correspond to the structural state and the microhardness value after $N = 8$ under deformation at 77 K [3.145]. In other words, higher values of deformation compensate for the temperature deficit during the diffusion processes of nanocrystallisation.

The study of the MPD of a partially crystallized alloy brought unexpected results (Fig. 3.76). It is easy to see from the presented X-ray diffraction pattern that the partially crystalline state formed

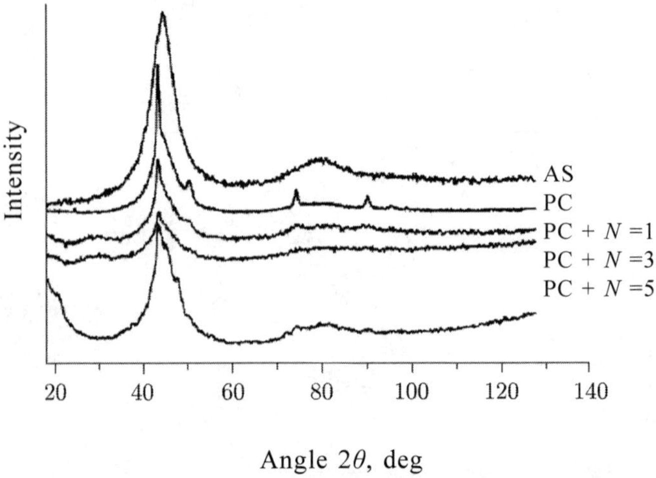

Fig. 3.76. Evolution of X-ray diffraction patterns of the partially crystallized (PC) state of $Ni_{44}Fe_{29}Co_{15}Si_2B_{10}$ alloy after MPD with different values of N (T = 293 K).

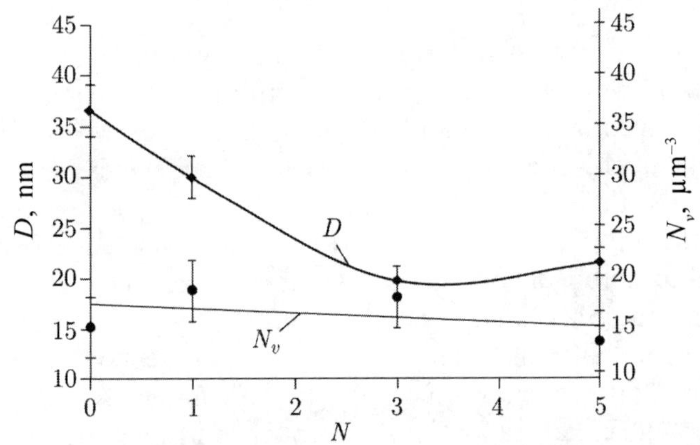

Fig. 3.77. Dependence of the average size (D) and the bulk density (N_v) of the nanoparticles of the crystalline phase in the original partially crystallized $Ni_{44}Fe_29Co_{15}Si_2B_{10}$ alloy on the N value at MPD (T = 293 K).

after annealing the amorphous alloy again becomes amorphous as the deformation magnitude increases to N = 5. Electron microscopic experiments unequivocally confirmed this trend: a sharp decrease in the size of nanocrystals occurs while their bulk density is retained. Such an unusual evolution of the amorphous–nanocrystalline structure with increasing N is clearly demonstrated by the dependence of the average size of nanoparticles and their bulk density with increasing N in a partially crystallized alloy (Fig. 3.77). With

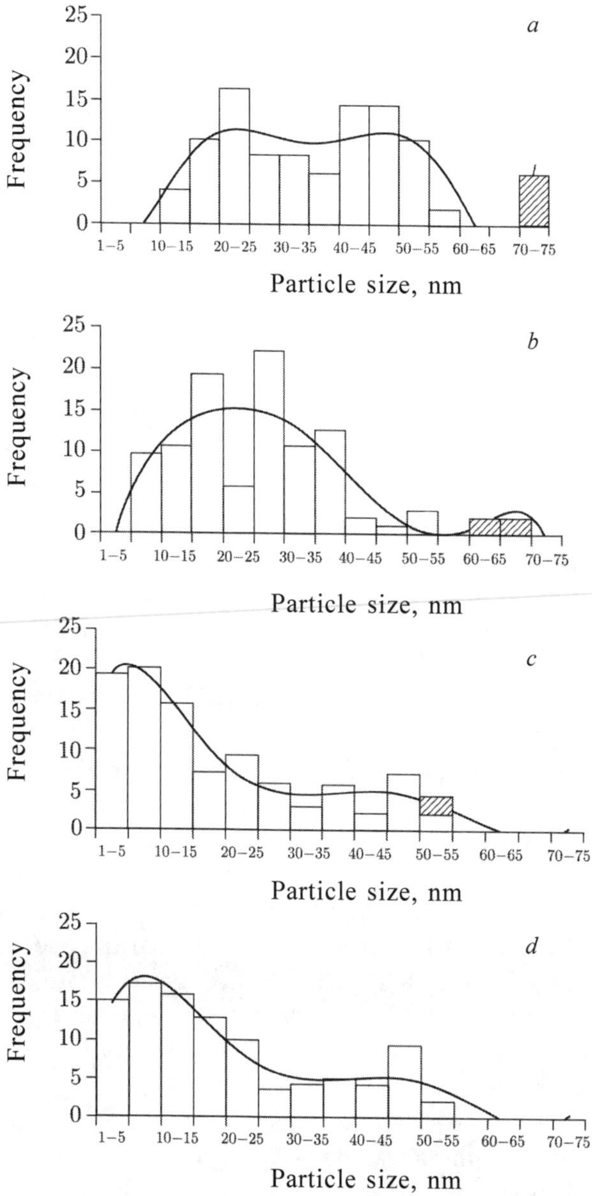

Fig. 3.78. Histograms of the size distribution of nanocrystals observed at different stages of the MPD of a partially crystallized $Ni_{44}Fe_{29}Co_{15}Si_2B_{10}$ alloy; $N = 0$ (*a*), 1 (*b*), 3 (*c*), 5 (*d*); aize fractions which disappear as the deformation increases are shaded.

practically unchanged amounts of nanoparticles per unit volume, the disappearance of the crystalline phase occurs due to a significant decrease in the size of the nanoparticles. In other words, due to their 'dissolution' in an amorphous matrix.

Even more clearly, the effect of 'dissolution' of nanoparticles in MPD can be recorded by comparing histograms of the distribution of nanoparticles of the crystalline phase obtained after different MPD regimes (Fig. 3.78). It is seen that as the *N* grows, the 'tail' corresponding to the largest nanoparticles disappears from each next histogram (shaded on the histograms shown in Fig. 3.78).

So, at first sight, obviously contradictory results were obtained. On the one hand, the MPD of the amorphous alloy $Ni_{44}Fe_{29}Co_{15}Si_2B_{10}$ leads to its partial transition to the crystalline (more precisely, nanocrystalline) state. On the other hand, the MPD of the same partially crystallized alloy leads to dissolution of the crystalline phase, i.e., a tendency to return to the original amorphous state is observed. The contradiction discovered is apparent and can be logically explained taking into account the specifics of the structural processes taking place at MPD.

In the MPD process, we introduce a significant elastic energy into the solid. As possible channels of dissipation in this case it is necessary to analyze plastic deformation, phase transformations and heat release. The reason for the crystallisation is both the local increase in temperature and the presence of significant local stresses in the amorphous matrix. The stresses stimulate the flow of processes that depend on temperature, and, the higher the stress, the lower the temperature of the thermally activated crystallisation process. In addition, one must take into account the fact that the value of the activation energy of the crystallisation process Q^* is lower than usual due to a significantly higher concentration of regions of excess free volume in the shear bands [3.147].

Finally, one more important detail must be taken into account: the local atomic structure of the amorphous matrix in the shear band can differ from the 'classical' for the amorphous state. It is quite possible that in the amorphous matrix even before crystallisation there exist regions induced by deformation with increased correlation in the arrangement of atoms – the nuclei of the crystalline phase with a markedly different degree of compositional and topological short-range order. This is indirectly evidenced by the results of [3.148], which show that the chemical composition of the crystalline phase in an amorphous aluminum-based alloy after normal annealing and after MPD is significantly different.

In this way,

$$Q^* = Q_k - G\tau - \Delta Q_{\text{fv}} - \Delta Q_{\text{sr}} \tag{3.17}$$

where Q^* is the effective activation energy of crystallisation in the shear band, Q_k is the activation energy of crystallisation due to thermal fluctuations, G is the shear modulus, and τ is the shear stress in the region of the plastic shear zone. ΔQ_{sr} is the contribution to the decrease in activation energy of crystallisation due to the presence of short-range order (topological and / or composite) in the zone of the shear band and ΔQ_{fv} is a contribution to the decrease in the activation energy of crystallisation due to the substantial enrichment of the shear band by excess free volume.

The appearance of crystals in the shear bands occurs during the MPD process, and not after its completion. This leads to the fact that the shear bands, newly arising in the amorphous matrix, begin to interact with the crystallites formed at the earlier stages of MPD. Such interaction can occur through several mechanisms (retardation of shear bands on particles of the crystalline phase, cutting or shearing of such particles by shear bands, as well as effects of primary and secondary accommodation). In any case, such an interaction will cause the appearance of dislocations in the crystalline particle itself. The highest density of dislocations will, obviously, be near the interphase boundary, where the effect of shear bands will be most effective. Finally, there comes a time when the density of dislocations in the boundary zone will be extremely high, and the boundary (or the entire crystal particle) will spontaneously transfer to an amorphous state, since the free energy of a region of a strongly defective crystal will be above the free energy of the amorphous state. In fact, this will be perceived as the 'dissolution' of crystals in an amorphous matrix under the action of shear bands active in the amorphous matrix when MPD is performed. It was this 'dissolution' process that was observed in the MPD of a partially crystallized amorphous alloy (Figs. 3.76 and 3.78). It should be borne in mind that the deformation 'dissolution' of crystals is unlikely to be realized until the end. It is known [3.149] that very small crystalline particles (less than 10 nm) can not accumulate dislocation-type defects due to the presence of very large image forces. In our case, this means that nanocrystalline particles of less than 10 nm in size in the amorphous matrix will not 'dissolve' in the MPD process simply because they will effectively push the dislocations to the interphase boundary and remain defect-free. In other words, nanocrystals of less than 10–20 nm in size formed in an amorphous matrix will be structurally stable and will persist throughout the long stages of the MPD. On the basis of this, it is easy to explain the fact that in all studies without exception, where the process of isolating crystals under MPD was studied, the crystals always had sizes less than 20 nm.

At the same time, as the experiment shows, cases are possible where the balance between the release of crystals on the shear bands and the subsequent deformation 'dissolution' can be violated. In this case, the MPD triggers the cyclicity principle, and structural states with large and small volume fractions of nanocrystals in the amorphous matrix periodically replace one another as the strain increases.

Figure 3.168 shows that at the stage preceding nanocrystallisation (N = 0.5) there is a sharp decrease in the *HV* value, the most significant in the case of MPD at 293 K. Essentially this means that the amorphous state in the initial stages of the MPD is structurally rearranged, that it facilitates the processes of plastic shear. The preliminary annealing of the amorphous alloy and its partial crystallisation, as we see, completely eliminate this effect (see Fig. 3.168). We can suggest the following structural model of the effect under consideration. In the process of application of hydrostatic pressure, the regions of free volume (probably, their coalescence) are redistributed in such a way that when a moderate shear stress is applied, the formation of shear bands is facilitated within which the concentration of free volume regions should be significantly (by several orders of magnitude) higher than in the surrounding matrix. This is facilitated by local atomic restructuring (a change in the topological and composite short-range order) as a result of the combined action of the hydrostatic and shear stress components. As a result, nanoclusters (associates) are formed with the predominance of the metallic and covalent nature of the interatomic interaction. Obviously, in the former, the process of plastic shear must be facilitated. Such transformations in an amorphous matrix are evidenced by the fact that at N = 0.5 there are significant changes in the width and position of the halo maximum on X-ray diffraction patterns. In addition, as we shall show later, in this region of deformation the magnetic characteristics change drastically. Since such rearrangements are to a certain extent thermally activated,

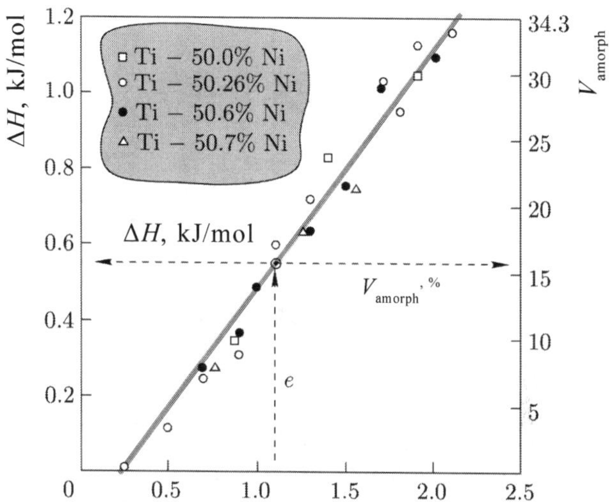

Fig. 3.79. Change in the amorphous phase amount (V_{amorph}) with an increase in deformation degree in cold rolling (e) in Ti–Ni Alloys (3.154).

their effect at low-temperature (77 K) deformation is much less pronounced (Fig. 3.168).

2. Amorphous metal–metal alloys

Recently, the focus of researchers has been on alloys based on titanium nickelide, which has the shape memory effect [3.150]. It is shown that the TiNi intermetallide in the MPD process under shear pressure in a Bridgman chamber or during cold rolling can partially or completely transform into an amorphous state [3.151]. Later, this effect, inherent in titanium nickelide, was repeatedly confirmed by other researchers [3.152, 3.153]. Figure 3.79 shows the linear dependence of the volume fraction of the amorphous phase formed in various alloys near the TiNi composition on the degree of deformation during cold rolling [3.154].

At the same time, there appeared works in which an alloy based on titanium nickelide $Ti_{50}Ni_{25}Cu_{25}$ was obtained by quenching from a melt in an amorphous state and then subjected to MPD in a Bridgman chamber and at a certain stage of deformation there was the transition from an amorphous structure to a nanocrystalline structure [3.143, 3.155]. So, on the one hand, MPD leads to the realization of the 'crystal ⇒ amorphous state' phase transition, and on the other – the 'amorphous state ⇒ crystal (nanocrystal)' phase transition. This obvious contradiction, which we discussed

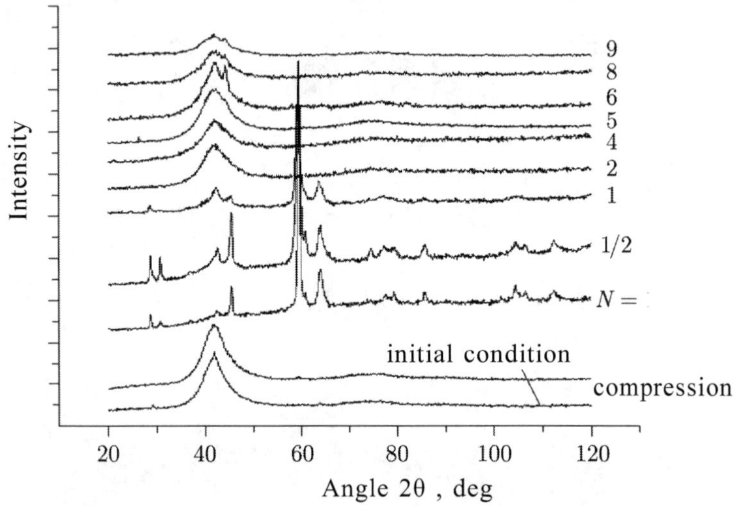

Fig. 3.80. The X-ray spectra of the initial amorphous $Ti_{50}Ni_{25}Cu_{25}$ alloy (in), after hydrostatic compression (P = 4) without shear (compression) and after shear under pressure with a different number of turns (N = 1/8, 1/2, 1, 2, 4, 6, 8).

earlier with reference to amorphous metal–metalloid alloys, led the authors of Ref. [3.156] to undertake a detailed and systematic study of structural–phase transformations in titanium nickelide-based alloys under the influence of MPD in the Bridgman chamber with varying chemical composition, the initial structure, as well as the temperature and magnitude of the MPD. In this case, $Ti_{50}Ni_{25}Cu_{25}$ alloy was chosen, which before deformation in the Bridgman chamber could be in both crystalline and amorphous states. The latter can be obtained by vacuum quenching from the melt by spinning at a rate of 10^6–10^7 deg/s [3.157].

The results obtained in Refs [3.156] unambiguously show that as a result of the MPD of the amorphous $Ti_{50}Ni_{25}Cu_{25}$ alloy obtained by quenching from the melt, phase transformations of various types occur in it (Fig. 3.80).

The application of only hydrostatic pressure without shear deformation already causes the appearance in the amorphous matrix of a small amount of a crystalline phase of the $B19$ type.

At the initial stage of the MPD (N = 1/8), the volume fraction of the crystalline phase (type $B19$ and type $B2$) increases substantially (to ~70%). Extremely high values of the volume fraction of crystals (~80%) were observed after N = 1/2, and then (after N = 1) there was a significant decrease in the fraction of the crystalline phase a

b

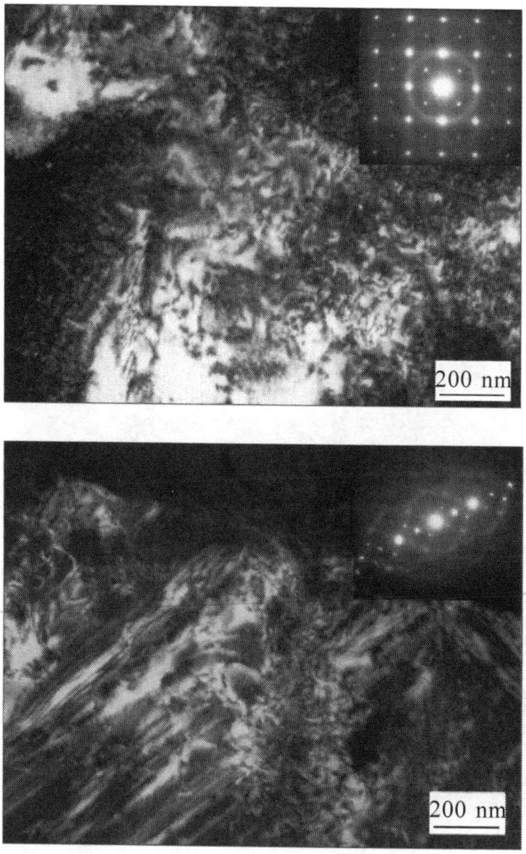

Fig. 3.81. Electron diffraction micropattern and dark field electron microscopic images of the structure of $Ti_{50}Ni_{25}Cu_{25}$ alloy after $N = 1$ in reflex $(110)_{B2}$; the axis of the zone $[001]_{B2}$ (*a*) and in the reflex $(100)_{B19}$; axis of the zone $[010]_{B19}$ (b).

(~30%) (Fig. 3.81). This tendency to a decrease in the fraction of the crystalline phase persists as the deformation increases, and after $N = 2$ the crystalline phase in the structure is almost completely absent (Figure 3.82 *a*).

This structure corresponds to the X-ray amorphous spectrum shown in Fig. 3.83.

It is typical for the amorphous state of a solid, and only the unusual nature of some diffraction patterns and high-resolution dark field electron microscopy reveal the presence of a small fraction of the nanocrystalline phase. Deformation at $N = 4$ leads to complete 'dissolution' of nanocrystals and to the formation of an amorphous state, which differs only in some details from the initial amorphous

Fig. 3.82. Electron diffraction pattern and dark field images of the structure of the $Ti_{50}Ni_{25}Cu_{25}$ alloy under the action of the main (first) diffraction halo corresponding to the MPD at $N = 2$ (*a*) and $N = 5$ (*b*).

state obtained by quenching from the melt. Further growth of MPD ($N = 5$) again leads to the appearance in the structure of a small fraction of the crystalline phase, which is detected only at the electron microscopic level (Fig. 3.82 *b*). But an even greater increase in deformation ($N = 6$) fixes the crystalline phase both radiographically and electron microscopically (Fig. 3.84). Theoretical analysis of X-ray spectra with a split halo in combination with electron microscopic data suggests that we are dealing with a two-phase amorphous–crystalline structure. A further increase in MPD leads to a new 'dissolution' of the crystalline phase ($N = 8$) and its subsequent appearance ($N = 9$).

All of the above is clearly demonstrated by the graph in Fig. 3.85, which shows the dependence of the volume fraction of the crystalline phase in the initially amorphous $Ti_{50}Ni_{25}Cu_{25}$ alloy on the number of

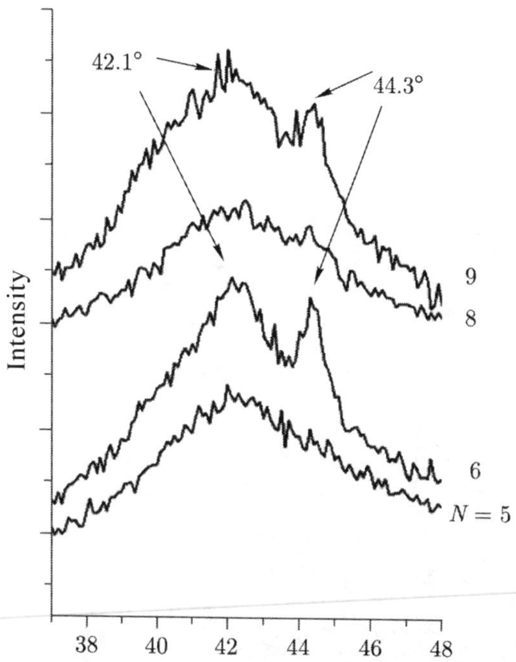

Fig. 3.83. Evolution of the main (first) halo of the X-ray amorphous spectra of the $Ti_{50}Ni_{25}Cu_{25}$ alloy at MPD with a different number of turns. The number of turns N in the Bridgman chamber is indicated in the figure.

turns N in the Bridgman anvil. The results shown in the graph were obtained both by X-ray diffraction analysis methods (a significant volume fraction of the crystalline phase) and by transmission electron microscopy (a small volume fraction).

In the study [3.156] it was apparently possible to discover for the first time practically three cycles of mutual amorphous–crystalline phase transitions. Earlier, a similar effect was observed in the process of mechanoactivation of the Co_3Ti intermetallide powder [3.120]. Such a cycle, that is, a tendency toward a phase transition to the crystalline state of the initially amorphous structure, and, on the contrary, the existence of a directly opposite trend at certain stages of the MPD, unambiguously explains, in our opinion, the apparent contradiction in the experimental results of [3.151–3.153], On the one hand, and [3.143, 3.155], on the other hand.

Let's try to find out what causes such a cycle. The 'initially amorphous state ⇒ crystal (nanocrystal)' phase transition is obviously connected with two reasons. First, as we noted above, the applied

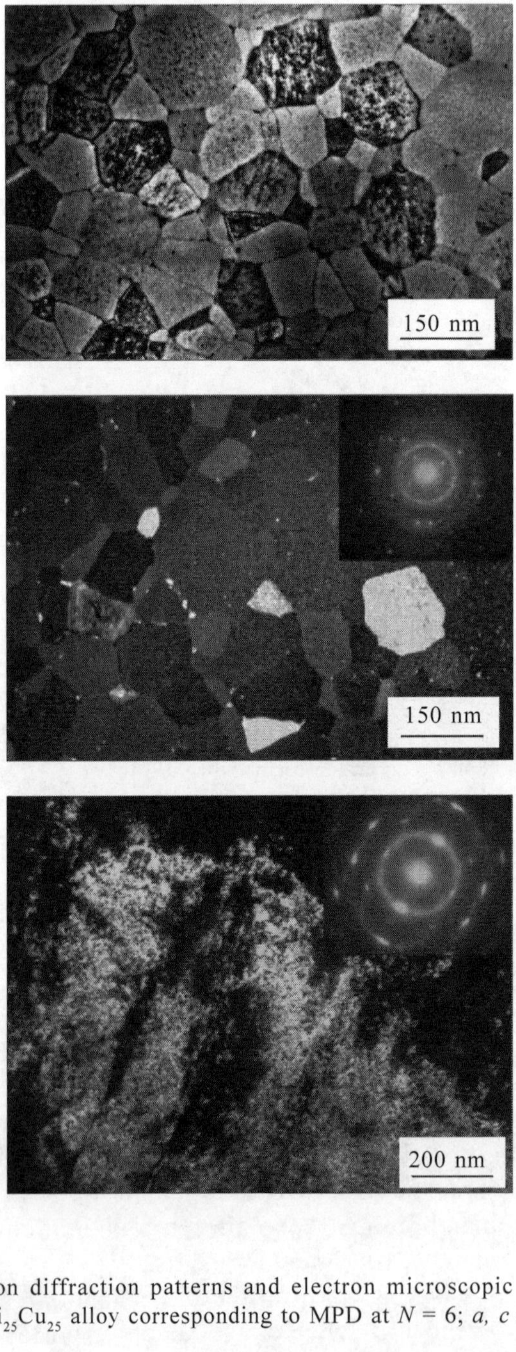

a

b

c

Fig. 3.84. Electron diffraction patterns and electron microscopic images of the structure of $Ti_{50}Ni_{25}Cu_{25}$ alloy corresponding to MPD at $N = 6$; *a, c* – bright field; *b* – dark field.

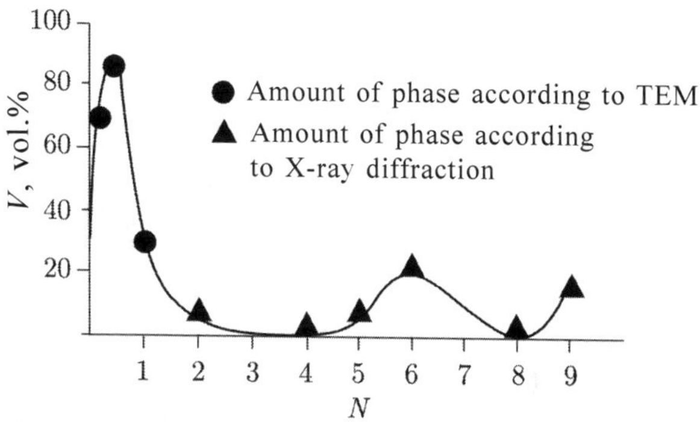

Fig. 3.85. Dependence of the volume fraction of the crystalline phase on the strain of an amorphous $Ti_{50}Ni_{25}Cu_{25}$ alloy.

hydrostatic pressure can stimulate a phase transition in which the equilibrium phase has a lower specific volume. The fact that the $Ti_{50}Ni_{25}Cu_{25}$ amorphous alloy partially passes into the crystalline state without shear deformation, but only due to hydrostatic pressure (P = 4 GPa), confirms this hypothesis. Secondly, the appearance of additional elastic energy dissipation channels, typical for MPD processes [3.158], leads under the conditions of the highly localized plastic flow in the shear bands to local heat energy release and to a corresponding local increase in the temperature of the amorphous matrix. Under these conditions, the shear bands, as we have already said, are able to crystallize.

After the formation of the crystalline phase in the MPD process, it undergoes very large plastic deformations. One of the effective channels of energy dissipation in the conditions of low dislocation mobility is amorphisation [3.128]. Such a phase transition (solid-phase melting) becomes possible, first of all, when the free energy of a crystal containing a colossal density of defects (vacancies, dislocations, disclinations, fragment boundaries, etc.) becomes higher than the free energy of the disordered state of the system. In this case, the negative volume effect departs, apparently, to the second plan and is not decisive. Such transitions have been observed experimentally most often in intermetallics and, in particular, in titanium nickelide [3.151]–[3.153].

Further, as the MPD develops, the picture is obviously repeated, but with certain features associated with the ever-increasing elastic energy and large (gigantic) internal stresses. Apparently, the described tendency to cyclic transformations is common to the behaviour of metallic materials prone to amorphisation during high-energy impacts, in particular, in MPD, carried out by various methods.

It is interesting to note that amorphisation of the crystalline phase is not the only additional channel for the dissipation of elastic energy in the MPD process. As shown in [3.129], these same functions can perform the processes of dynamic recrystallisation if the dislocation mobility proves to be sufficiently high. In our case, the process of amorphisation dominated, but sometimes, apparently, under conditions of the local temperature increase it was possible to observe in the structure the consequences of dynamic recrystallisation (Fig. 3.82 *a*).

An important feature of the processes occurring in MPD is their heterogeneity inherent in any kind of plastic deformation [3.159]. For this reason, the cyclic amorphous–crystalline phase transitions are 'mismatched' in the volume of the deformable material and proceed in different microvolumes at different deformation parameters. This can explain the fact that we extremely rarely can fix states with a limiting content of either an amorphous or a crystalline phase. At any stage of mega-deformation there are local parts of the matrix, which either 'outperform' neighboring regions, or 'lag behind' them. And the higher the value of the deformation, the more pronounced this effect of mismatch is. Finally, we can achieve a certain dynamic equilibrium when we are unable to fix any changes in the phase composition within the given averaging scale.

It is of undoubted interest to analyze the features of deformation amorphisation and crystallisation of the same material with a variation of its initial state, which would allow creating a unified structural model of cyclic phase transformations in the MPD process.

The structure–phase transformations were studied by varying the MPD value with the help of the Bridgman chamber in $Ti_{50}Ni_{25}Cu_{25}$ alloy, which, in contrast to the studies carried out in [3.160], was before the start of the deformation experiments, not in the amorphous but in the crystalline state. The complete X-ray spectra corresponding to all the investigated states are shown in Fig. 3.86.

The main results of the study are as follows.

• The investigated $Ti_{50}Ni_{25}Cu_{25}$ alloy has an initial crystalline structure, represented mainly by plate martensite *B*19. During the MPD, the plates unfold, crush, and finally disappear completely (Fig. 3.87).

• The alloy begins to amorphize already after $N = 0.25$ (Fig. 3.86). After deformation $N = 1$ ($e = 2.15$), the mass degradation

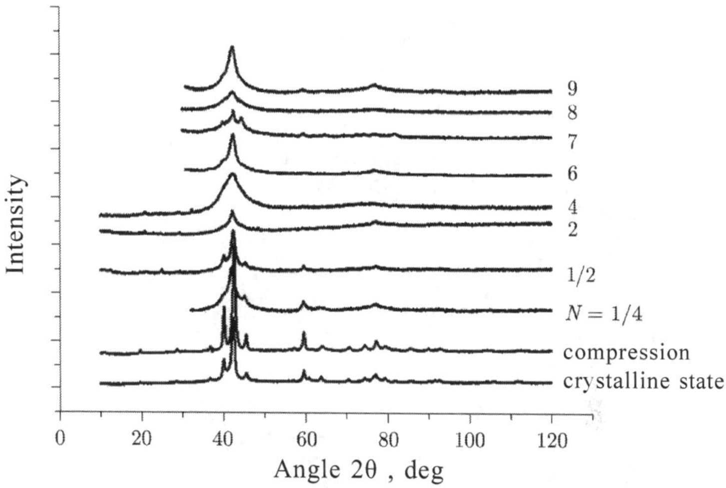

Fig. 3.86. The complete profiles of the X-ray diffraction patterns of the $Ti_{50}Ni_{25}Cu_{25}$ alloy at various stages of the experiments.

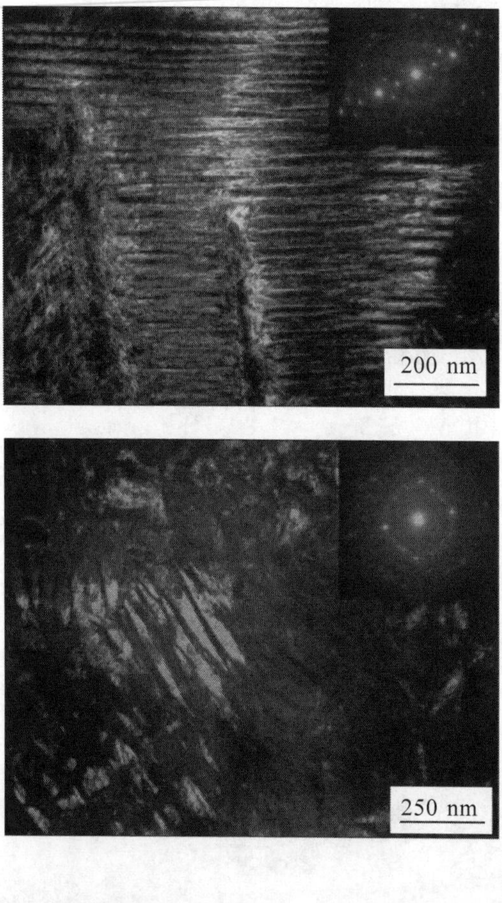

a

b

Fig. 3.87. Electron microscopic image of the structure (dark field) of martensite of phase *B*19 and corresponding electron diffraction patterns before MPD (*a*) and after MPD (N = 1) (*b*).

a

b

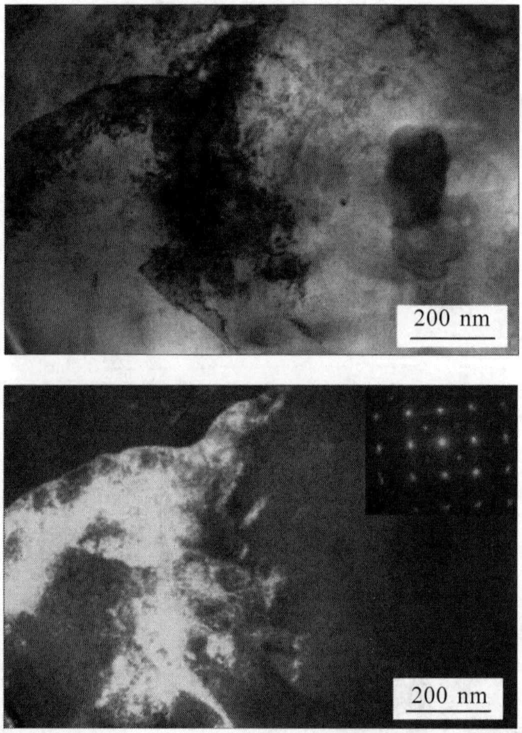

Fig. 3.88. Electron microscopic images (*a*, *b*) showing the presence of the *B*2-phase after MPD (N = 0.5); *a* and *b* are respectively bright and dark field in the phase *B*2 reflex.

of the plate structure and the transition to the amorphous state are clearly observed.

• Simultaneously with the degradation of the martensite plates, plastic deformation of the formed amorphous phase occurs, as a result of which, beginning with $N = 0.5$ ($e = 1.80$), the appearance of a nanocrystalline *B*2-phase with a particle size of up to 10 nm is electron microscopically recorded (Fig. 3.88). In addition, globular regions of phase *B*2 of about 300 nm in size are sometimes observed, containing a large number of defects of deformation origin.

• Further deformation $N = 2–4$ ($e = 2.51–2.90$) is structurally characterized by a superposition of the amorphous phase and *B*2-phase nanocrystals, which often appear in the shear bands of the amorphous matrix (Fig. 3.89).

• In the later stages of deformation, after $N = 6$ ($e = 3.5$), the local instability state of the *B*2-phase is electron microscopically revealed, which is an intermediate state of the martensitic transformation *B*2 $\Rightarrow$ *B*19 (Fig. 3.89).

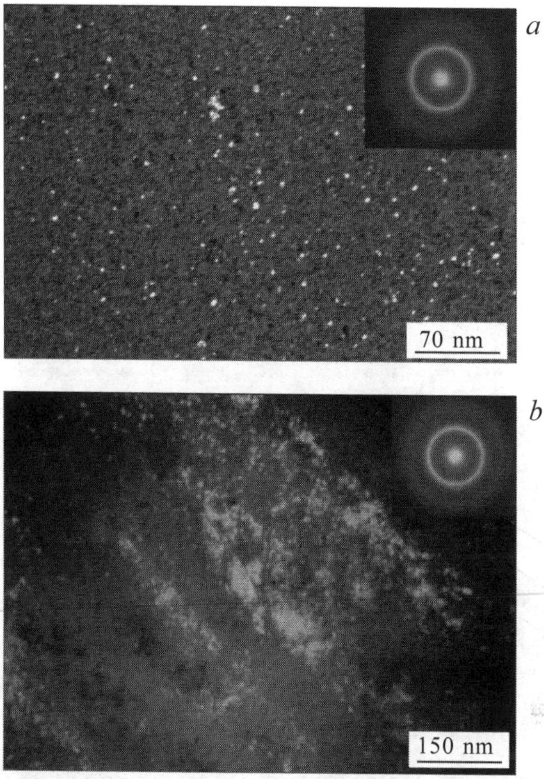

Fig. 3.89. Electron microscopic images (dark fields) of *B*2-phase nanocrystals distributed uniformly in volume (*a*) and in shear bands (*b*) after MPD ($N = 2$).

• The X-ray spectra of the later stages of deformation after $N = 7$ ($e = 4.0$) are already characterized by the presence of broad maxima of the *B*19 phase against the background of the X-ray amorphous state.

• X-ray spectra at the final stages of deformation ($N = 9$, $e = 5.3$) are again completely X-ray amorphous, and electron microscopy recorded the presence of nanocrystals of *B*2-phase (Figs. 3.86 and 3.90).

Thus, as in the case of the initially amorphous $Ti_{50}Ni_{25}Cu_{25}$ alloy [3.156], in the $Ti_{50}Ni_{25}Cu_{25}$ crystalline alloy, the following cyclic sequence of phase transitions occurs with increasing MPD values in the Bridgman chamber:

$$B19 \Rightarrow \text{AP} \Rightarrow B2 \Rightarrow B19 \Rightarrow \text{AP} \Rightarrow B2,$$

where AP is the amorphous state of a solid.

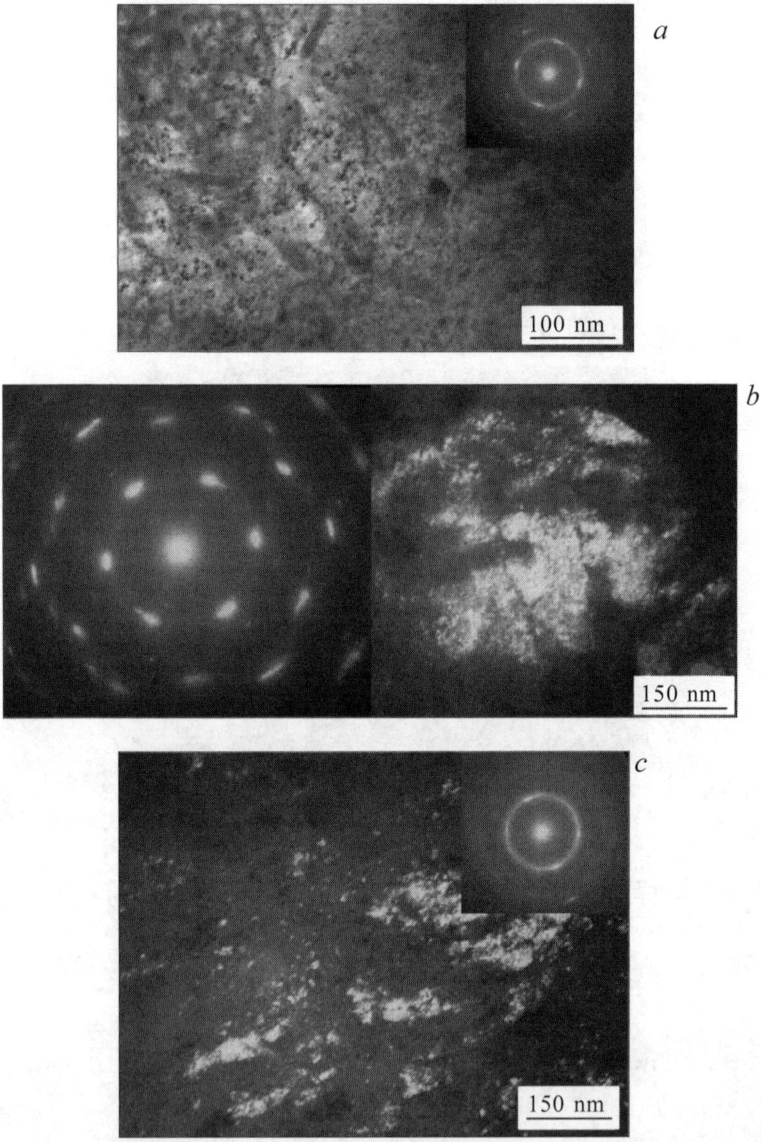

Fig. 3.90. Electron microscopic images (bright field (*a*) and dark field (*b, c*) under the action of reflections from crystalline phases) and the corresponding microdiffraction at different stages of the MPD; $N = 4$ (*a*), 6 (*b*) and 9 (*c*).

It is assumed in the literature that the periodicity of structural changes in MPD is generally determined by the activation of various channels of dissipation (relaxation) of elastic energy stored by the material during deformation [3.158].

It is obvious that the features of the structural change in the $Ti_{50}Ni_{25}Cu_{25}$ alloy in the MPD process found in [3.160] are related to the peculiarities of the flow of direct and inverse phase transformations of both diffusion and martensitic type. Figure 3.91 is a diagram that gives an idea of the nature of the cyclic transitions at MPD from the crystalline state to the amorphous state and then to the nanocrystalline state, followed by periodic repetition of the processes, but at the nanoscale level.

Let us dwell on the question of why in the course of the secondary crystallisation of the amorphous phase with MPD, phase *B*2 it the first phase to appear instead of the equilibrium phase *B*19 at room temperature. When heating, the process of crystallisation of an amorphous state in shear bands proceeds, obviously, through a diffusion mechanism at temperatures (500–510)°C, where the crystal phase of the *B*2 type is equilibrium [3.150]. Upon subsequent cooling to room temperature in the region of 50°C, a thermoelastic martensitic transformation takes place with the formation of phase *B*19 [3.150]. In our case, at certain stages of the MPD, the martensitic transformation is suppressed, and the *B*2-phase is stable at room temperature. In [3.161], a size effect was observed in the thermoelastic transformation $B2 \Rightarrow B19$ in the $Ti_{50}Ni_{25}Cu_{25}$ alloy. Nanoparticles smaller than 20 nm did not undergo conversion upon cooling to room temperature and had a high-temperature *B*2-phase structure.

In our case, nanocrystals formed in shear bands or otherwise and having a size of less than 10 nm are stable at certain MPD stages

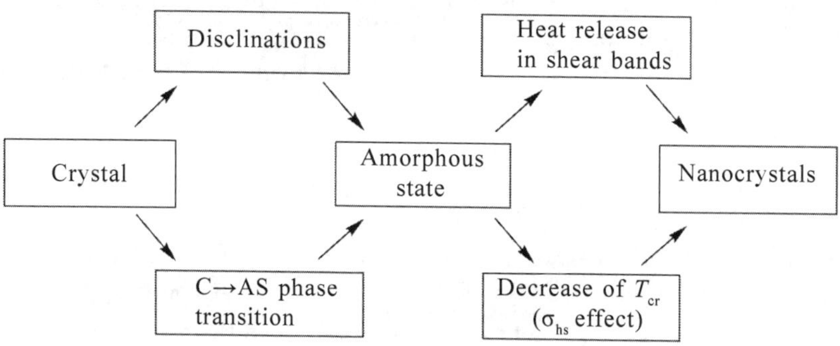

Fig. 3.91. Scheme of processes leading to the transition of the crystalline state to amorphous and, further, amorphous state to crystalline state during the course of the MPD; C is a crystal, AS is an amorphous state, T_{cr} is the transition temperature of an amorphous state to a crystalline state, and σ_{hs} is the stress of hydrostatic compression.

because of the small (nanoscale) particle size of the initial $B2$-phase. The shear transformation $B2 \Rightarrow B19$ is nevertheless realized in the later stages of MPD, apparently due to high operating stresses or coarsening of phase $B2$ particles.

The last fact is confirmed by the observation of microdiffraction patterns (Fig. 3.90 *c*), corresponding to the pre-martensitic state of $B2$-phases after $N = 6$ ($e = 3.5$), and the appearance of broad $B19$ lines on the X-ray spectrum with further increase of the strain to $N = 7$ ($e = 4.0$).

The most interesting, in our opinion, is the elucidation of the structural mechanism of the phase transition in the MPD process from the crystalline state to the amorphous one. In [3.160], rotations, distortion and crushing of the initially regularly located martensite plates of the $B19$ phase in the MPD process were experimentally observed (Fig. 3.87). However, the last act of amorphisation – the 'dissolution' of nanoscale 'debris' of martensitic plates remains speculative.

Apparently, one of the tools for testing this assumption is the experimental observation of the dissolution of nanosized crystals during deformation or computer simulation of the solid-phase dissolution process during shear deformation under the conditions of all-round compression.

Thus, we come to the conclusion that, both in the deformation of intermetallic compounds and complex phases prone to amorphisation at MPD, and during the deformation of amorphous alloys in the MPD process there are successive transitions from the amorphous state to the crystalline state, and, conversely, from crystalline to amorphous. As a result, a stable amorphous–nanocrystalline structure is formed, undergoing quantitative changes as the deformation continues to grow.

Theoretical consideration of nanocrystallisation at MPD of amorphous alloys

As we have already noted, one of the most significant features of structural transformations is the fact that the MPD at room and lower temperatures causes the formation of nanocrystals in a number of amorphous metal alloys in shear bands realizing plastic shear in an amorphous matrix.

Despite the fact that the nanocrystallisation effect at MPD of amorphous alloys was discovered about 15 years ago, this phenomenon, characteristic for amorphous alloys of various systems, has not as yet found a satisfactory, physically correct explanation. On a purely qualitative level, it has been suggested that nanocrystallisation is due to a local increase in temperature in the shear band or in the entire deformed sample, while it has been shown in a number of studies [3.142] that these effects are clearly not enough to explain nanocrystallisation. Earlier, we noted that the activation energy of crystallisation in the shear bands can be represented as an expression corresponding to equation (3.17).

Its essence lies in the fact that the activation energy of crystallisation at MPD decreases due to the local release of thermal energy as a result of the presence of a short-range order (topological and/or composite) in the zone of the shear band and due to substantial enrichment of the shear band by an excess free volume. In addition, the factors contributing to nanocrystallisation, in addition to a local increase in temperature, can be a high level of stresses, as well as the existence of a high concentration of the free volume and the short-range order in the shear band.

The purpose of this section is to evaluate theoretically the role of factors contributing to the appearance of nanocrystals in the shear bands of amorphous alloys. As a way to create MPD, we will consider torsion under the conditions of high hydrostatic pressures in the Bridgman chamber (THBC).

Estimation of the role of local temperature increase in shear bands

When using a Bridgman chamber, a sample having a disc shape is located between the loading punch and the anvil. Moving the punch creates the necessary pressure level, the shear deformation is due to the rotation of the anvil. When the anvil is rotated through a small angle $d\theta$, the shear strain $d\gamma$ is determined by the relation (3.1), which after some simplification has the form

$$\gamma = 2 \cdot \pi \cdot N \cdot \; r/h. \tag{3.18}$$

Since the yield stress of an amorphous state is determined from the von Mises law [2.122], the corresponding equivalent strain ε_{eq} equals

$$\varepsilon_{eq} = \gamma\sqrt{3}. \tag{3.19}$$

Taking into account the pressure as a result of which the thickness of the disk decreases from the initial value h_0 to h, ε_{eq} is determined, as shown in [3.162]:

$$\varepsilon_{eq} = 2 \cdot \pi \cdot N \cdot r \cdot h_0/h^2. \tag{3.20}$$

The rationale for the practical use of relation (3.20) in the form presented as a function of the number of turns N is given in [3.162]. The shear bands resulting from the torsion of a sample of an amorphous alloy having the shape of a solid cylinder are perpendicular to its axis [3.162]. They are areas of material with a thickness of less than 60–70 nm, enclosed between parallel planes. In the process of deformation in the plane stress state, several shear bands are formed; in our case of planar deformation only one, the main band, develops as a rule,

To calculate the energy stored in the main shear band, by two cross sections perpendicular to the axis of the sample, we separate out an element of length dz. From this element using two cylindrical surfaces having radii r and $(r + dr)$, we select the ring, Fig. 3.92. The right end section turns with torsion by an angle $d\varphi$, and the generating line of the cylinder by an angle γ and occupies the position AB′. The segment BB′ Is equal, on the one hand, to $r \cdot d\varphi$, on the other hand, to $\gamma \cdot dz$. Hence: $r \cdot d\varphi = \gamma \cdot dz$, and the magnitude of the angle γ is

$$\gamma = r \cdot d\varphi / dz. \tag{3.21}$$

The angle γ, as is clearly seen (Fig. 3.92), represents the angle of shear of a cylindrical surface. The quantity $d\varphi/dz = \theta$, where θ has the meaning of the angle of mutual rotation of sections, is related to the distance between them. θ is usually called the relative twist angle. For a purely elastic torsion

$$\gamma = r \cdot \theta; \ \tau = G \cdot r \cdot \theta, \tag{3.22}$$

where G is the shear modulus. The tangential stresses arising in the cross section are denoted by the symbol τ. The pair tangential stress also appears in longitudinal planes (axial sections) according to the law of reciprocity of the pair tangential stresses. Elementary forces dF cause the torque

$$M = \int_F \tau \cdot r \cdot dF. \tag{3.23}$$

In a purely elastic region, the expression for the torque has the form

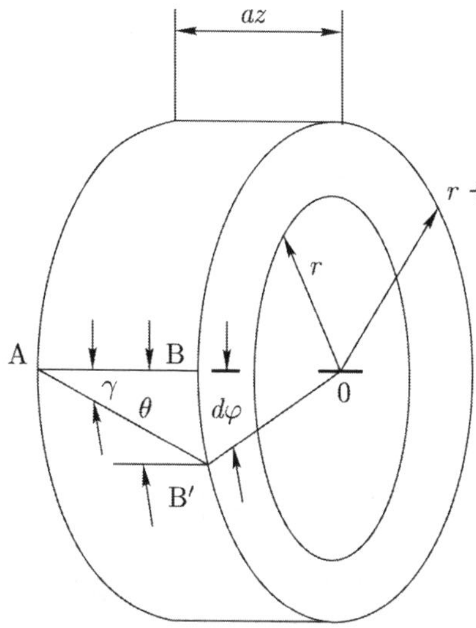

Fig. 3.92. Cylinder cut from the shear band. Designations are given in the text.

$$M = G \cdot \theta \int r^2 dF = 2 \cdot \pi \cdot G \cdot \theta \cdot R^3 / 3. \qquad (3.24)$$

The limiting value of the torque is obtained when the stresses in the entire section are equal to the yield stress is shear τ_p. It follows from the von Mises criterion [3.162] that $\tau_p = \sigma_s / 3^{1/2}$, where σ_s is the yield point for uniaxial tension, and the expression for purely plastic torque is given by

$$M^p = \int_F \tau_p r \, dF = \tau_p \cdot \theta \int_0^{2\pi} d\varphi \int_0^R r^2 \, dr = 2 \cdot \pi \cdot \sigma_s \cdot \theta \cdot R^3 / 3 \cdot 3^{1/2}. \qquad (3.25)$$

The value of the polar moment of inertia of the cross section is determined from expression

$$J = 2 \cdot \pi \int_0^R r^3 \, dr = \pi \cdot R^4 / 2. \qquad (3.26)$$

It is known that amorphous alloys are ideal elastoplastic bodies without any distinctive strain hardening. In [3.162] it is shown that after reaching the yield strength the work hardening coefficient $m \approx 0.025$ GPa is negligibly small. The value of m is determined by the additional stresses required by the moving shear band to overcome the shear bands formed earlier. Previous estimated [3.163] show that the value of m in practical calculations can be neglected since its inclusion is in the range of experimental errors.

We calculate the energy U spent on the formation of the main shear band. It is determined mainly by the quantity of plastic torque (3.25), since the elastic and elastoplastic deformation provide a negligible contribution to the energy U.

$$U=\int_0^t dt\int_0^h\left(\left(M^P\right)^2/2\cdot J\cdot\tau_p\right)\cdot dz=4\cdot\pi\cdot\omega\cdot\sigma_s\cdot R^2\cdot h\cdot t/\left(9\cdot 3^{1/2}\right),\quad(3.27)$$

where $\theta = 2\cdot\pi\cdot N$, $\omega = 2\cdot\pi\cdot N/t_0$ is the rotational speed of the anvil, t is the time of forming the main shear band, t_0 is the time during which the anvil makes one revolution. To formulate the boundary value problem of the theory of heat conductivity with a source, it is necessary:

1) set the initial and boundary conditions for the heat conductivity equation;

2) satisfy the condition for the distribution of the source in a section whose length is equal to the width of the main shear band.

The value of the thermal energy released in MPD in the main shear band as a function of time is found from the relationship

$$Q=\pi\cdot\rho\cdot h\cdot R^2\cdot\left(\Delta T/t_0\right)\cdot c_P\cdot t,\qquad(3.28)$$

where ρ is the density of the AMC material, ΔT is the temperature increase in the shear band during one revolution of the anvil; c_p is the specific heat, R is the radius of the sample. On the other hand, this thermal energy Q is equal to the mechanical work W during the formation time t of the shear band. W is given by

$$W=4\cdot\pi\cdot K\cdot\omega\cdot\sigma_s\cdot R^2\cdot h\cdot t/\left(9\cdot 3^{1/2}\right),\qquad(3.29)$$

where k is the ratio of the work expended to the work stored released as heat, ω is the rotational speed of the anvil, t is the duration of rotation.

To find the distributions of temperature and heat flow during the formation of the main shear band we consider an infinite rod of circular cross-section with the heat source placed in the centre of the origin. Its width is equal to the thickness of the shear band.

The temperature $T(x, t)$ in the rod satisfies the heat equation

$$\partial T/\partial t=\alpha^2\partial^2 T/dt^2,\qquad(3.30)$$

where α^2 is the heat conductivity coefficient. We reformulate the condition of the boundary value problem in the following way, using its symmetry. Instead of an infinite rod with an internal heat source of specific power q, we consider two semi-infinite rods, at the left end of each of which a heat source $q/2$ acts. By combining both bars at the origin, we obtain the original problem by virtue of symmetry.

Consider the temperature distribution as a function of the coordinate x in a semi-infinite rod $0 < x < \infty$ in the case when the change in temperature of its left end, located at the origin, is a linear function of time. For the sake of simplicity, the initial temperature of the rod will be assumed to be 0°C. The initial conditions for the left and right ends of the rod can be written in the form

$$T(x,0)=0;\ T(0,t)=A\cdot t;\ T(x,\infty)=0, \tag{3.31}$$

where $A = 8 \cdot \pi \cdot k \cdot \mathrm{N} \cdot \sigma_s / 9 \cdot 3^{1/2} = \rho \cdot \Delta T \cdot c_p$. We solve the problem by the Laplace transform method [3.164]. The formulation of the problem in this form completely corresponds to the experimental conditions, since the material of the disk made from the amorphous alloy has a heat conductivity coefficient α^2 comparable to α^2 of the steel anvil and the punch, the latter being connected with loading devices, taking into account the shape and dimensions of which is not possible. In addition, it is assumed that the lateral surface of the cylindrical rod is thermally insulated, and the isothermal surfaces at the initial instant of time coincide with its cross sections. In addition, this formulation of the problem assumes that the ends of the rod remain all the time isothermal surfaces, and therefore the isothermal surfaces in the rod will always coincide with the cross sections. Thus, the temperature will depend only on one spatial coordinate x. The absence of heat transfer on the lateral surface of the disk, taking into account the expected temperature of its heating, will apparently introduce small quantitative changes that do not play an appreciable role [3.165].

We apply the Laplace transform [3.164] to the heat conductivity equation (3.30), and also to the initial and boundary conditions (3.31). The images of functions have the form

$$\begin{aligned} LT(x,t)&=U(x,p),\\ LT_t(x,t)&=pU(x,p),\\ LT_{xx}(x,t)&=U_{xx}(x,p). \end{aligned} \tag{3.32}$$

The second of the formulas (3.32) is obtained with allowance for the first initial condition (3.31). Going to the images, instead of the problem for the function $T(x, t)$, we obtain the problem for the image $U(x, p)$:

$$\partial^2 U(x,p)/\partial x^2 - p \cdot U(x,p)/a^2 = 0. \quad (3.33)$$

$$U(0,p) = A/p^2;\ |U(x,p)| = M. \quad (3.34)$$

Equation (3.33) with conditions (3.34) is a boundary value problem for the ordinary differential equation, in which the variable p plays the role of a parameter. Its solution has the form

$$U(x,p) = (A/p^2)\exp(-p^{1/2} \cdot x/a). \quad (3.35)$$

To find the solution, we use the identity [3.164]:

$$L\left[\left(\exp\left(-a \cdot p^{1/2}\right)\right)/p\right] = 1 - \Phi\left(a/\left(2 \cdot t^{1/2}\right)\right) = \left(2/\pi^{1/2}\right) \int_{a/\left(2 \cdot t^{1/2}\right)}^{\infty} e^{-\eta^2} d\eta. \quad (3.36)$$

Therefore, representing $U(x, p)$ in the form $U(x, p) = (A/p^2) \exp [-p^{1/2}x/\alpha]$ and taking into account relation (3.36), and also the contents of theorems on the image of the derivative and convolution of functions, we obtain

$$LU(x,p) = T(x,t) = \left(x/\left(2 \cdot a \cdot \pi^{1/2}\right)\right) \int_0^t e^{-x^2/\left(4 \cdot a^2 \cdot [t-r]\right)} A \cdot \tau/(t-\tau)^{3/2} d\tau, \quad (3.37)$$

and after introducing the variable $\xi = x/ (2\alpha[t - \tau])^{1/2}$, the solution of the problem has the form

$$T(x,t) = 2 \cdot A/\pi^{1/2} \int_{x/\left(2 \cdot at^{1/2}\right)}^{\infty} \left(t - x^2\right)/\left(4 \cdot a^2 \xi^2\right) \cdot \exp\left(-\xi^2\right) d\xi. \quad (3.38)$$

Since the probability integrals are tabulated, we can use the corresponding tables to determine the integral entering into the solution of the heat conduction problem (3.38). On the other hand, the probability integrals rapidly converge [3.166], so we can use them to calculate the expansions of the integrands. Equation (3.38) can be simplified:

$$T(x,t) = \left(2 \cdot A / \pi^{1/2}\right) \left\{ t \left[1 - \int_0^{x/\left(2 \cdot a \cdot t^{1/2}\right)} e^{-\xi^2} d\xi \right] - \right.$$

$$\left. - \int_0^{x/\left(2 \cdot a \cdot t^{1/2}\right)} \left[x^2 / \left(4 \cdot a^2 \cdot \xi^2\right) \right] e -^{-\xi^2} d\xi \right\}. \qquad (3.39)$$

We estimate the temperature of the AMC material contained in the main shear band, in analogy with [3.167, 3.168], but taking into account the specificity of the MPD in the deformation of the amorphous alloy according to the torsion under pressure in the TPBC scheme. The temperature is found from the equality of the relations (3.28) and (3.29).

$$\Delta T = 8 \cdot \pi \cdot k \cdot \sigma_s \cdot N / \left(9 \cdot 3^{1/2} \rho \cdot c_p\right). \qquad (3.40)$$

Substituting the values of mass, thermophysical and mechanical characteristics for amorphous iron-based alloys into equation (3.40) and taking into account that σ_s = 1.5 GPa, ρ = 7 · 10^3 kg/m^3; c_p = 0.46 · 10^3 J/kg · deg, k = 0.2 according to [3.169] we obtain T = 170°C for N = 1, T = 320°C for N = 2, T = 470°C for N = 3 and T = 620°C for N = 4.

Thus, it can be concluded that the effect of the formation of nanocrystals in shear bands at MPDs only due to a local increase in temperature can occur for amorphous iron-based alloys between the second and fourth turns of the movable anvil.

Evaluation of the role of excess free volume and dilation in shear bands

During the MPD, the amorphous structure undergoes changes both in the shear bands and outside them, which must be taken into account when considering the process of nanocrystallisation. In [3.170], the effect of a high compressive pressure of 25 GPa on samples of an amorphous $Fe_{40}Ni_{40}P_{14}B_6$ alloy in a freshly quenched state was investigated. Samples were subjected to uniaxial compressive pressure in a diamond cell. Using the ruby-fluorescent method, it was possible to find the pressure distribution at various points on the surface of the sample, which was an essentially nonlinear function of the coordinate r representing the distance from the centre of the sample. As shown by theoretical estimates and experiments, the effect of pressure on the flow stress can be neglected, provided that its magnitude does not exceed 8–10 GPa [3.164, 3.170]. Nevertheless, the pressure has a significant effect, leading to dilation effects in the shear bands and to a corresponding change in the activation energy of nanocrystallisation.

At present, the effect of dilation in shear bands is actively involved in explaining a number of features of mechanical, thermal, electrical, and other properties of amorphous alloys. In fact, the only work in which the magnitude of the dilation in the shear bands was measured during mechanical testing of amorphous alloys for bending is work [3.148]. The maximum value of dilation (change in density) was 13–15%. In [3.171], based on the dilation effect in the shear bands, it was possible to explain the observed anisotropy of the resistivity of amorphous alloys after cold rolling with a reduction of ≈40–45%, and after mechanical testing of the ribbons for 180-degree free bending.

The defective structure of the amorphous state is characterized by the magnitude of the excess free volume (FV), which reflects the difference in the specific density of the amorphous and crystalline states by about 3%. Let us consider the role of FV in the development of the shear band and the associated dilation effect. From the physical point of view, the propagation of the shear band is analogous to the development of a crack with a 'filler', and the filler can be an arbitrary, including an amorphous, ideal elastic-plastic solid [3.172]. If elastic stresses can not be relaxed near the tip of such a crack, the stress concentration causes the dilation effect $\Delta V/V$, which is described by the equation [3.173]

$$\Delta V/V = (E \cdot \gamma_s/l)^{3/2}, \tag{3.41}$$

where γ_s is the surface energy per unit area, E is the Young's modulus, l is the crack length with the 'filler'. The authors of [3.146] suggested that in the amorphous state in the absence of a crystal lattice, such a dilation (micropore) can 'dissolve' to form an excess of the free volume. Within the framework of this model, on the one hand, dilation causes relaxation of elastic stresses, and on the other hand it serves as a source of a free volume that reduces the ductility of the amorphous alloy at the tip of the crack and promotes the plastic deformation (shear band) by the diffusion mechanism of the free volume. Since the concentration of free volume can not be measured by direct experimental methods, its value is estimated from a change (in this case, decrease) in viscosity. The viscosity is related to the strain rate $d\varepsilon/dt$ and stress σ by the Newton ratio

$$d\varepsilon/dt = \sigma/(3 \cdot \eta). \tag{3.42}$$

The relations obtained for vacancies in crystalline bodies are usually used [1.107] for analytical description of the diffusion coefficient of the excess part of the free volume, viscosity and deformation. In amorphous alloys, instead of vacancies, we use the free volume whose size is equal to V^*. Within the deformation model caused by the free volume [3.174], the ductility below the glass transition temperature T_c is determined by the relation

$$\eta = \left(k \cdot T / \left(v \cdot b^3\right)\right) \cdot \exp\left(-\delta V^* / V_f\right) \cdot \exp\left(G_m / kT\right), \tag{3.43}$$

where k is the Boltzmann constant, ν is the Debye frequency, b is the interatomic distance, δV^* is the probability of finding a pore or free volume of V^* size, V_f is some critical value that determines the development of the process [3.175], for example, the advance of a crack with a filler, ΔG_m – free energy of vacancy migration.

It is well known that in crystalline bodies at constant load, vacancies drift in the region where the compressive stresses act. By analogy with vacancies, the source of the flow of drifting FV regions is the entire volume of amorphous material. The trap for them, obviously, is the main sheat band, characterized by the dilation effect. In other words, there is a 'pumping' of the propagating shear band by an excess free volume, which leads to a decrease in the effective viscosity and, consequently, to a decrease in the activation energy of the formation of the nucleus of the crystalline phase. In addition, the effect of reducing the viscosity in the shear band is significantly enhanced (by several orders of magnitude) due to spontaneous release of heat and a local increase in temperature.

Generalized consideration of factors contributing to nano-crystallisation in the shear bands

A necessary condition for the formation of a crystalline phase in a shear band is, first of all, an increase in the temperature in the zone of plastic shear to values close to the temperature of the transition of the amorphous state to the crystalline state under conditions of ordinary heat treatment. From (3.37) and (3.38) it follows that the temperature in the shear band increases in proportion to the magnitude of the macroscopic deformation of the sample in the

Bridgman chamber, and after 2–3 turns at room temperature it reaches local values T = 600–750 K commensurate with the crystallisation temperature of most amorphous iron-based alloys [1.107]. In addition to increasing temperature, plastic deformation is accompanied by a sharp increase in the concentration of regions of excess FV that accumulate in the region of the shear bands. This is confirmed by independent dilatometric studies [1.107].

It has now been reliably established that the amorphous state formed during quenching from a melt contains clusters up to 1.5–2 nm in size [3.164], which are essentially subcritical nuclei of the crystalline phase. The rate of growth of nanocrystals in an amorphous matrix in this case is determined by the equation:

$$u = a_0 \cdot v_0 \left[\exp\left(-\Delta E / kT\right)\right]\left[1 - \exp\left(-\Delta F_v / kT\right)\right], \qquad (3.44)$$

where u is the diameter of the nanoparticle, ν_0 is the frequency of atomic jumps, Δ^E is the difference between the molar free energies of the amorphous and crystalline phases, and ΔF_v is the activation energy of the atom leaving the amorphous matrix and joining the growing crystalline phase. The crystallisation process proceeds through a diffusion mechanism, and its driving force is the difference in free energies ΔE. Obviously, this value is minimal for polymorphic crystallisation.

The local increase in temperature due to the dissipation of the accumulated energy of deformation ΔT_{dis} provides an increase in the driving force of crystallisation, an increase in the value of ν_0 and a decrease in the value of ΔF_v. The decrease in viscosity due to 'pumping' in the region of the shear band of an extremely high concentration of excess FV, which practically does not sinks, sharply increases the diffusion coefficient and reduces the value of ΔF_v in equation (3.44). This effect can be estimated as an effective temperature increase in the ΔT_{FV} shift band. In addition, under the action of very high acting shear stresses, the drift velocity of atoms increases according to the Eyring model [3.176], which we can estimate as ΔT_τ.

Thus, the total effective temperature increase in the shear band at MPD leading to nanocrystallisation is:

$$\Delta T_\Sigma = \Delta T_{dis} + \Delta T_{FV} + \Delta T_\tau. \qquad (3.45)$$

It is important to emphasize that, as follows from (3.45), crystallisation in shear bands can begin at a lower temperature in comparison with a similar characteristic in the usual thermal treatment of an amorphous alloy.

Let us formulate a number of questions arising from the experimental data that are being extracted and are important for understanding the phenomenon of nanocrystallisation in the shear bands of amorphous alloys, and we will try to give to them, if possible, exhaustive answers in accordance with the theoretical consideration given above.

1. *Why is nanocrystallisation in the shear bands observed only under MPD conditions and is absent in the course of ordinary macroscopic deformation?*

The main reasons for the crystallisation, as we have seen, are a local increase in temperature and the presence of a high concentration of excess free volume in the shear bands propagating in the amorphous matrix. Obviously, to some extent these effects are also present in the ordinary macroscopic flow, but they are markedly enhanced with the transition to the megaplastic deformation region. To effect a phase transition to the crystalline state, an increase in the temperature in the shear band by ΔT_Σ is necessary. As follows from expression (3.45), there are three physical phenomena that contribute to this. The first and foremost of these are the phenomena of the release of local heat, which, as follows from (3.40), depend strongly on the magnitude of the deformation. Apparently, the thermal effects observed during normal deformation are insufficient to achieve the effective crystallisation temperature inherent in each amorphous alloy.

On the other hand, the value of ΔT_Σ under the MFD conditions can turn out to be so great that crystallisation will occur at anomalously low MPD temperatures [3.144].

2. *Why in the process of crystallisation in the shear bands arise exactly nanocrystals, the size of which does not exceed several tens of nanometers?*

Essential for understanding the processes of structure formation in MPD is the fact that the appearance of crystals in the shear bands occurs in the process of the MPD, and not after its completion. This means that the crystals formed are constantly exposed to local plastic shear from the surrounding regions of the amorphous matrix. Such effects can lead to the effects of deformation dissolution of crystals, which is well known from the literature [3.120]. The reason for the formation of nanocrystals is also the fact that the true lifetime of relatively high temperatures in the plastic shear zone is too short to realize the formation of relatively large crystals. Finally, it should also be taken into account that the size of the crystals is limited to a region commensurate with the thickness of the shear band (up to 70 nm).

3.2.3. Structure when compacting powders

The density of nanocrystalline materials obtained by different compaction methods of nanopowders [3.177–3.184] is from 70–80 to 95–97% of the theoretical density. In the simplest case, a nanocrystalline material consisting of atoms of the same type contains two components that differ in structure [3.185]: ordered grains (crystallites) of 5–20 nm in size and intergranular boundaries up to 1.0 nm wide (Fig. 3.93).

All crystallites have the same structure and differ only in their crystallographic orientation and size. The structure of the interfaces is determined by the type of interatomic interactions (metallic, covalent, ionic) and the mutual orientation of neighboring crystallites. Different orientations of neighboring crystallites lead to a certain decrease in the density of matter within the boundaries. In addition, the atoms belonging to the interfaces have a different immediate environment than the atoms in the crystallites. Indeed, X-ray and neutron diffraction studies of nanocrystalline compacted nc-Pd [3.186, 3.187] show that the density of the interface material is 20–40% smaller than the density of the ordinary Pd, and the coordination number of the atom belonging to the interface is less than the coordination number of the atom in an ordinary crystal. The width of the interface, determined by different methods on various compact nanocrystalline materials, ranges from 0.4 to 1.0 nm [3.188–3.191].

According to the original model concepts [3.192, 3.181, 3.193], the structure of the intercrystalline substance is characterized by an arbitrary arrangement of atoms and the absence of not only long-range but also short-range order. Such a state was described by the authors of [3.192, 3.181, 3.193] as a gas-like structure, referring not to the mobility of the atoms, but only to their location (see Figure 3.93). The experimental evidence of some disorder in the intercrystallite matter in nanomaterials obtained by compaction was the results of diffraction studies [3.192, 3.193].

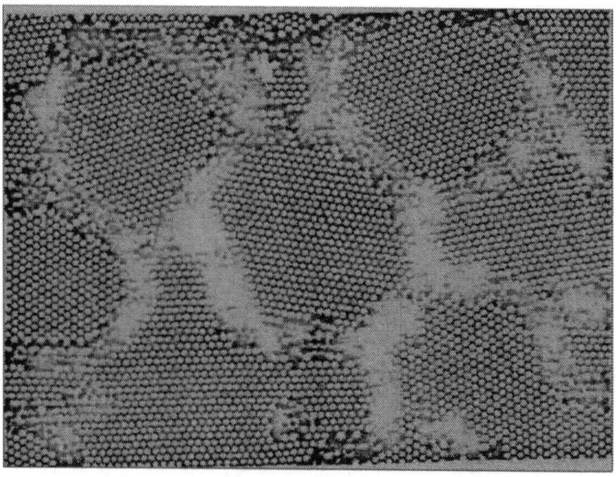

Fig. 3.93. A two-dimensional model of the atomic structure of a nanocrystalline material, calculated using the Morse potential: o – crystallite atoms; • – the atoms of the interface, displaced with respect to the nodes of the ideal crystal lattice by more than 10%; all atoms are chemically identical [3.185].

At the same time, according to the results of recent studies [3.194–3.198], the structure of the interfaces in nanomaterials is close to that in ordinary polycrystals, and the degree of order in the mutual arrangement of atoms in the boundaries is much higher than previously assumed. The use of high-resolution electron microscopy [3.199] has shown that in nanomaterials, as in ordinary polycrystals, the interface atoms are influenced only by two neighbouring crystallites. Pores were found only in triple junctions, and not along the entire length of the interface; the density of atoms in the intercrystalline boundaries turned out to be practically the same as in the crystallites.

The study of compact samples of nanocrystalline iron nc-Fe obtained in high vacuum with an average crystallite size of 10 nm [3.200] found that (95±5)% of all atoms are located at the sites of the bcc lattice. In an earlier paper [3.193], the authors did not find any appreciable short-range order in the arrangement of atoms at the grain boundaries of nc-Fe. It was shown in [3.200] that the unusual results [3.193] are associated with the oxidation of the surface of crystallites: in nc-Fe samples obtained in an insufficiently high vacuum with residual oxygen, only ±5% of the atoms are occupied by the bcc lattice sites of iron, most other iron atoms belong to amorphous oxide phase and only a small part (~5%) of Fe atoms is located not at the positions of the crystal lattice of iron.

An investigation of the short-range order in nanocrystalline compacted cobalt nc-Co [3.201] with an average crystallite size of 7 nm showed that the samples contained ~70% of the disordered amorphous phase and 30% of the ordered crystalline phase. The authors noted that the disordered phase is located along the grain boundaries and does not have specific features that are inherent in the disordered gas-like phase. The relative content of the disordered phase in nc-Co appears to be very high, since the nc-Co samples were partially oxidized (the authors themselves [3.201] report this); moreover, the processing of the experimental spectra did not take into account the presence of lattice defects and free volumes.

3.2.4. Structure under pulsed light annealing

Independent groups of researchers have been working on the effect of pulses of a powerful light source on the structure of amorphous alloys [3.202–3.207]. Since amorphous alloys are mainly obtained by spinning a melt in the form of thin bands (20–30 μm thick), irradiation of the ribbon surface with a powerful light pulse results in changes in the structure practically over its entire thickness.

The process of crystallisation under pulsed light annealing differs from isothermal annealing. A difference in the sequence of formation of crystalline phases under steady-state thermal and pulsed annealing was found. As a result of the powerful light pulsed annealing of Finemet alloy, no nanocrystalline structure is formed in one pulse at about 550°C for 1 h. At a fairly high energy input to the light source, the formation of the second crystalline hexagonal H-phase occurs simultaneously with the appearance of the phase α-Fe(Si), with the crystallites of the α-Fe (Si) phase being larger than necessary to achieve the optimum properties for magnetically soft materials.

Probably, this difference is due to the difference in the course of diffusion processes. As is known, at the first stage of annealing an amorphous alloy of the Fe–Si–B–Nb–Cu system, Cu atoms form into clusters, which are the basis for the nucleation and growth of α-Fe(Si) nanocrystals on their boundaries, and the Nb atoms, concentrating in an amorphous matrix, hinder the growth of α-Fe (Si) nanocrystals. The formation of Cu clusters requires some time, which depends on the diffusional mobility of Cu atoms. As shown in [3.208], copper is grouped at 400°C for 5 min, and the formed clusters are observed after annealing for 1 h. To form α-Fe(Si) crystals, according to [3.209], when annealing at 520°C at least 2 minutes are required.

In pulsed light annealing, the heating of the ribbon occurs very rapidly (about 50 μs), and Cu atoms, uniformly distributed in the amorphous alloy, do not have time to form clusters, hence, there are no conditions for the formation of α-Fe(Si) nanocrystals.

The α-Fe(Si) nanocrystallites in the Fe–Si–B–Nb–Cu alloy after pulsed annealing have a silicon content of about 17 at.%, which practically does not vary with the energy supplied to the radiation source. After isothermal annealing not higher than 600°C, the silicon content in crystallites can reach 20 at.%, and after annealing at a temperature above 800°C falls to 14 at.%. The constant concentration of silicon in crystallites during pulsed annealing can also be explained from the point of view of the limited diffusion of atoms due to the short duration of the action.

It is very likely that the annealing mechanism with pulsed light annealing is similar to the laser annealing mechanism: the energy of light radiation in a very short time (picoseconds), in comparison with the duration of the flare, is absorbed directly by the electronic subsystem. This energy is subsequently absorbed by the ion skeleton as a result of electron–electron, electron–phonon, and phonon–phonon relaxation, and the system acquires a certain temperature T_{ann}, which varies depending on the parameters of the sample and the external conditions in a finite time. In this case, pulsed annealing with incoherent optical radiation is advantageously distinguished from laser annealing in that a significant fraction of the ultraviolet radiation is present in the emission spectrum of the gas discharge, which is effectively absorbed by the amorphous metallic ribbon [3.210].

Crystallisation under pulsed light annealing takes place in the entire volume of the amorphous ribbon. Despite the fact that the irradiation occurs on one side of the sample, the unirradiated side of the tape is also crystallized. This fact can be caused by explosive crystallisation. With explosive crystallisation, latent heat of crystallisation is released, which leads to an increase in the temperature of the system, which further increases the activation process of latent heat release. This process can dramatically increase, and the crystallisation wave will pass through the entire volume of the thin band. This mechanism is possible in metastable systems such as amorphous glasses. The foregoing considerations are only one of the hypotheses, other mechanisms of crystallisation of amorphous alloys under irradiation with light are possible. For example, it is assumed in [3.211] that covalent bonds present in amorphous iron-based alloys with metalloids [3.211] are destroyed under the influence of a powerful light pulse, resulting in a shock wave initiating crystallisation of the alloy. Pulsed light annealing of amorphous alloys in the air atmosphere does not lead to oxidation until its complete crystallisation.

3.2.5. Changes of the structure under the action of pulsed laser radiation

As was discussed above (section 3.1.5.2), the effect of short pulses of laser radiation on massive samples with good heat removal from surface layers is accompanied by high rates of heating and cooling. The use of modern laser technologies can not only increase the efficiency of annealing, but also provide processing conditions that are unattainable by traditional methods, which makes it possible to obtain materials with new properties [3.212].

Using laser pulsed irradiation of amorphous 82K3KhSKhR metal alloy samples, it was established that pulsed laser action allows the annealing of an amorphous metallic alloy with a small boundary between the initial and processed material to be controlled in time and temperature distribution [3.213].

As a result of the action of focused pulsed laser radiation, local zones of the irradiated material are formed on the surface of metallic glasses. With a small area of the irradiated surface and sufficient radiation energy, a deposit is formed in the centre of the zone. The dimensions of the zones of melting and annealing vary depending on the energy of the action of the pulse and the area of the irradiated surface. After the action of laser irradiation in an amorphous matrix, the regions of the crystalline phase are nucleated. Crystallisation within the fused area leads to the formation of large grains and a change in the chemical composition due to the evaporation of some constituents of the alloys [3.214].

When studying the effect of laser annealing on the structure and properties of bulk amorphous Zr–Ti–Cu–Ni–Al (52.5% Zr) and Pd–Cu–Ni–P (40% Pd) alloys, it was established that the effect of laser pulsed radiation on the surface of the amorphous alloys is accompanied by structural transformations, which depend on the thermal properties of the alloys [3.215]. As a result of the action of pulsed laser radiation on the surface of the Zr-based alloy, a 'rosette' is formed, consisting of radially growing crystals (Fig 3.94*a*). The resulting crystals belong to the hexagonal close-packed (hcp) syngony, characteristic of crystalline zirconium. Areas of fusion and thermal influence [3.215] are distinguished. Behind crystallisation in the solid state (growth of crystals) is also observed within the limits of the fusion zone. The relief formed on the surface is due to the volume effect during crystallisation, which is confirmed by dilatometric studies.

In the region of laser action in an alloy based on Pd no visible structural changes are observed. The impact zone is a kind of 'lunar crater', the zone of thermal action is not metallographically detected, which may be due to high values of the alloy viscosity and the thermal conductivity coefficient (Fig. 3.94 *b*).

The analysis of the elemental composition determined at different points of the surface showed that no significant changes in the concentration of the constituents occur in the alloy based on palladium. In the alloy based on zirconium there is an increase in oxygen in the centre of the zone of action. Molten Zr reacts actively with oxygen, forming hardly soluble ZrO_2 oxides, which are the centres of crystallisation. The zone of thermal influence in the alloy based on Zr (275 μm) is greater than in the alloy based on Pd (110 μm) by approximately 2 times.

3.2.6. Structure of amorphous–crystalline films

With the growth of films from atomic or molecular fluxes for different substrate–film systems, it is possible to distinguish characteristic morphological, structural, and sub-structural transformations, which make it possible to separate these systems according to the types of growth. Separate signs of the corresponding species can be manifested in other processes of film synthesis.

Classification of growth types by structural–morphological features

The classification of growth types of crystalline films was based on qualitative morphological features that characterize the film at successive stages of its growth. At the present time, the accepted division into three species is retained according to the characteristic structural and morphological transformations occurring at all stages of growth.

1. *According to Volmer and Weber* (VW) [3.216], the growth of the film begins with the formation of discrete embryos–islands on the surface of a solid (substrate) (in the case of condensation in a vacuum these are clusters of several atoms); as the atoms

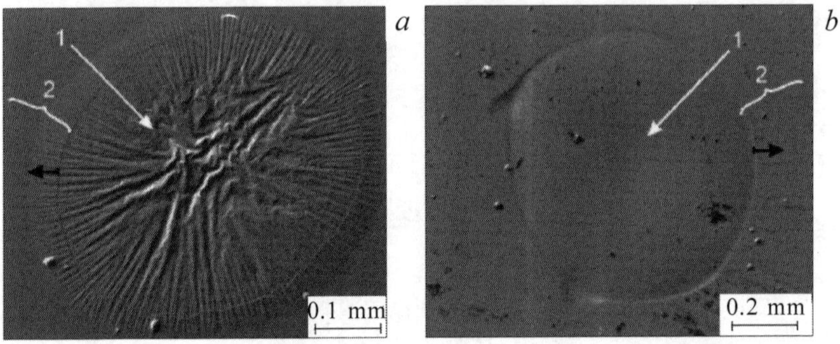

Fig. 3.94. Morphological differences between the zones of action of laser radiation: *a* – alloy based on Zr; 0 – alloy based on Pd; 1 – melting zone; 2 – zone of thermal influence (dark arrows – the beginning and direction of indentation).

enter the source (for example, the evaporator), the islands grow, with their intergrowth, the formation of a labyrinth and then a continuous coating (Fig. 3.95); the subsequent growth is actually the continuation of the normal growth of the crystal in the case of a single-crystal film and polycrystalline grains.

The average thickness of the film at which morphological transformations occur (coalescence, formation of a percolation microstructure, and the onset of continuity) depends on the film material, the interfacial interaction of the condensate with the substrate, and the process parameters: substrate temperature (T_s), condensation rate, expressed or through the flow of atoms entering the unit surface per unit time (J, $cm^{-2} \cdot s^{-1}$), or through the growth rate of the film (v_k, $nm \cdot s^{-1}$).

The growth according to the VW is generally irrelevant to the substrate structure and can be realized both on crystalline and amorphous substrates. On the surface of a single crystal, depending on the pair of substrate–film materials and growth conditions, it is possible to form both oriented (in the limiting case, single crystal) and non-oriented polycrystalline films, and for a number of multicomponent systems below certain substrate temperatures-amorphous films.

2. *According to Frank and Van der Merv*e (FM) [3.217, 3.218], the growth of a film begins with the formation of two-dimensional nuclei and occurs by successive buildup of monoatomic layers (Fig. 3.96). In this case, the morphology of the growth front can be developed to different degrees, which is determined by the parameters of the process.

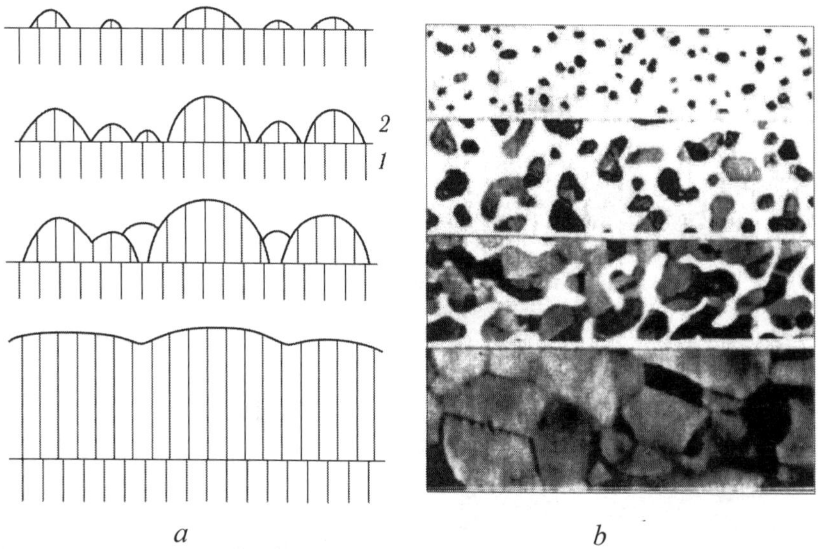

Fig. 3.95. Morphological changes in the growth of films according to the Volmer and Weber mechanism: *a* – scheme of successive stages of transition from the islet to a continuous structure; *b* – structural morphological changes with increasing Mo film thickness on the fluorite at $T_p \approx 1170$ K; the fourth stage corresponds to a thickness of about 50 nm; 1 – substrate, 2 – film.

Moreover, as follows from the experimental studies, the formation of the next layer does not necessarily require the filling of the previous layer. In the initial stages, the FM growth can be considered as an extension of the substrate crystal (at least in the basal plane). In the absence of pronounced morphological changes, the growth of the FM film is accompanied by characteristic structural transformations. Up to a certain critical thickness t_k controlled by the elastic strain energy, a layer is formed which is accommodated by means of elastic deformation (ε_0) until the mismatch $f_0 = (a_2 - a_1) / a_2$ of the parameters a_1 and a_2 of the crystal lattices of the substrate and the film ($\varepsilon_0 = f_0$). In this case, the crystal lattice of a film changes almost always, and the phenomenon itself is called the pseudomorphism. At a thickness $t > t_K$, the film relaxes to the normal structure of the given material.

3. *According to Stranski and Krastanov* (CK) [3.219], layered growth occurs first on the surface of the substrate crystal, with the formation of two-dimensional crystals from a fraction of the monolayer to several atomic layers, depending on the substrate–film system, and on (or) this two-dimensional crystal discrete islets-

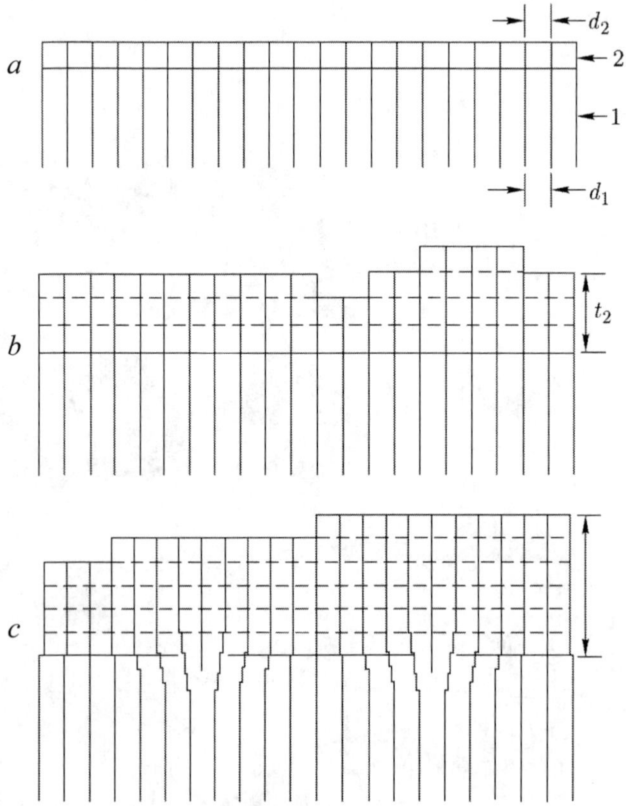

Fig. 3.96. Diagram illustrating the growth of films by the Frank and van der Merwe mechanisms: *a, b* – $t_2 < t_k$, $d_2 = d_u$ ($\varepsilon_0 = f_0$); *c* – $t_2 > t_k$, $d_2 < d_1$, d_1 and d_2 are the interplanar distances for the interfaces between the planes of the crystal lattices of the substrate and the film.

embryos are formed, and the subsequent growth of the film occurs as in the first variant (Fig. 3.97).

With this growth mechanism, one can observe the sequence of structural transitions as the number of adsorbed atoms increases, even before the filling of one monolayer (0 < 1, 0 is the ratio of the number of atoms in the film to the number of possible adsorption sites – the minima of the substrate potential).

Unlike the first, the second and third growth mechanisms, of course, are realized only on the surface of the crystals, and oriented film crystallisation necessarily takes place.

Oriented crystallisation of films on amorphous substrates

The problem of oriented growth and especially the production of single-crystal films on amorphous substrates is relevant in two aspects.

First, the nature and mechanism of spontaneous orientation of crystallites for growth on amorphous substrates have long been of interest. If for epitaxy, based on a large number of different substrate-film systems, some criteria have already been worked out that allow predicting expected orientations and, to some extent, manage the process of oriented growth, the problem under consideration is still under development today. Secondly, the development of methods that allow growing single-crystal films on amorphous substrates is stimulated by the possibility of three-dimensional integration of semiconductor devices: for example, in silicon-based VLSI

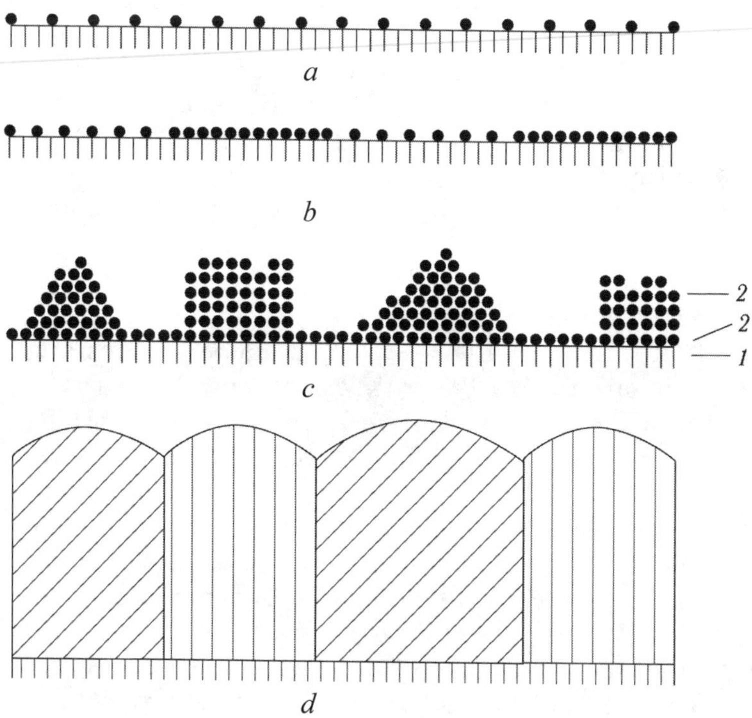

Fig. 3.97. Illustration of the structural and morphological transformation according to Stranski and Krastanov: *a*, *b* – structural transitions for $\Theta < 1$; *c* – the formation of three-dimensional islands on (or in) the layered cover; *d* – polydomain (polycrystalline) film with $\Theta \gg 1$. 1 – the substrate; 2′ – two-dimensional (monolayer) coating; 2 – three-dimensional islets on (in) a two-dimensional layer.

technology, this is achieved with the possibility of growing single-crystal Si films on amorphous SiCb.

The nature of textures with the growth of films on amorphous substrates

Experimental studies of the growth of films of different materials on amorphous substrates (mainly glass, metal oxides and semiconductors) have shown that polycrystalline films are always formed in which, under certain conditions, preferential grain orientations can be formed. For such films, one or more uniaxial textures are characteristic.

In the case of an uniaxial (axial) texture, most crystallites have only one common crystallographic axis. As a rule, this axis coincides with the direction of film growth and is often located perpendicular to the surface of the substrate. In the plane of the substrate, the grains are oriented arbitrarily. As we saw in the previous sections, in biaxial textures, the crystallites are oriented in such a way that two general directions can be distinguished, including the direction of the majority of the grains in the plane of the substrate.

Orientational changes with the growth of films have served as the basis for isolating some characteristic textures corresponding to the successive stages of growth of relatively thick films on the surface of amorphous materials [3.220].

The nature of the possible textures formed by the growth of films on non-orienting substrates is the following. The preferred orientation of the embryos–islets (the nucleation texture) is due to the minimization of the energy of the interfaces (free surface, the interfacial film–substrate surface), i.e., the desire of the islet to take the form corresponding to the minimum of free energy. With the growth of films by the vapour–crystal mechanism without coalescence, when the supersaturation of the vapour is not very high, individual crystallites form at the earliest stage of nucleation. Under such conditions, the forms of their growth are close to the equilibrium form of the crystal. Each crystal is placed parallel to the substrate of one of the equilibrium faces. In the case of a strongly anisotropic crystal lattice, clearly defined axial textures are usually formed. In the case of less anisotropic lattices (for example, FCC or BCC), it is possible to form two or more axial textures in accordance with the number of different faces of the equilibrium crystal. The equilibrium shape of the microcrystals on the substrate (Fig. 3.98) satisfies Wolf's law:

$$\sigma_i / r_i = \text{const}, \qquad (3.46)$$

where σ_i is the specific free energy of the *i*-th face of the crystal, r_i is the distance from the centre of the free crystal of the equilibrium shape to the corresponding surface.

Proceeding from this condition, one should expect a preferential orientation by the most densely packed face parallel to the substrate, and this orientation is more probable at lower interface energy. For example, for materials with FCC and diamond structures, the equilibrium shape of the crystal is a cuboctahedron, in which the faceting occurs mainly on the {111} and {001} planes. On amorphous substrates for vacuum-condensed metal films with a FCC lattice, the formation of a ⟨111⟩ texture is typical, for BCC ⟨110⟩, HCP ⟨0001⟩. Cubic ionic crystals, for example MgO or ZrO_2, are characterized by ⟨001⟩ textures.

The orientations accompanying the main one can disappear in the first stage of coalescence, thus already during the period of 'horizontal' growth the share of the basic texture (nucleation) in the volume of the film can increase.

The result of the continuing nucleation after the film becomes continuous can be the orientation of nucleation–growth.

The formation of the coalescence texture is due to the fact that the stability of the islets of the condensed phase is determined not only by their dimensions, but also by orientation. In the presence of complete or partial disorder in the orientation of the islets, only those that survive, orientations and facings of which ensure a minimum of free energy of the surfaces. If two particles of different orientations merge, the resultant single-crystal particle will inherit the one of the two orientations which was more profitable [3.221]. The reorientation process is the migration of the intercrystalline boundary toward the unfavorably oriented grain. A feature of the coalescence texture

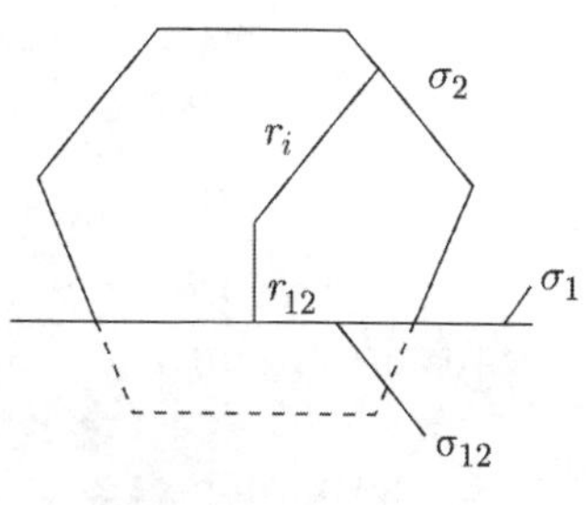

Fig. 3.98. Scheme for determining the shape of an islet of a film on a Wolf substrate; σ_1, σ_2 and σ_{12} are the specific free energies of the surfaces of the substrate, the *i*-th edge of the islet and the interphase boundary, r_i is the distance from the centre of the free crystal of the equilibrium shape (denoted by a dashed line) to the corresponding surfaces.

is that the location of its axis does not change when the angle of incidence of the molecular beam changes: the texture axis is always perpendicular to the substrate. This is explained by the fact that the coalescence process does not depend on the direction of the molecular beam, but is determined by substructural transformations in directions parallel to the surface of the substrate.

Figure 3.99 shows the electron diffraction pattern of a polycrystalline Pt film on amorphous SiO_2. The anomalous intensity ratio of the diffraction rings (increased intensity of the ring 220) indicates a preferential orientation of the grains by the (111) plane parallel to the substrate (texture ⟨111⟩) and to a lesser extent – the plane (112).

After the substrate is completely covered by a growing multi-orientational film (stage τ_1), a new stage of growth begins, in which competition between adjacent crystallites of significantly different orientations plays an important role (stage τ_2). At this stage of growth, a growth texture is formed as a result of geometric selection of the first layer crystallites. Due to the anisotropy of the growth rate the grains the growth rate of which is maximal in directions close to the normal to the surface of the substrate (stage τ_3) survive. This process is called evolutionary selection [3.222]. As a result of selection, a small number of favorable orientations remains, and the growth of the remaining grains is suppressed in accordance with the scheme depicted in Fig. 3.100.

In the case of thermal evaporation and condensation in vacuum or ion-plasma sputtering under conditions of low adatom mobility, the shading effect of slowly growing grains with grains with a high growth rate can contribute to the selection process.

As an example of the evolutionary selection process, let us consider the texture of thick Si films on SiO_2 formed in hydrogen reduction of silicon tetrachloride [3.223, 3.224]. Figure 3.101 shows electron microscopic images of replicas from the surface of poly-Si films with a thickness of 0.1 to 100 μm obtained in one regime. Comparing the images of the replica (a) and the structure of the same film (b), one can see that the replica is a good representation of the grain substructure of the film. Therefore, we can conclude that with increasing thickness, a continuous increase in the lateral grain size occurs in the near-surface region, and in a thick film it is two orders of magnitude larger than in the labyrinth morphology stage.

Table 3.1 shows, according to a number of papers, the dependence of the orientation of Si films on SiO_2 on the deposition temperature

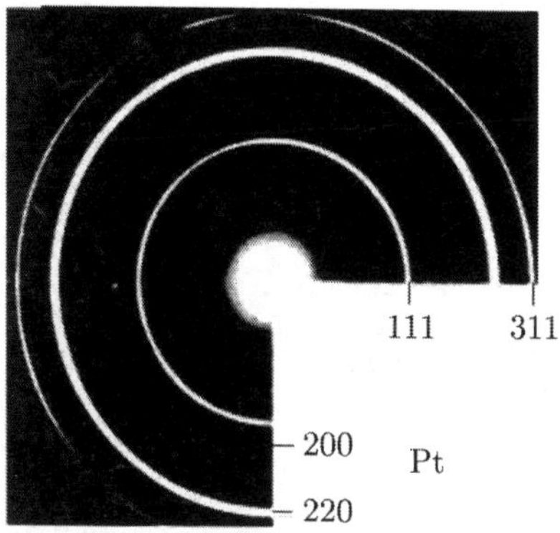

Fig. 3.99. Electron diffraction pattern of a polycrystalline Pt film condensed in a vacuum on amorphous SiO_2.

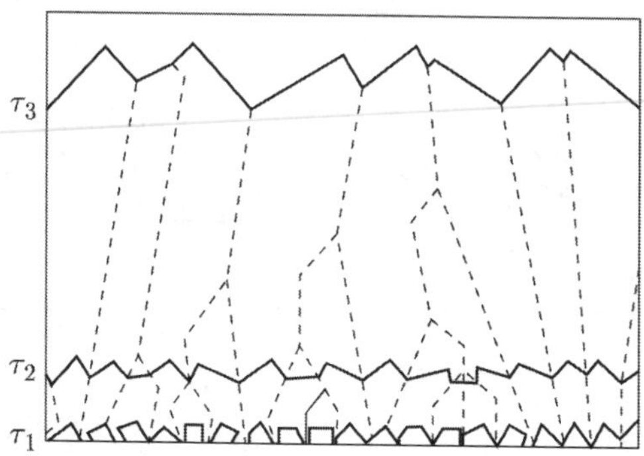

Fig. 3.100. A diagram illustrating the evolutionary selection of grains with the growth of the film (τ_1, τ_2, τ_3 – the growth stages of film grains) (according to Van der Drift, 1967).

in pyrolysis of SiH_4. It can be concluded that the main texture of the films for low temperatures is ⟨110⟩, for high (more than 1000°C) ⟨111⟩ The transition between them is the texture ⟨001⟩.

To explain the observed pattern of the appearance of textures ⟨110⟩ and ⟨111⟩ one can proceed from the following. First, the direction of favorable growth of crystallites should correspond to the favorable orientation of the surface (in this case, the {111} surface) to the flow of atoms. Secondly, it is necessary to take into account the peculiarity of the kinetic characteristics of film growth in the pyrolysis of SiH_4, from which it follows that in the region of low T_s the growth rate is controlled by the surface reaction. Thus, the

Table 3.1. Preferential orientations of unalloyed Si films in dependence of the deposition temperature on amorphous SiO in pyrolysis of SiH_4 [3.220]

t, μm	T_s, K	Texture
0.3	923 973 1073	Amorphous (110)w (100)a
0.5	923 973 1000 1023 1073 1100 1123	Amorphous (110)a (100)w (110) a, (100)a (100)a (100)a, (111)a (100)w, (111)s
0.6	1023–1073 1123 1173	(100)s (100)a, (111)a (111)s
0.8	923 973 1073 1123	Amorphous (110)s (100)s (100)a, (111)w
Comment: w – weak preferential orientation; a – average; s – strong		

texture ⟨110⟩ which ensures the perpendicularity of the substrate surface and, consequently, the surface flow of the adatoms of the two planes {111} of the ⟨110⟩ zone, will be optimal.

For $T_s > 1170$ K, the enhancement of the homogeneous decay of SiH_4 in the near-surface region of the gas phase is characteristic, and the growth rate is controlled by mass transfer. Homogeneous decay of SiH_4 in the near-surface region provides a normal component of the flux of silicon atoms and, consequently, the texture of ⟨111⟩becomes optimal. Formation of the texture ⟨111⟩ in the case of vacuum condensation of Si also supports the proposed explanation.

The realization of the considered principle of geometric selection in the formation of preferential orientations in thick films is confirmed on other materials. The structure and orientation of the films depend on the parameters of the process. The dependence of the normal axial texture on the direction of the condensed stream explains the improvement of the texture ⟨0001⟩ in ZnO films during magnetron sputtering of Zn in an oxygen atmosphere, when the distance between the target and the substrate became less than the free-flight length of the atoms [3.225]. In addition, the texture is

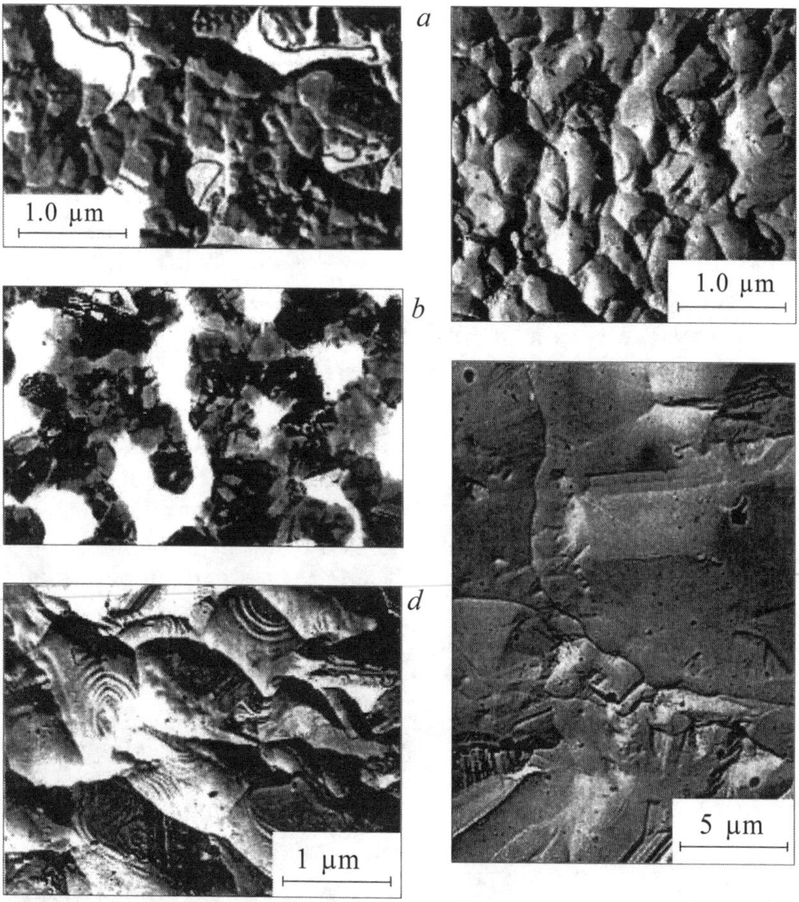

Fig. 3.101. Electron diffraction patterns of replicas taken from the surface of polycrystalline Si films on SiO_2 produced by reduction with $SiCl_4$ at 1400 K, thickness *t*: *a* – 0.1, *c* – 0.7, *d* – 2.5 100 μm, *b* – electron microscopic images of the substructure of the film *a*.

improved and the size of the blocks is increased many times when they exit to the near-surface region (selective growth) as compared to the boundary layer to the substrate.

With the growth of thick films in the process of thermal evaporation and condensation in vacuum, the recrystallisation, which is activated by the energy released during the phase transition of the crystal pair, and also by the energy of the light from the source, also affects the substructural and orientational inhomogeneity in thickness. Figure 3.102 shows electron microscopic images of the structure of a Pd film 4 μm thick obtained by electron-beam evaporation and condensation in vacuum on an unheated substrate (an oxidized silicon wafer). The region bordering the substrate has a nanocrystalline structure with a grain size of about 20 nm inherent in thin films. In the near-surface region, the lateral grain size is already several micrometers. Recrystallisation of metal films always leads to a texture: ⟨111⟩ for FCC, ⟨110⟩ for BCC. Thus, we can conclude: thick polycrystalline condensates always have a gradient substructure.

The effect of irradiation concomitant with condensation in the oriented crystallisation of films

As numerous studies have shown (review [3.226]), irradiation of growing films with low-energy ions can have a significant effect on many parameters that are important for their practical use: adhesion, density, structure dispersion, surface morphology, macrostresses, microhardness, and orientation. Already in the first studies changes in the texture of the films were detected as a result of ion bombardment, accompanying the growth of metal films (Al, Ni, Cu, Nb, N), metallic alloys (for example, Ni–Fe), compounds (AlN, TiN). At present, the realization of the effect of ion-beam processing in condensation

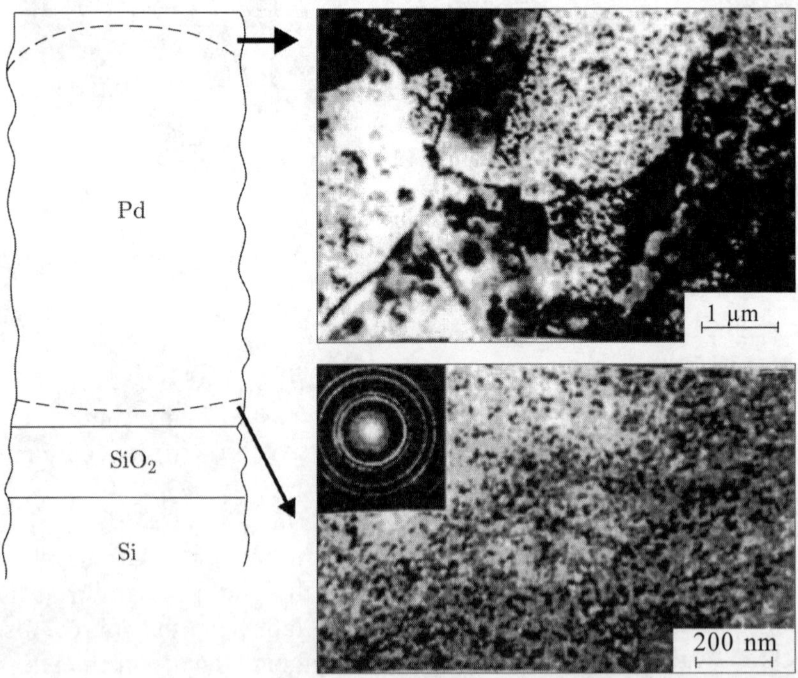

Fig. 3.102. Illustration of substructural heterogeneity in the thickness of polycrystalline Pd film.

is separated into a special method – ion-beam assisted deposition (IBAD). The results of investigations of the ion irradiation effect during film growth show that the IBAD method is promising for improving the process of oriented crystallisation during vapour phase condensation.

Under conditions of 'natural' oriented crystallisation of films on amorphous substrates uniaxial textures form in the best case In a number of practical applications, it is necessary to obtain biaxial textures. In particular, it is known that high parameters of films of high-temperature superconductors (HTSCs) can be achieved only in epitaxial samples, for the preparation of which they can be used as substrates, for example, single crystals of MgO or yttrium stabilized ZrO_2 (YSZ).

The mechanism for the formation of biaxial textures of films on amorphous or polycrystalline substrates has not yet been completely solved. The original technical solution was obtained with the help of IBAD. The development of the texture at IBAD of Cu was associated with the channeling of ions in the crystalline lattice of the condensate: in a polycrystalline film, crystallites with directions of light channeling along the ion beam will be less affected by ions (remain the coldest) and serve as embryos of recrystallisation of the surrounding matrix. The model explaining the development of the preferred orientation in IBAD with low-energy ion bombardment is based on the difference in the sputtering coefficient of grains of different orientations. In the model it is assumed that the crystallographic axis normal to the substrate remains constant, and in the plane of the substrate it is arbitrary. With a certain agreement between the rate of condensation and ion sputtering, it is possible to grow only those crystallites that are oriented in the direction of minimal ion sputtering. The directions of light channeling in perfect FCC, BCC and HCP crystals are respectively:

$$\langle 110\rangle,\ \langle 100\rangle,\ \langle 111\rangle$$
$$\langle 111\rangle,\ \langle 100\rangle,\ \langle 110\rangle$$
$$\langle 1120\rangle,\quad \langle 0001\rangle.$$

Thus, for example, in the case of growth of a film with an FCC structure, the natural normal texture $\langle 111\rangle$ at normal incidence of the ion beam should degenerate into a $\langle 110\rangle$ texture. The orientation of the ion beam at an angle to the growth front of the film should not only change the texture, but also lead to an advantageous azimuth orientation, that is, to a biaxial texture. This is the most significant effect of IBAD, and it can be formulated as follows: with IBAD, the overall azimuthal orientation is given by a set of selectively growing crystallites oriented with respect to the ion beam, which is favourable for its channeling through crystallites. Channeling of ions through the growing crystallites (grains), which are oriented relatively to the ion beam in turn reduces the probability of their sputtering by ions. At the same time, the growth of unfavourably oriented crystallites should be suppressed by ion sputtering, subsequent shading and occlusion by growing crystallites. The diagrams in Fig. 3.103 depict the stages of 'selection' of channeling-favouring orientations (dark grains) by suppressing growth by spraying and then sequestering the unfavorably oriented grains (white).

Consider the effect of IBAD in more detail on the growth of YSZ films. In [3.227], the texture of YSZ films formed at 600°C on pyrolytic glass was studied in detail, depending on the IBAD parameters: the condensation rate, ion beam density and its orientation relative to the substrate plane, ion energy, ion/atom (η) ratio, which is one of the essential parameters of the process, since the optimal biaxial texture is achieved at a certain value of η, the value of which depends on the evaporation (sputtering) method. The value of η is greater the smaller is the energy of the condensed atoms. In the case of electron-beam evaporation, $\eta = 4$, with laser radiation <0.5, with ion-beam evaporation 2–3. The main results of the study are as follows. On an unheated substrate and without IBAD amorphous films are formed, at 600°C – polycrystalline films with a uniaxial normal $\langle 001 \rangle$ texture which is the result of selective growth. A selective dependence of the texture on the parameters of the ionic accompanying of the growth process is observed, the possibility of the formation of biaxial textures is shown. At a beam incidence angle of 48° a not very pronounced biaxial texture was detected [001]. The formation of a biaxial texture is not the result of repeated nucleation, but a consequence of selective growth.

In [3.228], the results of a systematic investigation of the IBAD effect are shown with the growth of YSZ films on different substrates.

The scheme for implementing the method is shown in Fig. 3.104. One source of ions (Ar, energy 1500 eV, current 200 mA) serves to sputter the material, the second (100–550 eV, 80 mA) is used for IBAD.

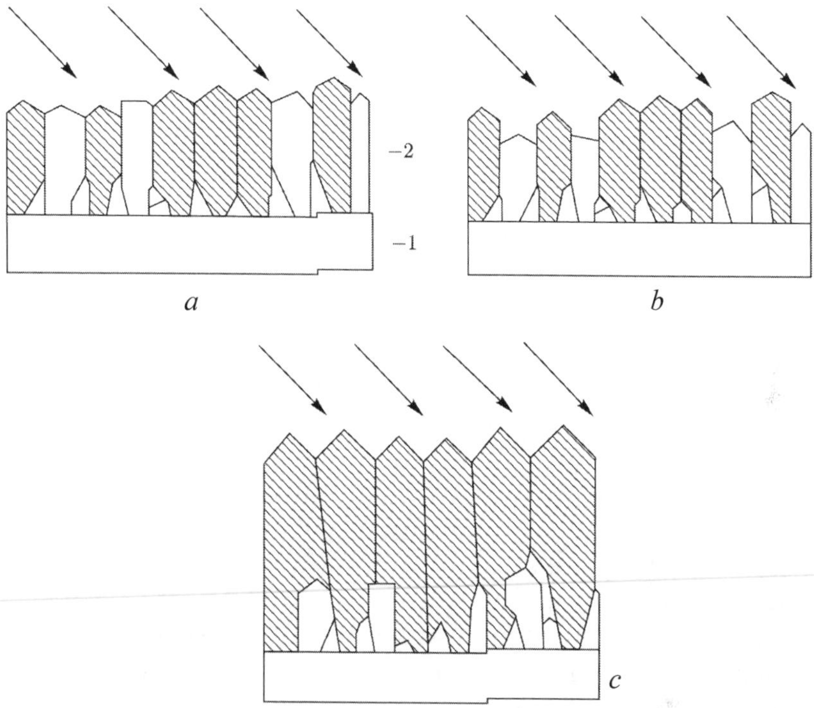

Fig. 3.103. Selection of growing grains of the same orientation in ion irradiation of the growing film (1 – substrate; 2 – the film).

Artificial epitaxy

One of the interesting and practically important directions in the oriented crystallisation of films is artificial epitaxy. The method is based on the orienting effect of an artificially created lattice of a certain symmetry on the surface of an amorphous substrate. The lattice period can exceed the period of the crystal lattice of the film material by several orders of magnitude. The effect of steps on the surface of an amorphous substrate in the oriented crystallisation of films was first demonstrated by the growth of Ag on carbon prints from the stepped (001) surface of a NaCl crystal [3.229]. At the same time, the concept of 'artificial epitaxy' as a method was formulated in our country by N.N. Sheftal [3.230] and confirmed by experiments on the crystallisation of ammonium iodide from an aqueous solution onto a glass substrate with diffraction gratings. As applied to other materials, the method was developed in the works of E.I. Givargizov [3.231] and other researchers.

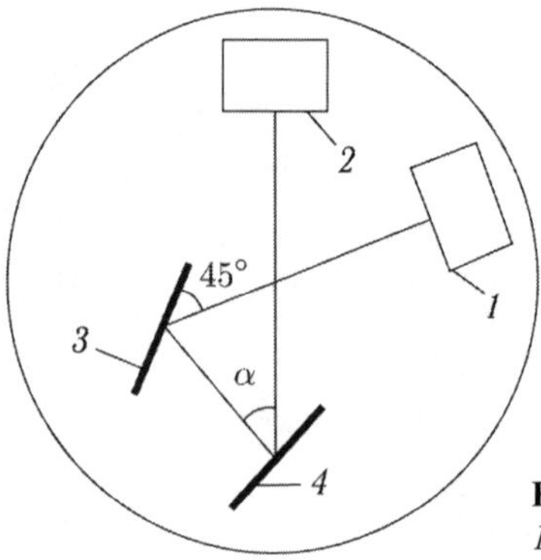

Fig. 3.104. Diagram illustrating the IBAD method: *1, 2* – ion sources; *3* – target; *4* – substrate.

The basic starting position is that the symmetry of the artificial lattice created on the substrate must correspond to the symmetry of the crystalline lattice of the crystallized material. Thus, the symmetry of the artificially created periodic pattern on the substrate is given in accordance with the symmetry of the corresponding lattice plane of the crystallized material, taking into account the necessary orientation of the film, that is, the starting criterion for choosing the orienting relief is symmetry. As a rule, when crystallizing materials, only two faces become dominant. For example, the combination of the {111} and {100} faces is characteristic for metals with an FCC lattice and semiconductors with a diamond lattice, {110} and {100} for metals with a BCC lattice, for materials with a hexagonal lattice {0001}, {1120} and/or {1010}. Thus, not many variants of symmetry of the surface relief are required to ensure the oriented growth of the corresponding crystallites.

Consider, for example, a film of a material with an FCC lattice in the (111) orientation. There are two possible microrelief schemes with symmetry providing a given orientation of the film: in the first case we have surface projections corresponding to the lines in Fig. 3.105 *a*, in the second – a system of triangular depressions (iris), corresponding to the light triangles in Fig. 3.105 *b*. Provided that the film originates at the two-sided corners (shown by hatching), the relief of the first type admits twin positions and, accordingly, the formation of twin boundaries, and a second type relief is more advantageous for the formation of a single-crystal film.

To ensure the orientation (001) of a film of a material with a cubic lattice, it is expedient to form a microrelief with a symmetry axis of the fourth order (Fig. 3.105 *c*).

For better integration of the microcrystals–film embryos, taking into account the given orientation and the equilibrium shape, it is ideal to set the corresponding corners of the side faces relative to the 'bottom' of the cells (relief elements). For example, for the orientation (111) of the FCC lattice, the dihedral angle is 109°28′ (Fig. 3.106 *a*), for the orientation $(11\bar{2}0)$ of the HCP 120° (Fig. 3.106 *b*), (001) of the HCP 90°, (001) crystals with the NaCl lattice 90°. In practice precise values of angles are difficult to realize, but minor deviations do not have a significant effect on the orientation of the film.

To ensure the orienting influence of the artificial relief, the period of the resultant lattices should not exceed a certain value depending on the crystallized material and the conditions of crystallisation. The height of the lateral faces of the lattices is limited by subsequent

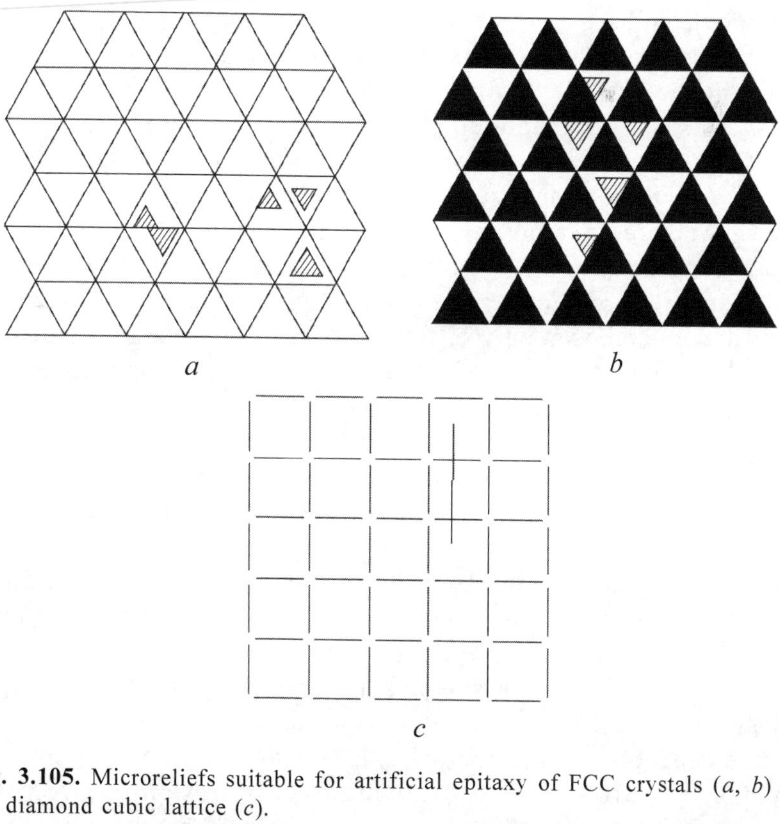

Fig. 3.105. Microreliefs suitable for artificial epitaxy of FCC crystals (*a*, *b*) and the diamond cubic lattice (*c*).

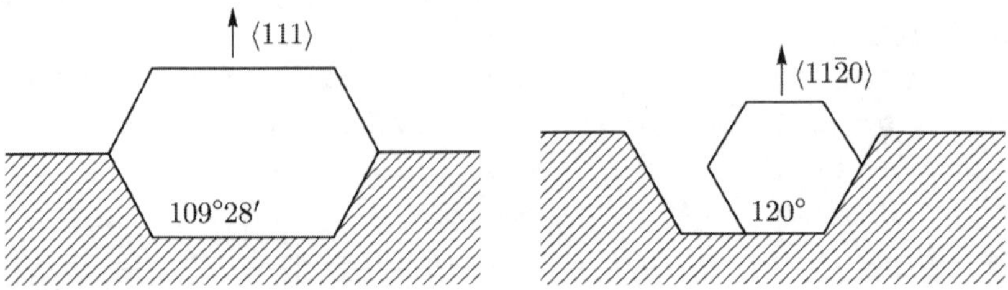

Fig. 3.106. Microreliefs with inclined walls, variants of crystallisation on them of FCC (*a*) and HCP (*b*) materials in the indicated orientations.

technological operations when creating instruments and decreases with decreasing period of the lattice being created.

Taking into account the mechanism of film growth, the period of the lattices should be selected, proceeding from the possibility of nucleation and growth outside their walls, where the orientation of embryo-islets can be arbitrary. Therefore, under the conditions of low mobility of adatoms, the effect of artificial relief is in principle very small, and its greatest manifestation is possible in substrate-film systems, for which the bands of capture of adatoms by embryos formed at the opposite walls overlap.

In [3.232], the graphoepitaxy method was implemented for texturing thick films of multicomponent functional materials. The influence of more than a dozen variants of the artificial relief and surface structures was investigated, optimal parameters and synthesis modes were determined, which ensure biaxial texturing of HTSC films. The main requirements for artificial relief are formulated:

- it should consist only of the most effective elements that cause the oriented growth of crystallites, excluding simultaneously inefficient elements;
- to provide the desired geometric restrictions on the growth of crystallites in accordance with the symmetry of the arrangement of the effective elements of the symmetry relief (habitus) of crystallites;
- provide the conditions in which the formation of crystallite growth centres will occur near the elements of the artificial relief, since otherwise their orienting efficiency is low. \

As a result of the experiments, it was established that a simple banded relief consisting of only narrow parallel grooves is effective for orienting the crystallites of HTSC materials. Such geometry provides: effective geometric constraints for crystallite growth in the form of parallel walls of narrow grooves; the required symmetry corresponding to the habit of growing lamellar crystals of rare-earth element–barium cuprates (symmetry axes of the fourth or second orders); the density of the ensemble of crystallites almost independent of the cooling rate, since the size of the crystallites is physically limited by the width of the channels. This made it possible to obtain oriented layers on substrates with such an artificial relief, in which 75–85% of the sublimate-sized crystallites are oriented with a misorientation angle of less than 10°.

The method of solid-state SiC epitaxy on silicon was developed, consisting in the treatment of single-crystal silicon with carbon monoxide (1100–1400°C) [3.233]. The SiC films obtained by the methods of chemical and physical deposition are mostly amorphous. For their crystallisation it is either necessary to increase the temperature of the substrate or use additional annealing. Figure 3.107 [3.234] shows images of high-resolution transmission electron microscopy (HRTEM) of cross sections and IR spectra of SiC films deposited by the magnetron sputtering method in hydrogen–argon plasma at various temperatures of the silicon substrate. It is seen that crystallisation of the SiC nanocomposites progresses with increasing temperature.

Directional crystallisation from the melt

Back in the 1960s, a 'printing' technique was proposed, the essence of which was that a drop of melt on the substrate was covered with a piece of glass and heating of such a system was stopped system. Crystallisation resulted in the formation of plates with large crystals. The possibility of multiple nucleation during the crystallisation of such a layer makes it difficult to produce single-crystal films in this variant. One of the first inventions that modify this method consists in directional crystallisation of the melt [3.235]. A liquid layer, for example Si on an amorphous SiO_2 film, can be crystallized as a monocrystal from one single-crystal seed nucleus (Fig. 3.108) of the desired orientation, if spontaneous emergence of uncontrolled embryos within the liquid layer is avoided. This is achieved by creating a radially symmetric thermal field with a minimum temperature in the centre of the layer (Fig. 3.108 *a*), where a seed is placed in the form of a single-crystal needle or a floating crystal-embryo (Fig. 3.108 *b*). Crystallisation goes from top to bottom and from the centre to the edges of the layer, and the substrate in this case plays a passive role.

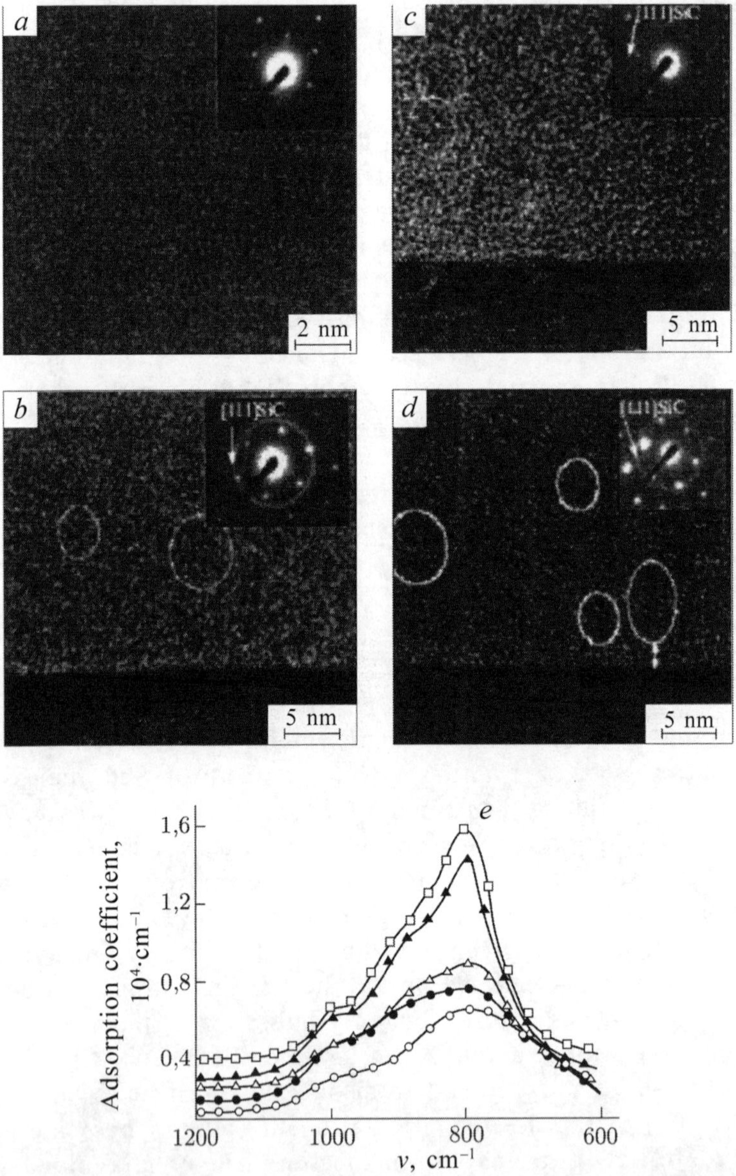

Fig. 3.107. Influence of the temperature of a silicon substrate on the crystallisation of an amorphous SiC film. HRTEM data: *a* – 200°C; *b* - 300°C; *c* - 400°C; *d* – 600°C; *e* – IRS: o – 200°C; • – 300°C; Δ – 400°C; ▲ – 500°C; □ – 600°C.

As an embryo of recrystallisation during the formation of a single-crystal film on the Si–SiO_2 heterosystem, it is possible to use open 'windows' in a SiO_2 film, as shown in Fig. 3.109, which illustrates one of the options for the implementation of the system, the so-called lateral (i.e., 'side') epitaxy [3.236]. Here, the oriented crystallisation of amorphous or recrystallisation of polycrystalline silicon comes from the substrate and propagates in a Si film on SiO_2. Subsequent processing 'cuts off' areas the junction of the recrystallized Si film with the single crystal.

When the system of individual islets on an amorphous substrate formed by photolithography is recrystallized, the probability of multiple nucleation can be reduced with a decrease their size. As a result, single-crystal islets can be obtained. Optimization of the recrystallisation process can be achieved by setting the appropriate shape of the islets and the temperature gradient within the area of the islet. The shape effect is investigated on islets narrowed on one side (Fig. 3.110) [3.237].

If the laser beam is scanned along two polycrystalline (amorphous) islets of silicon of this shape on SiO_2 misoriented to 180°, then an islet is recrystallized in a single crystal along which the beam moves with a narrowed side; the second islet remains polycrystalline. The effect is explained by the fact that only one embryo is formed in the narrowest part of the islet, which grows into the rest of the islet (similar to the growth of massive single crystals by the Bridgman-Stockbarger method).

The temperature gradient within the recrystallized island depends both on the properties of the heterostructure being processed and on the emission spectrum, the influence of which is also manifested

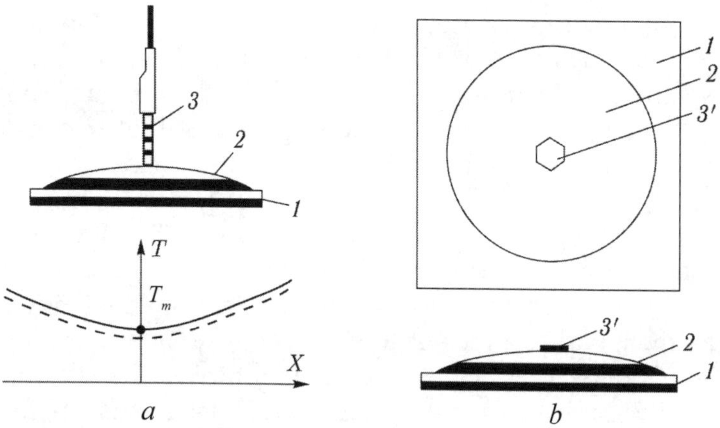

Fig. 3.108. Variants of the method of oriented crystallisation of films on an amorphous substrate: 1 – substrate, 2 – melt, 3 – nucleation of crystallisation in the form of a needle, 3′ – in the form of a floating crystal.

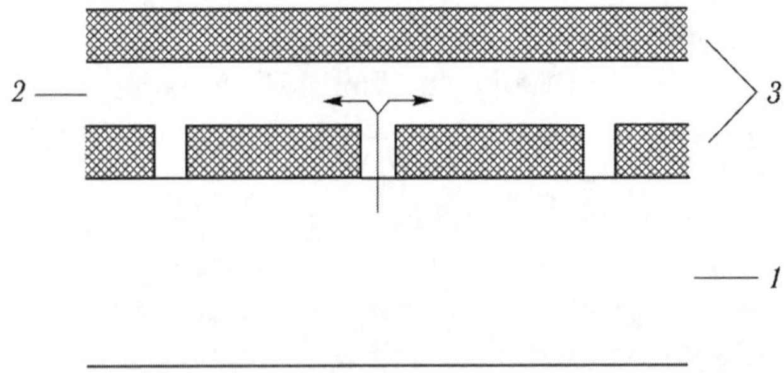

3.109. One of the possible variants of structures for 'lateral' epitaxy: 1 – mono-Si, 2 – poly-Si, 3 – SiO_2; the arrows mark the directions of the recrystallisation fronts.

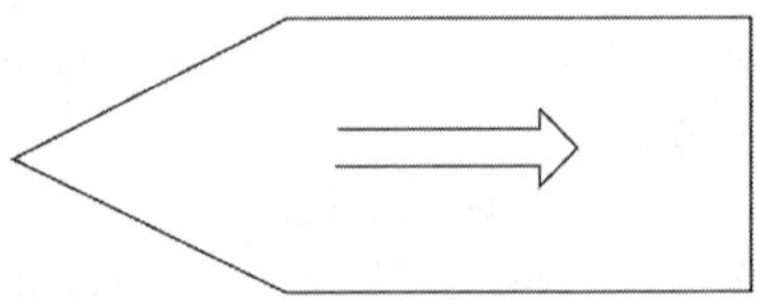

Fig. 3.110. The shape of the film islet for recrystallisation (the arrow indicates the direction of the process).

through the properties of the components of the heterostructure. For recrystallisation of islets it is desirable to have a concave profile of temperature distribution within the islet, which is achieved, for example, for silicon on quartz glass using a CO_2 laser (Fig. 3.111). Such a profile is achieved due to the self-limitation of the heating of the recrystallized silicon islets due to an increase in the reflectivity of the melt compared to crystalline silicon at a higher substrate temperature maintained in the spaces between the islets. This method produced single-crystal islands with a width of up to 20 μm [3.238].

The properties of amorphous–nanocrystalline films have been studied to the greatest extent for nanocomposites based on an amorphous carbon matrix and nanocrystalline inclusions of TiC. Such films are produced by various methods: laser ablation of graphite and graphite–polymer targets [3.239], electric arc plasma sputtering of titanium targets in hydrocarbon media [3.240], magnetron sputtering of graphite and titanium targets [3.241, 3.242]. The participation of hydrogen in these processes leads to saturation of carbon matrices with this element; in the western literature, such matrices were designated H:DLC or *a*-C:H (diamond-like carbon saturated with

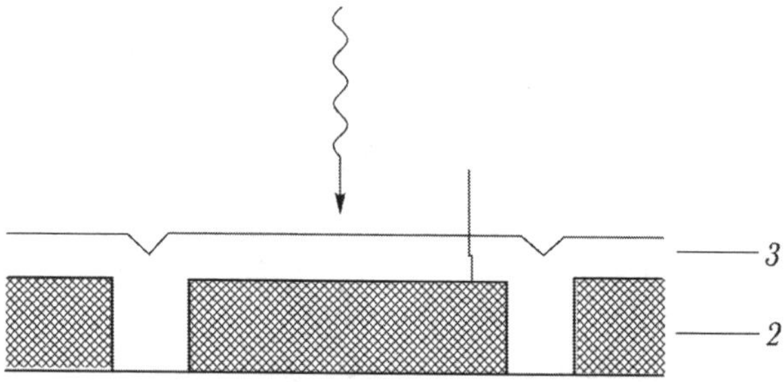

Fig. 3.111. The formation of a concave crystallisation front in Si strips surrounded by a moat under the action of a CO_2 laser: 1 – substrate, 2 – poly-Si, 3 – SiO_2 or Si_3N_4.

hydrogen or amorphous carbon saturated with hydrogen, the content of the latter can reach up to 25 at.% [3.239].

The authors of [3.127], acting on chromium by plasma obtained by an arc discharge in low-pressure argon, applied a chromium film with a mean crystallite size of approximately 20 nm to the copper substrate; a film with a thickness of less than 500 nm had an amorphous structure, and at a greater thickness was in the crystalline state. High hardness (up to 20 GPa) of the film was due to the formation of supersaturated solid solutions of interstitial impurities (C, N) in chromium.

In [3.243], intermetallic Ni_3Al films with an average crystallite size of about 20 nm were obtained by magnetron sputtering a $Ni_{0.75}Al_{0.25}$ target and depositing metallic vapours on an amorphous substrate.

Film amorphous-nanocrystalline composites, due to their high wear resistance, various tribological and physicomechanical properties, have become widespread in the manufacture of coatings for friction units and tool products, as well as in the technology of other functional nanomaterials.

In Fig. 3.112 is a schematic of the nanostructure of a tribological coating of the type YSZ / Au / MoS2 / DLC (YSC-yttria-doped zirconium oxide, DLC-diamond-like carbon) [3.244]. This multiphase nanocomposite structure in which, on the one hand, oxide (carbide, nitride) nanocrystals, due to its high hardness, provide high wear resistance and amorphous inclusion disulfide (WS2 or MoS2) are functioning as a solid lubricant and vacuum conditions of dry friction. On the other hand, the carbon-based amorphous matrix (with predominant sp2 coordination) is characterized by a low coefficient

Fig. 3.112. Schematic representation of the nanostructure of the tribological coating: 1 – solid nanocrystals of oxides (nitrides, carbides); 2 – amorphous or nanocrystalline inclusions of Mo or W sulphides; 3 – amorphous matrix (oxides, gold or carbon) [3.244].

low coefficient of friction (0.1–0.15) and is designed for humid conditions, while oxides and gold additives facilitate operation at high temperatures. To this we can add that the disulphides of tungsten and molybdenum can undergo phase transformations, passing from the amorphous to nanocrystalline state; the coefficient of friction can then decrease to ~0.01. High adaptability of such coatings with a thickness of 1–2 μm to the operating conditions (temperature range from cryogenic to 800°C with varying humidity) and their wear resistance are effectively used for coatings in the frictional units of space technology, and the coatings themselves have received the conditional name of chameleons [3.244].

The method of ion implantation mainly relates to the production of light-emitting semiconductors with regard to problems of opto- and nanoelectronics (creation of LEDs, lasers, non-volatile memory, etc.) [3.46]. Most of the work in this field is related to the formation of Si and SiC nanocrystals in amorphous SiO_2 layers (see, for example, [3.245, 3.246]). Thus, the double implantation of silicon and carbon ions in a layer of amorphous silica and additional annealing at 1000°C leads to the synthesis of luminescent SiC nanocrystals about 5 nm in size [3.246].

Multiphase nanostructures

Unlimited possibilities of vacuum technologies were manifested in the creation of various types of film composites: metal, metal-dielectric, etc; with arbitrary arrangement of the phases or layers; with an arbitrary or regular orientation of crystalline phases; with a combination of crystalline and amorphous or only amorphous phases. Low-temperature condensation in vacuum due to the high rate of nucleation of the condensed phase and insufficient activation of the processes controlling coalescence and recrystallisation during the growth process makes it possible to grow compact film nanostructures even for single-phase materials, including pure metals.

Single-layer film nanostructures are formed naturally during the process of formation under appropriate conditions of a single-component or multicomponent vapour, and as a result of subsequent processing of the condensed structures.

The multilayered film nanostructures, considered in this section, are purposefully created multilayer heterostructures that are formed by alternating the necessary phases of nanometer thickness during the film growth. The artificially formed periodicity, defined by the double layers, leads to changes in the band structure, the phonon spectrum, and affects many physical properties. Interest in such heterostructures is associated with the possibility of realizing the dimensional effect of the physical properties already in a relatively 'thick' object.

Multiphase nanostructures with arbitrary spatial distribution of phases

It has long been noted that the formation of heterogeneous (granular) structures [3.247] due to the trapping of gas molecules by the condensate and the formation of intergranular layers, in particular, oxides, occurs under the vacuum condensation of certain metals under the conditions of a relatively high oxygen partial pressure. This inhibits the process of recrystallisation, which occurs with the growth of the pure condensate. In particular, at a constant condensation rate of Al, an increase in the partial pressure of oxygen in the vacuum chamber from 1.3×10^{-4} to 10.4×10^{-3} Pa led to a decrease in the grain size from 100 to 4 nm [3.248]. The chemisorption of oxygen by metal clusters increases the set of orientations during epitaxial growth, which also contributes to the high dispersity of the substructure [3.247].

The specific nature of vacuum technological processes allows for the formation of films of a rather complex composition, and in many cases the problem is reduced, first of all, to ensuring the specified ratios of the film components. Therefore, in the deviation from the stoichiometry of the expected crystalline phases, the excess concentration of a component can be rejected during the crystallisation process, accumulate at the front of growing nanocrystals and slow down their growth, forming disordered phases at the interfaces between crystalline nanophases. Such a scenario can develop, in particular, in the crystallisation of films of amorphous alloys.

The least studied are *amorphous* film nanocomposites. One of the possible ways of their formation is the joint condensation of a multicomponent system, prone to amorphisation (metallic glass), and amorphous oxide. For example, amorphous granular nanocomposites with a thickness from fractions of a few microns were obtained by ion-beam sputtering in vacuum of composite targets from the corresponding alloys and single-crystal quartz $(Co_{41}Fe_{39}B_{20})_x\ (SiO_2)_{100-x}$, $(Co_{86}Ta_{12}Nb_2)_x(SiO_2)_{100-x}$ [3.249], $(Co_{45}Fe_{45}Zr_{10})_x(SiO_2)_{100-x}$ [3.250]. In a wide range of variation of x the samples are characterized by a granular substructure formed by nanometer-sized metallic granules in the oxide matrix. The size of the pellets decreases with increasing proportion of the oxide. The concentration dependence of the electrical resistivity of composites is typical for percolation systems. The percolation threshold corresponds to x = 49% for the first and 46% for the second, about 43% for the third system. The maximum value of the magnetoresistance of the samples corresponds to a composition close to the percolation threshold.

Molecular–dynamic modelling of the atomic structure of amorphous binary metal alloys (AA) predicts the nanocomposite nature of their substructure, which manifests itself in the possibility of forming fractal clusters of second-metal atoms [3.251]. In the structural aspect, until recently the problem of amorphous metallic materials was limited mainly to the consideration of the topological short-range order, i.e., in the form of averaged characteristics (structural factor, pair function of atomic distribution) calculated or determined by diffraction methods. As shown in [3.251], the main parameters of topological short-range order for the amorphous Re–Tb, Re–Ta alloys do not depend on the type and concentration of atoms of the second component. At the same time, computer modelling of the atomic structure of AA Re_{100-x}–Tb_x ($x \approx$ 10, 12, 13, 15, 20 and 30 at.%) showed the concentration dependence of the sizes of clusters formed from Tb atoms. The percolation transition corresponds to a concentration of Tb atoms of about 13%. At this concentration, the cluster formed by the Tb atoms first connects the opposite faces of the simulated cube. The percolation cluster is fractal: it contains pores of all sizes – from the diameter of the atom to the size of the main cube with the dimension 2.51 averaged over many implementations of the model. Thus, the approach developed gives an idea of the atomic substructure of binary AAs, which is important for explaining qualitative and quantitative changes in the physical properties of the AAs. For example, a magnetic phase transition is observed in alloys containing more than 20% of terbium atoms, i.e., the magnetic ordering occurs only above the percolation threshold in the given system.

The structural model of vitrification of pure metals proposed in [3.252] predicts the possibility of forming a percolation cluster at high cooling rates from interpenetrating and contacting edges, icosahedron vertices at the time of transition from a liquid to a solid amorphous state. The energy advantage of the icosahedral organization and the presence of fifth-order axes that make them incompatible with translational symmetry create the prerequisites for stabilizing the amorphous structure. Thus, the fractal cluster of icosahedra, which includes half of all atoms, plays the role of a restraining crystallisation of the framework and characterizes the fundamental difference between the structure of a solid amorphous state and the melt.

The most flexible way of creating amorphous–crystalline nanostructures is joint condensation, for example, of metallic and oxide phases. The resulting granular metal–insulator nanocomposites with metallic nanoparticles distributed in the dielectric matrix can be obtained (depending on the initial composition) with different particle size and density of the metal phase particles, which makes it possible to significantly change the electrical conductivity of the system. Composites based on crystalline Co, Fe, and Ni nanoparticles or their alloys are of interest as materials exhibiting giant magnetoresistance, as well as the ability to control the transmission spectra of films in the IR region (magneto-refractive effect) [3.253]. The possibility of creating on the basis of these materials magnetically sensitive elements explains such a high interest of researchers in such systems.

Various variants of the methods for creating heterostructures with discrete nanoparticles of the magnetic phase uniformly distributed in the matrix of the film are possible. Metal–dielectric granular heterostructures are mainly obtained by electron-beam evaporation and condensation in a vacuum simultaneously of a magnetic material and an oxide matrix, for example, aluminum oxide characterized by thermal stability and a relatively high dielectric constant, as well as by magnetron sputtering of composite targets. The stratification of the dielectric and metallic phases occurs during the growth of the film even at relatively low substrate temperatures. For example, in [3.253], films of the system $(Co_{50}Fe_{50})_x(Al_2O_3)_{1-x}$ with the size of metallic granules from 1 to 3 nm for concentrations below the percolation threshold ($\chi_P \approx 0.17$) were created by the first method at T_s = 200 °C. In the region of the percolation threshold, the magnetoresistive effect reached 6%, and also the magnetorefractive effect in Al_2O_3 was revealed. Replacing the dielectric can significantly change the properties of the composite both quantitatively and qualitatively. Therefore, the list of initial dielectrics used is quite wide. One should note the incompetence of use for the resulting composites of the formula representation corresponding to the original system. For example, for the initial system $Co_x(LiNbO_3)_{100-x}$ [3.254], the actual structure of the condensate consists of an amorphous matrix with the inclusion of metal nanoparticles (Fig. 3.113) or (after heat treatment) of additional crystalline oxide phases, for example $Co_{0.33}Nb_{0.67}O_2$, in the absence of the $LiNbO_3$ phase.

The observed physical effects show that for their manifestation it is necessary to have objects with a large ratio of the interface area to the volume. This is achieved either by creating short-period (up to units of nm) multilayer composites or composites forming a mixture of nanophases. In particular, metallic nanocomposites with the giant magnetoresistance effect are formed either in the form of granular spatial distribution of nanoparticles of the ferromagnetic phase (for example, [3.255–3.257]), Co–Ag, Fe–Ag, Ni–Ag, Co–Au, etc., or multilayer heterostructures from alternating corresponding metallic films (for example, [3.258–3.261]). Even for couples that are insoluble in the solid and liquid state, co-condensation in vacuum in the low-temperature region makes it possible to form quasi-alloys or quasi-eutectics, which are a mixture of nanocrystals of two phases. The first way to form nanostructures should be recognized as more technological than the creation of multilayer heterostructures.

Systematic studies of the structure and properties of granular metallic films were performed on the Ag–Co system. It is shown that by changing the concentration of the magnetic component it is possible to control the structure and properties of the composite, to obtain different variants of magnetic states.

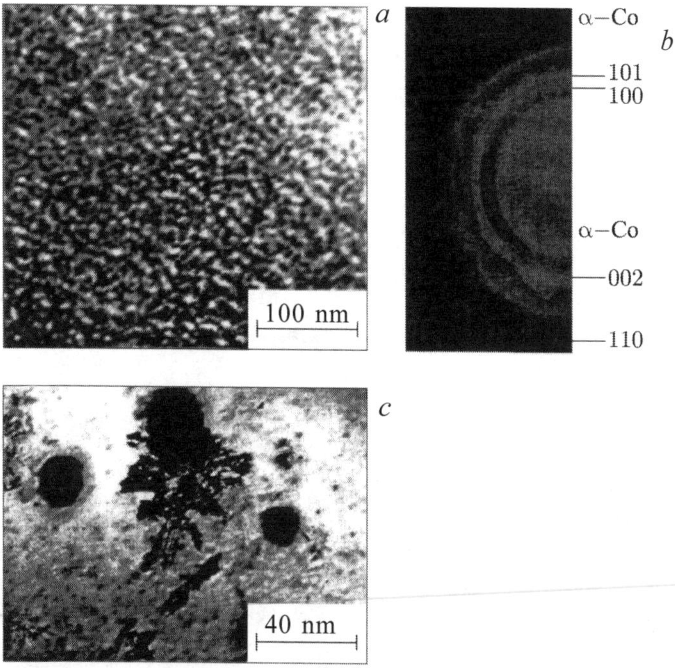

Fig. 3.113. TEM image (*a*) and fragment of the electron diffraction pattern (*b*) of the film composite obtained by ion beam sputtering of the target of the composition $Co_x(LiNbO_3)_{100-x}$ (x = 15%); *c* – after annealing at 500°C (segregations of Co and $Co_{0.33}Nb_{0.67}O_2$ phases are visible).

The original method for obtaining nanocrystalline condensates of Ag–Me (Me: Co, Fe, Ni) systems was proposed in [3.258]. Joint metal vapour condensation was carried out in a continuous flow of an inert gas in the ultrahigh vacuum chamber so that a convection flow can carry Me nanoparticles through the flux of Ag atoms, to thereby coating them with silver. For the corresponding inert gas pressures heterostructures with the magnetoresistive effect in a field H = 1.5 T 26% (for Co), 8% (Fe) and 7% (Ni) were obtained.

It should be noted that the methods for creating film granular nanostructures are not limited to vacuum technologies alone. In particular, the mechanism of formation of Co-containing structures during electrolytic deposition is discussed using examples of Re–Co [3.262] and Cu–Co [3.263] systems. Of course, the questions of structure formation in such systems, the kinetic regularities of the process of phase separation during the growth of metal films and under natural and artificial ageing remain to be answered, and this problem is especially difficult for polycomponent systems.

In [3.264, 3.265] it was shown that in relatively simple metallic systems with limited solubility *orientational nanocrystallisation* can take place during film growth from the two component vapour phase. Let us consider, as an example, the growth of films of the Ag–Ni system on the (001) NaCl surface [3.265]. Figure 3.114 shows an electron diffraction pattern and a micrograph of a film with a thickness of about 100 nm.

Oriented nanostructures are formed at a substrate temperature of about 350°C. No noticeable changes in the parameters of the crystal lattices were detected. Well expressed reflexes of double diffraction on thin nanograins of both nanocrystalline phases reflect their layered mutual arrangement. Thus, with the growth of films, predominantly non-lateral phase separation occurs, but in the direction of growth, i.e., automodulation with respect to composition in the direction of growth takes place, while the nanocrystalline structure as a whole is preserved. In this case, irregular layered nanocomposites are formed from alternating very thin, mutually oriented plates of both phases, up to several nanometers in diameter. This is confirmed by the conservation of the relative concentrations of the components integrally measured by Auger electron spectroscopy with the layerwise ion etching a heterostructure [3.264]. The temperature range of the realization is quite strict (as in Fig. 3.114) the parallel orientation is only a few tens of degrees. A further increase in T_s leads to a deterioration in the quality of the biaxial texture, as well as to the through growth of the grains of each phase, and the grains become monoblock (in this case, monophase) in the thickness of the sample.

When condensing from a two-component vapour phase, it is possible to separate the condensed-phase islet clusters by composition: for different values of the activation energy of the surface diffusion of the components at the time of the stage preceding the coalescence, the film should consist of a highly dispersed 'matrix' formed by clusters of compositions *A*, *B* and *AB* with including a small number of relatively large islets of composition A. Calculations for model pairs with similar values of the energy E_d show that in this case the distributions with respect to the dimensions of clusters of composition *A* and composition *B* practically coincide, and the agreement is better the greater the value of E_d of components.

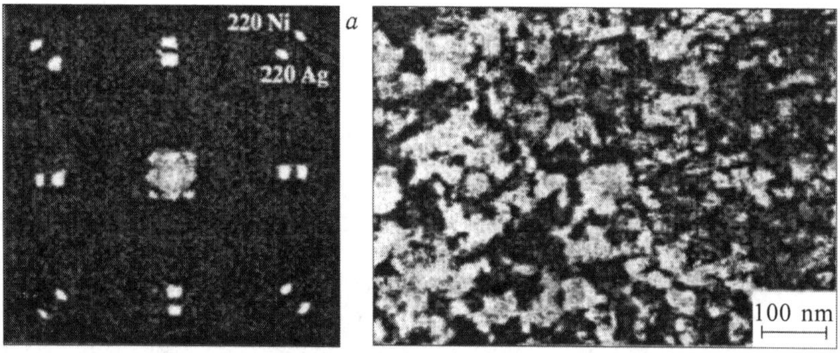

Fig. 3.114. Electron diffraction pattern and TEM image of nanocrystalline oriented multilayer heterostructure Ag–Ni.

Thus, when considering the mechanism of formation of such a two-phase nanostructure, it is assumed that atoms A and B with equal probability can both form a crystalline phase and act as impurity atoms with respect to the phase of atoms of another kind (Fig. 3.115*a*).

At the initial stage of condensation, these roles are distributed by random fluctuations in the density of fluxes of atoms from the vapour phase. Atoms A, falling on the surface of a cluster of B type atoms, are embedded in the crystal lattice of their phase, the 'impurity' B atoms migrate along its surface to the boundary and are integrated into its phase. Similarly, the process develops on the surface of the second phase. At certain values of the condensation temperature, it is possible to establish a dynamic equilibrium between the flows of 'impurity' atoms leaving the 'matrix' phase as a result of surface diffusion and the flow of atoms coming to its surface from the vapour. In this case, the phase will grow, having on its surface a constant number of 'impurity' atoms.

At low condensation temperatures the surface diffusion slows, the resulting mixture will decompose due to the almost complete insolubility components to form a two-phase non-oriented nanodisperse heterostructure (Fig. 3.115 *b*). As the temperature of the substrate increases, the 'impurity' B atoms, along with the escape from the surface of the epitaxial phase from the A atoms, accumulate on the front of its growth in an amount sufficient for self-organization of the epitaxial phase from several monolayers, i.e., a changing phase occurs. The process is periodically repeated, as a result of which a two-phase epitaxial nanocrystalline heterostructure is formed. At a higher T_s, the grains of both phases grow through the entire thickness of the condensate, 'tearing off' impurity atoms to neighbouring phases due to the activation of surface heterodiffusion. An estimate of the model parameters at which the diffusion stratification is realized shows that when E_d varies from 0.7 to 0.9 eV, the expected temperature range of the oriented stratification varies from (305...330 K) to (385...410 K), remaining very narrow (25–30 K) [3.267].

When estimating the temperature range of the formation of the modulated structure, depending on the activation energy of surface diffusion, it is limited from above by a temperature above which the concentration of *B* atoms in the condensation process does not reach one monolayer at the centre of the surface of one phase (*A*): the crystal grows through the entire thickness of the film displacing these atoms to the adjacent crystals of phase *B*. The lower boundary of this region is found from the condition that during the time of increase in the concentration of atoms of the second grade in the centre of the surface of the crystal to two monolayers, the average diffusion path length is less than the radius of the crystal. This means that in the process of condensation there is an accumulation of *B* atoms, and the stratification occurs at the growth front of phase *A*, and the dispersion of the structure increases with decreasing temperature.

The diagram of possible structures shown in Fig. 3.115 *b* qualitatively agrees with the results of electron microscopic studies of Ag–Cu and Ag–Ni films.

Attention is drawn to the preservation of the parallelism of the orientation of the Ag and Ni layers, while the orientation of (111) Ag on (001)Ni and (110)Ni on (001)Ag is predicted from the size mismatch of the lattice parameters ($\alpha \sim 0.15$). The observed orientation relationship can be explained on the basis of the peculiarities of the growth process manifested in the formation of a mixture of Ag and Ni atoms at the stage of nucleation of each grain of the new phase, as well as the effect of the third layer [3.266].

In the Ag–Cu system, the parameters of the crystal lattices were found to converge mainly due to a decrease in a_{Ag} to values of 0.402-0.405 nm (based on the results of electron diffraction measurements) at a substrate temperature of up to 300°C, which corresponds to dissolution in Ag nanocrystals of up to 17% Cu. This pattern is also characteristic of thick films in which the lattice parameter of Ag decreases to 0.406 nm (corresponds to a Cu concentration in Ag grains of about 7%) and the lattice parameter of copper increases to 0.362 nm (Ag concentration in Cu grains is about 1%). For

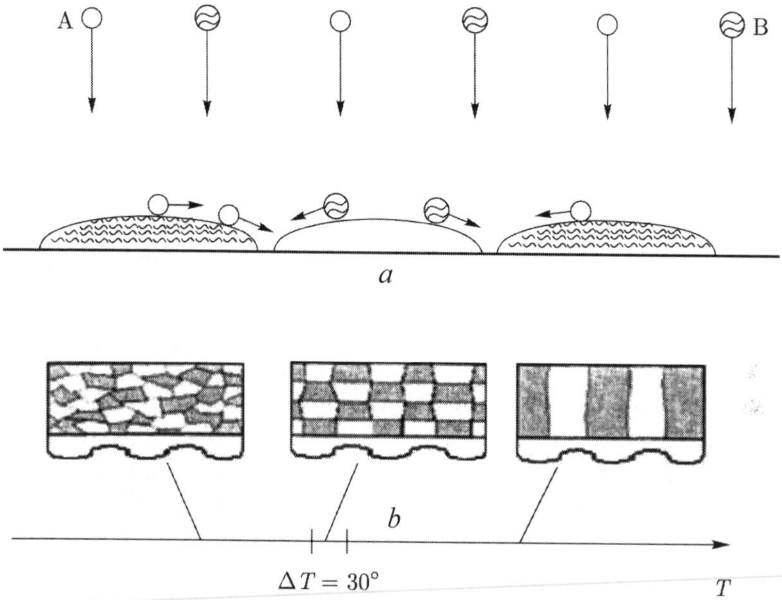

Fig. 3.115. *a* – flow diagram for neighbouring formed nanophases from components *A* and *B*; *b* – schemes of heterostructures of two-component metal films with limited mutual solubility of components; the regions of their formation temperature are shown.

the process parameters used, the change in the lattice constants is completely explainable by the condensation-stimulated excess mutual dissolution of Cu and Ag, which is not taken into account by the equilibrium diagram for this system.

In the films of the Ag–Ni system, no noticeable changes in the parameters of the crystal lattices of the corresponding phases were detected.

Multilayer oriented nanostructures

Composites can be a volumetric compact mixture of discrete phases of different degrees of dispersion, and a system of alternating layers of phases *A* and *B* with a period determined by the sum of their thicknesses $t_{AB} = t_A + t_B$. In the form of multilayered composites (MC), it is also possible to create such heterostructures that naturally can not form in principle. In short-period MCs, representing on the whole are already a relatively 'thick' object, it is possible to realize both the actual size effect of many physical properties of very thin films and the effect of a large specific volume of internal interface boundaries as a special phase. Interest in multilayer composites in recent years has grown so much that large international conferences are devoted to discussing the problems of their growth, structure and physical properties.

A rather large group of metallic systems (Cu–Ni, Au–Ni, Ru–Jr, Co–Cr, Ni–Mo, Ni–V, Mo–V, Nb–Cu, Nb–Al, Nb–Ti, Nb–Zr, Nb–Ta, Ag–Ti, etc.), semiconductor compounds (A_3B_5 and solid solutions based on them, metal chalcogenides, etc.) has been studied.

Depending on the choice of materials forming the MC, and the growth conditions, it is possible to distinguish the following variants of the multilayer nanostructures formed.

1. From alternating amorphous and (or) polycrystalline films. The clarity (structural thickness) of the interphase boundary in this case is determined by the mutual solubility of the components of the composite. This kind of MC can be defined as compositionally modulated in elemental composition (Fig. 3.116 *a*).

2. From single-orientation coherent-conjugate films of related materials with close parameters of crystal lattices (Fig. 3.116 *b*). Coherence can be ensured by elastic deformation of each film (long-range coherence in perpendicular and parallel film directions and a sharp interphase boundary (IB)) or by partial mutual dissolution of components, which determines the non-sharp character of the concentration profile.

As follows from the dimensional and orientational dependences of the IB energy, at $0.9 < \alpha < 1.1$ in the general case and for any α, parallel conjugation is realized for conjugating planes of the densest packing of the same type of crystal lattices, that is, a single-crystal structure from the layer to layer. Considering the features of this variant of heterostructures, we can note the following. With a parallel mutual orientation of the layers to a certain thickness of each layer coherent coupling is possible with compensation of the discrepancy $f_0 = (a_B - a_A)$ by means of elastic deformation leading to transformation of, for example, a cubic lattice into a tetragonal (pseudomorphism). The critical thickness to which coherence of conjugation is maintained depends on the elastic characteristics of the material and the value of f_0, as well as the relaxation mechanism controlled by the growth conditions. For example, in [3.269], by the X-ray method of the (001)Pd–Co MC, it was established that at $t_{Co} = 0.9$ nm and 1.2 nm, the lattice periods in the basal plane become the same (0.271 nm), while the Co tetragonality reaches 0.1 and that of Pd = 0.015.

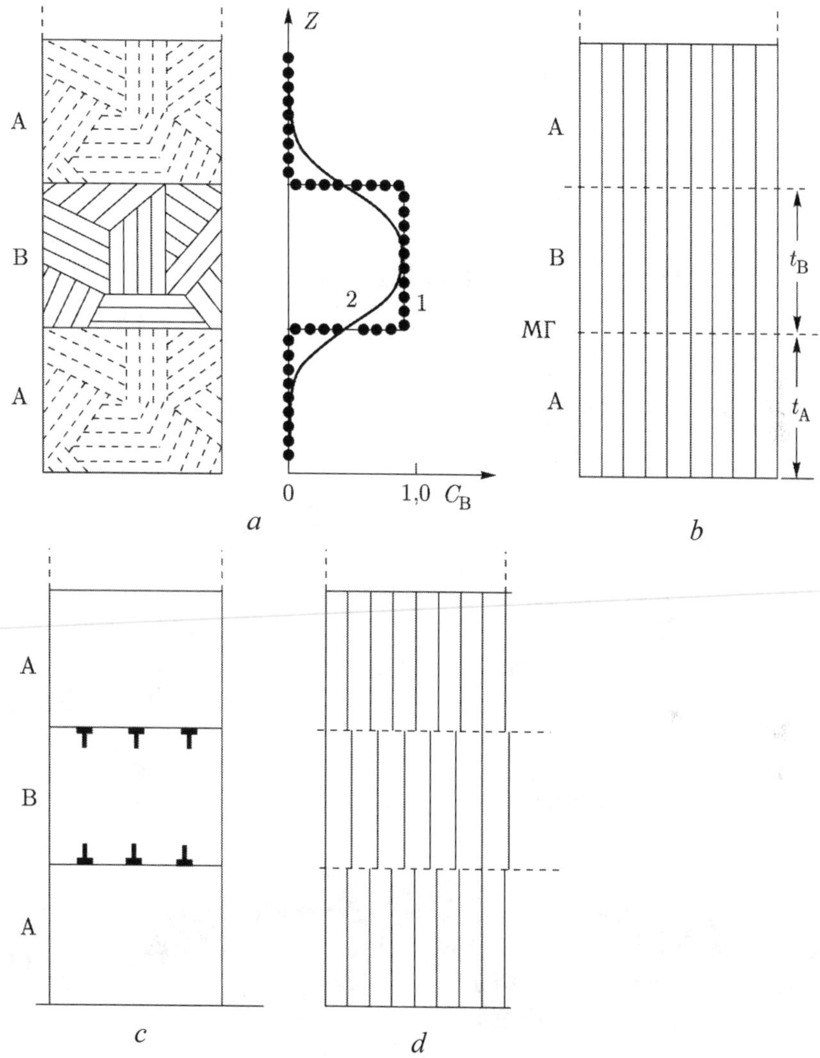

Fig. 3.116. Examples of multilayer film composites: *a* – from alternating amorphous and (or) polycrystalline films (the graph shows conditionally the concentration distribution in the absence of diffusion mixing of components (1) and with mixing (2), *b, c, d* – single-orientation MC with coherent, partially coherent and incoherent conjugation on IB.

Since, in accordance with the diagrams, the dimensional dependence of the orientation of the films upon conjugation of close-packed planes of the same type of lattices should not be manifested in principle, the orientation (111) is most advantageous for the FCC–FCC systems, and the orientation (111)||(001) is the most advantageous for FCC– HCP systems. This is confirmed by the relatively good quality of multilayer epitaxial heterostructures, for example, (111)Au–Ni ($f_0 \sim 0.15$) [3.268].

3. For partially coherent conjugation of films with the formation of a regular dislocation substructure of the IB (Fig. 2.106 *c*). As a result, nanostructures characterized by three-dimensional modulation are formed: in composition along the *z* axis and along the substructure through misfit dislocations (MD) in the IB plane [3.270].

The possibility of creating a periodic modulation of the crystal lattice by the formation of regular grids of MD in IB is demonstrated by electron diffraction [3.271, 3.272] on the corresponding periodic substructure. Depending on the size of the discrepancy (it should be large enough to ensure strict periodicity), the period of the dislocation superlattice can be varied within wide limits. The effect of dislocation superlattices on the properties of this kind of nanostructures is associated with a periodic stress field created by dislocations modulating the electron density in the near-boundary region approximately equal to the dislocation network period.

The most systematic studies of the growth, substructure, and properties of multilayer heteroepitaxial structures of a wide range of metal chalcogenides were performed in [3.273]. For metal chalcogenide pairs with a dimensional mismatch from 0.02 (EuTe-PbTe) to 0.13 (YbS–PbTe), t_k varies from 5 to 1.2 nm, and P_w is from 23 to 3.3 nm. It is established that combinations of layers from narrow-band, wide-band, ferromagnetic and diamagnetic materials broaden the possibilities of manifesting the effects of compositional and substructural organization of film nanostructures in optical, transport, magnetic, and other properties.

To form regular dislocation gratings, it is necessary to ensure the layer growth of the corresponding materials. Of fundamental importance is the choice of a single-crystal substrate: as shown in [3.273], for chalcogenides of lead and tin – KCl and BaF_2, for chalcogenides of rare-earth elements – Si, chalcogenides of lead and tin.

4. One-orientation incoherently conjugated films (Fig. 3.116 *d*): systems with large mismatch and (or) relatively weak interfacial coupling.

5. Multi-orientational heterostructures, when the system allows more than one regular orientation relationship.

Proceeding from the consideration of the above variants of multilayer nanostructures, one can say that the problem of their structure is reduced to the establishment of regularities of conjugation, reliable criteria that predetermine the type of film growth (*B* on the surface *A* and *A* on *B*) and the orientation relations between them, taking into account the sequence of growth, and the structure (substructure) of interphase boundaries. The criteria of oriented crystallisation of films considered at the beginning of Section 3.2.6 can in principle be used as a basis for predicting the orientation relationships and the nature of the substructure of the films forming the multilayered composite.

The authors of [3.274] carried out an electron microscopic analysis of the structure formation of $Fe_{80-78}Zr_{10}N_{10-12}$ films 0.7 and 1.8 μm thick, obtained by reactive magnetron sputtering, and its evolution after annealing at T = 200, 500, 600 and 650°C. holding time 1 h. The phase composition of the films after deposition is represented by a supersaturated solid solution of Zr and N in α-Fe with a BCC lattice and in films 1.8 μm thick, ZrN nitride with an FCC lattice, and in films with a thickness of 0.7 μm – a cluster phase of the composition ZrN. In this case, in a film with a thickness of 0.7 μm the concentration of Zr and N in the solid solution is higher. It was found that annealing with increasing temperature decreases the degree of supersaturation of the solid solution, with zirconium forming the compound ZrN (200, 500°C), and nitrogen – Fe_xN (600, 650°C) and diffusing into the substrate. Thus, a decrease in the nitrogen concentration in the film at high temperatures is due to the interaction with the substrate and the formation of a diffusion layer with an increased concentration of this element. It was revealed that in films annealed at 600°C, due to the counter diffusion of nitrogen from the film into the substrate and oxygen from the substrate into the film at the film–substrate interface a region is formed consisting of several diffusion layers of different phase composition, the formation of which is possible in the Fe–Zr–N–Si–O system. The α-Fe, ZrN, ZrO_2, γ-Fe, and Fe_xN phases are found in the near-boundary region of a 1.8-μm film. It is shown that the structure of the investigated films is represented by nanoscale grains with an average size of 3–5 nm, some of which form columnar clusters formed in the direction of film growth, and is characterized by heterogeneity. The average grain size, the degree of heterogeneity and the number of columnar clusters are larger in films with a thickness of 1.8 μm. The obtained results indicate that the observed differences in the structure and phase composition of the 0.7 and 1.8 μm films in the initial state and after annealing are mainly due to the existence and difference of the temperature gradient arising in the sputtering process and providing conditions for more complete diffusion processes in thicker films.

The authors of [3.275] investigated promising thin-film magnetically soft alloys Fe–Zr and Fe–Zr–N. Sputtered films of compositions Fe–8 at.% Zr and (Fe–8 at.% Zr)–17 at.% N were obtained by specially developed techniques. X-ray diffractometry showed that during 1 hour of annealing the initially amorphous materials at 400°C, crystallisation proceeds and a BCC phase is formed – a solid solution of iron-based alloying elements. X-ray patterns show lines corresponding to the BCC-crystalline phase based on iron, the amount of which increases with increasing annealing temperature up to 550°C. This is evidenced by an increase in the intensity of these lines (see Fig. 3.117).

In this case, the material sputtered in a nitrogen-containing medium exhibits a higher thermal stability of the amorphous state – in it and after annealing at 550°C a significant amount of the amorphous phase remains. This is evidenced by the fact that on the X-ray diffraction pattern of the nitrogen-containing sample, in addition to the BCC-solid solution lines, there was also a halo corresponding to the amorphous phase. In a nitrogen-free sample after annealing for 1 hour at 550°C the material completely converted to the crystalline state – the halo does not form. In addition, the nitrogen-free material annealed at 550°C for 1 hour, showed the formation of the the Fe_2Zr intermetallic. In the case of annealing the nitrogen-containing material the thermal stability of the amorphous phase ws very high and retained to some extent up to 550°C. X-ray spectrum microanalysis showed that the vacuum (residual pressure of 1 MPa) annealing of the nitrogen-containing material leads to a reduction of the nitrogen concentration (17 ± 2) to (9 ± 1) at.%.

3.3. Mechanical properties

The problem of strength occupies one of the leading places in the development of constructional and functional materials of the new generation, since the reliable operation of the latter requires the provision of sufficient load-bearing capacity and a certain margin of safety and resistance to catastrophic failure.

The properties of amorphous–nanocrystalline materials are largely determined by the conditions under which a crystalline phase is formed, since this determines the morphology, phase

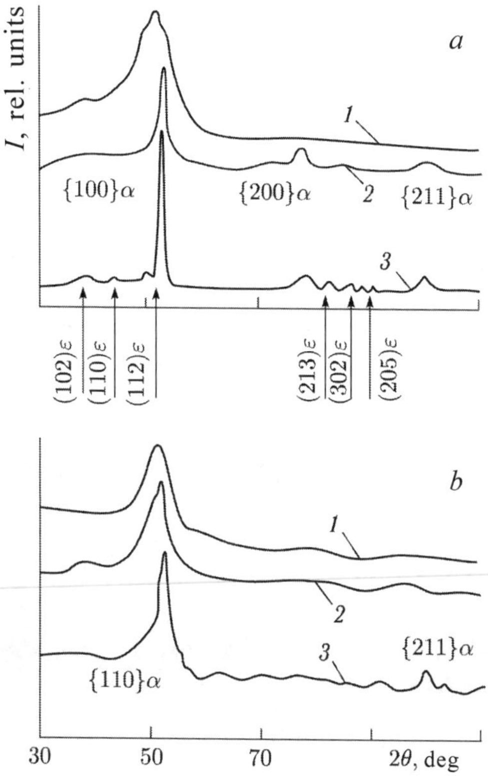

Fig. 3.117. X-ray diffraction patterns of the investigated material sputtered in the Ar (*a*) and Ar + N_2 (*b*) atmosphere in the initial state (1) and after annealing for 1 hour at 400 (2) and 550°C (3); α – iron-based BCC phase; ε – Fe_2Zr.

composition, and the number of structural components in the amorphous–nanocrystalline state. Intensive developments in the field of nanostructured material science were accompanied by extensive studies of the strength of amorphous–crystalline materials, since immediately a significant increase in strength (hardness) and a reduction in the plasticity of these objects was observed.

Microhardness measurements were carried out both for single-component Se [3.276] nanocrystals, and for single-phase ($NiZr_2$ [1.6]) and multiphase (Ni–P [3.277], Fe–Si–Me–B [3.278]) systems. The general rule is that after the formation of nanocrystals in an amorphous matrix the microhardness usually grows. When comparing the amorphous and nanocrystalline states it should be borne in mind that crystallisation leads to a significant (up to 50%) increase in elastic moduli [1.1], which may be the main reason for the increase in hardness.

In section 3.1, the methods of obtaining materials with an amorphous–nanocrystalline structure are discussed in detail; in this section, the mechanical properties of alloys with an amorphous-nanocrystalline structure obtained in various ways are generalized.

Earlier it was shown (section 3.2) that conditionally we distinguish two basic structural states formed in amorphous–nanocrystalline alloys (ANA): 1) alloys with statistically distributed nanoparticles in an amorphous matrix; and 2) alloys containing predominantly isolated nanocrystals separated by amorphous interlayers . The first group of alloys corresponds to the initial stages of the transition from the amorphous to crystalline state in the process of controlled thermal effects, and the second group corresponds to the later (final) stages of this process. In the sections 3.3.1 and 3.3.2 of this chapter, the features of mechanical behaviour for each of these states are considered.

3.3.1. Mechanical properties of amorphous alloys in early stages of crystallisation

$Fe_{70}Cr_{15}B_{15}$ alloy. A detailed study of the mechanical behaviour of the $Fe_{70}Cr_{15}B_{15}$ alloy having an amorphous–crystalline structure, which is a two-phase mixture of an amorphous phase and a eutectic, was conducted in [3.92] and contains a volume fraction of eutectic not more than 0.5. In this alloy, the microhardness *HV* always increases with increasing annealing parameters – time and temperature (Fig. 3.118).

In order to establish the effect of the structure on the mechanical properties of the alloy, the measured value of the microhardness *HV* and the calculated values of the structural parameters of the crystalline phase were compared for each thermal treatment regime and then the parametric dependences of the microhardness *HV* were plotted against the structural parameters D^e, N^e_v, and V^e_v. Since these structural parameters are related (see expression (3.16)), only the dependences $HV(D^e)$ and $HV(V^e_v)$, which, from our point of view, are most visible from the physical point of view, were analysed.

Figure 3.119 shows graphs of the $HV(D^e)$ dependence for a fixed annealing time and for a fixed temperature. There is a normal increase in *HV* with an increase in the size of the eutectic.

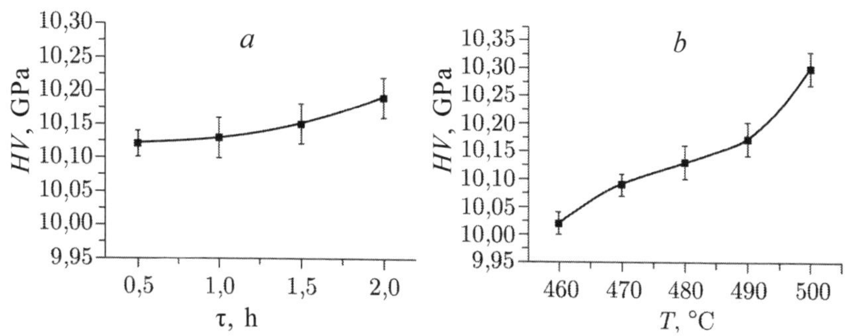

Fig. 3.118. The dependence of HV on the annealing time τ at a fixed temperature (480°C) (*a*) and on the annealing temperature T at constant exposure (1 h) (*b*) for the $Fe_{70}Cr_{15}B_{15}$ alloy.

Figure 3.120 shows graphs of $HV(V_v^e)$ at a fixed annealing time and at a fixed temperature. There is an increase in HV with an increase in the volume fraction of the eutectic. It is known that the presence of crystalline phases (especially carbides or borides) leads, as a rule, to a significant increase in Young's modulus of the amorphous alloy, the dependence of this quantity on the volume fraction of the high-modulus crystalline phase being linear [3.279]. In turn, the yield stress (microhardness) of alloys also increases with increasing volume fraction of crystalline particles in the amorphous matrix in accordance with the rule of additive addition of Young's moduli of the amorphous and crystalline phases [3.279]:

$$HV = HV_0^M\left[1+V_v\left(\frac{E_K}{E_M}-1\right)\right], \qquad (3.47)$$

where E_M and E_K are the Young's moduli of the amorphous and crystalline phases respectively; HV_0^M is the microhardness of the amorphous matrix.

Equation (3.47) analytically analyzes the possible strengthening effect which can be achieved by eutectic crystallisation of the amorphous $Fe_{70}Cr_{15}B_{15}$ alloy. In this case, a crystalline eutectic is formed in the amorphous matrix, which includes not one but two crystalline phases (boride Fe_3B and α-(Fe–Cr)), and therefore formula (3.47) should be somewhat transformed. To do this, instead of E_K

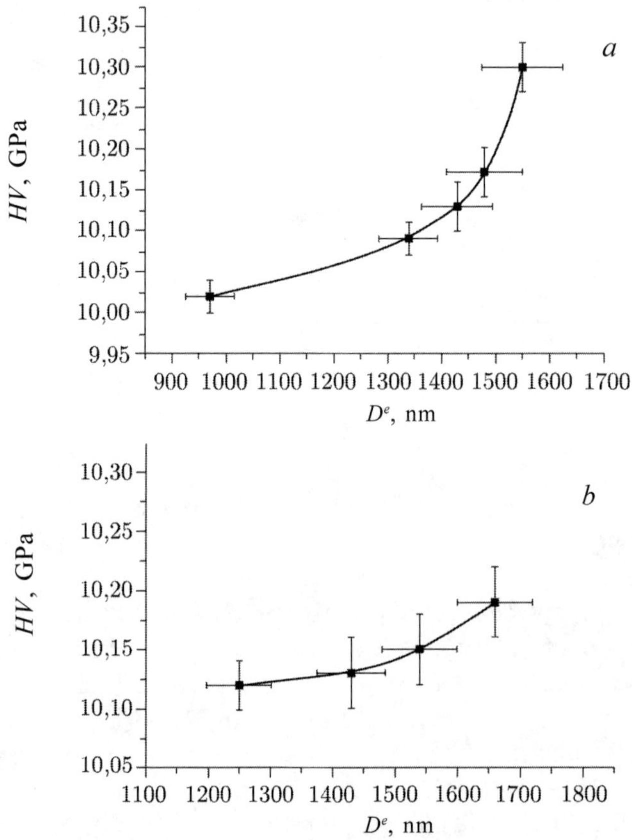

Fig. 3.119. Dependence of the microhardness *HV* of the $Fe_{70}Cr_{15}B_{15}$ alloy on the average size of the eutectic regions D^e at a fixed annealing time (a – τ = 1 h) and at a fixed temperature (b – T = 480°C).

(the Young's modulus of the crystal), the effective value E_E (Young's modulus of the composite eutectic phase) was used for calculations. It was assumed that the components of the eutectic make an additive contribution to the effective value of the Young's modulus:

$$E_E = V_1E_1 + V_2E_2, \tag{3.48}$$

where V_1 and V_2 are the volume fractions, E_1 and E_2 are the Young's moduli of the first and second crystalline phases, respectively, which are part of the eutectic.

Substituting into (3.48) the experimental values of V_1 and V_2, equal to 3/4 and 1/4, and the values of E_1 and E_2, taken from the literature data [1.107, 3.280], we obtain E_E = 240 GPa.

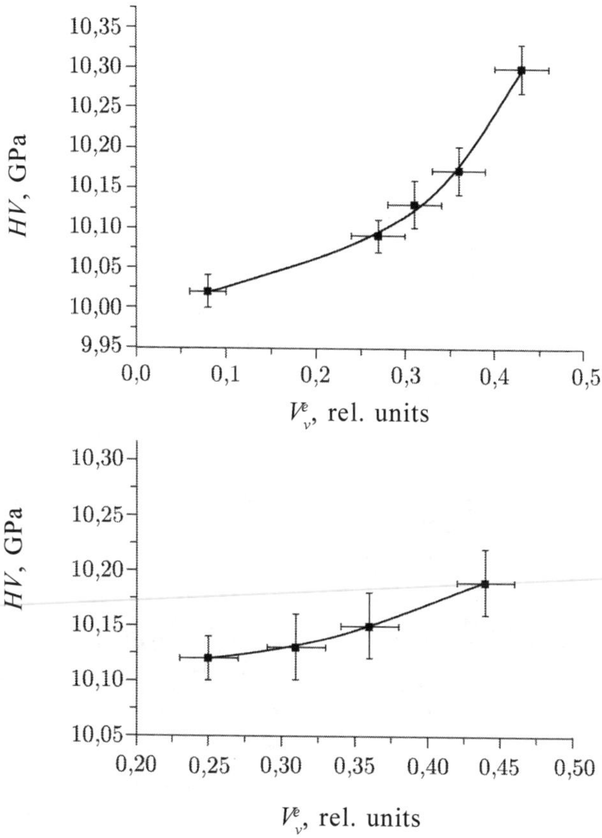

Fig. 3.120. Dependence of the microhardness HV of the alloy $Fe_{70}Cr_{15}B_{15}$ on the volume fraction of the eutectic V_v^e for a fixed annealing time (a – τ = 1 h) and at a fixed temperature (b – T = 480°C).

Figure 3.121 shows the linear dependence $HV(V_v^e)$ calculated at the eutectic crystallisation stage of the amorphous $Fe_{70}Cr_{15}B_{15}$ alloy, using Eq. (3.47), assuming the orrect use of the effective Young's modulus E_E obtained with the aid of expression (3.48). The experimental values of HV and V_v^e measured on samples corresponding to different stages of eutectic crystallisation and obtained after thermal treatments selected in accordance with the temperature–time diagram (see 3.2.1) are plotted on the same graph. It was assumed that the structural states formed after different annealing regimes differ only in the average size, bulk density and volume fraction of the eutectic phase,

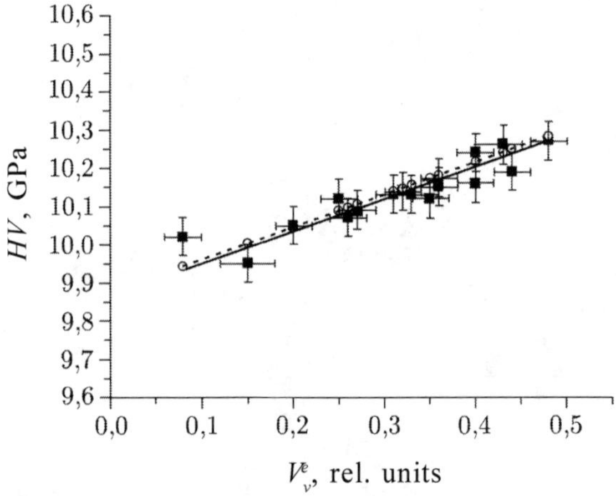

Fig. 3.121. The graph of the dependence of microhardness *HV* on the volume fraction V_v^e of the crystalline phase for all modes of annealing of the $Fe_{70}Cr_{15}B_{15}$ alloy (■ – experimental values, O – values calculated on the basis of equation (3.46))

while the structure of the amorphous matrix, as well as the shape and internal substructure of the eutectic, practically do not change.

As can be seen in Fig. 3.121, the experimental points are in satisfactory agreement with the theoretical dependence. This means that, in this case, the hardening due to the higher values of the elastic moduli of particles of the crystalline phase precipitated in the amorphous matrix (the 'modular' hardening factor) contributes to the hardening of the amorphous $Fe_{70}Cr_{15}B_{15}$ alloy in the process of eutectic phase precipitation.

In addition to the relationship between the elastic moduli of nanoparticles and the amorphous matrix, the structural parameters of the nanocrystalline phase themselves must play an important role in the change in strength during nanocrystallisation: the particle size, their bulk density, volume fraction, type of crystal lattice, texture, distribution by the size and volume of the amorphous matrix, etc.

$Fe_{58}Ni_{25}B_{17}$ alloy. In the $Fe_{58}Ni_{25}B_{17}$ alloy, *HV* decreases with increasing *T* and τ, but the *HV* of the two-phase structure still exceeds the *HV* of the initial amorphous state (Fig. 3.122). With other parameters of heat treatment the dependences are of a similar nature.

In order to understand the physical nature of the change in the strength characteristics in the crystallisation of an amorphous state, it is important to consider how the microhardness varies with the

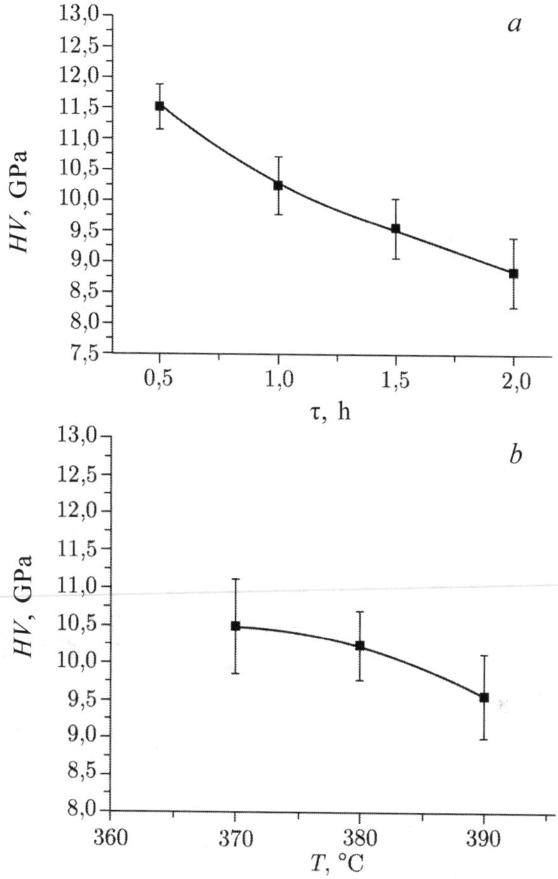

Fig. 3.122. The dependence of *HV* on the annealing time τ at a fixed temperature (380°C) and on annealing temperature *T* at constant exposure (1 h) for the $Fe_{58}Ni_{25}B_{17}$ alloy.

structural parameters of the nanoparticles. For this, parametric dependences of the microhardness *HV* on the structural parameters *D*, N_v and V_v were constructed for each heat treatment mode.

In the $Fe_{58}Ni_{25}B_{17}$ alloy, a noticeable drop in the microhardness *HV* is observed with the growth of all structural parameters (Figs. 3.123, 3.124) at a fixed annealing time (1 h) and at a fixed temperature (380°C). The dependences after annealing at other temperatures and with constant annealing times look similar. Dependences $HV(V_v)$ were built analytically on the basis of the experimental data presented in Fig. 3.123, 3.124 (*a*, *b*), taking into account the fact that the volume fraction $V_v = \pi \cdot N_v \cdot D^3/6$.

For the values of $V_v \leq 0.1$ for a fixed $T = 380°C$ there is a noticeable decrease in the microhardness, and at higher values of V_v the dependence $HV(V_v)$ goes to saturation. The dependences $HV(V_v)$, corresponding to other temperatures and fixed annealing durations, although differ in detail, in principle have a similar character. Such an unusual, falling character of *HV* dependences with the increase of all three structural parameters analyzed by us is connected apparently with large (more than 100 nm) average crystal sizes formed after annealing in the given used regimes.

$Fe_{50}Ni_{33}B_{17}$ and $Ni_{44}Fe_{29}Co_{15}B_{10}Si_2$ alloys. For $Fe_{50}Ni_{33}B_{17}$ alloy, as a result of heat treatment, the *HV* value always increases linearly (Fig. 3.125) with respect to the initial amorphous state (8.71 GPa), reaching a maximum value (10.30 GPa) after annealing at 390°C for 2 hours.

Due to the fact that in the $Fe_{50}Ni_{33}B_{17}$ alloy the size of the nanoparticles did not change during annealing ($D = 20$ nm) (see Fig. 3.40), it was excluded from consideration as a parameter influencing the measurement of *HV*. In this case, only the dependence $HV(N_v)$ and the $HV(V_v)$ dependence analogous to it were analyzed in this case (Figs. 3.126, 3.127).

The $HV(V_v)$ dependences for the $Fe_{50}Ni_{33}B_{17}$ alloy after annealing at a constant annealing time of 0.5 h (Fig. 3.126) or at a constant temperature of 370°C were close to linear (Fig. 3.127). For other annealing parameters they turned out to be similar. In some cases, the dependence $HV(V_v)$ could be described not by a linear but by a power law with the output of the *HV* value for saturation.

In the $Ni_{44}Fe_{29}Co_{15}B_{10}Si_2$ alloy, as in the $Fe_{50}Ni_{33}B_{17}$ alloy, an increase in the annealing temperature and duration always leads to an increase in the microhardness (from 8.80 GPa in the initial amorphous state and up to 12.10 GPa after at annealing 440°C, 2 h) (Fig. 3.130). Since this alloy exists in a two-phase state in a wider (in comparison with other alloys) temperature range, it was possible to obtain a larger number of experimental values and, consequently, to increase the accuracy of the experimental results.

Since the average size of the nanoparticles at constant annealing parameters in the alloy $Ni_{44}Fe_{29}Co_{15}B_{10}Si_2$ is constant (20 nm) (see Fig. 3.46), then, just as for the $Fe_{50}Ni_{33}B_{17}$ alloy, only the dependence $HV(N_v)$ and the identical $HV(V_v)$ dependence, monotonically

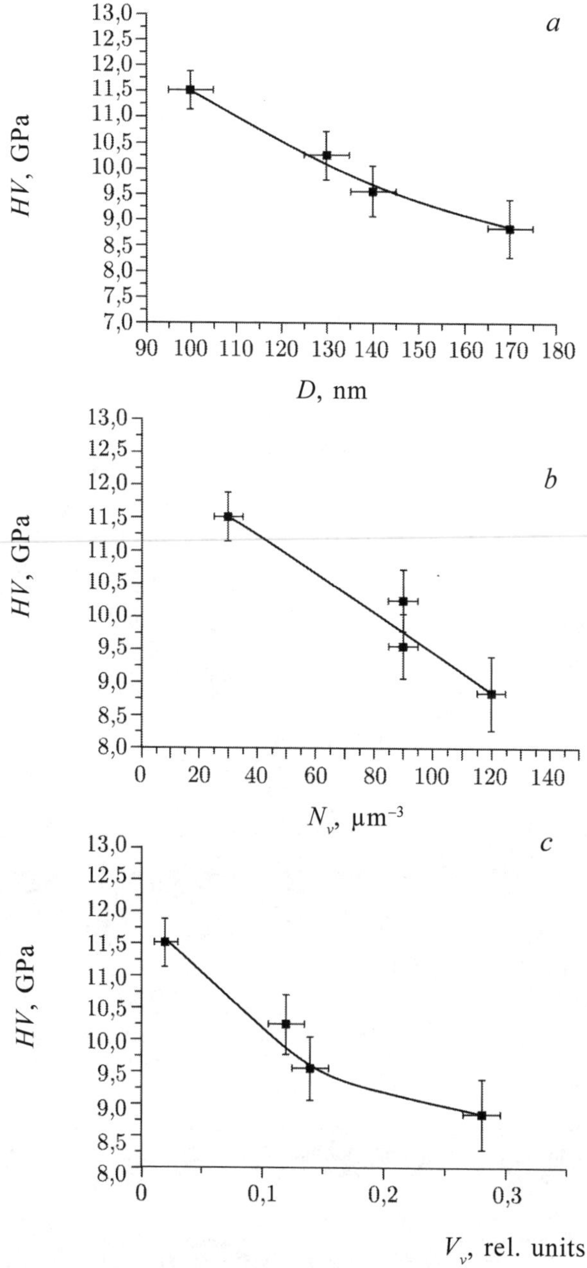

Fig. 3.123. The dependence of microhardness of *HV* of $Fe_{58}Ni_{25}B_{17}$ alloy on D, N_v and V_v of nanoparticles at a fixed annealing temperature (380°C) .

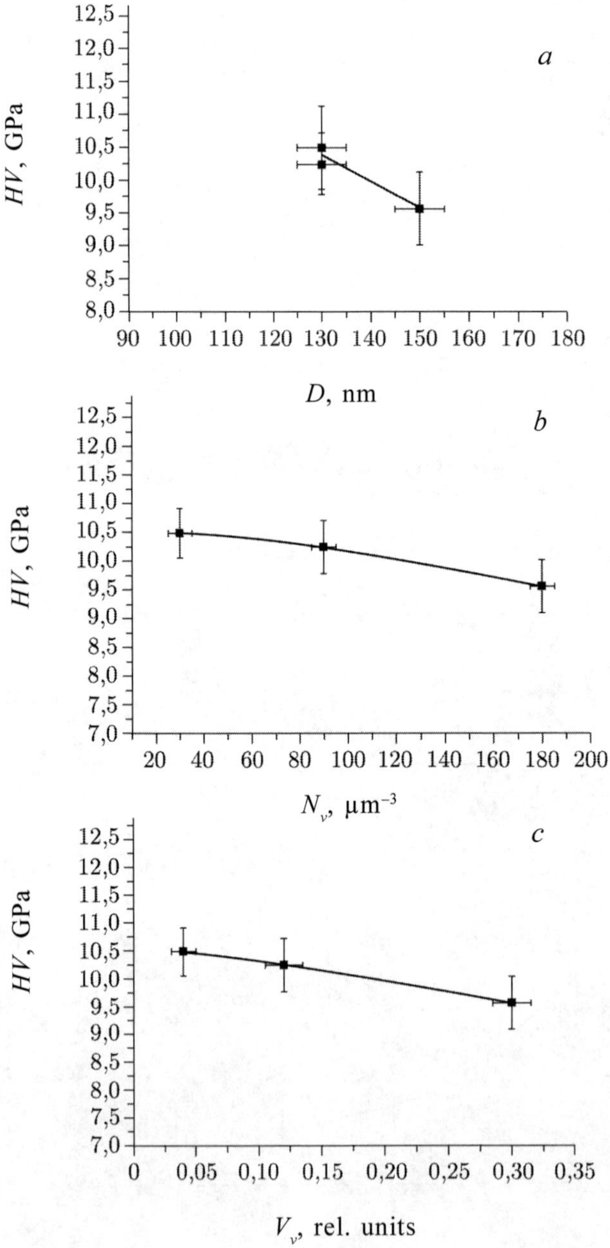

Fig. 3.124. TThe dependence of microhardness of *HV* of $Fe_{58}Ni_{25}B_{17}$ alloy on *D*, N_v and V_v of nanoparticles at a fixed annealing time of 1 h.

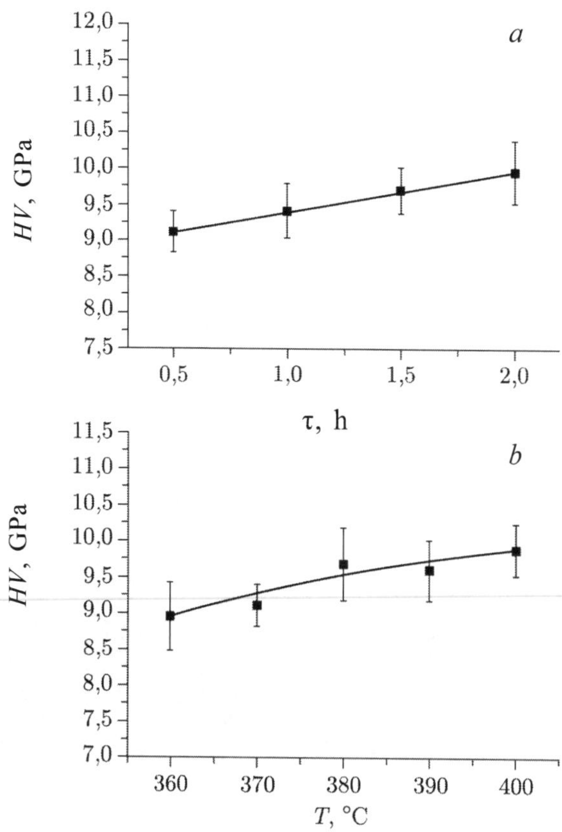

Fig. 3.125. The dependence of *HV* on the annealing time τ at a fixed temperature (380°C) and on annealing temperature *T* at constant exposure (1 h) for the $Fe_{58}Ni_{25}B_{17}$ alloy.

increasing with an increase in V_v (Fig. 3.131) were analysed. This dependence is fixed for all annealing regimes.

As is known, a possible cause of the influence of the nanoparticles on the strength of amorphous alloys is the higher value of their Young modulus [3.279].

In the amorphous alloys $Fe_{50}Ni_{33}B_{17}$ and $Ni_{44}Fe_{29}Co_{15}B_{10}$Si2, the nanoparticles that are BCC or FCC solid substitutional solutions based on iron and nickel precipitate during crystallisation (see section 3.2.2). The boron present in large amounts in the alloys (17 at.% in the $Fe_{58}Ni_{25}B_{17}$, $Fe_{50}Ni_{33}B_{17}$ allloys as well as 10 at.% in $Ni_{44}Fe_{29}Co_{15}B_{10}Si_2$ alloy) practically does not dissolve in nanoparticles (less than 0.1%) [3.281]. This means that all boron atoms remain in the amorphous matrix during the nanocrystallisation process. In this

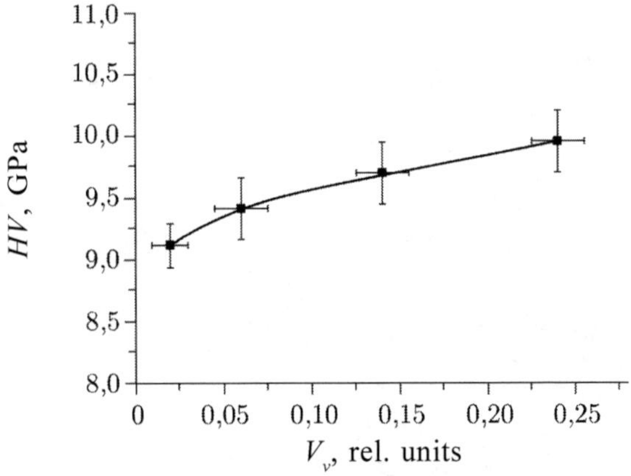

Fig. 3.126. Dependence of the microhardness of $Fe_{50}Ni_{33}B_{17}$ alloy on the volume fraction of nanoparticles V_v for a fixed annealing time (τ = 0.5 h).

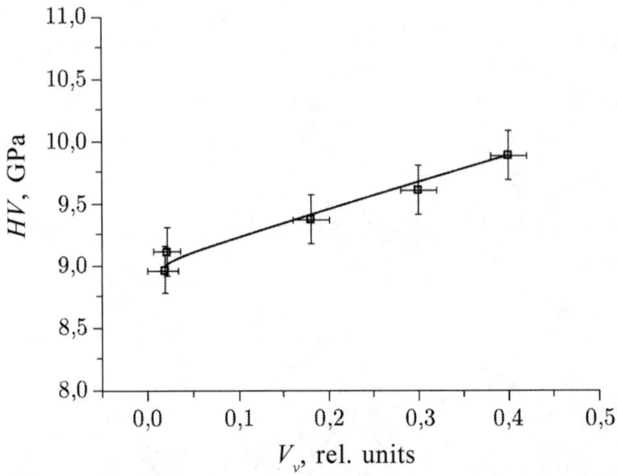

Fig. 3.127. Dependence of the microhardness *HV* of the $Fe_{50}Ni_{33}B_{17}$ alloy on the volume fraction of nanoparticles V_v at a fixed temperature (T = 370°C).

case, the sharp difference between the E values of the amorphous and crystalline phases, which can reach 30–50% for the same chemical composition of the alloy [1.107], is noticeably levelled. In fact, for the nanocrystals precipitated in the studied alloys, the value of E is in the range 200–210 GPa [3.280], and the E value for an amorphous matrix retaining a high boron concentration varies within 195–200 GPa [3.281]. Thus, the maximum possible value of E_K/E_M is an insignificant value of 1.076, and hence the hardening due to the difference in the elastic moduli of the amorphous and crystalline phases (equation (3.47)) can not be considered for these alloys as

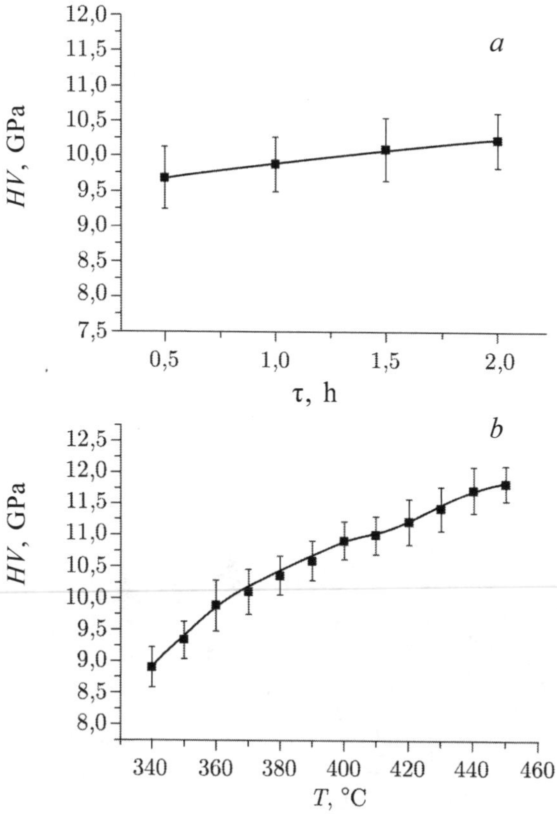

Fig. 3.128. Dependence of the microhardness *HV* on annealing time τ at a fixed temperature (360°C) and on annealing temperature *T* at constant holding time (1 h) for $Ni_{44}Fe_{29}Co_{15}B_{10}Si_2$ alloy.

the main reason for the change in the strength of amorphous alloys in the stage of their crystallisation.

Another possible reason for the increase in strength as the volume fraction of nanocrystalline particles increases is the retardation of crystal nanoparticles propagating in the amorphous matrix (the 'structural' hardening factor). The process is similar to the 'retardation' of moving dislocations in a crystal containing coherent or incoherent particles of the second phase. Such effects were indeed observed experimentally by electron microscopy (see section 3.3.1.2) and calculated theoretically [3.282].

Thus, for the $Fe_{50}Ni_{33}B_{17}$ alloys and $Ni_{44}Fe_{29}Co_{15}B_{10}Si_2$ seems possible to analyze in pure form (without additional significant influence the effect of the elastic moduli) the hardening associated with the interaction of propagating in the amorphous matrix by

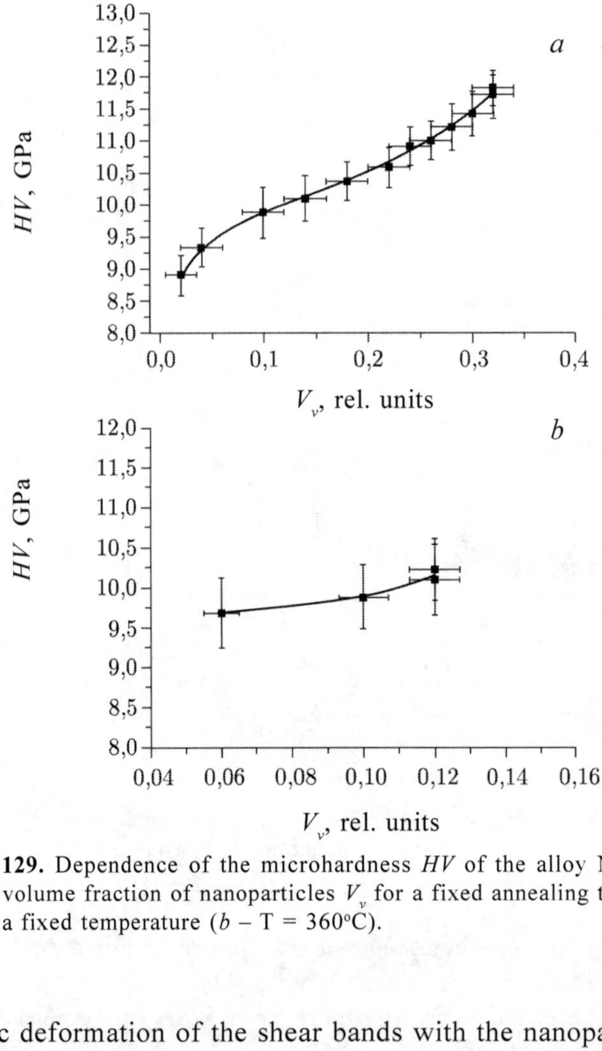

Fig. 3.129. Dependence of the microhardness *HV* of the alloy $Ni_{44}Fe_{29}Co_{15}B_{10}Si_2$ on the volume fraction of nanoparticles V_v for a fixed annealing time (a – τ = 1 h) and at a fixed temperature (b – T = 360°C).

plastic deformation of the shear bands with the nanoparticles formed in an amorphous matrix, depending on their volume density and size.

Figures 3.130 and 3.131 show for the $Fe_{50}Ni_{33}B_{17}$ and $Ni_{44}Fe_{29}Co_{15}B_{10}Si_2$ alloys the total dependences $HV(V_v)$ for all heat treatment modes used in the experiments and, consequently, for all realized two-phase states. When analyzing them, it is necessary to take into account two circumstances. First, the comparison of the structural states obtained at different temperatures and annealing times permits a certain error due to the fact that the structural states of the amorphous matrix and, possibly, of the nanocrystalline particles may differ somewhat. Secondly, since the alloys $Fe_{50}Ni_{33}B_{17}$ and $Ni_{44}Fe_{29}Co_{15}B_{10}Si_2$ are characterised by the effect of stabilizing

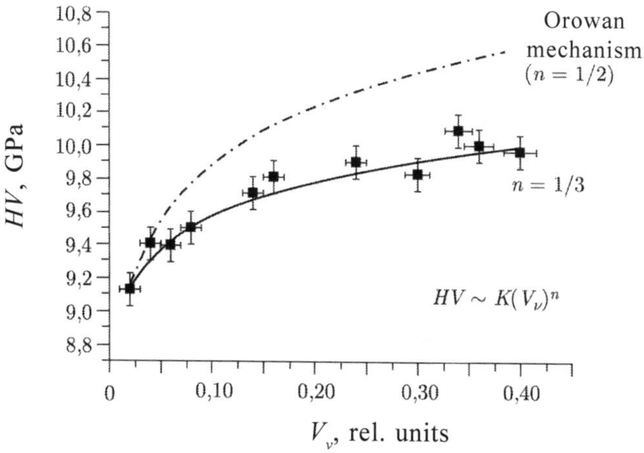

Fig. 3.130. Graph of the dependence of the microhardness *HV* on the volume fraction V_v of the crystalline phase for all annealing modes for $Fe_{50}Ni_{33}B_{17}$ alloy.

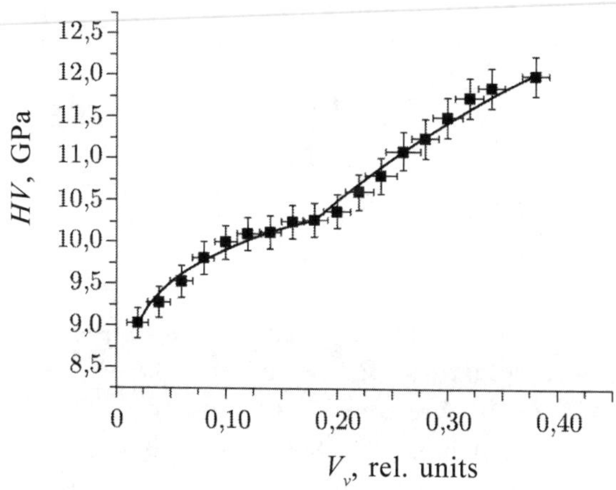

Fig. 3.131. The graph of the dependence of *HV* on the volume fraction V_v of the crystalline phase for all annealing regimes of the $Ni_{44}Fe_{29}Co_{15}B_{10}Si_2$ alloy.

the nanoparticle sizes, the graphs in Figs. 3.130 and 3.131 essentially characterize the dependence $HV(N_v)$, since $V_v \sim N_v$.

The nature of the change in the microhardness of the $Fe_{50}Ni_{33}B_{17}$ alloy in relation to the volume fraction of the nanoparticles (Fig. 3.130) can be described by the dependence

$$HV \sim K(V_v)^n, \tag{3.49}$$

where $n = 1/3$.

The dependence $HV(V_v)$ in Fig. 3.131 for the $Ni_{44}Fe_{29}Co_{15}B_{10}Si_2$ alloy is divided into two sections ($((V_v)_{cr} = 0.2)$, each of which is analogous to that obtained for the $Fe_{50}Ni_{33}B_{17}$ alloy. One can find an analogy between the hardening in question in the case of nanocrystallisation and the hardening due to the retardation of dislocations sliding in a crystal on incoherent and non-intersected particles, described by the Orowan mechanism [3.283]. This mechanism is proposed for calculating the hardening by the non-coherent particles located at a distance much greater than their radius. The mechanism is applicable to the shear modulus of the particles much larger than the shear modulus of the matrix. According to Orowan's model, the dislocations in motion are held on the particles until the applied stress is sufficient to cause the dislocation line to bend and pass between the particles, leaving a dislocation loop around them. In accordance with the Orovan mechanism, the Ashby theory developed on the basis of it, and the modified theory of planar clusters of the Orowan loops on particles, which satisfactorily describe the experimental results, we obtain [3.283]

$$\sigma = \sigma_0 + AG\left(\frac{V_v b}{D}\right)^{1/2} \varepsilon, \tag{3.50}$$

where σ is the deforming stress, σ_0 is the deforming stress in a crystal that does not contain particles, c is a constant equal to 0.1-0.6, G is the shear modulus, V_v is the volume fraction of particles, b is the Burgers vector of sliding dislocations, D is the particle size, and ε is the degree of plastic deformation.

Thus, there is a noticeable similarity between the influence of the volume fraction of particles and the bulk density of the particles on hardening in crystalline materials and in amorphous alloys (Fig. 3.130). The difference lies in the fact that the value of the exponent n in the crystals is 1/2, and in the amorphous materials it is 1/3. This analogy is not unexpected, since the shear bands realizing the plastic shift in the amorphous state are at the mesolevel essentially effective dislocations the Burgers vector of which is not exactly defined. In this case, the shear strain capacity in such a band is thousands of percent [1.124]. A lower value of n indicates that the retardation of the shear bands by particles in the amorphous alloys is less effective than in the crystals by the Orowan mechanism. Apparently, the nanocrystalline particles of about 20 nm in size are partially cut, or there is another mechanism, which is more effective than in crystals, for overcoming the shear bands of particles of the nanocrystalline phase.

$Fe_{73.5}Si_{13.5}B_9Nb_3Cu_1$ alloy. Figures 3.132 and 3.133 show the dependences of the measured microhardness *HV* on the annealing time τ (Fig. 3.132) and on annealing temperature *T* (Fig. 3.133), respectively. It can be noted that as the two heat treatment parameters increase, the *HV* value is constantly increasing. The dependences of *HV*(τ) are typical of the kinetic curves, which go to saturation after τ = 1 h at higher *HV* values which increase with increasing *T* (Fig. 3.132).

The *HV*(*T*) dependences contain three growth stages: at $T \leq 500$ °C there is a complex course determined by τ; in the temperature range 520–560°C the dependences are close to linear and do not depend on the annealing time; finally, at $T \geq 580$°C the dependences go to saturation, and the values of *HV* are very close to each other.

The three stages on the *HV*(*T*) curves (Fig. 3.133) can be explained if we compare the dependences in Figs. 3.52 and 3.54. In the first stage of annealing (up to 520°C), both the average size of the nanocrystals *D* and their bulk density N_v increase, while in the second stage (annealing at 520–560°C) only N_v grows at a constant parameter *D*, and in the third stage (560–580°C), both structural parameters of nanocrystals are practically constant. Accordingly,

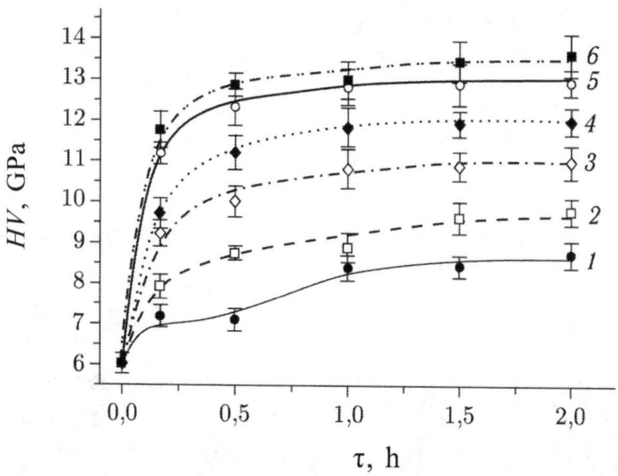

Fig. 3.132. Dependences of the microhardness *HV* on τ at a fixed temperature for the $Fe_{73.5}Si_{13.5}B_9Nb_3Cu_1$ alloy (1 – 480, 2 – 500, 3 – 520, 4 – 540, 5 – 560, 6 – 580°C).

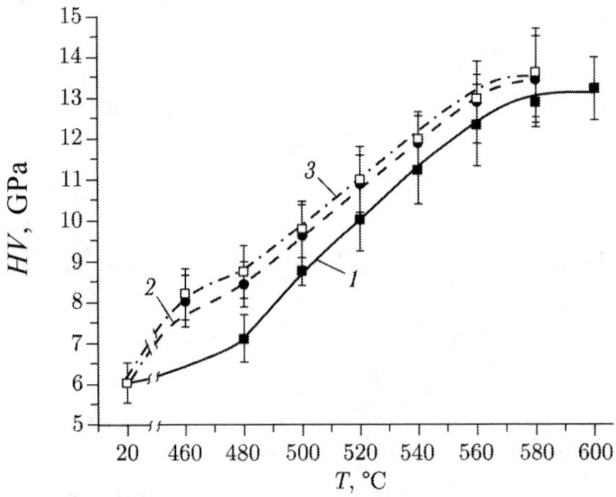

Fig. 3.133. Dependences of microhardness *HV* on annealing temperature *T* at different annealing times for the alloy $Fe_{73.5}Si_{13.5}B_9Nb_3Cu_1$ (1 – 0.5, 2 – 1.5, 3 – 2 h).

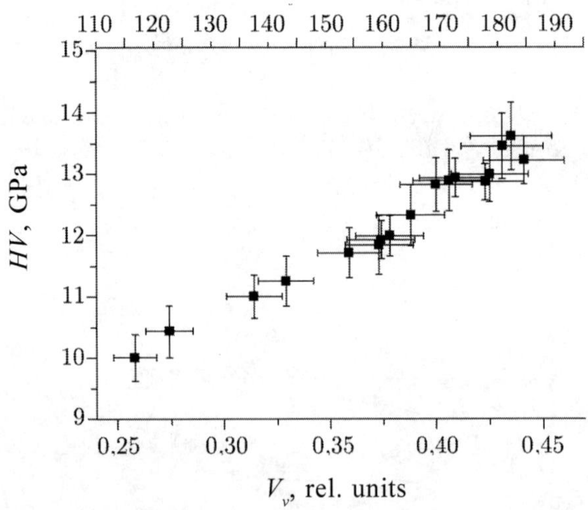

Fig. 3.134. The experimental dependence of microhardness *HV* on the bulk density N_v (volume fraction V_v) for all annealing modes of $Fe_{73.5}Si_{13.5}B_9Nb_3Cu_1$ alloy at a constant D = 16 nm.

HV in the first stage is determined by two independent structural parameters (D and N_v), and in the second stage – only one (N_v).

Let us use the fact that after the heat treatments of the the $Fe_{73.5}Si_{13.5}B_9Nb_3Cu_1$ alloy at 520–560°C only one parameter of the nanoparticles (bulk density N_v) changes and we analyze in pure form the dependence $HV(N_v)$ at D = const = 16 nm.

For this purpose, Fig. 3.134 shows all the values of *HV* obtained by us after certain heat treatment regimes and the N_v values corresponding to these states, for which D = const = 16 nm. We assume that changes in the strengths (microhardness) of structural states in the amorphous–nanocrystalline state are determined only by a change in the bulk density of the nanoparticles, while all other possible differences in the structure can be neglected. In addition, since the volume fraction of the nanocrystalline phase V_v for the spherical nanoparticles is determined by the relation [3.93] $V_v = \pi \cdot N_v \cdot D^3/6$, then, therefore, for D = const, each value of N_v in Fig. 3.134 corresponds to a certain value of V_v, and the dependences $HV(N_v)$ and $HV(V_v)$ are of a similar nature.

As already shown for the $Fe_{70}Cr_{15}B_{15}$, $Fe_{50}Ni_{33}B_{17}$ and $Ni_{44}Fe_{29}Co_{15}B_{10}Si_2$ alloys, there are two main reasons for the increase in strength (microhardness) as N_v and V_v grow in the nanocrystallisation stage: a higher value of Young's modulus of the nanocrystalline particles evolved compared to the amorphous matrix ('modular' hardening factor) (equation (3.47)); the interaction of shear bands propagating during plastic deformation through the amorphous matrix and the nanoparticles of the crystalline phase encountered in their path ('structural' hardening factor) (equation (3.49)).

Substituting into the equations (3.47) and (3.49) the experimental value of the bulk density N_v of the crystalline particles obtained under all conditions of controlled heat treatment of the alloys $Fe_{70}Cr_{15}B_{15}$ and $Fe_{50}Ni_{33}B_{17}$, and taking the Young's modulus of the matrix equal to the Young's modulus of the amorphous alloy $Fe_{78}B_{10}Si_{12}$ (E = 120 GPa), and Young's modulus of the crystalline particles of the α-phase (Fe–16 at.% Si) equal to E = 170 GPa [3.15], we obtain (by two calculations based on the equations (3.47) and (3.49)) two dependences of various contributions to hardening $HV^{(1)}(N_v)$ (Fig. 3.135 *a*) and $HV^{(2)}(N_v)$ (Fig. 3.135 *b*) due to the above reasons.

The overall dependence (obtained by summing) of the microhardness $HV^{(\Sigma)}(N_v) = HV^{(1)}(N_v) + HV^{(2)}(N_v)$ (dependence 3 in Fig. 3.135 *c*) describes satisfactorily the experimental results (Fig. 3.134). This means that in this case both factors contribute to the hardening of the amorphous alloy in the process of primary nanocrystallisation: the hardening associated with the higher Young's modulus of crystalline particles (the 'modular' factor), and the hardening associated with the interaction of shear bands that realize

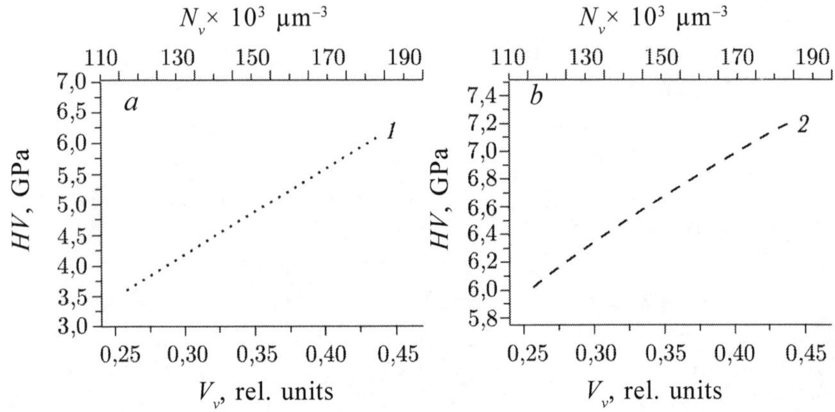

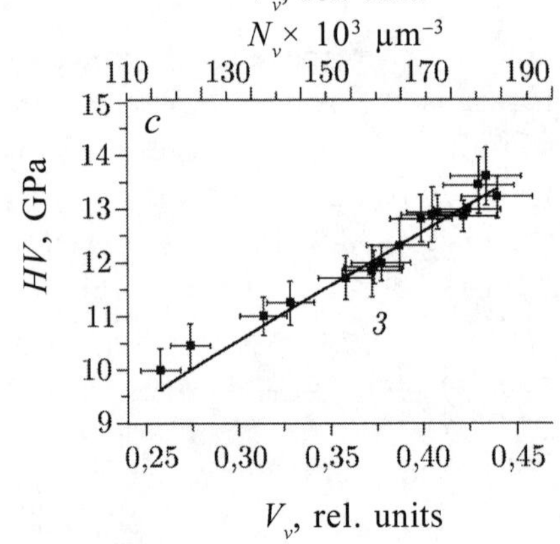

Fig. 3.135. Dependences of microhardness *HV* on the volume fraction $V_v(N_v)$ for all annealing modes of $Fe_{73.5}Si_{13.5}B_9Nb_3Cu_1$: *a*, *b* – theoretical dependences (1 – 'modular' and 2 – 'structural' hardening factors), *c* – experimental values (■) (3 is the total dependence).

the deformation of the amorphous–nanocrystalline alloy, with particles of the nanocrystalline phase ('structural' factor).

As a result of crystallisation of rapidly solidifying amorphous Al–Cr–Ce–M (M–Fe, Co, Ni, Cu) aluminum alloys with a content of more than 92 at.% Al, a structure is formed in which there is an amorphous phase and icosahedral nanoparticles enriched in Al (Fig. 3.136) [3.285].

These alloys have a high tensile strength (up to 1340 MPa), close or superior to the strength of steels. The main causes of high tensile strength are the spherical morphology of icosahedral nanoparticles and the presence of a thin layer of aluminum around them.

Consequently, an amorphous–crystalline composite can, with an optimal combination of volume fraction, structure and morphology of its constituents, possess higher mechanical properties than the crystalline and even amorphous materials. The problem is only to achieve an optimal combination of strength and plasticity when using

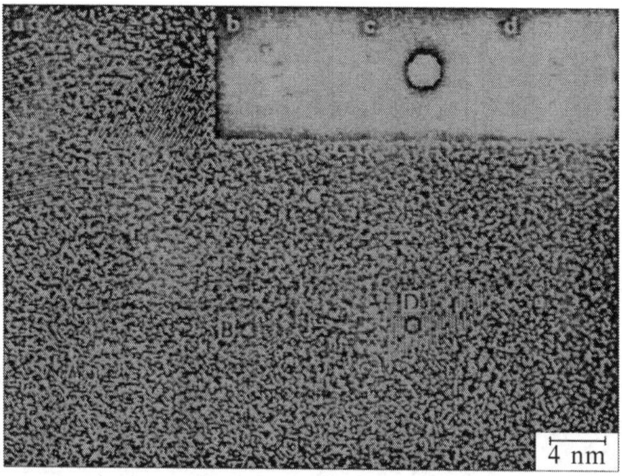

Fig. 3.136. Micrograph of a rapidly hardening $Al_{94.5}Cr_3Ce_1Co_{1.5}$ alloy [3.285]: icosahedral nanoparticles *B*, *D*, etc. with an average size of ~5–10 nm are distributed in the amorphous matrix C; *b*, *c* and *d* are diffraction patterns from 1 nm diameter sections marked by circles and belonging to regions *B*, *C* and *D*, respectively..

a fairly easily reproducible technology for obtaining the material and its subsequent heat treatment.

3.3.1.1. Influence of the size of nanocrystalline particles on strength

Let us turn to the question of how the size of nanocrystals (size effect) affects the hardening of an amorphous matrix.

In the $Fe_{50}Ni_{33}B_{17}$, $Ni_{44}Fe_{29}Co_{15}B_{10}Si_2$, the effect of retardation of the growth of the nanocrystalline phase at the crystallisation stage (see section 3.2.1) is observed and, therefore, it is not possible to determine the size effect.

In the $Fe_{58}Ni_{25}B_{17}$ alloy, after the selected heat treatment conditions, the average crystal size changes from 100 to 170 nm, and in principle it is possible to follow the size effect for the *HV* value. But such an analysis will be incorrect, since when the average particle size in this alloy changes, their bulk density N_v also changes. To exclude the influence of the parameter N_v, the following procedure was used.

As can be seen from Figs. 3.130 and 3.131, the dependence $HV = f(V_v)$ in the alloys $Fe_{50}Ni_{33}B_{17}$ and $Ni_{44}Fe_{29}Co_{15}B_{10}Si_2$ has the same form: $HV \sim K(V_v)^{1/3}$. It can be assumed that the same dependence will also be valid for the $Fe_{58}Ni_{25}B_{17}$ alloy. In this case, we can bring all the *HV* values obtained for the $Fe_{58}Ni_{25}B_{17}$ alloy to some values corresponding to the same constant value $(N_v)_0$. For the standard value, we have chosen $(N_v)_0$ = 100 μm^{-3}, since it is in the middle of the interval of all the obtained N_v values for the $Fe_{58}Ni_{25}B_{17}$ alloy after the heat treatment modes used (50–160 μm^{-3}).

The given values $(HV)_0$, corresponding to $(N_v)_0$, were determined for different states using expression

$$HV_0 \cong HV_i + K_{\text{cor}}\left[(N_v)_0 - (N_v)_i\right], \tag{3.51}$$

where $(HV)_0$ and $(HV)_i$ are respectively the reduced and true values of the microhardness of the given structural state of the amorphous–nanocrystalline alloys (ANA), $(N_v)_0$ and $(N_v)_\text{i}$ are respectively the reduced (100 μm^{-3}) and the true value of the bulk density of nanocrystals of the given structural state of ANA, and K_{cor} is the correction parameter, which was calculated as the average value of the coefficients for the $Fe_{50}Ni_{33}B_{17}$ and $Ni_{44}Fe_{29}Co_{15}B_{10}Si_2$ alloy and was K_{cor} = 38.5 MPa · μm^3.

Figure 3.137 shows the dependence of the corrected value of the microhardness HV_{cor} for the $Fe_{58}Ni_{25}B_{17}$ alloy on the average size of the nanocrystals *D*. Dark circles denote the values of the HV_{cor} corresponding to the initial amorphous state (D = 0) and to the states obtained after the thermal treatments (D = 100–170 nm). It can be seen that for D > 100 nm the microhardness *HV* decreases with the average size of nanoparticles *D*. At the same time, comparing the

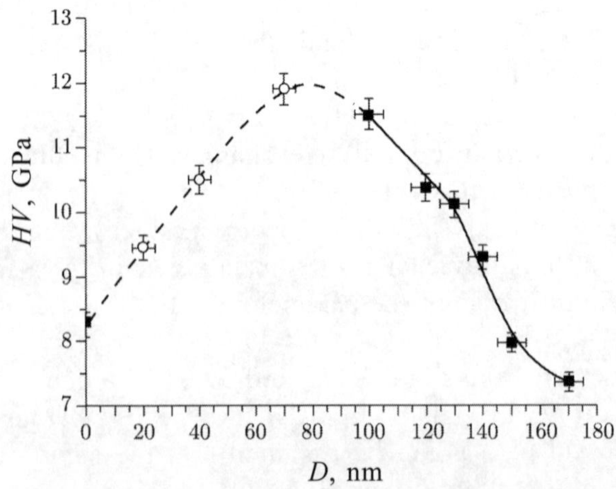

Fig. 3.137. Dependence of the corrected value of microhardness HV_{cor} on the average size *D* of nanocrystals for the $Fe_{58}Ni_{25}B_{17}$ alloy for all the investigated states.

values of HV_{cor} for $D = 0$ (amorphous state) and for $D > 100$ nm, we can assume that the dependence $HV_{cor} = f(D)$ is generally complex and non-monotonic. In order to reproduce this dependence in its full form, we lack the values of the HV_{cor} in the range $D < 100$ nm.

Additional experiments were carried out to obtain these data. Samples of $Fe_{58}Ni_{25}B_{17}$ alloy were annealed isothermally at 400°C for very short time intervals (3, 5 and 10 min) in order to fix the earliest stages of nanocrystallisation. For the values of the average size of nanocrystals located in the amorphous matrix, the corrected microhardness values at $Nv = 100\ \mu m^{-3}$ were calculated and are plotted on the graph shown in Fig. 3.137 in the form of light circles. In a more complete form, the dependence $HV_{cor} = f(D)$ is a curve with a maximum corresponding to $D = 70$–80 nm.

It is known that for single-phase nanocrystals the dependence of the strength on the grain size is anomalous, which for ordinary polycrystals is usually described by the Hall–Petch relation [3.12] (Fig. 2.4)

$$HV(\sigma_T) = HV_0(\sigma_0) + k_y D^{-1/2}, \tag{3.52}$$

where HV is the hardness, σ_T is the yield strength, HV_0 is the hardness of the body of the grain, σ_0 is the internal stress, preventing the propagation of plastic shear in the body of grain, k_y is the coefficient of proportionality and D is the average grain size.

In many experimental studies [2.11, 3.286-3.289] devoted to the study of the mechanical properties of nanocrystalline materials, it was found that in the nanometer grain size range there are significant deviations from the standard function (1 in Fig. 2.4.) In the range $D < 30$–50 nm noticeable deviations from the Hall–Petch ratio begin (Fig. 3.138).

The apparent similarity of the dependences $HV_{cor} = f(D)$ in Fig. 3.137 and $HV = f(D)$ in Fig. 3.138 is clearly visible. It is important to keep in mind that the dependences on the average size of nanocrystals for two completely different structural states are compared. The dependences in Fig. 3.138 were obtained for a single-phase polycrystalline (nanocrystalline) state. In Fig. 3.137 is a two-phase amorphous crystal structure when the crystalline phase is distributed in the form of statistically located equiaxed nanoparticles with a volume fraction not exceeding 0.5.

We shall conventionally assume that the dependence on the right of the region of the maximum of HV in Fig. 3.137 is 'normal', since it smoothly passes into the region of the crystalline particles of

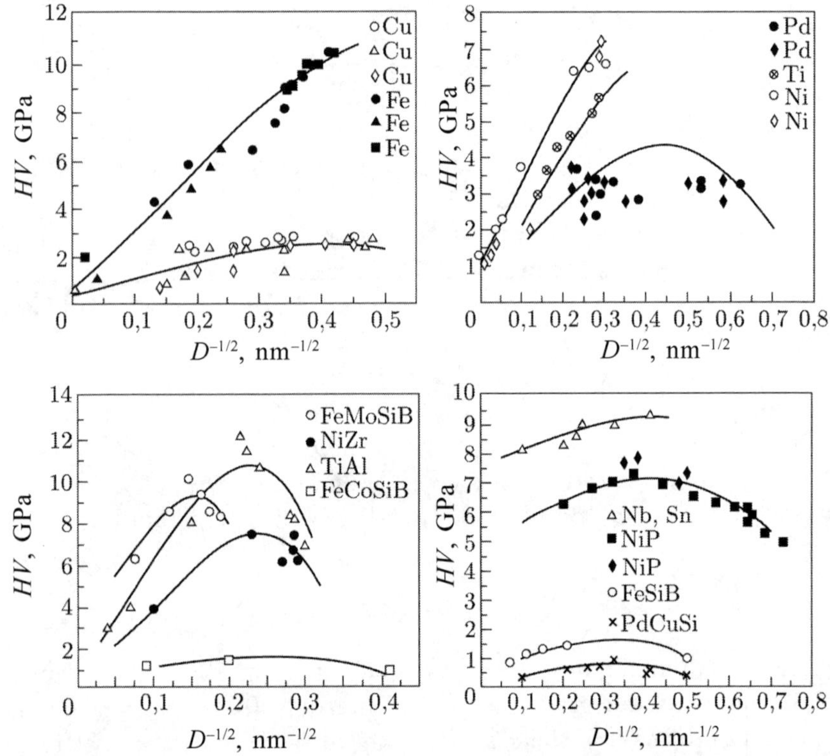

Fig. 3.138. Dependence of hardness on the grain size of nanocrystals for various alloys [2.11].

ordinary size far from the nanometer range. This dependence differs somewhat from those that occur in crystals containing particles of the second phase [3.283], but it is physically easy to understand from the following reasoning: as the size of the particles of the second phase grows, the shear bands on the crystalline particles become less effective due to the more active flow of deformation processes in the particles themselves. With a decrease in their size and a transition to the region of the nanoparticles, dislocation sliding inside them becomes less active, and the moment comes when the nanocrystalline phase behaves like rigid, non-intersected particles of the second phase, in which the dislocation and other relaxation processes are completely suppressed. At the same time, at $D < 70$ nm, anomalous processes begin to occur that do not fit into the structural model developed above. The overcoming of nanoparticles by shear bands is considerably facilitated and, in the limit, reaches the stress level, at which the shear bands propagate in an amorphous matrix containing no nanoparticles ($D = 0$). It should be emphasized that the anomalous behaviour of the microhardness (strength) manifests itself precisely when the size of the nanoparticles of the crystalline phase becomes smaller than the thickness of the shear bands (60–70 nm [1.124]) propagating in the amorphous matrix during plastic deformation.

In the $Fe_{73.5}Si_{13.5}B_9Nb_3Cu_1$ alloy, the average crystal size is also changed from 10 to 17 nm after the selected heat treatment conditions. However, in this case, as well as for the $Fe_{58}Ni_{25}B_{17}$ alloy, it is not possible to correctly determine the size effect in pure form, since the size of the nanoparticles does not change in any interval of thermal treatments (Fig. 3.53). At the same time, assuming, as for the amorphous alloy $Fe_{58}Ni_{25}B_{17}$, that the influence of the bulk density of nanoparticles on the value of the microhardness $HV(N_v)$ is the same in the entire range of structural states of the $Fe_{73.5}Si_{13.5}B_9Nb_3Cu_1$ alloy, regardless of the heat treatment regime, we consider in more detail the temperature interval for the change in the average crystal size from 10 to 17 nm, and, therefore, estimate the size effect of HV, eliminating the contribution from N_v due to friction. To this end, we, by analogy with the amorphous $Fe_{58}Ni_{25}B_{17}$ alloy, reduce all the HV values obtained for the interval $10 < D < 17$ nm, to a certain constant value $(N_v)_0$. The value $(N_v)_0 = 80 \times 10^3$ μm^{-3} was chosen, since it is in the middle of the range of N_v values, after which the values of D and N_v ($20–140 \times 10^3$ μm^{-3}) simultaneously changed. The given values of $(HV)_0$, corresponding to $(N_v)_0$, were determined for different states using the expression (3.51).

The correction factor in this case was defined as $K_{cor} = d(HV)/d(N_v)$ for the $HV(_{Nv})$ function shown in Fig. 3.134 in the interval $T = 520–560$°C. A value of $(HV)_0$ for the structural states of the amorphous–nanocrystalline alloy formed at $T < 520$°C was calculated using a K_{cor} value of 46 MPa × μm^3 obtained for the annealing conditions $T = 520–560$°C, and expressions (3.51).

Figure 3.139 shows the dependence $(HV)_0 = f(D)$ calculated using the above described procedure, which gives a representation (albeit not quite accurate) of the effect of the value of D of the nanocrystals of the α-phase in the range 10–17 nm on the HV value of the $Fe_{73.5}Si_{13.5}B_9Nb_3Cu_1$ alloy. As can be seen in Fig. 3.139, there is a linear decrease in the values of $(HV)_0$ as D decreases. In other words, as in the case of the $Fe_{58}Ni_{25}B_{17}$ alloy a dependence is observed, which is opposite in character to the Hall–Petch relation (equation (3.52)). Note that in the $Fe_{58}Ni_{25}B_{17}$ alloy (Fig. 3.137), the dependence $(HV)_0 = f(D)$, anomalous from the point of view of the

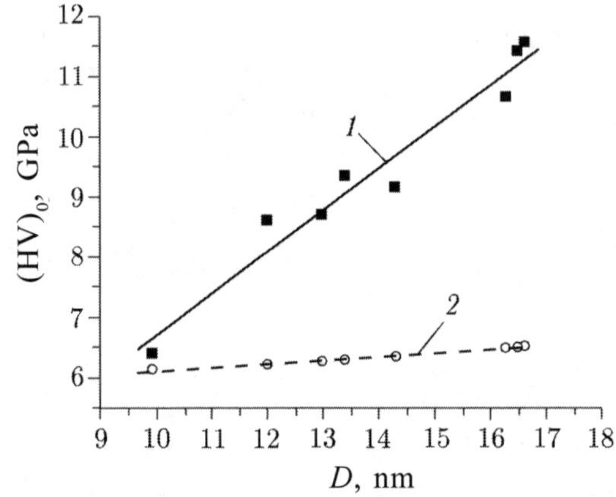

Fig. 3.139. The dependence of the reduced value of microhardness $(HV)_0$ on the average size of α-phase nanocrystals for $Fe_{73.5}Si_{13.5}B_9Nb_3Cu_1$: *1* – experimental data, *2* – theoretical dependence for the 'modular' hardening factor.

Hall–Petch ratio, also showed an almost linear character in the range of D = 20–70 nm. At higher values of D, the dependence $(HV)_0 = f(D)$ had a 'normal' form $(HV)_0 = AD^{-1/2}$, where A is a constant.

The nature of the change in HV as a function of D in Fig. 3.139, as well as the nature of the dependence of $HV(N_v)$ in Fig. 3.134, can be determined by two factors: 'modular' and 'structural'. The effect of hardening due to the growth of the size of nanoparticles with a constant value of N_v is easily calculated using expression (3.47), taking into account that $Vv = \pi \cdot N_v \cdot D^3/6$. The corresponding theoretical dependence is shown by the dashed line in Fig. 3.139. It can be seen that the main contribution to the anomalous growth of microhardness as the average size of the nanoparticles increases is the retardation of the shear bands realizing plastic deformation on the crystalline particles, despite the fact that the thickness of the shear bands δ (about 70–80 nm) exceeds the diameter of the nanocrystalline phase D. For $\delta < D$, structural mechanisms of strengthening due to the interaction of nanoparticles with shear bands will be effective, the study of which will be the subject of further theoretical and experimental consideration.

3.3.1.2. Structural mechanisms of plastic deformation

In clause 3.3.1, which was devoted to the investigation of the mechanical properties of alloys with an amorphous–nanocrystalline structure in the early stages of crystallisation, it was shown that as the volume fraction of nanocrystalline particles in the amorphous matrix increases, the strength of alloys increases both due to the 'modular' factor and also the 'structural' hardening factor. The latter, apparently, is mainly due to the interaction of shear bands propagating in the amorphous matrix and the nanocrystalline particles.

The amorphous matrix is the basic structural component of the amorphous–nanocrystalline under consideration and the features of the electron microscopic studies of shear bands in the amorphous alloys were considered in Sec. 3.1.

V.A. Pozdnyakov theoretically analyzed the conditions for the development of shear bands in amorphous alloys (see Chapter 1, section 1.3.3), as well as the heterogeneous plastic flow of alloys with an amorphous–nanocrystalline structure [3.282] when the structure is an amorphous matrix and nanocrystalline particles, distributed in it. The characteristic size of nanocrystalline inclusions is much smaller than the thickness of the shear band. We shall analyze the conditions for the development of shear bands in such a material. The volume fraction of inclusions $f_K \ll 1$. The effect of rigid dispersed particles on the shear flow stress for uniform plastic deformation ε of a material and a small volume fraction of inclusions in accordance with [3.290, 3.291] can be represented in the form

$$\tau_y = \tau_{y0} + A^* E \varepsilon f_K, \tag{3.53}$$

where τ_{y0} is the shear stress of the flow of a homogeneous material (matrix); A^* is a constant of the order of unity; E is the Young modulus. When the elastic interaction between the inclusions is taken into account in the second term, an additional factor $(1 - f_K)$ appears in (3.53).

Assuming the smallness of the particle size compared with the thickness of the shear band, one can estimate the effect of nanocrystalline inclusions on the shear stress in amorphous alloys. Relation (3.53) is valid under the condition that there is no relaxation of the stresses arising in the particles during the plastic deformation of the matrix. This condition is satisfied only immediately behind the front of the shear band, i.e., with deformation γ^* corresponding to relative displacement u^*. So the presence of dispersed particles will lead to an increase in the stress of initiation of the plastic flow $\tau_m = \tau_{m0} + A^* E q_1 \gamma^* f_K$, where q_1 is a numerical parameter approximately equal to unity. The stationary stress τ_0 in the main part of the shear band, where all the relaxation processes are facilitated, is not affected by the presence of particles, unless through a change in the viscosity of the material. Thus, the dependence of the critical stress of development of the shear band on the volume fraction of dispersed crystalline inclusions will have the form

$$\tau_p - \tau_0 \geq \left\{4\mu\left(\tau_{m0} - \tau_0 + A^* E\varepsilon^* f_K\right)u_e / \left[\pi(1-\nu)L\right]\right\}^{1/2}. \quad (3.54)$$

When the dimensions of the crystalline particles are commensurable with the thickness of the shear band, processes of bending the front of the strip between the particles and changing the trajectory of its development are possible.

The flow stress of the amorphous–nanocrystalline alloys can be defined as the average value of the propagation stress of shear bands in a given sample, if the stress of the formation (nucleation) of the bands is less than their propagation stress. If the distribution function of the propagation stresses τ_p of the shear bands in the amorphous–nanocrystalline alloys is $\Phi(\tau_p)$, then the flow stress is

$$\sigma_y = \eta \int \Phi(\tau_p)\, d\tau_p, \quad (3.55)$$

where η is the orientation factor.

The total deformation of the material is equal to the sum of the elastic and plastic ε_p deformations, and the plastic deformation in the case of heterogeneous flow is

$$\varepsilon_p = C\langle\gamma\rangle V n_b, \quad (3.56)$$

where $\langle\gamma\rangle$ is the mean shear deformation in the strip; n_b is the concentration of the shear bands; V is the average volume of the bands; C is a numerical coefficient, including the orientation factor. The development of shear bands will lead to the formation and growth of shear steps of height H on the surface of the sample. In the two-dimensional approximation with the linear density of the activated shear bands along the length of the sample, $N = 1/L$, where L is the average distance between the bands of one system, the magnitude of the plastic deformation is

$$\varepsilon_p = C_1 N H \cos\theta, \quad (3.57)$$

where θ is the slope angle of the strip to the face of the sample (loading axis).

The resulting shear steps are stress concentrators, and when they reach a certain critical value, microcracks will form on them. The length of a stable microcrack opening on a concentrator of power H is [3.292]

$$L_{TP} = B(\mu/\gamma)H^2;\ B = (1/8\pi)\ (1 + v/1 - v), \qquad (3.58)$$

where γ is the specific energy of fracture; μ is the shear modulus of the amorphous material; v is Poisson's ratio of the amorphous matrix.

The condition for unrestricted propagation of such a crack (destruction of the material): $\sigma > \gamma/H$. For a thin sample of thickness t (tape), as a fracture condition we can take $L_{TP} > \alpha t$, where $\alpha < 1$.

The deformation before destruction of εf is then equal to

$$\varepsilon_f = (\sigma_y/E) + C_1N[\alpha\gamma t/(B\mu)]^{1/2}. \qquad (3.59)$$

According to this consideration, under the condition σ_y = const the ductility of the material can be changed either by changing the fracture energy or by varying the density of the resulting shear bands. The higher the concentration of the resulting shear bands, the greater the strain before fracture. The nature of the interaction of propagating shear bands and nanoparticles located in the amorphous matrix is obviously influenced by the following factors: the shear band power, their propagation velocity, the mutual orientation of the shear bands and nanoparticles, the size, shape and crystal structure of the nanoparticles, the difference in the coefficients of thermal expansion and elastic moduli of an amorphous matrix and nanoparticles, the chemical composition of the precipitating phases.

The interaction of shear bands and crystalline particles has been studied only is a few investigations, since the setting of experiments is accompanied by great methodological difficulties – the creation of shear bands directly on the finished foils for electron microscope research.

The effect of nanocrystallisation on the geometric parameters of the shear bands was investigated in [3.293] for an amorphous and partially crystallized Zr-based alloy. For bulk amorphous alloys, the average size of the shear bands was 0.3 μm. As the volume fraction of the crystalline phase increases, the average length and power (the height of the steps) of the shear bands decrease, and the density of the shear bands on the surface of the sample, which was tested by scanning electron microscopy, increases. When the volume fraction became greater than 42%, no shear bands were observed at all. The effect of nanocrystallisation on the geometric parameters of the shear bands has been revealed: a decrease in the average length and power (step heights), as well as an increase in the density of shear bands on the surface of the sample, which was tested by scanning electron microscopy.

R.D. Conner and colleagues [3.294] investigated the mechanical properties of a massive metal alloy based on Zr, which is a composite of a matrix and particles. The particles of W, WC, Ta or SiC were introduced into the amorphous matrix. The compressive strength before fracture increased by more than 300% compared to an amorphous alloy in which these particles were not introduced. Increase in viscosity occurs, in the opinion of the authors, due to the fact that the particles inhibit the propagation of shear bands and stimulate the formation of complex shear bands. Unfortunately, the authors in this and subsequent works did not confirm their hypothesis experimentally.

V.A. Pozdnyakov [3.295] theoretically analysed possible mechanisms for the interaction of nanoparticles with a shear band moving in the amorphous matrix. They showed that in the case when the dimensions of the crystalline phases are commensurable with the thickness of the shear band (40–60 nm), processes of buckling of the front of the band between the nanoparticles and, as a consequence, changes of its trajectory are possible. The variant of the cutting of nanocrystals by a shear band can apparently occur provided the size of the nanoparticles exceeds a certain critical value, promoting activation of the development of the slip band in the crystalline region under the action of stress concentrators. In [3.149], taking into account the effect of configuration forces (image forces reflecting the effect of interfaces and free surfaces) and lattice friction forces, the characteristic size of a free nanocrystal L was determined, below which the probability of the existence of mobile dislocations inside it decreases markedly (for iron, $L \approx 2–5$ nm, for nickel $L \approx 10–15$ nm).

On the basis of theoretical studies [3.296] it was shown that the cutting of the nanoparticles by shear bands results in the formation of new interfaces, since the cutting process is associated with the separation of nanoparticles into two (or more) parts, which requires additional energy ΔE:

$$\Delta E \approx 2\pi (r_i)^2 \, \gamma_f + kP, \qquad (3.60)$$

where r_i is the characteristic particle size at the place of cutting it by a shear band; γ_f is the specific energy of the interface between the crystal and the amorphous phase; k is the coefficient of 'dry' friction of the shear band on the surface of the particle; P is the force of normal pressure, and in the case of deformation in the steady state, $P \approx 4\sigma_T (\pi r_i)^2$.

Thus, the cutting of the particle must be accompanied by an increase in the stress necessary for further advancement of the shear band.

It follows from (3.60) that the magnitude of the additional stress is of the order of magnitude determined from the relationship:

$$\sigma_{\text{add}} \approx [2\pi(r_i)^2 [\gamma_{\text{f}} + 2k\sigma_{\text{T}}]] / h, \qquad (3.61)$$

where h is the thickness of the shear band.

Based on the theoretical analysis of the implementation of a specific mechanism of interaction of the shear band with crystalline nanoparticles, a scheme was proposed that describes the possibility of the occurrence of a particular interaction mechanism depending on the size of the crystalline particles (Fig. 3.140) [3.296].

When analyzing the scheme it is seen that for a small amount of crystalline particles (30–40 nm), the interaction of the shear bands and nanoparticles occurs by the 'absorption' mechanism. At a particle size of 40–60 nm mechanisms of 'bypassing', 'cutting' and, to a lesser extent, 'accommodation' can be realized; In the range of 60–100 nm, the most likely interaction is the 'cutting' mechanism,

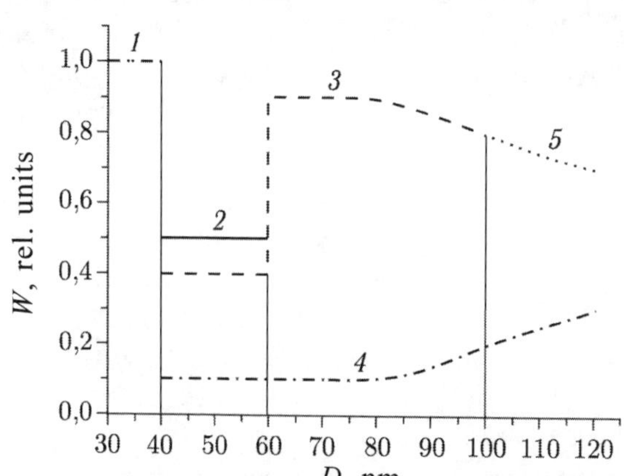

Fig. 3.140. Theoretical dependence of the possibility of realizing (W) the mechanisms of interaction of the shear bands with crystalline particles on their size D: 1 – absorption; 2 – bypassing; 3 – cutting; 4 – accommodation; 5 – retardation.

but there is the possibility of implementing the mechanism of 'accommodation'. In the size range of crystalline particles greater than 100 nm, the mechanism of 'retardation' is most effectively realized, and the probability of the mechanism of 'accommodation' increases with the growth of the average particle size.

Experimental electron microscopic studies of the deformation behaviour were carried out on the $Fe_{58}Ni_{25}B_{17}$ alloy [3.296], in which the structural state (ANA of the first type) was realized by controlled thermal treatment in the 380°C mode, and a nanocrystalline phase was uniformly distributed in an amorphous matrix and its share of did not exceed 0.3.

It should be noted that any previously formed deformation bands (more precisely, steps) in an amorphous sample will be destroyed during electropolishing in the manufacture of foils for electron microscopic studies. Therefore, the deformation shear bands were created in fully prepared foils up to 0.1 μm in thickness by the action of a diamond pyramid – the indenter of a microhardness measuring device. Its loading system allows one to select a load in the range of 0.001–0.5 N. The precision guidance system of the indenter allows one to apply deforming 'pricks' at a certain distance along the entire perimeter of the central hole of the foil. As a result, a large number of shear bands were formed in each foil, located at a convenient location for electron microscopic examination.

In the study of electron microscopic images, attention was focused on the morphology of the shear bands during passage through the nanoparticles or in close proximity to them. More than a hundred cases of interaction of shear bands with nanoparticles were analysed. At the same time, special attention was paid to unambiguous identification of the shift bands in electronic images, since the absorption contrast from them in some cases was similar to the absorption contrast from microcracks. In these cases, detailed analysis of the contrast was carried out at different positions of the defect relative to the incident electron beam and, in addition, a number of contrast features from the shear bands and microcracks were taken into account (the intensity of the absorption contrast from the shear bands is substantially lower than the contrast intensity of the microcracks).

All the observed acts of interaction can be conditionally divided into five large groups [3.296]:

Group I. The shear band absorbs small nanoparticles, without changing the trajectory of its motion in the amorphous matrix

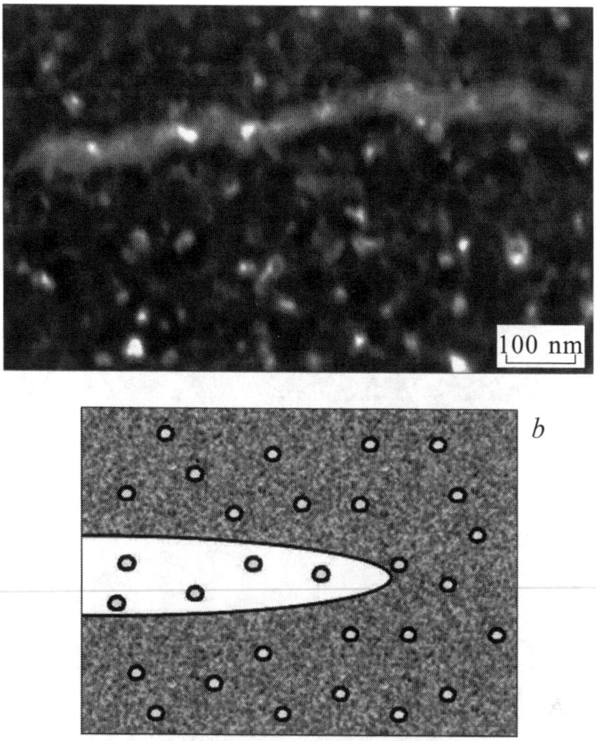

Fig. 3.141. Structure (TEM) (*a*) and scheme (*b*) of the interaction of the shear band and nanocrystals by the 'absorption' mechanism.

(the 'absorption' mechanism). Figure 3.141 *a* shows an electron microscopic image and in Fig. 3.141 *b* – diagram of the 'absorption' mechanism.

Group II. The shear band bypasses the nanoparticle on the path of its advance, changing the trajectory of motion in the amorphous matrix (the 'bypassing' mechanism). The motion of the shear band at the same time resembles the process of double transverse sliding of a dislocation that overcomes a rigid barrier. Figure 3.142 shows, as an example, the corresponding electron microscopic image (*a*) and the bypassing scheme (*b*) [3.295].

Group III. The shear band passes through the nanoparticle, 'cutting' it. A typical case of the action of the 'cutting' mechanism is presented on an electron microscopic image in Fig. 3.143 *a*, and its scheme is shown in Fig. 3.143 *b*. This interaction variant can be realized if the shear band propagating in the amorphous matrix can stimulate the dislocation flow inside the nanoparticle [3.149].

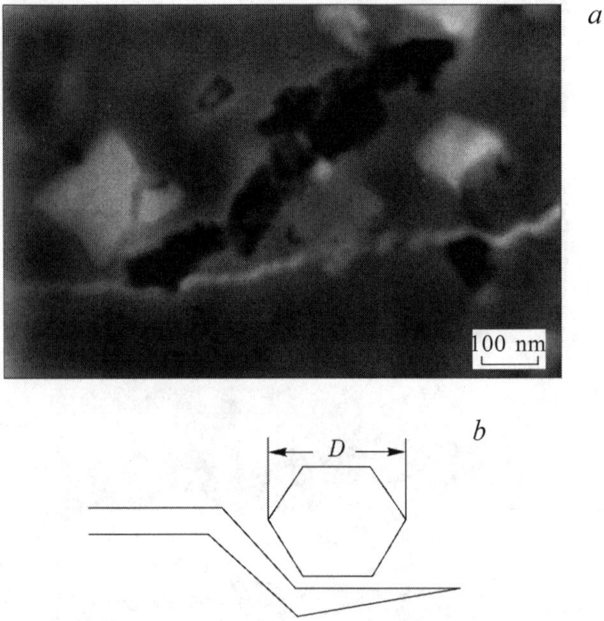

Fig. 3.142. The structure (TEM) (*a*) and the scheme (*b*) of the interaction of the shear band and nanocrystals by the 'bypassing' mechanism.

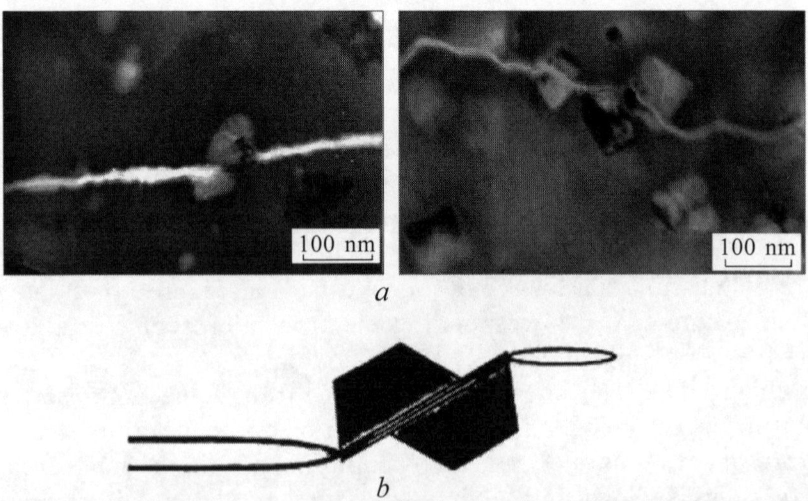

Fig. 3.143. Structure (TEM) (*a*) and scheme (*b*) of the interaction of the shear band and nanocrystals by the 'cutting' mechanism.

Group IV. The shear band, resting on the nanoparticle, causes in it very large elastic stresses, which in turn initiate a new shear band in the amorphous matrix on the other side of the nanoparticle (the 'accommodation' mechanism). As a rule, the trajectory of the secondary accommodation band coincides with the trajectory of the primary shear band. A typical electron microscopic photograph (*a*) and scheme (*b*) are shown in Fig. 3.144.

It was extremely rare to observe multiple secondary accommodation effects when an elastically stressed nanocrystal initiated several secondary shear bands into an amorphous matrix. Figure 3.145 *a* demonstrates this rather rare case: the emission into the amorphous matrix of several new shear bands from the interphase boundary (indicated by arrows in the photo). The corresponding scheme is shown in Fig. 3.145 *b*.

Group V. The shear band is retarded and stops at the interphase boundary or inside the nanoparticle (the 'retardation' mechanism) (Fig. 3.146). It should be noted that this phenomenon was observed much less frequently than the alternative interaction options, assigned to the groups II–IV.

Based on the study of the mechanisms of interaction of particles and shear bands, the results were used to construct the histograms corresponding to the relative frequency of the interaction mechanisms:

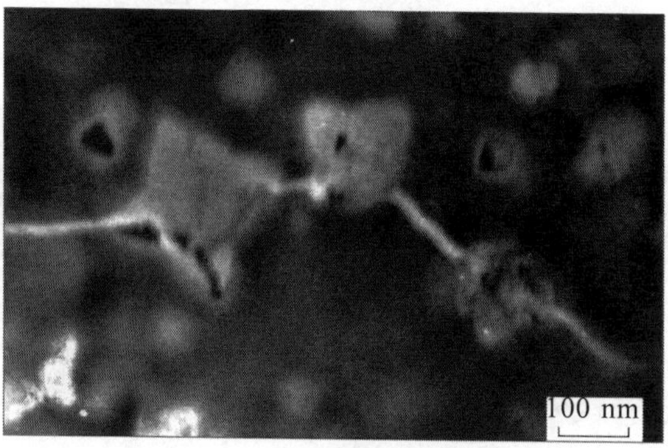

a

b

Fig. 3.144. Structure (TEM) (*a*) and scheme (*b*) of the interaction of the shear band and nanocrystals by the 'accommodation' mechanism.

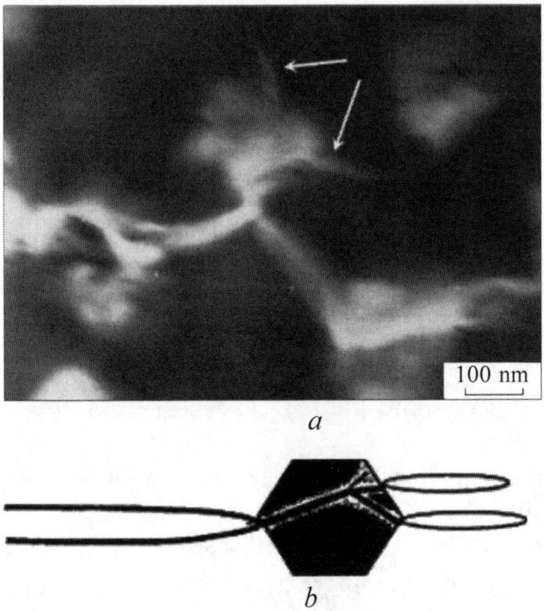

Fig. 3.145. The structure (TEM) (*a*) and the scheme (*b*) of the action of the shear band and nanocrystals by the mechanism of the secondary 'accommodation' (the secondary slip bands are indicated by arrows).

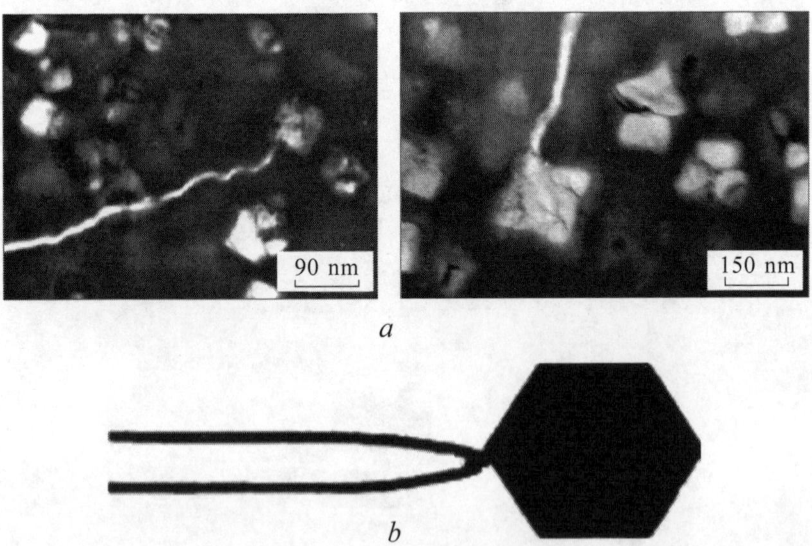

Fig. 3.146. Structure (TEM) (*a*) and scheme (*b*) of the interaction of the shear band and nanocrystals by the 'retardation' mechanism.

'bypassing'(2), 'cutting' (3), 'accommodation' (4) and 'retardation' (5), depending on the size range of nanoparticles: 35–65 nm (*a*), 65–95 nm (*b*), 95–125 nm (*c*) and 125–160 nm (*d*) (Fig. 3.147). The mechanism of 'absorption' was not considered, since such cases were relatively rare (with a very small particle size) and, as a result, did not yield to statistical estimation.

Figure 3.147 compares the relative frequencies of the implementation of the interaction mechanisms – 'bypassing', 'cutting', 'accommodation' and 'retardation', depending on the size interval of the nanoparticles. In the nanoparticle size range 35–65 nm it is evident from the histogram presented that the 'bypassing' mechanism is most often realized, the mechanisms of 'cutting' and 'accommodation' are much more rare, and the 'retardation' mechanism is not realized at all (Fig. 3.147 *a*) . In the range 65–95 nm (Fig. 3.147 *b*), the 'cutting' mechanism becomes dominant with the complete absence of the 'retardation' mechanism. For a larger nanoparticle size fraction of 95–125 nm (Fig. 3.147 *c*), the effect of all mechanisms is approximately equally probable, and the mechanism of 'retardation' dominates for the largest particle size fraction of 125–160 nm (Fig. 3.147 *d*).

Figure 3.148 shows the distribution of the relative frequency of realization of each of the interaction mechanisms, depending on the size interval of the nanoparticles. The mechanism of 'bypassing' most often occurs at D = 35–65 nm with a noticeable decrease in the relative frequency as the particle size increases (Fig. 3.148 *a*). The mechanism of 'cutting' is most likely at D = 65–95 nm and is much less frequently observed both with decreasing and with increasing particle size (Fig. 3.148 *b*). The relative frequency of the 'accommodation' mechanism depends little on the value of D (Figure 3.148 *c*), and the action of the 'retardation' mechanism is most often observed with $D \geq 100$ nm (Fig. 3.148 *d*).

In the size range of nanoparticles 35–65 nm, the dominant interaction mechanism is the 'bypassing' mechanism, in the size interval 65–95 nm – the 'cutting' mechanism, and in the size range above 100 nm – the 'retardation' mechanism. At a particle size smaller than 35 nm, the 'absorption' mechanism dominates.

The frequency of realization of the 'accommodation' mechanism depends little on the size of the nanoparticles, however, in the range of 95–125 nm, its large realization is observed in comparison with other dimensional intervals.

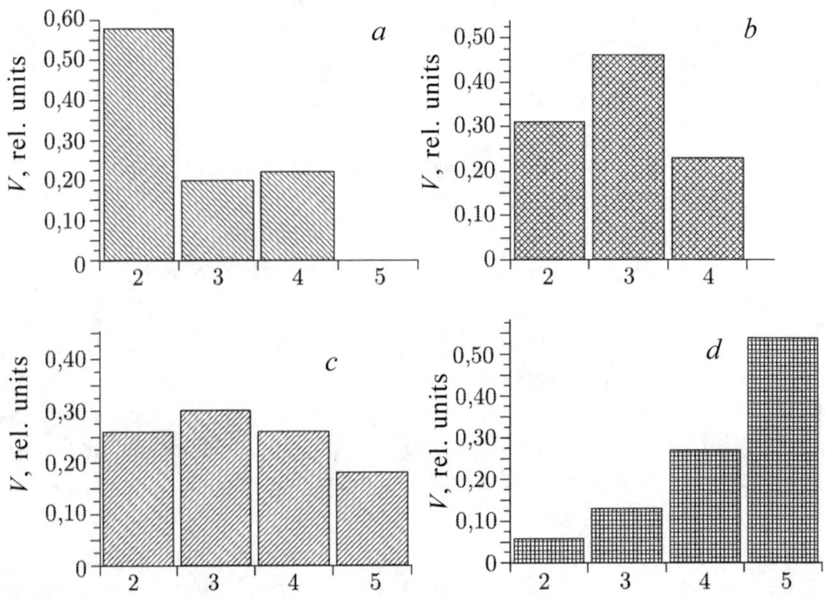

Fig. 3.147. The relative frequency of realization of the interaction of the shear bands (*V*) by mechanisms (2 – 'bypassing', 3 – 'cutting', 4 – 'accommodation', 5 - 'retardation') with nanoparticles of various sizes: *a* – 35–65 nm, *b* – 65-95 nm, *c* – 95–125 nm, *d* – 125–160 nm.

In section 3.3.1.1, the dimensional effect of nanocrystals of hardening of an amorphous matrix was investigated. The dependence of the microhardness *HV* on the average size *D* of nanocrystalline particles for the $Fe_{58}Ni_{25}B_{17}$ alloy, was obtained at the constant bulk density of the particles (Fig. 3.137).

In Fig. 3.149 the $HV = f(D)$ dependence obtained for the $Fe_{58}Ni_{25}B_{17}$ alloy at N_v = const is combined with the regions of dominant mechanisms of interaction of nanoparticles with shear bands, established on the basis of the results of the investigation of this section (Figs. 3.147 and 3.148).

For a size fraction of 95–125 nm where there is no clear leader in implementing the mechanism (Fig. 3.147 *c*), two mechanisms are the dominant ones and their relative frequency of realization exceeds 0.5. A 'normal' $HV = f(D)$ dependence is observed in the D = 80–170 nm region, which is in many respects similar to the Hall–Petch relation for polycrystalline materials: as the *D* decreases a noticeable increase in *HV* occurs. With $D \leq 80$ nm, the 'anomalous'

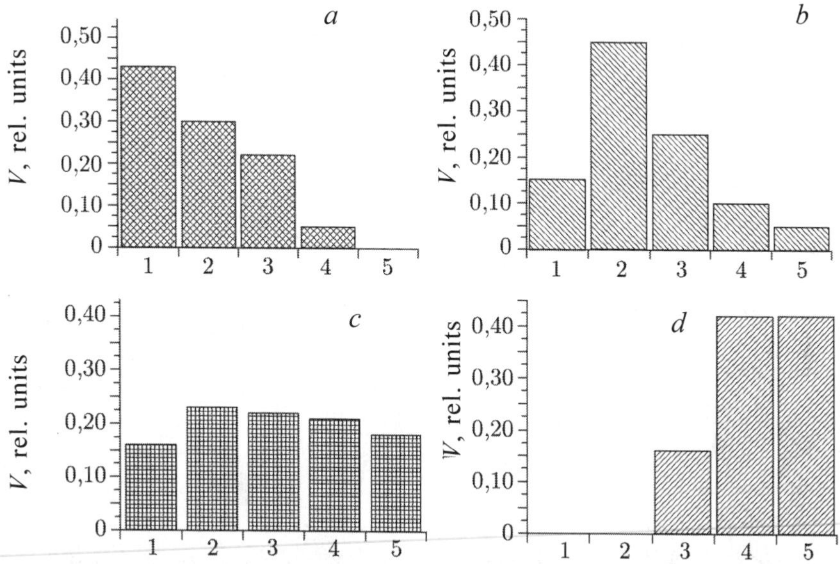

Fig. 3.148. The relative frequency of realization of the interaction of nanoparticles (V) of 35–65 nm (1), 65–95 nm (2), 95–125 nm (3), 125–160 nm (4), 160–190 nm (5) with shear bands along different mechanisms: *a* – 'bypassing', *b* – 'cutting', *c* – 'accommodation', *d* – 'retardation'.

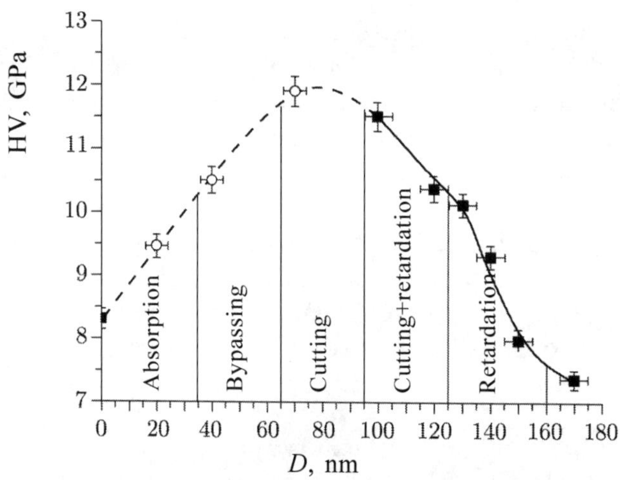

Fig. 3.149. Dependence of the microhardness *HV* on the average size of the crystal particles *D* for the $Fe_{58}Ni_{25}B_{17}$ alloy at a constant bulk particle density N_v.

dependence $HV = f(D)$ is observed: as D decreases, a noticeable decrease in HV occurs. On the basis of Fig. 3.148 we can assume that the 'normal' ratio $HV = f(D)$ is mainly due to the mechanisms of 'cutting' and 'retardation', and for 'anomalous' – the mechanisms of 'bypassing' and 'absorption'.

3.3.1.3. Plasticity

When carrying out research in the field of nanostructured materials science, it was established that after the formation of nanocrystals in an amorphous matrix, the strength (hardness) of the amorphous-crystalline materials tends to increase (see section 3.3.1). The problem of giving plasticity to bulk metallic glasses as materials with high strength and corrosion resistance was discussed a long ago (see, for example, [3.297]).

The authors of [3.298] established that the dependence of the ductility of a material on the controlled crystallisation regime (isothermal or thermocyclic annealing) is characterized by kinetic curves $\ln t - 1/T \cdot 10^3$ and can be described by an exponential function:

$$\varepsilon = \varepsilon_0 \exp\,[W_0 - W(\sigma)/RT], \tag{3.62}$$

where ε is the deformation under thermal action; ε_0 is a quantity that depends on the annealing regime and characterizes the structural state of the material; W_0 is the activation energy of the isothermal annealing process, depending on the degree of ordering of the structural structure and composition of the alloy; $W(\sigma)$ is the activation energy of the process due to the action of the thermocyclic annealing regime; σ are the stresses caused by the temperature and structural–phase state of the material; R is the gas constant; T is temperature.

The value of ε_0 is estimated by extrapolating the dependences $\ln \varepsilon - 1/T$. Taking into account that the parameter ε_0 in equation (3.62) depends on the stresses arising during thermal cycling, this value should also be estimated taking into account the structural and stressed state of the material, for example [3.299]:

$$\varepsilon_0 = \varepsilon'_0 \exp\,[\gamma\sigma], \tag{3.63}$$

where ε'_0 is a constant characterizing the structural state of the material; γ is a constant having the dimension of the reciprocal dimension of the stress; σ are the internal stresses caused by the thermal exposure regime.

In the work [3.300], attention was paid to the manifestation of plasticity in the case of nanosized brittle crystalline nanomaterials based on refractory compounds. In this connection, it seems quite probable that the effect of the size effect on the manifestation of plasticity in brittle amorphous and nanocrystalline objects has approximately the same character.

It should be noted that some cases of the plasticity of metallic glasses as a result of annealing and crystallisation of nanoparticles characterized as 'regions with a high (relative to the amorphous matrix) correlation in the arrangement of atoms' have been noted for a long time (see, for example, [1.1]). The results of recent experiments extend this information. Thus, a deformation of about 10% induced by crystalline nanoparticles was noted in compression tests of Zr–Cu-based metal bulk glasses at room temperature [3.299]. The appearance of inclusions (~2 nm in size) during annealing was recorded every 3 seconds with the help of synchrotron radiation. It is assumed that when plasticity develops, the shear bands generate a free volume.

High hardness (about 100 GPa) and thermal stability (up to 1100-1200°C) are characteristic for two-phase composites based on TiN nanocrystals 3–4 nm in size and an amorphous matrix consisting of silicon nitride and titanium disilicide [3.301, 3.302].

High mechanical and thermal properties of bulk amorphous steels $Fe_{50}Cr_{14}Mo_{14}C_{14}B_6M_2$ (M – Y or Dy) have also been reported; annealing (T = 600–850°C) of these steels resulted in the formation in the amorphous matrix of nanoinclusions of austenite and chromium carbide [3.303]. The hardness HV and the modulus of elasticity E values of the annealed samples studied by the nanoindentation method were at the level of 21–22 GPa and 320–340 GPa, respectively; elastic recovery (W; at W = 100% – complete elastic recovery, with W = 0 – total plasticity) was 29–32%. It is noted that these characteristics obtained on samples 3–5 mm in diameter are much higher than those for many bulk metal glasses. The temperature range of supercooling (the range from the crystallisation temperature to the glass transition temperature) turned out to be particularly significant for steels doped with dysprosium (ΔT = 70 K), which makes it possible to effectively perform thermoplastic processing for shaping in this range.

Investigation of the influence of the concentration of silicon and tin in Ti–Zr–Cu–Ni–Sn–Si glasses on deformation by compression revealed the optimal composition for which the deformation was at 5%, and the fracture stress was 2 GPa (at ΔT = 35 K) [3.302]. It was noted that there were many shear bands on the surface of these samples, and nanocrystalline inclusions of ~5 nm in size were detected in the nanostructure by high-resolution TEM [3.303].

It is interesting to compare the bending behaviour of amorphous, nanocrystalline, and amorphous–nanocrystalline samples of the Zr–Cu–Al–Ni–Nb alloy, which revealed a great advantage in the strength and plasticity of the composite containing about 30 vol.% of Zr_2Cu-type nanoinclusions [3.304].

The author of [3.305] developed amorphous matrix–dendrite composites that possess both high strength and high viscosity. The mechanical properties of such composites after uniaxial loading are studied in detail in the article. Under quasi-static compression at room temperature almost all such composites exhibit increased ductility due to the high volume fraction of dendrites and, as a result, the effective blocking of the propagation of the shear bands. At quasistatic tension at room temperature, although composites exhibit ductility under tension, inhomogeneous deformation and the associated softening predominate. Also, the amorphous–dendritic composites have increased maximum strength also at cryogenic temperatures.

The microstructure of the $Zr_{58.5}Ti_{14.3}Nb_{5.2}Cu_{6.1}Ni_{4.9}Be_{11.0}$ composite is shown in Fig. 3.150 [3.306]. Dendrites are evenly distributed in an amorphous matrix, the distance between dendritic branches is 2–4 μm, the volume fraction of dendrites is ~58%. The X-ray spectrum shows the presence of a two-phase structure – the peaks of the BCC lattice of the β-Zr solid solution are superimposed on the broad diffuse scattering of the amorphous maximum. A high volume fraction of dendrites makes such composites sufficiently plastic.

In addition, the fatigue limit for $Zr_{58.5}Ti_{14.3}Nb_{5.2}Cu_{6.1}Ni_{4.9}Be_{11.0}$ amorphous matrix/dendrite composites rises to 473 MPa and fatigue limit at four-point bending up to 567 MPa [3.305].

In work [3.307] two-phase structures consisting of an amorphous matrix and a crystalline phase in the form of dendrites are considered. The authors of [3.308] showed that in the Vitreloy1 alloys ($Zr_{41.2}Ti_{13.8}Cu_{12.5}Ni_{10}Be_{22.5}$) two-phase composites based on an amorphous matrix and dendrites with a volume fraction of 25% are formed, the barrel length reaches 50–150 μm, the barrel

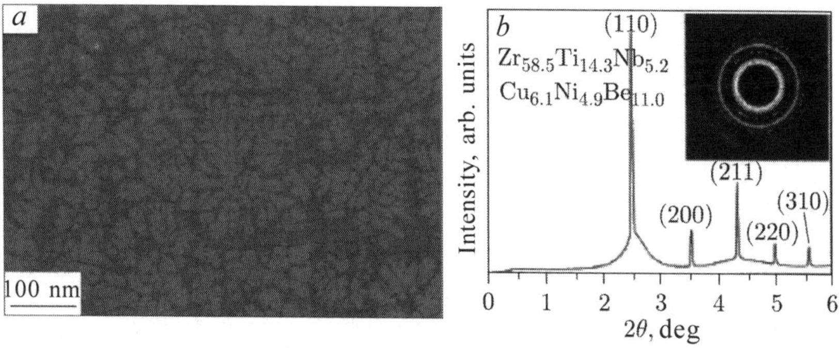

Fig. 3.150. The microstructure and radiograph of the $Zr_{58.5}Ti_{14.3}Nb_{5.2}Cu_{6.1}Ni_{4.9}Be_{11.0}$.

radius is 1.5–2.0 μm. Compared with the fully amorphous Vitreloy1 alloy, the two-phase alloy showed an improvement in flexural and compression strength: the tensile strength reaches 1.7 GPa, and the total deformation (elastic + plastic) was more than 8% [3.308]. In uniaxial tension, the two-phase amorphous–crystalline composite showed an elongation of about 5% (average along the length of the sensor) and up to 15% of the neck region of the sample, in contrast to the zero plasticity of the completely amorphous alloy. Also, the Charpy impact toughness of the composite was significantly higher (200 kJ · m^{-2}) than that of the fully amorphous Vitreloy1 alloy (80 kJ · m^{-2}).

Hays et al. [3.308] concluded that the ductility of amorphous-dendritic composites is facilitated by a soft β-phase of dendrites, which facilitates the appearance of shear bands and simultaneously blocks their spread beyond the domain. It is assumed in [3.307] that strengthening by the β-phase can play a key role in increasing the plasticity of composites.

Not all the experimental facts described above have been unambiguously explained, which is connected with the general state of the theory of the structure and mechanical properties of metallic amorphous alloys [3.309, 3.310]. The basic models of the deformation of these objects, based on the representations of the zone of shear transformations and on local atomic displacements due to free volume, continue to be actively discussed and compared with numerous (and often ambiguous) experimental data.

Nevertheless, it is quite obvious that the formation of nanocrystals in amorphous objects is in many cases accompanied by an increase in their strength and plasticity, but the specific mechanism of this phenomenon requires additional theoretical and experimental studies. In this connection, attention should be paid to the recently proposed composite model of the plastic flow of amorphous covalent materials, in which the homogeneous nucleation and growth of liquid-like nanostructures that initiate the development of homogeneous and inhomogeneous deformation of the entire composite are theoretically considered [3.311].

3.3.2. Mechanical properties in the late stages of crystallisation

As already mentioned above (Fig. 3.19), the late stages of crystallisation are characterised by the formation of an amorphous-nanocrystalline structure, containing predominantly isolated nanocrystals separated by amorphous interlayers. Moreover, the higher the volume fraction of the crystalline phase V_v, the thinner the amorphous interlayers.

In chapter 2 we briefly discussed the characteristics of the mechanical properties and mechanisms of plastic deformation of alloys with the nanocrystalline structure, since nanocrystals are the main structural component, and amorphous interlayers can be considered as 'diffuse' grain boundaries in the late stages of crystallisation.

Nanocrystalline materials with amorphous thin interlayers along the grain boundaries (with $V_v \geq 0.5$), formed at the final stages of crystallisation, are the subject of research in this section. One such representative is a shape memory alloy based on the Ti–Ni–Cu system which, due to the presence of copper, can be obtained in the amorphous state [3.312]. The transition to nanoscale sizes of crystallites in these alloys, which is difficult to achieve by other methods, improves the service characteristics of alloys and increases their efficiency and durability.

At the initial stage of crystallisation, separate single-crystal inclusions of the round form of a high-temperature austenite phase *B*2 with a bcc lattice ordered in the CsCl type based on TiNiCu with a lattice period $a = 0.30$ nm appear in the amorphous matrix. The average diameter of the crystals is about 0.5 μm and practically does not change with a change in the heating temperature and the residence time of the sample at this temperature.

X-ray and *in situ* electron microscopy experiments showed that when heated, the ternary amorphous $Ti_{50}Ni_{50-x}Cu_x$ alloys with a copper content of $x \leq 25$ at.% crystallize with formation of the $B2$-phase throughout the entire volume [3.313].

The resistometry method [3.314] established that under isochronous annealing conditions (30 min) the crystallisation of the amorphous $Ti_{50}Ni_{25}Cu_{25}$ alloy occurs in the temperature range 425-430°C.

As is known, the grain boundaries are the most important element in the structure of nanocrystalline materials which determine their strength properties. Influencing them, one can control the physical and mechanical characteristics of nanomaterials. As discussed in detail in the previous chapter (section 2.4.2), noticeable deviations from the Hall–Petch ratio occur in the nanometer range of grain size (D < 30–50 nm). The observed anomaly leads to a decrease in strength with a decrease in the grain size and, obviously, does not make it possible to achieve strength values of nanocrystals close to the limiting ones, and thereby realize in practice the physical potential of the strength that is embedded in the solid. The anomalous Hall–Petch dependence is a direct consequence of the change in the structural mechanism of the plastic flow from the classical dislocation to grain boundary microsliding (GBMS). This process is realized due to the formation of shear microregions in the grain boundaries [3.133].

The most effective GBMS is realized in nanocrystals in the late stages of controlled annealing of the amorphous state where amorphous grain boundary layers are formed. This is due to the formation in the late stages of crystallisation of an amorphous matrix of thin grain-boundary layers with increased thermal stability. Such layers can, as we have already noted, be regarded as 'diffuse' grain boundaries. The deformation of a material with a similar structure is carried out by GBMS on amorphous intercrystalline interlayers by the formation of localized shear bands [3.133].

Inhibiting the GBMS, it is possible to suppress the effects of an abnormal decrease in the strength of nanocrystals and, consequently, approach the level of ultimate (theoretical) strength (hardness) (Fig. 2.4) [3.315].

One of the ways to suppress the process of GBMS is to create effective structural barriers to the propagation of deformation in amorphous intercrystalline interlayers. In section 3.3.1, devoted to the study of the mechanical behaviour of materials having an amorphous crystal structure (an amorphous matrix and individual crystalline particles) in the early stages of crystallisation, it was shown that the interaction of propagating shear bands in the amorphous matrix with crystalline nanoparticles can lead to substantial hardening of these alloys. By analogy, it can be assumed that in the presence of nanoparticles the plastic deformation processes in amorphous intercrystalline interlayers will be hampered (GBMS processes). Thus, by purposefully introducing nanoparticles into amorphous intercrystalline interlayers of nanocrystals at the stage of transition from dislocation to grain boundary sliding, we get a chance to approach the ultimate strength values.

In work [3.316], in order to inhibit GBMS, the specimens of the rapidly quenched amorphous alloys of the Ti–Ni–Cu system ($Ti_{50}Ni_{25}Cu_{25}$ and $Ti_{49}Ni_{24}Cu_{24}B_3$ alloys) were studied. Boron atoms are, in this case, 'useful' impurities, which are introduced into the amorphous intergranular interlayers to inhibit the GBMS processes by means of the principle of grain boundary engineering (the purposeful introduction of 'useful' impurities into grain boundaries). Samples of these two alloys underwent isothermal annealing to form a nanocrystalline structure in the temperature range 440–470°C for 10–60 min.

When the annealing temperature rises to the temperature of the start of crystallisation, depending on the annealing time, a crystallisation process is observed in the structure of the alloys. Crystallisation of amorphous $Ti_{50}Ni_{25}Cu_{25}$-based alloys at thermal exposures above 400°C takes place in a single step with the formation of a BCC phase ordered in type *B*2 [3.312].

Electron microscopic experiments showed that an amorphous-nanocrystalline state forms in the early stages of crystallisation of the $Ti_{50}Ni_{25}Cu_{25}$ alloy. When annealing at 450°C, 10 min, the *B*2-phase nanocrystals are uniformly distributed in an amorphous matrix and have the shape close to spherical (Figure 3.151*a*).

The features of diffraction contrast at the interphase boundaries (amorphous phase–crystal) were not detected. In the later stages of annealing: 450°C, 1 hour a microcrystalline state with equiaxed grains of the *B*2-phase was observed and no diffraction effects at the grain boundaries (Fig. 3.151 *b*).

The crystallisation of the $Ti_{49}Ni_{24}Cu_{24}B_3$ alloy also begins with the precipitation of crystalline *B*2-phase particles in the amorphous matrix, which leads to the formation of an amorphous-nanocrystalline structure. At the same time, on the surface of growing nanocrystals located in an amorphous matrix, the effects of deformation contrast associated with segregations or nanoparticles

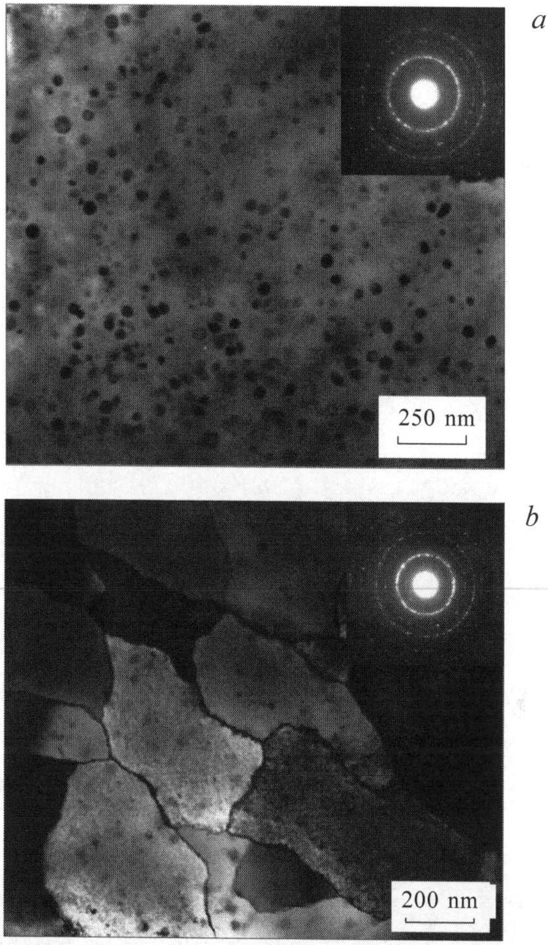

Fig. 3.151. The structure of $Ti_{50}Ni_{25}Cu_{25}$ alloy after annealing at 450°C; τ = 10 min (*a*) and τ = 1 h (*b*) (TEM, bright field).

of phases are clearly observed (Fig. 3.152). In the later stages of annealing, a nanocrystalline state is formed, for which the existence of nanoparticles located directly in amorphous intercrystalline interlayers is characteristic (the nanoparticles are indicated by the arrows in Fig. 3.152 *b*).

In Fig. 3.153 the dark-field image is represented in the reflexes of the grain boundary phase. Nanoparticles have a size of about 5 nm and form a characteristic 'skeleton' in accordance with the location along the grain boundaries of the *B*2-phase. As shown by the interpretation of the diffraction pattern (Figure 3.153*b*), these nanoparticles are a mixture of two boride phases: Ti_2B and TiB_2 [3.317].

Using the TEM, the average size *D* of the *B*2-phase crystals was determined in the temperature range of the beginning and the end of crystallisation.

Figure 3.154 shows the dependence of *D* on the annealing temperature *T* at τ = 10 and 30 min for the two amorphous $Ti_{50}Ni_{25}Cu_{25}$ and $Ti_{49}Ni_{24}Cu_{24}B_3$ alloys studied.

The change in the average size *D* of the *B*2-phase of the nanocrystals as a function of *T* for both alloys is non-monotonic. In the $Ti_{50}Ni_{25}Cu_{25}$ alloy, with increasing temperature from 440 to 470°C at τ = 10 min, the *D* value rises from 50 to 195 nm. In the first stage (440-450°C) a very slow growth is observed up to 55 nm, and then in the second stage (over 450°C) – faster (one and a half time) increase in *D* from 60 to 140 nm. At τ = 30 min, the values of *D* for all values of *T* are much higher (260–580 nm), and the dependence $D = f(T)$ is monotonic.

In the $Ti_{49}Ni_{24}Cu_{24}B_3$ alloy, the dependence $D = f(T)$ is characterized by two stages. In the first stage (440–450°C), *D* is practically constant at τ = 30 min and there is even a slight decrease in *D* at τ = 10 min. With a further increase in *T* (from 450 to 470 °C), there is a significant increase in *D* from 20 to 60 nm (τ = 10 min) and from 33 to 120 nm (τ = 30 min).

It should be noted that the growth of the grains in the $Ti_{49}Ni_{24}Cu_{24}B_3$ alloy is much weaker than in the $Ti_{50}Ni_{25}Cu_{25}$ alloy. Also for the $Ti_{49}Ni_{24}Cu_{24}B_3$ alloy, there is a slight decrease in *D* as *T* increases in the range 440–460°C, which is apparently due to the specificity of the transition from a two-phase amorphous-nanocrystalline state to a nanocrystalline state.

Thus, using structural treatment in the $Ti_{49}Ni_{24}Cu_{24}B_3$ alloy, structural states were obtained, including a nanocrystalline matrix with *B*2-phase grains and grain-boundary precipitates of ultradisperse boride phases (Ti_2B and TiB_2). A similar structure was observed at $T \geq 450$°C (except for 450°C mode – 10 minutes). At a lower *T* value, including 450°C – 10 min, a two-phase amorphous-nanocrystalline structure is formed (Fig. 3.152). In the $Ti_{50}Ni_{25}Cu_{25}$ alloy, nanocrystalline structures without grain boundaries are formed at all annealing temperatures for 30 min and at T > 450°C (τ = 10 min).

The mechanical properties were measured on a scanning nanohardness meterr NanoScan-3D using the method of dynamic

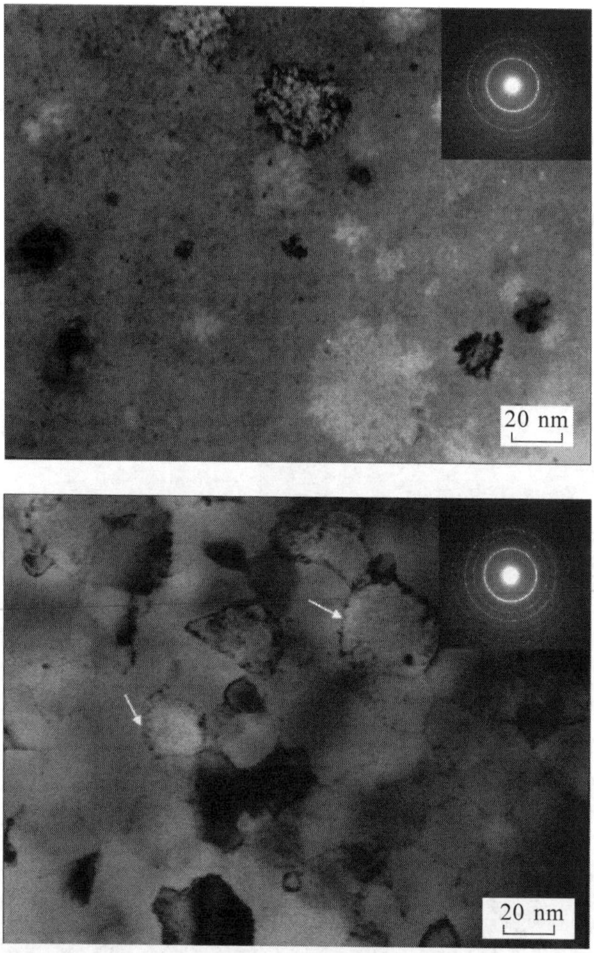

Fig. 3.152. The structure of the $Ti_{49}Ni_{24}Cu_{24}B_3$ alloy after annealing at 450°C; τ = 10 min (*a*) and τ = 1 h (*b*) (TEM, bright field).

dynamic indentation. The method is based on measuring and analyzing the load dependence when the indenter is pressed into the surface of the material on the depth of introduction of the indenter.

The mechanical tests are carried out using an indentor of the Berkovich type, which is a triangular diamond pyramid with an apex angle of about 142°. The method of measuring dynamic indentation is as follows: the indentor is pressed into the surface of the sample at a constant speed and when the specified load is reached the indenter is retracted in the opposite direction. During this test, the values of

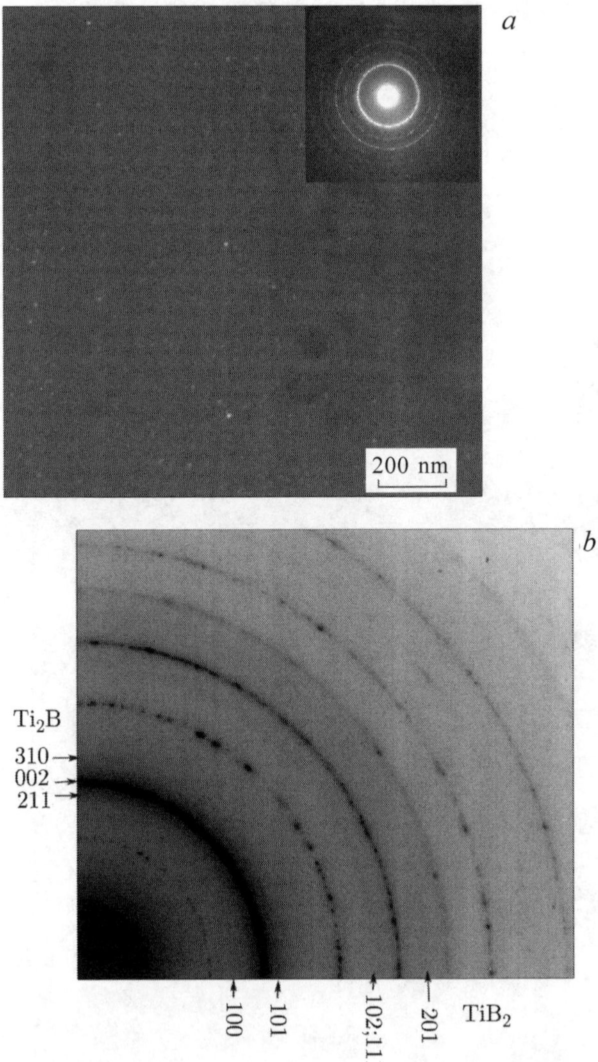

Fig. 3.153. Structure (*a*) and microdiffraction pattern (*b*) of the $Ti_{49}Ni_{24}Cu_{24}B_3$ alloy after annealing for 1 h at 450°C, TEM. (*a* – dark field in boride reflections, *b* – electron microdiffraction pattern containing reflections from borides and its interpretation).

the load *P* and the corresponding displacement of the indenter *h* are recorded.

An experimental curve typical for this method in the form of a graph of the dependence of the load (*P*) on the depth of indentation (*h*) is shown in Fig. 3.155. The resulting *P–h* diagram is composed of two branches: loading and unloading.

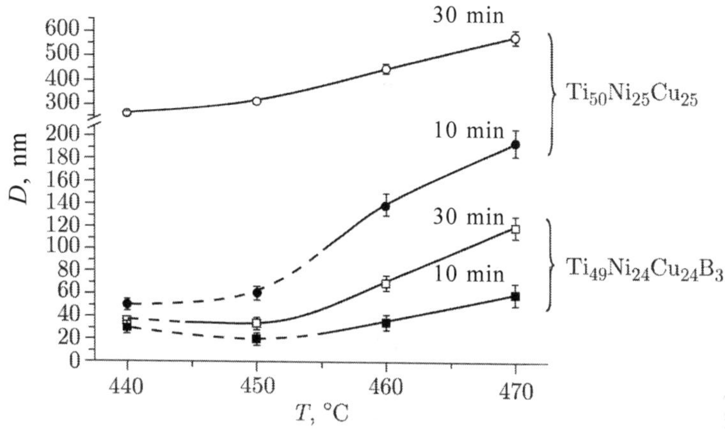

Fig. 3.154. Dependence of the average size of the *B*2-phase *D* nanocrystals on the annealing temperature *T* of the amorphous alloys studied (the solid lines correspond to the nanocrystalline state, the dashed lines to the amorphous–nanocrystalline state).

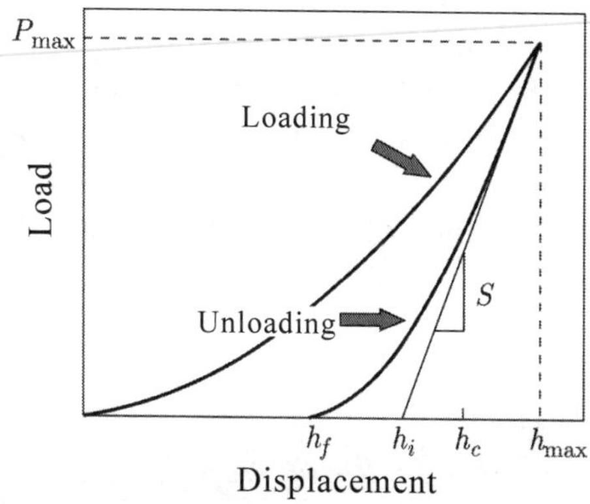

Fig. 3.155. General view of the *P–h* loading diagram (for the notations h_f, h_c, h_{max} see Fig. 3.156).

The loading curve characterizes the resistance of the material to the introduction of a rigid indentor and reflects both the elastic and plastic properties of the material under study. The unloading curve is mainly determined by the elastic restoration of the indentation pattern.

In the framework of this method, the hardness of the HV_{IT} sample is determined by the equation

$$HV_{IT} = \frac{P_{max}}{A_c}, \tag{3.64}$$

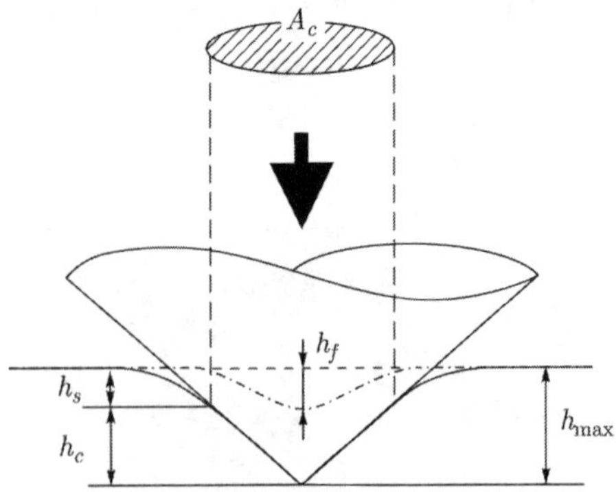

Fig. 3.156. The scheme of contact with the designations of the quantities used in the calculation of the modulus of elasticity and hardness

where A_c is the area of the projection of the indentation at the maximum value of the applied load P_{max}.

Figure 3.156 shows a contact scheme for nanoidentification, where h_c is the depth of penetration during which the full contact between the indenter and the material is made after full loading (up to P_{max}). At the same time, part of the depth of introduction of the indenter, during which contact with the material is absent as a result of the indentation formed around the indenter, is h_s.

The value of the reduced elasticity modulus E^* is calculated as follows:

$$E^* = \frac{1}{\beta} \cdot \frac{\sqrt{\pi}}{2} \cdot \frac{S}{\sqrt{A_c}}, \quad (3.65)$$

where the constant β depends on the shape of the indenter, and the stiffness of the contact S is determined from the slope of the tangent to the unloading curve at the point P_{max} (Fig. 3.155):

$$S = \left(\frac{dP}{dh} \right)_{P=P_{max}}. \quad (3.66)$$

The quantity E^* (reduced modulus of elasticity) takes into account the elastic interaction of the material with the indenter.

The contact area at maximum load A_c is determined by the geometry of the indenter and the contact depth h_c and is described by the so-called needle-shaped functions $A_c = f(h_c)$.

Figure 3.157 shows the dependence of the nanohardness HV_{IT} measured on the scanning nanoscale device NanoScan-3D by the method of dynamic indentation, on the annealing time for both alloys. As τ increases, the first noticeable increase in HV_{IT} relative to the initial amorphous state (τ = 0) is observed in the $Ti_{50}Ni_{25}Cu_{25}$ alloy. The maximum value of HV_{IT} was recorded during annealing for 10 min (HV_{IT} = 6.5 GPa); with further growth of τ, the HV_{IT} falls to 5 GPa.

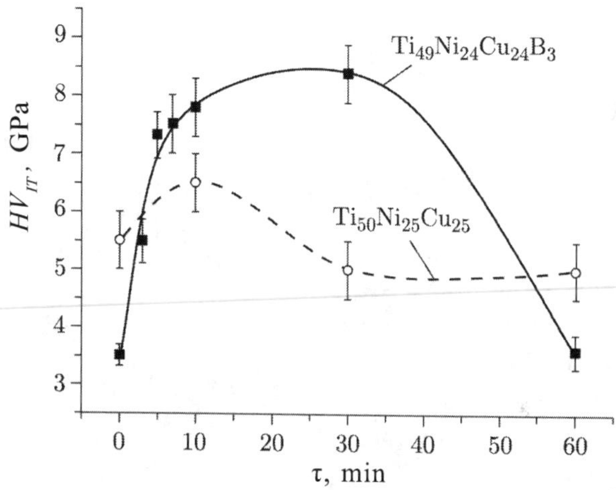

Fig. 3.157. The dependence of the nanohardness HV_{IT} on annealing time τ at an annealing temperature of 450°C.

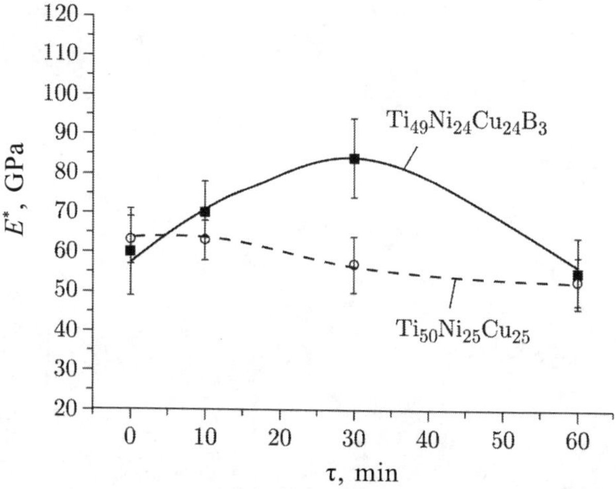

Fig. 3.158. Dependence of the reduced Young's modulus E^* on annealing time τ at T = 450°C.

For the $Ti_{49}Ni_{24}Cu_{24}B_3$ alloy, the $HV_{IT} = f(\tau)$ dependence is a curve with a maximum: with increasing τ, the growth of HV_{IT} in relation to the amorphous state is observed; At $\tau > 30$ min there is a noticeable drop. The maximum value of $HV_{IT} = 8.4$ GPa is reached at $\tau = 30$ min, significantly exceeding the maximum value of HV_{IT} for the $Ti_{50}Ni_{25}Cu_{25}$ alloy.

Figure 3.158 shows the dependence of the reduced Young modulus E^* of the $Ti_{50}Ni_{25}Cu_{25}$ and $Ti_{49}Ni_{24}Cu_{24}B_3$ alloys as a function of the value of τ at $T = 450$°C. The value of E^* in the initial amorphous state ($\tau = 0$) is 60 GPa for both alloys. With increasing τ in the $Ti_{50}Ni_{25}Cu_{25}$ alloy, the E^* value is insignificant and decreases monotonically. The value of E^* in the $Ti_{49}Ni_{24}Cu_{24}B_3$ alloy first grows and reaches its maximum value (84 GPa) at $\tau = 30$ min, and then decreases to a value of 55 GPa corresponding to the E^* value for the $Ti_{50}Ni_{25}Cu_{25}$ alloy.

In order to analyze the strength properties of the studied alloys depending on their structural state formed in isothermal annealing, the dependences $HV_{IT} = f(D)$ were constructed using the results presented in Fig. 3.154 and 3.157.

Figure 3.159 shows the dependences $HV_{IT} = f(D)$ for $Ti_{50}Ni_{25}Cu_{25}$ and $Ti_{49}Ni_{24}Cu_{24}B_3$ alloys annealed at 450◦C (τ = 10–60 min). In the $Ti_{50}Ni_{25}Cu_{25}$ alloy at $D > 70$ nm, the HV_{IT} value increases with a decrease in D in full accordance with the Hall–Petch ratio. At $D < 70$ nm, anomalous behavior (decreasing hardness values with decreasing grain sizes) is revealed, which is well known from the

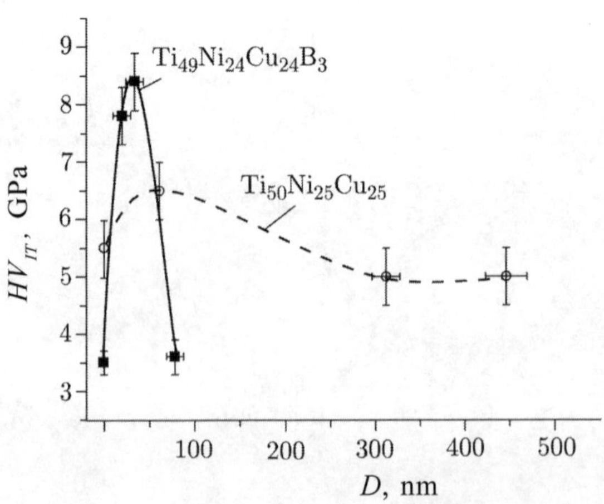

Fig. 3.159. Dependence of the nanowhardness of HV_{IT} on the size of *B*2-phase *D* of the nanoparticles.

literature [2.5, 3.288, 3.318]. The maximum value of HV_{IT} (6.5 GPa) corresponds to D = 60–70 nm.

The dependence $HV_{IT} = f(D)$ for the $Ti_{49}Ni_{24}Cu_{24}B_3$ alloy has a markedly different character. Although in this case a curve with a maximum is observed, however, the maximum value of HV_{IT} (8.4 GPa) considerably exceeds the analogous characteristic for the $Ti_{50}Ni_{25}Cu_{25}$ alloy and is achieved at a much lower D = 30 nm. In addition, the dependence $HV_{IT} = f(D)$ in the alloy $Ti_{49}Ni_{24}Cu_{24}B_3$ is much 'steeper' than in the $Ti_{50}Ni_{25}Cu_{25}$ alloy.

Using the method of grain boundary engineering (a focused introduction of 'useful' impurities (boron atoms) in the grain boundaries), the authors of [3.316] succeeded in obtaining a nanocrystalline state of the material under study with an average size of nanocrystals in the interval $30 \leq D \leq 120$ nm and with grain boundary precipitates of boron nanoparticles about 5 nm in size. The main idea of this approach was that a certain amount of boron atoms was introduced into the $Ti_{50}Ni_{25}Cu_{25}$ alloy, which can completely dissolve in the amorphous matrix and dissolve very insignificantly in the crystalline BCC type *B*2-phase formed in the crystallisation process.

Figure 3.160 schematically represents the mechanism of formation of such a structure. In the initial stages of crystallisation, boron atoms are displaced to the interphase boundary of the amorphous and crystalline phases, since boron is practically insoluble in the BCC lattice of the *B*2-phase of nanocrystals (Fig. 3.160 *a*). This slows down the growth of crystalline nanoparticles noted earlier (Fig. 3.154), since the motion of the interphase boundary is inhibited by boron atmospheres in the amorphous matrix encircling the nanocrystals. This effect is clearly visible in the corresponding electron microscopic images (Fig. 3.152 *b*). As the degree of crystallisation increases, a nanocrystalline structure with thin amorphous interlayers is formed along the boundaries of crystalline particles which are characterized by an elevated boron content (Fig. 3.160 *b*). As the crystallisation process develops, the formation of boride phases in the form of Ti_2B and TiB_2 nanoparticles about 5 nm in size takes place in amorphous interlayers and in the grain boundaries (Fig. 3.160 *c*).

The strength characteristic was the value of HV_{IT} nanohardness for which the Hall–Petch ratio (3.52) is valid. In [3.319], S.A. Firstov and T.G. Rogul' introduced the concept of the theoretical hardness HV^*_{IT} – the maximum hardness of the material, which can be achieved

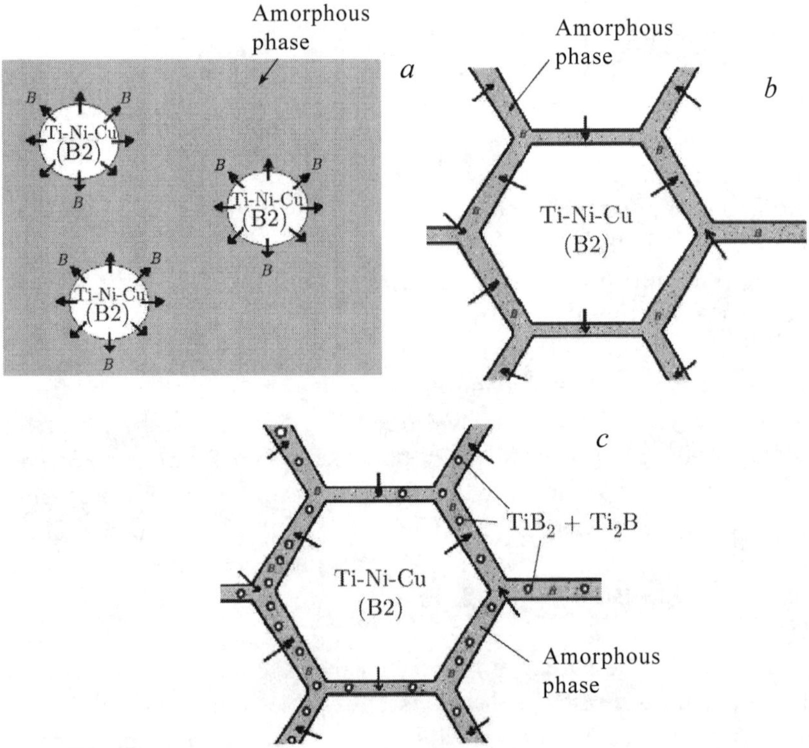

Fig. 3.160. Stages of formation of the nanocrystalline structure during annealing of the amorphous $Ti_{49}Ni_{24}Cu_{24}B_3$ alloy: *a* – primary crystallisation with separation of *B*2-phase, displacement of boron atoms into an amorphous matrix; *b* – formation of nanocrystalline structure, individual crystals separated by thin amorphous interlayers with high boron content; *c* – formation of boron nanophases Ti_2B and TiB_2 in amorphous interlayers and grain boundaries.

under the condition that the stress of the beginning of the plastic flow under the indenter corresponds to the theoretical shear strength of the material.

This characteristic is very important for evaluating the extremely hardened state of materials and is

$$HV_{IT}^{*} = \frac{\beta E}{\alpha(1+\nu)} = \frac{E^{*}(1-\nu)\beta}{\alpha}, \qquad (3.67)$$

where E^* is the reduced Young modulus equal to $E/(1 + \nu)$, ν is Poisson's ratio, α is the proportionality coefficient between Young's modulus and theoretical shear strength τ_{theor} ($\alpha = E/\tau_{\text{theor}}$) and β is the proportionality coefficient between HV and σ_y, ($\beta = HV/\sigma_y$).

The value of β is in the range 1.5–3.3, and the values of α in the range 5–30.

Later, a number of refinements were introduced into expression (3.67) [3.320].

The obtained HV_{IT} values for both alloys are analyzed from the position of the concept of theoretical hardness. For this purpose, instead of the dependence $HV_{IT} = f(D)$, shown in Fig. 3.159, the dependences $(HV_{IT}/E^*) = f(D)$ for the alloys $Ti_{50}Ni_{25}Cu_{25}$ and $Ti_{49}Ni_{24}Cu_{24}B_3$ were constructed. The value of HV_{IT}/E^* (normalized hardness) determines, obviously, the physical hardness of the real material. The experimental values of the normalized hardness HV_{IT}/E^* are estimated in comparison with the theoretical normalized hardness HV^*_{IT}/E^*. Figure 3.161 shows the dependences of the normalized hardness (HV_{IT}/E^*) on the crystal grain size D for $Ti_{50}Ni_{25}Cu_{25}$ and $Ti_{49}Ni_{24}Cu_{24}$B3 alloys.

Expression (3.67) can not give a sufficiently accurate value for HV_{IT}/E^*, since the coefficients β and α can vary within very wide limits (1–3 and 5–30, respectively [3.320]). Another way to estimate HV_{IT}/E^* is to obtain indentation [3.320]:

$$HV^*_{IT} / E^* = 0.386 \text{ ctg } \varphi, \tag{3.68}$$

where φ is the angle of sharpening of the Berkovich indenter.

Strictly speaking, equation (3.68) describes an 'instrumental' limitation of the maximum hardness value, determined by the angle of sharpening of the indenter, and does not contain information about the material itself. In principle, the theoretical normalized hardness of the material HV_{IT}/E^* can be higher, but it can not be measured using the dynamic indentation method, since the load and unloading curves will coincide. The theoretical normalized hardness of the material HV_{IT}/E^* in [3.316] is the 'instrumental' hardness (3.68) since it can be determined much more accurately than the 'physical' hardness (3.67).In these experiments, $\varphi = 71°$, therefore, in accordance with (3.68) $HV_{IT}/E^* = 0.134$. It is this value of the theoretical normalized hardness that is shown in Fig. 3.161. Note that the complete coincidence of the 'instrumental' and 'physical' normalized hardness HV^*_{IT}/E^* is achieved for completely real values of $\beta = 3$ and $\alpha = 20$ in the equation (3.67).

The limiting value of HV_{IT} in the $Ti_{49}Ni_{24}Cu_{24}B_3$ alloy is 20% higher than the limiting value of HV_{IT} in the $Ti_{50}Ni_{25}Cu_{25}$ alloy. It is important to note that the maximum HV_{IT} in the $Ti_{49}Ni_{24}Cu_{24}B_3$alloy

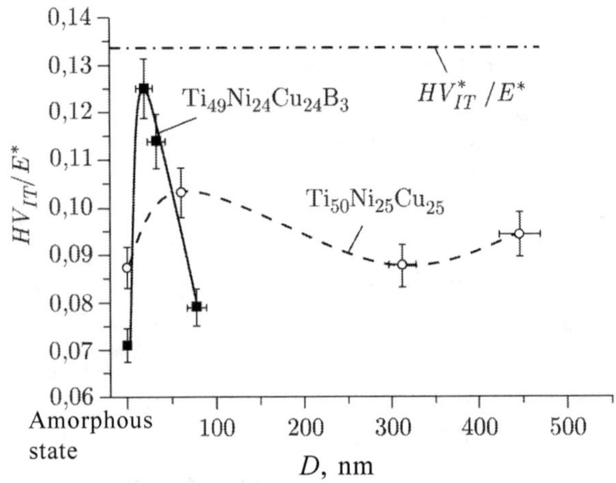

Fig. 3.161. Dependence of the normalized hardness of HV_{IT}/E^* on the size D of nanoparticles of $B2$-phase.

corresponds to the lower values of D at which an anomalous decrease in hardness is already observed in the $Ti_{50}Ni_{25}Cu_{25}$ alloy. In other words, the transition from the dislocation deformation mode to deformation by the mechanism of the GBMS with which the Hall–Petch anomaly is associated in the $Ti_{49}Ni_{24}Cu_{24}B_3$ alloy is suppressed due to the presence of high-density boron nanoparticles within the grain boundaries. The value of normalized nanohardness HV_{IT}/E^* is substantially closer to the theoretical value of HV^*_{IT}/E^*.

If we introduce the coefficient of 'proximity to the physical limit', equal to

$$\Delta = \frac{HV^*_{IT}/E^* - HV_{IT}/E^*}{HV^*_{IT}/E^*}, \tag{3.69}$$

then in accordance with Fig. 3.161 $\Delta_1 = 0.22$ for the $Ti_{50}Ni_{25}Cu_{25}$ alloy and $\Delta_2 = 0.07$ for the $Ti_{49}Ni_{24}Cu_{24}B_3$ alloy.

Thus, when the nanoparticles of boride phases are created in the grain boundaries by grain boundary engineering, the GBMS processes and the manifestation of the anomaly of the Hall–Petch ratio to essentially smaller D ($\ll$60 nm) are shifted and thus the nanohardness of the material approaches the physical limit of the HV_{IT} existing in the crystals [3.316]. By optimizing the size and surface density of the nanoparticles in the grain boundaries, perhaps in the future we will be able to approach the theoretical hardness even further.

3.3.3. Mechanical properties during quenching from a melt at a critical rate

The specificity of the amorphous–crystalline state and the resultant mechanical properties during quenching from the melt with a rate close to the critical one, when in the process of a sharp drop in the temperature of the melt only crystals of submicroscopic size can be formed, has been studied very little to date. Let us turn to the results obtained in [3.321, 3.322] on Fe–Cr–B alloys. Figure 3.162 shows the change in the microhardness *HV* as the effective cooling rate decreases during quenching from the melt in the region of the critical cooling rate v_{cr} for the $Fe_{70}Cr_{15}B_{15}$ alloy.

At cooling rates close to v_{cr}, a sharp maximum is observed corresponding to the transition of the alloy to the crystalline state. At $v > v_{cr}$, the alloy is in the amorphous state, which has characteristic diffuse rings on electron diffraction diagrams, and for $v < v_{cr}$ – in the crystalline state formed by several phases. It is characteristic that the values of *HV* in all the states obtained by quenching from the melt significantly exceed the values corresponding to this alloy but obtained by the 'usual' technology of melting and heat treatment (8 GPa). In the region of the *HV*(v) curve related to the amorphous state, there is an increase in electron microscopic effects associated with the presence of regions of increased correlation in the arrangement of atoms (see Chapter 1). This essentially determines the smooth increase of *HV* as v decreases and approaches to v_{cr}.

In the region of the maximum on the curve *HV*(v) (21–22 GPa), corresponding to the transition from the amorphous to crystalline

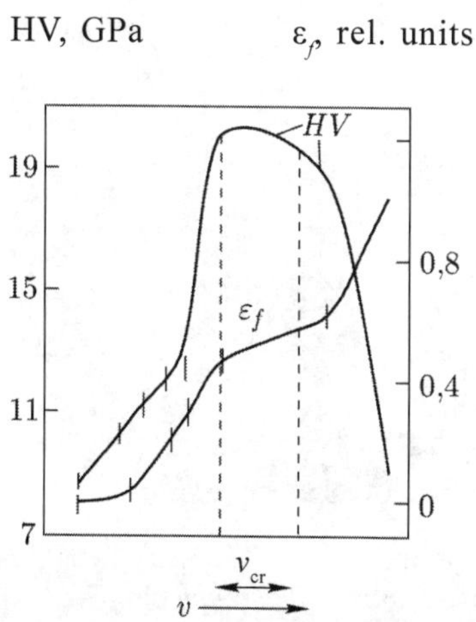

Fig. 3.162. Change in microhardness *HV* and ductility ε_f, depending on the effective cooling rate during melt quenching v.

state, in addition to the small volume fraction of the amorphous phase there is an ultradispersed crystalline phase (Fig. 3.163 *a*).

Attempts to identify microelectron diffraction pattern, similar to that shown in Fig. 3.163 *a*, led the authors of [3.321] to the assumption that in the structure there exist basically submicrocrystallites with a BCC lattice or a lattice close to it with a parameter close to 0.285 nm (in the case of a BCC lattice) and an appreciable scatter of the lattice parameter reaching several percent. The presence of the scatter in the lattice parameter causes the appearance of a 'cloud' of point reflexes on the microelectron diffraction pattern of Fig. 3.163, and with an obvious variation in the length of the effective reflection vector for each of the reflections of the azimuthally disoriented particle reflections. The average size of individual crystals is 8–10 nm (the minimum size is 1–2 nm). The morphology of submicrocrystals is clearly visible on dark-field images under the action of one or several point reflexes (Fig. 3.163*b*). Frequently observed blurring of reflexes on diffraction patterns and the reduction of the intensity of the primary diffraction contrast in the peripheral sections of the dark-field images of individual crystals suggest that the crystal lattice parameter can in this case vary not only from one microcrystal to another, but also from the central part of each of the crystallites to the peripheral one. Moreover, it is likely that at the earliest stages of the formation of the ultradispersed structure ($v \approx v_{cr}$), the border areas are partially amorphous. This is evidenced, in particular, by the fact that in the transition amorphous–

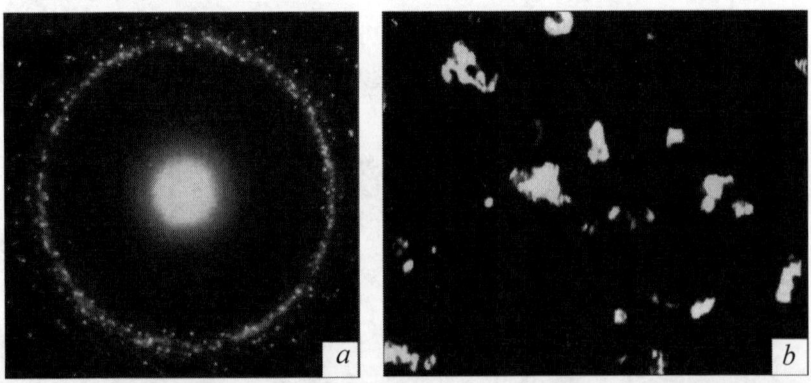

Fig. 3.163. Microelectron diffraction pattern (*a*) and dark-field image (*b*) in one of the reflections of the first ring of the Fe–Cr–B alloy after quenching from the melt with $v \approx v_{cr}$.

crystalline state there is no characteristic banded contrast from grain boundaries, which, apparently, is simply not present.

On the basis of the foregoing, the following structural model of an anomalously strong amorphous–crystalline state obtained by quenching from a melt can be assumed. Microcrystals that form a homogeneous conglomerate are characterized by a smoothly varying degree of crystalline order: in the central part of each crystallite formed at high temperature, there is an ideal crystal structure that approaches the periphery (i.e., essentially as the solidification temperature decreases sharply), Gradually turns into an amorphous structure (Fig. 3.164).

The observed interlayers of the amorphous phase are enriched with metalloid (boron) atoms and do not have clear interphase boundaries with the crystalline phase. The same structure can be imagined as microcrystalline, in which the grain boundaries between individual crystals are 'blurred' to such an extent that they are sufficiently extended regions of the amorphous phase.

Let us pay attention to one very important circumstance inherent in the amorphous-crystalline state we are considering: the boron concentration in microcrystallites exceeds the equilibrium value for α-iron. This is indicated by the dependence of the values of the coercive force H_c and the saturation induction B_s of the $Fe_{70}Cr_{15}B_{15}$ alloy in the quenched state on the effective quenching rate (Fig. 3.165).

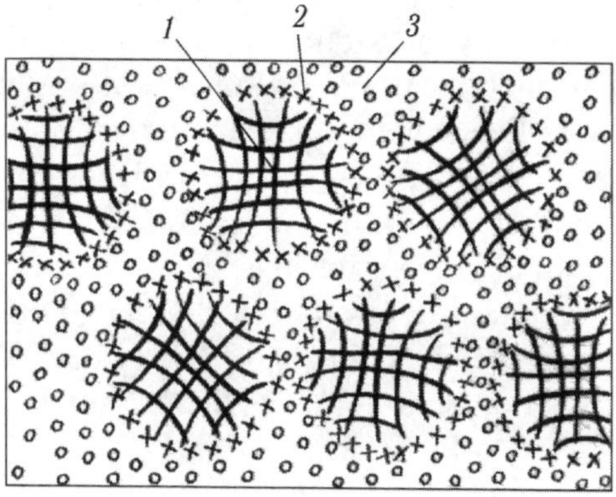

Fig. 3.164. The structural model of the transition amorphous–crystalline state which is realized upon cooling from the melt with the velocity $v \approx v_{cr}$: 1 – area of a microcrystal with a variable crystal parameter of the lattice; 2 – region of transition from crystalline to amorphous state; 3 – thin interlayers of amorphous phase.

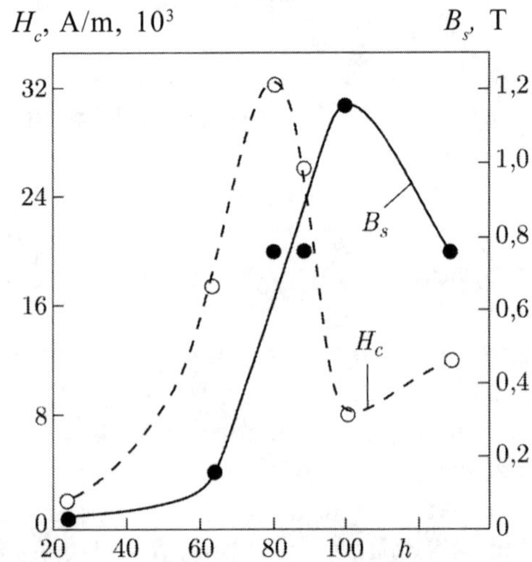

Fig. 3.165. Dependence of the coercive force H_c and saturation induction B_s on the effective quenching rate from the melt (ribbon thickness *h*).

The point is that in an amorphous state the alloy is paramagnetic: the saturation induction at high values of v is close to zero. The appearance of the crystalline phase with decreasing v leads to an increase in both saturation induction and coercive force. The peak of H_c coincides, as experiments have shown, with the peak of *HV*, i.e., it corresponds to the amorphous–crystalline state of interest to us. It can be seen that the maximum of H_c corresponds to a value of B_s that did not reach a maximum value equal to the saturation induction of α-Fe–Cr. This can be explained only by the fact that in the amorphous–crystalline state, a large amount of boron is contained in the crystallites. The subsequent decrease leads to the formation of crystalline α-Fe–Cr crystallites not containing boron (saturation induction reaches a maximum), but the hardness and coercive force corresponding to such a structure are no longer optimal.

Static bending tests have established [3.321] that the plasticity of the alloy in the transition state is lower than in the amorphous state, but at the same time it is far from zero (Fig. 3.162). Thus, it can be stated that the investigated state of Fe–Cr–B alloys quenched from a melt has not only a unique high strength, but also sufficient ductility. Studies of the features of the structure of plastically deformed ultradispersed alloys have shown that the process of plastic flow has features inherent in the deformation of amorphous alloys. Thus,

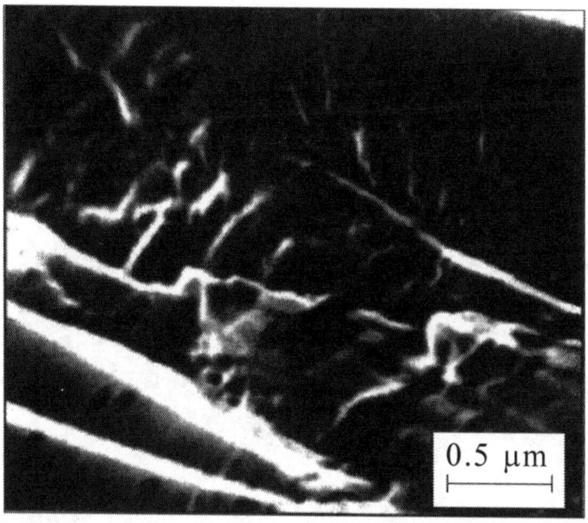

Fig. 3.166. Slip lines in a bent ribbon specimen of the $Fe_{70}Cr_{15}B_{15}$ alloy, obtained by quenching from a melt with a velocity $v \approx v_{cr}$. Scanning electron microscopy in secondary electrons.

for example, a study was made of the pictures of shear bands by transmission scanning electron microscopy (Fig. 3.166).

In this case, systems of highly localized bands were detected (the height of the slip steps reaches 0.3–0.4 μm), realizing the degree of deformation of several hundred percent, which is typical for amorphous materials. Since the authors of [3.322] could not detect electronically signs of the existence of dislocations in the shear bands [3.322], they assumed that the process of plastic deformation in the ultradispersed state is localized in amorphous intercrystalline interlayers and resembles, to a certain extent, the process of slipping along the grain boundaries.

The question of how thermally stable the ultradispersed state obtained by quenching from the melt is of unconditional interest. To some extent, the *HV* dependences on the annealing temperature for 1 h for the $Fe_{70}Cr_{15}B_{15}$ alloy, melt quenched with different rates (Fig. 3.167) correspond to some extent.

Curve 1 refers to an alloy quenched at a rate $v > v_{cr}$ and, therefore, in an amorphous state. It is characterized by a slight increase in *HV* due to low-temperature relaxation hardening, which is inherent in virtually all amorphous metal–metalloid-type alloys. Curve 2 corresponds to the transition amorphous–crystalline state obtained by quenching from the melt with a rate $v \approx v_{cr}$. In this case, at an annealing temperature of slightly above 200°C, a sharp decrease in *HV* occurs followed by a reduction, but in an absolutely fragile state.

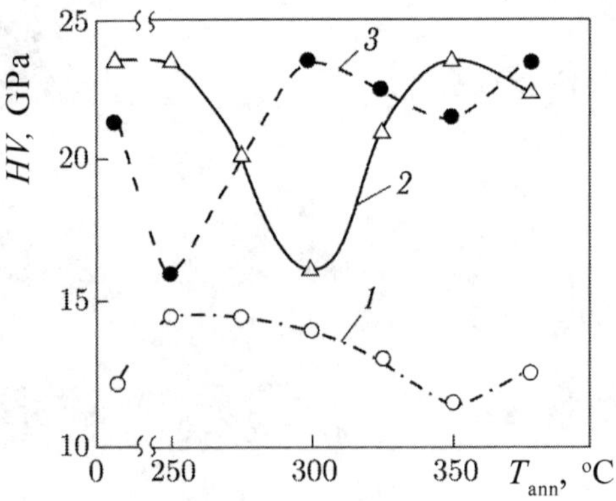

Fig. 3.167. Dependence of the microhardness *HV* on annealing temperature *T* for 1 hour of the $Fe_{70}Cr_{15}B_{15}$ alloy obtained by quenching from the melt with a rate of $v > v_{cr}$ (1), $v \approx v_{cr}$ (2) and $v < v_{cr}$ (3).

As shown by structural studies, such a relationship is associated with the simultaneous occurrence of two processes: decrease in boron concentration in the matrix and intensive boride formation. The separation of boride occurs mainly in amorphous interlayers along the crystal boundaries and blocks the propagation of plastic shear over them.

Curve 3 corresponds to the crystalline state obtained during quenching from the melt with a rate $v < v_{cr}$. It is the least stable state, although the processes occurring in it during tempering are, as it were, shifted to a phase relative to those that occur in the amorphous-crystalline state. Some paradox, still requiring its explanation, lies in the fact that the initial structure is non-equilibrium, the more it is stable with respect to thermal effects: the most stable amorphous state, and the least thermally stable – the crystalline one.

Briefly summarizing, it should be noted that there are several reasons for the anomalously high strength and hardness of the amorphous–crystalline state obtained by quenching from the melt and having a structure schematically shown in Fig. 3.163. First, alloys in this state are characterized by a very small size of microcrystallites, supersaturated with boron atoms. Secondly, an amorphous interlayer along the boundaries of crystallites completely excludes the dislocation mechanism of the transfer of deformation from one crystallite to another. And, finally, a high concentration of boron in an amorphous interlayer creates additional conditions for the realization of high stresses of the onset of plastic flow.

It should also be noted that the structural state of Fe–Cr–B alloys quenched from a melt at a critical rate described in [3.321, 3.322] can apparently be realized in other amorphous systems. Indeed, recently we were able to observe the same state, accompanied by a very high strength (σ_p = 6 GPa) in the Co_4B alloy.

3.3.4. Mechanical properties under deformation effects

In the work [3.145] an attempt was made to analyze in detail the features of the structure and properties under the influence of MPD on a series of amorphous alloys of the metal–metalloid type. The $Ni_{44}Fe_{29}Co_{15}Si_2B_{10}$, obtained by melt quenching, was mainly studied. But in addition, the $Fe_{74}Si_{13}B_9Nb_3Cu_1$ (Finemet), $Fe_{57.5}Ni_{25}B_{17.5}$, $Fe_{49.5}Ni_{33}B_{17.5}$, and $Fe_{70}Cr_{15}B_{15}$ were studied in [3.145].

As it was shown in section 3.2.2, after MPD crystallisation processes begin in the alloys, expressed more noticeably after deformation at room temperature. For example, for the number of turns N = 4 and T = 293 K, the fraction of the crystalline phase α and the average crystallite size D are 8% and 3 nm, respectively. The authors note that the values of these parameters almost completely correspond to those that are observed after N = 8, but at T = 77 K (α = 6% and r = 2 nm). Electron microscopic observations (Fig. 3.70 in section 3.2.2) confirmed these results qualitatively and quantitatively.

Figure 3.168 shows the change in the microhardness HV of the amorphous and partially crystallized $Ni_{44}Fe_{29}Co_{15}Si_2B_{10}$ alloy (preliminary annealing of the amorphous state at a temperature above T_{cr}), depending on the magnitude of the deformation (number of turns) N in the Bridgman chamber at various temperatures.

Attention is drawn to the noticeable difference in the character of the curves in the initially amorphous and initially partially crystalline states. In the first case, there is a significant decrease in HV in the initial stages of the MPD (N = 0.5) and subsequent monotonous growth, with both the drop effect and the subsequent growth effect being significantly more pronounced at T = 293 K (–2.8 and +2.1 GPa at 293 K and –0.8 and +1.1 GPa at 77 K). As for the initially partially crystalline alloy, its HV value was found to be 0.75 GPa higher than that of the initially amorphous alloy, but then, first noticeable a nd then weak decrease in the HV value without any extreme manifestations to a value almost coinciding With an HV value for an amorphous undeformed state.

In the region of a sharp drop in HV, inhomogeneous plastic deformation is observed with the formation of coarse shear bands,

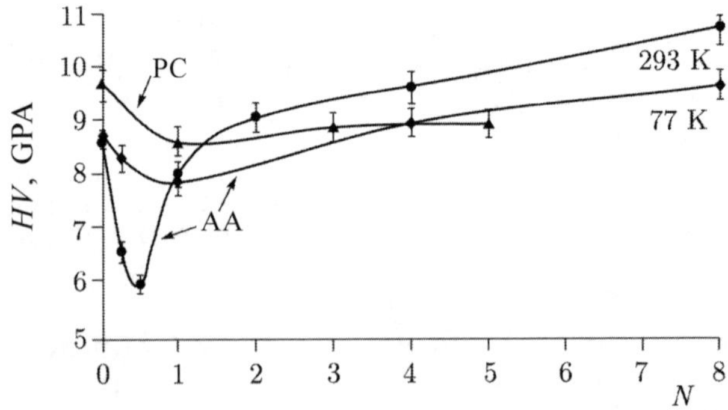

Fig. 3.168. The change in the microhardness *HV* amorphous (AA) and partially crystallized (PC) $Ni_{44}Fe29Co_{15}Si_2B_{10}$ alloy, depending on the value of *N* after MPD at temperatures of 293 and 77 K.

which is inherent in all amorphous alloys at temperatures well below the transition point to the crystalline state [1.107]. The phenomenon of nanocrystallisation in MPD is considered in detail in section 3.2.2. The MPD of annealed copper samples with an initial grain size of 100–250 μm was carried out in [3.323] by dynamic compression at the liquid nitrogen temperature. During deformation, the formation of a matrix structure consisting of fragmented nanograins and twins was observed. According to the TEM data, the average grain size was ~68 nm, the yield strength reached 600 MPa, the tensile strength 633 MPa, the elongation to break 11%. According to the strength and ductility indexes, these samples are somewhat inferior to the data [3.324–3.326] for samples with a purely twin structure.

Non-equilibrium thermodynamics for the description of MPD
As discussed in section 3.2.2, megaplastic deformation (MPD) is a complex multilevel process in which several deformation mechanisms can participate simultaneously.

A theoretical description of such a process is a rather difficult task. Therefore, the attraction of new ideas, views and approaches, the search for more adequate theoretical models are highly desirable for its solution. At present, there are several theoretical approaches aimed at solving this problem. This is the engineering theory [3.327, 3.328], which proceeds from simple mechanical concepts of the process of deformation. The system of kinetic equations is constructed on the basis of generalization of experimental results and regularities. In the framework of this theory it is not possible to explain many important features of the MPD process, and, in particular, it can not describe the formation of the 'limiting' structure of metals – the minimum average grain size. To somehow describe the fragmentation of grains, some attraction attractor is introduced into the theory by 'hands', forcing the system to strive for a state with the desired (prescribed) grain size.

The theory of material dispersion in the MPD process [3.329-3.331] is based on another specific deformation mechanism, which is based on the concepts of interboundary slip and anomalously high diffusion along the grain boundaries. In addition, the theory is based on the concepts of free volume borrowed from the theory of amorphous materials. At the same time, experimental work [3.332, 3.333] casts serious doubts on the fact that the free volume can be used as a determining thermodynamic parameter for describing amorphous alloys, at least in the classical interpretation that is accepted in the vast majority of works.

The theory of fragmentation of dislocation structures [3.334-3.336] describes the formation of high-angle grain boundaries for large plastic deformations. It is based on the idea of self-organization of dislocations. Flexural and twisting deformations and stresses arising in the process of equal-channel angular pressing lead to the generation of a large number of geometrically necessary dislocations whose behaviour is no different from the behaviour of chaotically arising dislocations. In fact, the ideas that previously developed to describe the formation of a cellular structure [3.337] were extended to describe the formation of a larger defect, the high-angle grain boundary. In the subsequent stages of hardening of the material [3.334], the newly formed structure must cross several elementary dislocation cells. It can be assumed that it is more advantageous for dislocations to leave the volume of a dislocation cell or grain and 'settle' at the boundaries, thereby increasing the level of their excess energy (non-equilibrium), and only then to form a new boundary as a process of relaxation of the excess energy of the boundaries. At the same time, it is logical to assume that the achievement of the ultimate grain size as a result of dispersion in the MPD process is a general property of obtaining a stationary state by the material and can be investigated in the most general form without concretizing the deformation mechanism in the framework of mesoscopic non-equilibrium thermodynamics [3.338–3.344]. The peculiarity of this approach is that whereas ordinary thermodynamics is an unstructured theory, here the defective structure is taken into account by the explicit introduction of the corresponding terms in the basic thermodynamic relationships. Structural elements of the medium in such an analysis can be quite large formations of the grain type, grain boundaries, dislocation clusters, etc. While ordinary thermodynamics in the study of a finely dispersed nanostructured medium loses its validity, the applicability of mesoscopic non-equilibrium thermodynamics becomes more acceptable due to the larger number of elements of the medium in a representative volume.

The basic thermodynamic relationship in this case is a certain thermodynamic identity or equality, which includes the first and second laws of thermodynamics [3.338, 3.341, 3.342]. We write it here in the form

$$du = \sigma_{ij} d\varepsilon_{ij}^{e} + Tds + \chi ds' + \sum_{l=1}^{N_d} \varphi_1 dh_1, \tag{3.70}$$

where u is the internal energy density, σ_{ij}, ε_{ij} is the stress tensor and the elastic part of the strain tensor, T, s is the temperature and the total (from external and internal sources) equilibrium entropy, χ, s' are the non-equilibrium temperature and non-equilibrium entropy only from internal sources, φ_1, h_1 are the energy and density of structural defects of l-type. Summation is carried out for all types of structural defects N_d present in the bulk of a solid, but in specific problems it may be sufficient to take into account one or two types of defects, $N_d + l$ is the total number of dissipation channels for the irreversible part of the work of external forces, including the thermal one.

The first two terms in (3.70) describe the equilibrium part of the system, the latter – non-equilibrium. A full differential means that the variables under it are functions (more precisely, arguments) of the state – the variables ε_{ij}^{e} and s in the classical sense of equilibrium thermodynamics, the variables s' and h_l in the generalized sense of non-equilibrium thermodynamics, since they are understood in the Landau–Khalatnikov equations, for which the generalized thermodynamic forces are calculated as derivatives of internal energy:

$$\chi = \frac{\partial u}{\partial s'}, \quad \varphi_l = \frac{\partial u}{\partial s'}. \tag{3.71}$$

Due to the formal equality of the thermodynamic variables entering into the relation (3.70), the variables φ_l and h_l can be given the meaning of static temperature and static entropy, since they are caused by a static (potential) field of defects, in contrast to the usual (dynamic) temperature due to atomic (kinetic) movement. The evolution of non-equilibrium parameters in the general case is proposed to be described by a system of equations, each of which is an equation of the Landau–Khalatnikov type in the case of homogeneous or Ginzburg–Landau in the case of a non-homogeneous problem:

$$\frac{\partial s'}{\partial t} = \gamma \frac{\partial u}{\partial s'} = \gamma \chi,$$
$$\frac{\partial h_l}{\partial t} = \chi \left(\frac{\partial u}{\partial h_l} - \varphi_l \right) = \chi_{h_l} \left(\varphi_l - \overline{\varphi}_l \right), \tag{3.72}$$

where γ, γ_l are the kinetic coefficients of the rate of the relaxation process of the corresponding dissipation channel. The parameter $\varphi_1 > 0$ describes the positive actually observed value – the average energy of the l-type defect. In contrast to the classical analogue, internal, rather than free, energy appears here. The stationary point of the second equation in (3.72) is shifted relative to the extremum of the internal energy, so that the thermodynamic force in the steady-state regime will not be zero. The latter circumstance may be due to the manifestation of the structural viscosity of the defective material – defects can not escape due to their low (zero) mobility. In addition, thanks to this displacement, it is possible to construct a consistent closed system of thermodynamic potentials obtained from the internal energy by means of the Legendre transformation [3.345]. Finally, unlike the classical analogue, the plus sign stands in the right-hand sides of equations (3.72) which indicates that the stable state of the system corresponds to those states that are located near the maximum of the internal energy. This seems somewhat paradoxical in comparison with equilibrium thermodynamics, where the minima of all thermodynamic potentials (except entropy) correspond to stable states. Such an inversion for internal energy is associated with the energy pumping process due to the irreversible part of the work of external forces.

Equations (3.72) form a connected system of equations, which must be considered together. However, for simplicity, consider them separately. The first equation in (3.72) can really be considered separately in the case when the production of new defects terminates, for example, when the system as a result of dispersion reaches a minimum grain size. If we confine ourselves to the quadratic approximation of the internal energy with respect to the non-equilibrium entropy

$$u = \chi_0 s' - \chi_l (s')^2, \tag{3.73}$$

Then the evolution equation can be written in an explicit form:

$$\frac{\partial s'}{\partial t} = \gamma(\chi_0 - 2\chi_1 s'). \tag{3.74}$$

The first term describes the constant sources of non-equilibrium entropy, which arise from the irreversible part of the work of external forces. The second term is the ordinary relaxation term describing the transition of the non-equilibrium entropy to the equilibrium state. Due to this, the equilibrium (s) part of the entropy will constantly increase which will lead through the equation of state to a constant increase in temperature, while the fraction of the non-equilibrium part in the steady-state regime $s' = \chi_0/2\chi_1$ will remain constant.

Thus, the equilibrium entropy s will vary in time in accordance with the evolution of the non-equilibrium entropy:

$$\frac{\partial s}{\partial t} = 2\gamma\chi_1 s'. \tag{3.75}$$

At the initial stage of deformation, the non-equilibrium part of the entropy will change as quickly as possible, and the equilibrium entropy is the slowest, while in the stationary stage, on the contrary, the non-equilibrium part of entropy will remain constant, and the equilibrium entropy will change as quickly as possible. If the kinetic constant γ is large, then the stationary state will be reached quickly (the adiabatic approximation), and then the change with time of the equilibrium entropy in the absence of heat outflow through external boundaries can be written directly in the form $\partial s/\partial t = \gamma\chi_0$.

In this case, the first equation in (3.72) can be ignored. It should be noted that such a regime can not be considered strictly stationary, since an increase in temperature will lead to a change in the expansion constants of the thermodynamic potential, and in the second order the system will continue to evolve. The regime will be strictly stationary only in the case when all the heat released from internal sources will completely leave the heat through its external boundaries.

The second equation in (3.72) can be considered separately in the adiabatic case considered above, or in the case when thermal effects can be neglected. In the case of MPD, this can be done (possibly only for some materials) at the stage of intensive production of boundaries, when the main channel for dissipation will be the production of defects. In the latter case, it is convenient to go over to the free-energy potential $f = u - T_s$ and assume that the process is carried out at a constant temperature. Separately, the equation for one type of defect can be considered in the case when the evolution of other types of defects is completed. In the case of MPD at the stage of direct fragmentation, it can be assumed that the process of formation of a dislocation structure has already reached the stationary regime and is reduced to providing boundaries with a new 'building' material. Then the main material defect, which is in a state of evolution, has the form of grain boundaries. In the quadratic approximation with respect to the density of defects, the evolution equation can be written in explicit form:

$$\frac{\partial h_g}{\partial t} = \gamma_g \left(\varphi_{g0} - \overline{\varphi_g} - 2\varphi_{g1} h_g \right), \tag{3.76}$$

where h_g is the boundary density (the total area of boundaries per unit volume), γ_g, φ_{g0}, φ_{g1} are the kinetic constant and the coefficients of internal energy expansion, and φ_g is the stationary value of the boundary density.

The coefficient of decomposition φ_{g0}, on the one hand, is responsible for the connection of grain boundaries with defects of deeper structural levels, such as dislocations, cellular structures, subgrain boundaries, etc., as a building material for boundaries. In addition, the value φ_{g0} determines the energy redundancy (non-equilibrium) of defects of deeper structural levels, as a result of which they tend to 'settle' at the boundaries. On the other hand, it determines an irreversible part of the external work, that is, the energy received from the work of external forces is directed to the production of new boundaries not directly, but indirectly through the production of smaller defects. The coefficient φ_{gl} determines the relaxation rate of non-equilibrium boundaries, the limit of which is given by the constant $\overline{\varphi}_g$.

The presence of the displacement $\overline{\varphi}_g$ will lead to the fact that a stable stationary point will correspond not even to the maximum of the internal energy, but to some constant slope of the tangent to its graph.

The value $\overline{\varphi}_g$ can depend on the parameters of the problem in a complex manner, and at this stage of understanding the problem must be determined experimentally. It can be quite definitely stated that a higher level of non-equilibrium of stationary boundaries [3.340, 3.341] should correspond to a finer dispersed material (large values of h_g).

The free energy for the isothermal problem can be expanded in a series of material defectiveness with allowance for contributions up to 4 degrees:

$$f(h_g) = f_0 + \varphi_{g0} h_g - \frac{1}{2}\varphi_{g1} h_g^2 + \frac{1}{3}\varphi_{g2} h_g^3 - \frac{1}{4}\varphi_{g3} h_g^4, \qquad (3.77)$$

where the expansion constants depend on the first and second invariants of the elastic part of the strain tensor:

$$f_0 = f_0^* + \frac{1}{2}\lambda\left(\varepsilon_{ii}^e\right)^2 + \mu\varepsilon_{ij}^e\varepsilon_{ji}^e,$$
$$\varphi_{g0} = \varphi_{g0}^* + g\varepsilon_{ii}^e + \frac{1}{2}\lambda\mu\left(\varepsilon_{ii}^e\right)^2 + \overline{\mu}\varepsilon_{ij}^e\varepsilon_{ji}^e, \qquad (3.78)$$
$$\varphi_{g1} = \varphi_{g1}^* + e\varepsilon_{ii}^e,$$

where λ, μ are the elastic Lamé constants, $f_0^* = 2 \cdot 10^8$ Pa, λ = μ = 20.8 GPa, $\varphi_{g0}^* = 0.8 \cdot 10^{-3}$ Pa·m, $\varphi_{g1}^* = 0.8 \cdot 10^{-12}$ Pa·m², $\varphi_{g2} = 2.25 \cdot 10^{-23}$ J; $\varphi_{g3} = 1.2 \cdot 10^{-34}$ J·m, g = 0.12 Pa·m, $\overline{\lambda}$ = 1000 Pa·m, $\overline{\mu}$ = 2500 Pa·m, e = 3.6 · 10⁻¹² Pa · m², $\varepsilon_{ji}^e = -0.002$.

The free energy diagrams corresponding to the values of the parameters are shown in Fig. 3.169 for different values of the second invariant $\varepsilon_{ij}^e\varepsilon_{ji}^e$.

Smaller values of defectiveness correspond to a more coarse-grained structure of the material. For small values of the parameter ε_{ij}^e (low stresses), there is one maximum of free energy in the region of a coarse-grained structure. It corresponds to the usual plasticity of the material, and the coarse-grained structure can therefore be considered as one of the limiting structure types.

With the growth of stress there is another maximum, which at high stresses becomes the only one. At this point, the system under the first-order phase transition scenario turns into a fine-grained state, into the next stationary or limiting structure. Such a transition is the essence of the MPD.

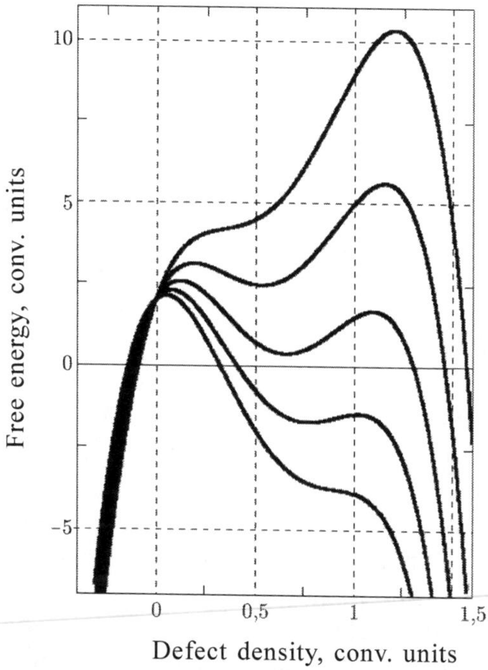

Fig. 3.169. Free energy graphs for different values of shear stresses. Here, the values of the second invariant correspond to curves from below upwards $\varepsilon^e_{ij}\varepsilon^e_{ji}$ = {1,7; 2.7; 3.9; 5.2; 6.8} · 10^{-6}.

It should be noted that, because of the greater number of thermodynamic pairs of variables compared with the equilibrium case, there will be more variants of the Legendre transformation. For example, by performing the Legendre transformation directly on the thermal (T, s) variables describing the defectiveness of the material (φ, h), one can obtain free energy of the form:

$$F = u - Ts - \varphi h. \tag{3.79}$$

For a free energy of this type, the stable state of the system will be near its minimum [3.340, 3.341], and not the maximum, as it was for the potentials considered above. A more detailed exposition of the problem of the connection between different types of thermodynamic potentials can be found in [3.345].

In conclusion, we note that the approach of mesoscopic non-equilibrium thermodynamics presented here is a kind of attempt to correctly reconcile two extreme points of view, one of which is based on the postulation of certain macroscopic constitutive relations, which often include the large volume of empirical data, and the second is based on a microscopic (structural) description . The first approach is much simpler and allows us to obtain foreseeable solutions, but at times it allows inaccuracies that they seem to follow, but the way to obtain macroscopic consequences from microscopic equations is so complicated, and to obtain more or less visible solutions it takes so many additional assumptions that the advantages of the approach become not obvious. The proposed approach here intelligently combines the advantages of both points of view. On the one hand, the simplicity of the constitutive approach is achieved, but the working relations are not postulated, but are derived from general principles, on the other hand, the structure of the material is fully taken into account here. In addition, the approach also provides a natural way of how to truncate the system of equations, leaving one or two types of defects in the field of view that are most significant when describing a particular process.

Since in describing the fragmentation of the grain structure in the MPD process, the most significant defect is the grain boundary, all other defects are effectively taken into account in the absence of the individualization of their individual behavior by the coefficients of the theory. Of course, when describing the second and third stages of material hardening, the set of defects will be different, and here it becomes necessary to consider the various systems and subsystems of the dislocation organization. This task, in view of its special complexity, has so far not been resolved within the framework of the proposed approach, and it still has to be addressed. In particular, within the framework of this approach, it is necessary to take into account the nature of the evolution of dislocation loops as it occurs in real material [3.346–3.351]. According to [3.346], macroscopic deformation consists of events occurring in regions of loss of stability. Its minimal structural element is dislocation loops, and not straight-line infinite dislocations, as is customary in most plasticity theories.

Thus, the author of [3.345] proposed a mesoscopic non-equilibrium thermodynamics variant for describing the fragmentation of a material during metal processing by MPD methods, which takes into account the structural structure of the material. The coarse-grained structure within the framework of the proposed consideration can be treated as one of the models of the stationary (limiting) structure. In the next stationary (limiting) finer-grained state with an increase in the level of external shear stresses, the system quite sharply goes over the scenario of a first-order phase transition. This type of transition justifies the application to this process of the term of severe plastic deformations, and not just large deformations, as is often the case in the literature.

3.3.5. Mechanical properties after pulsed treatment

In [3.215], the effects of pulsed laser annealing on the structure and properties of bulk amorphous Zr–Ti–Cu–Ni–Al (52.5% Zr) and Pd–Cu–Ni–P (40% Pd) alloys was studied and it was established that the effect of pulsed laser radiation on the surface of amorphous alloys is accompanied by structural transformations that depend on the thermal properties of the alloys. The nanoindentation method shows that in the zone of action of laser radiation, the values of nanohardness and modulus of elasticity of amorphous bulk alloys on the basis of zirconium and palladium change in comparison with the initial state (Table 3.2). In the Zr-based alloy, a significant decrease in the nanohardness and Young's modulus is due to the structural transformation of the amorphous matrix → HCP crystal, and in the Pd-based alloy this decrease is 10 and 5%, respectively, and is associated with the secondary glass transition processes occurring on the surface.

When using pulsed laser irradiation of 82K3KhSKhR amorphous metal alloy [3.213], the regions of the crystalline phase are nucleated in the amorphous matrix. In the crystallisation of an amorphous substance, in the course of the appearance and 'optimal' distribution over the volume of an amorphous matrix of metallic glasses of finely dispersed crystalline particles, the hardness increases. The greatest increase in microhardness is observed near the fusion boundary and is associated mainly with stresses that arise as a result of the formation of a transition zone from an amorphous matrix containing

Table 3.2. Mechanical characteristics of alloys in the zones of effect and in the initial state

Amorphous alloy		Nanohardness HV_{IT}, GPa	Young modulus E, GPa
Based on Zr	Inside the effect zone	0.249	4.598
	In the initial state	12.215	150.492
Based on Pd	Inside the effect zone	12.111	193.47
	In the initial state	13.396	202.888

finely dispersed inclusions to regions containing a crystalline phase. Due to the action of a relatively powerful and short laser pulse, leading to local heating of the material, a change in the mechanical characteristics at the boundary of the zone of thermal influence is noted.

In [3.352] the effect of the pulsed photon irradiation (PPR) on the structure and properties of the amorphous alloy $Al_{87}Ni_{10}Nd_3$ is examined It is established that under the influence of PPR, a nanocrystalline structure with a grain size of less than 50 nm can be obtained. After the effect of PPR with an energy of radiation E_r = 1–10 J/cm^2, nanocrystalline phases (Al, Al–Nd) with an average matrix phase size (Al) ~ 2.5 nm (E_r = 5 J/cm^2) to 5 nm (E_r = 10 J/cm^2) start to precipitate. At E_r > 15 J/cm^2, the alloy becomes amorphous–crystalline, and at E_r = 17 J/cm^2 it completely crystallizes with a particle size of 50–150 nm. The results of measuring the microhardness of this alloy after the PPR with an energy of E_r = 15 J/cm^2 show an increase in this characteristic from 3.7 GPa to 4.8 GPa, while the hardening HV/HV_{amorph} was 1.34–1.36. It should be noted that the maximum hardening while maintaining the ductility of the alloy was achieved after the PPR with E_r ~ 12 J/cm^2 (HV/HV_{amorph} = 1.31). After the PPR with E_r = 10 J/cm^2, the alloy is retained in the plastic state, but it is not prone to hardening, at E_r = 15 J/cm^2, maximum hardening is achieved with loss of plasticity, and after PPR with E_r = 17 J/cm^2, it softens ($HV < HV_{amorph}$). The mechanism of hardening is associated with the formation of nanocrystalline phases (Al and Al_3Ni), in the case of PPR with E_r = 15 and 17 J/cm^2, the size of aluminum nanoparticles is 50 and 84 nm, respectively.

It is reported in [3.353] that when PPR affects alloys of the Fe–P–Mn–V systems structural inhomogeneities (clusters) are formed in the amorphous matrix. Two types of clusters form in the $Fe_{74}P_{18}Mn_5V_3$ alloy that are unstable near the crystallisation temperature, and in $Fe_{73}P_{18}Mn_5V_4$ alloy the formation of clusters of the same composition was observed which are resistant to a wide range of temperatures and are characterized by a greater chemical bond strength, resulting in a greater degree of reinforcement in the transition to nanocrystalline state. PPR with E_r <16 J/cm^2 for 2 s while maintaining ductility leads to hardening of alloys in comparison with the initial amorphous state [3.354]. Under the influence of PPR with E_r <20 J/cm^2 for 2 s, a nanocrystalline structure with nanocrystal dimensions of 2–3 nm is formed in these alloys, while plasticity is preserved. After the effect of PPR with E_r = 21 J/cm^2, the $Fe_{73}P_{18}Mn_5V_4$ alloy begins to crystallize to form nanocrystalline phases of ~70 nm in size in the $Fe_{73}P_{18}Mn_5V_4$ alloy, which leads to a decrease in the microhardness.

3.3.6. Mechanical properties of amorphous–nanocrystalline films

Film technology is very effective for obtaining high-strength silicon–titanium nitride nanocomposites. Record hardness values (80-100 GPa) are described for nanostructured films (Ti, Si) N sputtered in plasma-activated regimes by the method of chemical deposition [3.355]. Such high-hard and highly wear-resistant nanocomposites $(Ti_{1-x}Al_x)N/a\text{-}Si_3N_4$ have an amorphous matrix of silicon nitride in which nanoinclusions of titanium–aluminum nitride (3–4 nm in size) are located. Comparison of the wear of the incisors with a coating from this nanocomposite and standard coatings such as TiN, Ti (C, N), and (Ti, Al) N has shown the advantage of nanocomposite coatings (Fig. 3.170).

Brittle materials, such as nitrides, carbides, oxides, etc., have high hardness values in the ordinary coarse-grained state (up to 20–30 GPa or more). This property served as a good motive for creating on their basis new types of superhard materials using the nanostructural approach. Several scientific groups in the United States, Sweden, Germany, the former USSR, and Austria independently and practically simultaneously produced single- and multilayer films based on nitrides, carbides and borides with a hardness of 50–80 GPa.

Figures 3.171 and 3.172 show the effect of the number of layers in multilayer nitride films (the total thickness of all films was the same and was 2 μm) on the hardness of the samples [3.356].

With an increase in the number of layers (and, correspondingly, with a decrease in their thickness), a considerable increase in hardness is observed, which is explained by the difficulty in spreading the propagation of dislocations and cracks due to the increase in the number of interfaces.

The properties of amorphous–nanocrystalline films have been studied to the greatest degree for nanocomposites based on an amorphous carbon matrix and nanocrystalline inclusions of TiC, ie, TiC–*a*C films (*nc*-TiC–*a*C:H – the British designation for hydrogen in the carbon matrix) . Such films are obtained by various methods: electric arc plasma spraying of titanium targets in hydrocarbon

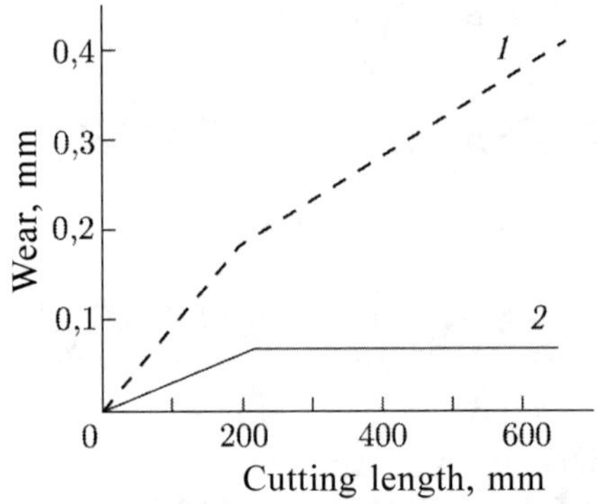

Fig. 3.170. Change in wear of the cutters as a function of the cutting length during processing of the refractory nickel alloy of the Inconel 718 type for standard (1) and nanocomposite coatings (2).

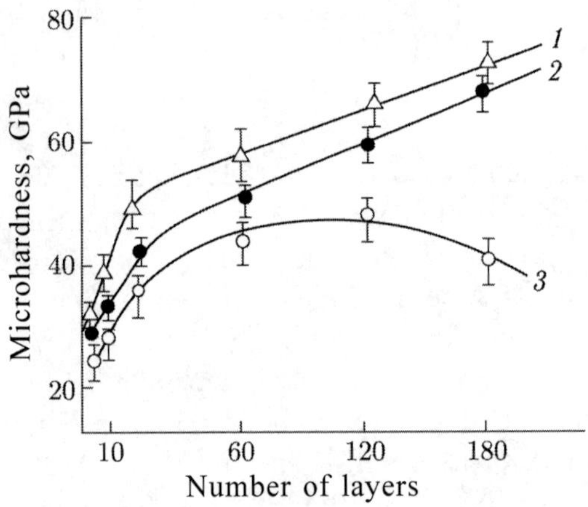

Fig. 3.171. A change in the hardness of multilayer nitride films with an increase in the number of layers: 1 – TiN–NbN, 2 – TiN–ZrN, 3 – TiN–CrN.

media, magnetron sputtering of graphite and titanium targets. The participation of hydrogen in these processes leads to the saturation of carbon amorphous matrices (in the foreign literature they were designated *a*-C:H, i.e., amorphous carbon saturated with hydrogen, whose content can reach up to 25 at.%) [3.357].

Table 3.3 shows some mechanical properties of amorphous-nanocrystalline composites of the TiC–*a*C type, studied using the nanoindentation method. For comparison, data on the properties of single-phase single-layer and multilayer carbide films are included in the table. Although the comparison of the results of nanoindentation

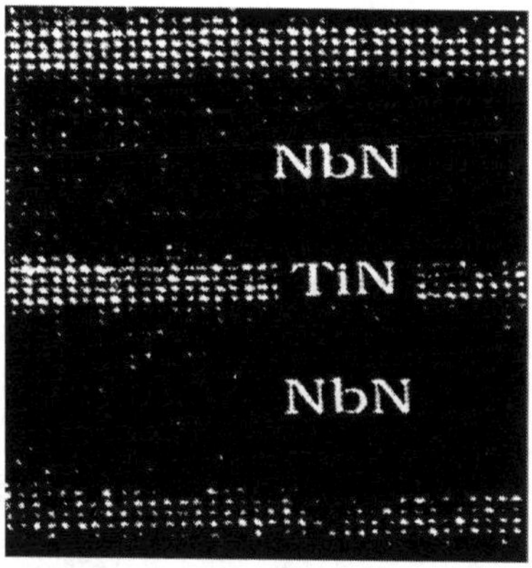

Fig. 3.172. The image obtained in a high-resolution transmission microscope of a TiN–NbN film (layer thickness: TiN ~ 1 nm, NbN ~ 4 nm).

Table 3.3. Mechanical properties of amorphous–nanocrystalline film composites

Object	Thickness, μm	TiC nanocrystal size, nm	Hardness *HV*, GPa	Young modulus *E*, GPa
Single-layer TiC–*a*C film	0.5	–	32	370
Multi-layer TiC–*a*C film	3.2	–	30	350
Single-layer TiC–*a*C film (22 at.% Ti)	3.0	1	18.2	150.8
Single-layer TiC–*a*C film (29 at.%)	3.1	2	21.8	162.8
Single-layer TiC film (40 at.% Ti)	2.7	12	25.6	191.0
Single-layer TiC film (54 at.% Ti)	2.7	35	16.4	230.6

[3.357–3.359], carried out by different authors, is very difficult due to methodological features (for example, unequal loads in measurements), however, one can draw some conclusions from the data of the table.

1. The hardness and modulus of elasticity of amorphous-nanocrystalline TiC–*a*C film composites do not follow the additive rule [3.357] in comparison with the analogous results for the initial components.

2. An attempt to explain a number of data (for example, the effect of the size of TiC nanocrystals) leads to contradictions, which

is obviously connected with the influence of many factors on the measured properties [3.359].

More informative was the study of deformation of single- and multilayer films with nanoindentation *in situ* in a high-resolution transmission microscope [3.360]. Investigation of the morphology of films and the type of prints (the presence and absence of cracks) revealed a viscous behavior and the presence of plastic deformation. The identification of multilayer films causes a homogeneous deformation followed by the generation of shear bands. The presence of TiC nanocrystals in an amorphous carbon matrix effectively promotes dissipation and deflection of cracks. All this leads to the appearance of plasticity while maintaining high values of hardness and modulus of elasticity [3.360].

In work [3.361], film coatings based on TiN are widely used as hard wear-resistant coatings for cutting tools, diffusion barriers in electronics, etc. It turns out that when doping nitride coatings with elements such as Si, B, Al, Ni, Y, there are significant changes in the structure and properties of such coatings. The high efficiency of alloying coatings with such elements, insoluble under equilibrium conditions, is due to their thermodynamically controlled segregation along the grain boundaries of TiN with the formation of a grain boundary phase limiting grain growth at a level of less than 10-15 nm. For example, in a coating of Ti–Si–B–N, a two-phase amorphous–nanocrystalline structure is formed consisting of titanium nitride sized about 20–25 nm and amorphous grain-boundary phase thickness, according to the estimates of the authors [3.351], 5–8 nm, which makes it possible to achieve superhardness (hardness more than 40 GPa) of such coating.

3.4. Magnetic properties

The study of the magnetic characteristics of nanocrystals is stimulated by significant applied successes in the creation of new high-efficiency magnetically soft and magnetically hard materials in the nanocrystalline state [3.362–3.366].

In principle, partial or complete crystallisation of amorphous ferromagnetic materials, as a rule, leads to a high coercive state [3.367]. In heat treatments above T_K, nanocrystalline phases with high magneto-crystalline anisotropy can be emitted in an amorphous matrix. For example, the coercive force H_c of an amorphous $Fe_5Co_{70}Si_{15}B_{10}$ alloy after annealing at 723 K for 1 hour increases to

24 kA/m, which is due to the release of the nanocrystalline cobalt phase [3.368]. With partial crystallisation of initially paramagnetic at amorphous alloys, a high coercive force value can also be achieved [3.368]. In this case, nanoparticles of the crystalline phase with a high Curie point T_c are formed in the amorphous matrix.

It was shown in [3.369] that, depending on the thermal treatment regime, two completely different structural states of partially crystallized amorphous $Fe_{85-x}Cr_xB_{15}$ alloys (x = 10, 12, 15, 18, 20) can be achieved, which are characterized by high H_c values. It turned out that for each of the alloys there exists an annealing temperature T^*_{ann} at which the H_c value passes through the aximum. Figure 3.173 shows a characteristic hysteresis loop for an alloy with x = 18 annealed in the optimal regime. Here, for comparison, a loop of a completely crystallized alloy of the same composition is shown.

The maximum value of H_c increases as the chromium content increases from 16 kA/m to 40 kA/m, while the saturation magnetization I_s decreases from 1 to 0.65 T. The value of T^*_{ann} also decreases as x increases from 950 to 870 K [3.369].

The dependence of the magnetic properties on the annealing temperature obtained for the $Fe_{67}Cr_{18}B_{15}$ alloy is shown in Fig. 3.174. Two H_c peaks of approximately equal magnitude (35–36 kA/m) are observed. The first peak is very close to T_K, and the second peak is about 120 K higher. The structural state of the amorphous-nanocrystalline alloy corresponding to the second maximum is more stable and less sensitive to the time of heat treatment.

The I_s value in Fig. 3.174 first increases noticeably, but then experiences a local minimum corresponding to the second peak H_c. On the basis of a detailed analysis, the authors of [3.369] concluded that the first maximum of H_c is due to the predominance of processes of rotation of the magnetization vector of single-domain α-FeCr nanoparticles in a paramagnetic amorphous matrix. The second maximum is due to the delay in the displacement of the domain boundaries of α-FeCr on the precipitates of the $(Fe, Cr)_3B$ phase, with magnetization fluctuations being the main factor in this process.

In recent years a new class of magnetic materials with a mixed amorphous–nanocrystalline structure and a higher level of static and dynamic magnetic characteristics has been discovered compared with similar crystalline and amorphous alloys [3.362]. For example, the characteristics of magnetically hard materials such as Fe–Nd–B (coercive force, residual induction and maximum magnetic energy) exhibit a clear tendency to increase when a nanocrystalline structure

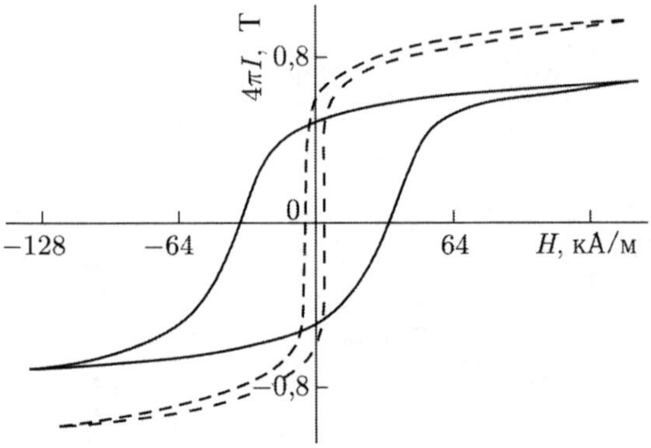

Fig. 3.173. Magnetic hysteresis loops for the $Fe_{67}Cr_{18}B_{15}$ alloy in amorphous nanocrystalline (solid curves) and nanocrystalline (dashed curves) states.

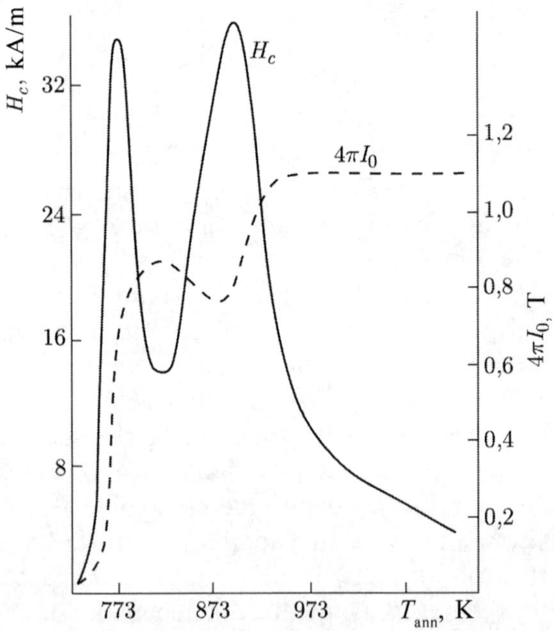

Fig. 3.174. Dependences of the coercive force and saturation magnetization on the pre-annealing temperature for $Fe_{67}Cr_{18}B_{15}$ alloy.

is formed. Magnetically hard nanomaterials of this type, obtained by melt quenching, are acquiring an increasingly applied value [3.370].

The paradoxicality of the situation is that in a two-phase amorphous–nanocrystalline state (and not in a single-phase state), a unique high level of magnetic parameters inherent in magnetically soft ferromagnetic alloys is realized, which at first glance contradicts the generally accepted views on the nature of ferromagnetism. Nevertheless, the amorphous–nanocrystalline magnetic materials possessing high values of magnetic induction and permeability (initial and maximum) successfully compete with well-known Fe–Si and Fe–Co alloys both in the amorphous state and in the crystalline state.

Figure 3.175 clearly shows the advantages of the amorphous-nanocrystalline alloys obtained by controlled annealing of the amorphous state, in comparison with traditional soft magnetic materials, including the amorphous ones. To an even greater extent, these advantages become apparent in the high-frequency range of magnetization reversal.

3.4.1. The theory of magnetism in nanocrystals with a strong intergranular interaction

The fact that the measure of magnetic hardness – coercive force (H_c)

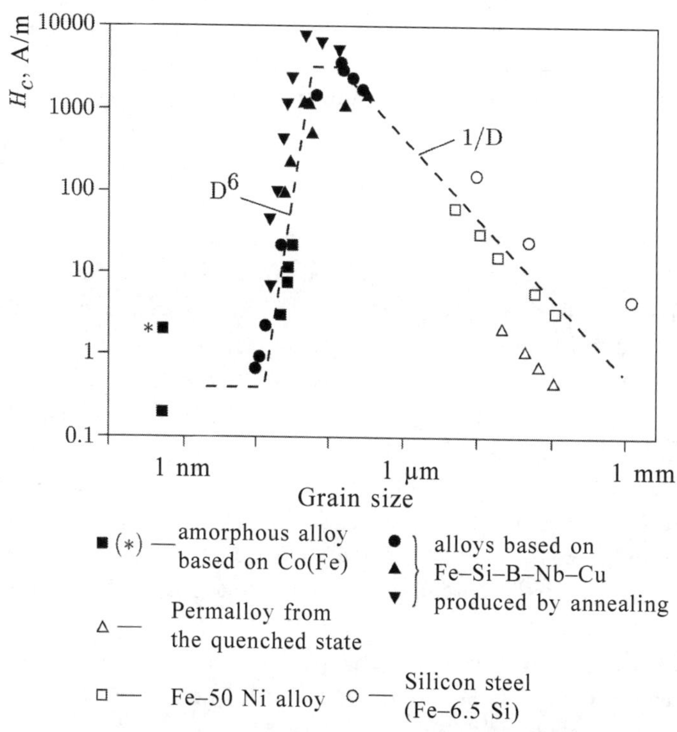

Fig. 3.175. A diagram illustrating the dependence of the coercive force on the grain size of magnetic materials [3.371].

– is inversely proportional to the grain size in the interval 0.1–1 mm in which D exceeds the thickness of the domain (Bloch) wall δ_w is the determining factor in understanding the optimal parameters of the nanocrystalline soft magnetic state; $D \gg \delta_w$. In such cases, the grain boundaries act as an obstacle to the motion of the domain walls, and hence the fine-grained materials are usually more magnetically harder than the coarse-grained ones. Recent progress in understanding the nature of coercive force led to the conclusion that a very small grain size $D \ll 100$ nm leads to a sharp decrease in H_c [3.371–3.376] (Fig. 3.175). This is due to the fact that D becomes substantially smaller than δ_w ($D \ll \delta_w$). In this case, the wall covers several grains, so that the fluctuations of the magnetic anisotropy in the scale of one grain do not lead to the retardation of the entire domain wall. This important concept suggests that nanocrystalline alloys have a significant potential as a soft magnetic material, since its properties require that the nanocrystalline grains are a single unit in the magnetic aspect. Similar ideas were expressed for the so-called 'spring-exchange' magnetically-hard materials [3.377–3.379].

In principle, the reduced coordination number of atoms in the surface should affect the Curie point of ferromagnetic nanocrystals, where the share of the grain boundaries is large. However, in most of the observations in the literature, the value of the Curie point T_c does not deviate strongly from those values that are characteristic of massive materials. For example, the value of T_c for Ni is 360°C in both coarse-grained and nanocrystalline states [3.380]. The value of $T_c = 1366$ K for Co nanoparticles and $T_c = 1388$ K for the massive state were obtained in [3.381].

The amorphous–nanocrystalline materials are in a two-phase state and have two Curie points (nanocrystalline and amorphous phases). Both are important parameters in the description of magnetic properties. The amorphous phase in which nanograins formed during crystallisation is usually enriched in non-magnetic atoms and, consequently, has lower magnetic ordering parameters and a lower Curie point. The same is true for a mixture of two crystalline phases, for example, for α-Fe as the main phase and the second phase in the form of carbides with a lower value of T_c.

When discussing the advantages of nanocrystalline alloys as soft magnetic materials, first of all, one should consider such properties as coercive force and permeability. The reduction in coercive force and the associated increase in permeability turned out to be those desirable phenomena that allow preferring amorphous and nanocrystalline alloys. When considering magnetic anisotropy in magnetically soft nanocrystals, such a characteristic as the length of the magnetic exchange interaction (exchange length) and its relation to the width of the domain wall and to the size of the monodomain [3.382] becomes very important.

These parameters can be determined using the following relationships [3.377]:

$$\delta_w = \pi\sqrt{\frac{A}{k}} \text{ and } L_{ex} = \sqrt{\frac{A}{4\pi M_s^2}}, \tag{3.80}$$

where δ_w is the thickness of the domain wall, L_{ex} is the exchange length, A is the rigidity of the exchange interaction, K is the magnetic anisotropy constant, and M_s is the saturation magnetization.

The model of randomly distributed anisotropy, proposed by Hertzer [3.372–3.376, 3.383, 3.384], was a prerequisite for explaining the soft magnetic properties of ferromagnetic nanocrystals. Within the framework of this model, the consideration of effective anisotropy in nanocrystals is based on the concept of randomly distributed anisotropy with respect to amorphous alloys. In particular, the notion of the characteristic volume is introduced, whose linear dimensions correspond to the characteristic exchange length $L_{ex} \approx \left(\frac{A}{K}\right)^{1/2}$ (Fig. 3.176).

N grains with chaotically distributed light magnetization axes in a volume of L_{ex}^3 having an exchange interaction were considered. Since the easy magnetization axes are randomly distributed, it is possible to carry out statistical averaging over all N grains, and the effective anisotropy will be $K_{ef} = K/(N)^{1/2}$, where K is the anisotropy constant for any of the grains. The number of grains in the exchange interaction is $N(L_{ex}/D)^3$, where D is the average diameter of the individual grain. Converting the expression to K_{ef}, we get

$$K_{ex} \cong KD^{3/2} \approx \left(\frac{K_{ex}}{A}\right) \approx \left(\frac{K^4 D^6}{A^3}\right). \tag{3.81}$$

Since the coercive force is proportional to the effective anisotropy, this analysis leads to the conclusion that the effective anisotropy and, consequently, the coercive force should grow as the sixth power of the grain size:

$$H_c \sim D^6. \tag{3.82}$$

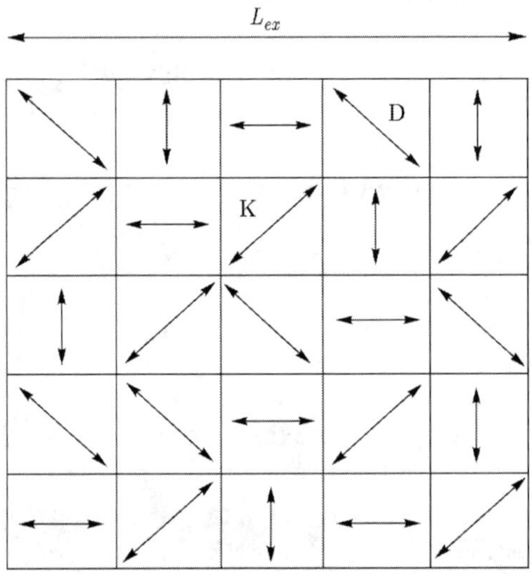

Fig. 3.176. Schematic model of chaotic anisotropy. The arrows denote randomly oriented axes of easy magnetization of the magnetocrystalline anisotropy [3.375].

An important condition for this relationship is that nanocrystalline grains must necessarily have an exchange interaction. It is not valid for non-interacting particles, which are characterized by an interaction parameter commensurate with the diameter of each particle, and sensitive to a superparamagnetic susceptibility.

Another dependence of the coercive force on the grain size was proposed in the literature for systems with reduced dimensionality and was subsequently confirmed experimentally. In [3.385], the contribution of domain walls to the coercive force was calculated, H_w. It was shown that it depends on the fluctuation parameter and the magnitude of equiaxial magnetic anisotropy. The authors concluded that the coercive force H_c should in principle be determined by one of three different parameters:

1) the inverse nucleation field of the domain H_N;
2) the field at which the domain walls begin to grow, H_G;
3) the field at which the domain walls become mobile, H_W.

The coercive force is determined by the greatest of these three quantities. The initial model of randomly distributed anisotropy was addressed to a homogeneous amorphous or nanocrystalline phase. The next step was the extension of this model to two-phase amorphous-nanocrystalline systems, characteristic of soft magnetic nanocrystals of the third type (for example, for the Finemet alloy) [3.374, 3.384].

For multiphase materials, the effective anisotropy is

$$D_{\text{ef}} = \left(\sum_i \frac{X_i D_i^3 K_i^2}{A^{1/2}} \right)^2 \tag{3.83}$$

where the sum corresponds to i phases in the material. In alloys of the Finemet type, $K_{\text{am}} \ll K_{\text{BCC}}$ (K_{am} and K_{BCC} are the constants of the magnetic anisotropy of the amorphous and nanocrystalline BCC phase, respectively), and a simple two-phase nanocrystalline material has an effective magnetic anisotropy:

$$K_{\text{ef}} \approx \left(1 - X_{\text{am}}\right)^2 \frac{K_i^4 D^6}{A^3}, \tag{3.84}$$

assuming that the volume fraction of the amorphous phase (H_{am}) is small. This simple expression predicts the 'dilution' effect associated with an increase in the relative volume of the amorphous phase.

In the literature, a somewhat different, power-law character of the dependence of D^n for H_c was also noted for $n < 6$ [3.386-3.388]. Unlike the Finemet alloys, the coercive force in the Fe–Zr–B–(Cu) [3.389] and Fe–P–C–Ga–Si–Cu alloys [3.390] obeys the simpler law D^3. Such a dependence was explained in [3.388, 3.391, 3.392] as a special case of the Hertzer model in the presence of an additional effect of a more long-range equiaxial anisotropy (K_u), i.e., with a parameter of the exchange interaction parameter much larger than L_{ex}. Under these conditions, the effective anisotropy can be represented in the form

$$K_{\text{ef}}^{\text{total}} = \sqrt{\left(K_u^2 + K_{\text{ef}\,nc}^2\right)}, \tag{3.85}$$

where $K_{\text{ef}\,nc}$ is the effective anisotropy of the nanocrystal in the absence of additional equiaxial anisotropy. Substituting in Eq. (3.85) the expression for the effective magnetic anisotropy from the model of randomly distributed anisotropy, taken from equation (3.81), we obtain

$$K_{\text{ef}}^{\text{total}} = \left(K_u^2 + \frac{K_j^2 D^3 \left(K_{\text{ef}}^{\text{total}}\right)^{3/2}}{A^{3/2}} \right)^{1/2}. \tag{3.86}$$

Equation (3.86), unfortunately, can not be solved analytically, but in the limiting case, when $K_u \gg K_{\text{ef}\,nc}$, it simplifies:

$$K_{\text{ef}}^{\text{total}} = K_u + \frac{1}{2}\left(\frac{\sqrt{K_u}K_i^2D^3}{A^{3/2}}\right) = a + bD^3, \tag{3.87}$$

which was observed experimentally.

The macroscopic characteristics of equiaxial magnetic anisotropy may be due to the induced anisotropy, which is caused by the domain structure arising during annealing or as a result of magnetoelastic interaction. In [3.389], a modification of the randomly distributed anisotropy model for systems with reduced dimensionality is proposed, where the coercive force is expressed as

$$H_c \approx K\left(\frac{D}{\sqrt{A/K}}\right)^{2n/(4-n)}, \tag{3.88}$$

where A is the rigidity of the exchange interaction and n is the dimension of the exchange interaction region.

The two-phase model of effective anisotropy deals with the case when the rigidity of the exchange interaction of two phases (amorphous (AM) and nanocrystalline (NC)) are comparable quantities. In [3.393], a two-phase model of randomly distributed anisotropy was extended to a more realistic case when $A_{\text{am}} < A_{\text{nc}}$. In this model, the effective anisotropy is:

$$K_{\text{ef}} \approx \frac{1}{\phi^6}\left(1 - X_{\text{am}}\right)^4 K_1^4 D^6 \left(\frac{1}{A_{\text{nc}}^{1/2}} + \frac{\left(1 - X_{\text{am}}\right)^{-1/3} - 1}{A_{\text{am}}^{1/2}}\right)^6, \tag{3.89}$$

where ϕ is the coefficient reflecting the symmetry K_{ef} and the spin rotation angle along L_{ex}. Note that in the classical model, $\phi \approx 1$. The nature of saturation magnetostriction for nanocrystalline alloys is discussed in [3.383]. It suggests that λ_s is determined by the balance between the contributions of nanocrystallites and the amorphous matrix. When $\lambda_{\text{nc}} < 0$ and $\lambda_{\text{am}} > 0$, a very small value of λ can be achieved for an amorphous–nanocrystalline composite. A simple two-phase model was proposed for induction and magnetostriction of saturation, which follows from the simple rule of mixtures [3.372, 3.373].

As we have already said, the Hertzer model [3.374–3.376, 3.383, 3.384] very well describes the coercive force data for many ferromagnetic nanocrystals. However, to meet this theory, two important conditions are required:

1) the grain size should be less than the characteristic parameter of the exchange interaction;

2) the grains must retain the ferromagnetic interaction.

In alloys with a single nanocrystalline phase with the conditions 1 satisfied the model operates at temperatures below the Curie point. but this is not required in the case of multiphase systems, which we encounter in the case of magnetic alloys obtained by crystallisation from an amorphous state.

For two-phase microstructures with ferromagnetic amorphous interlayers (AI) and the only ferromagnetic nanocrystals (NC), which are inherently intrinsic in nature, the character of the exchange interaction 'NC–AI–NC' is of paramount importance for the formation of the magnetic properties of these materials. This interaction depends on the size of nanocrystals and, more importantly, on the chemical composition, size and volume fraction of AI. The best properties obviously correspond to the state when both criteria are satisfied at a temperature below the Curie point of the amorphous phase, which, as a rule, is in turn below the Curie point of the nanocrystals.

The deviations of the reduced H_c from the values predicted by the model of randomly distributed anisotropy were first measured at temperatures near or higher than the Curie point of the amorphous phase [3.372]. The decrease in the interaction between ferromagnetic NC particles via AI directly correlates with the growth of H_c. This discovery became the basis for many studies in which the limiting parameters of alloys were established, which contain a sufficient number of very small nanocrystals or isolated nanoparticles in an amorphous matrix.

A temperature-dependent magnetic susceptibility was observed in [3.394, 3.395] in partially and completely nanocrystallized Fe–Si–B based alloys. It was concluded that for small NCs with a sufficient volume fraction of AI, it is possible to minimize or completely eliminate the exchange interaction between NCs and observe the superparamagnetic behaviour of the material. In the work [3.396] it was shown that the magnetic interaction increases with the increase in the volume fraction of the NC. These interactions tend to suppress superparamagnetism. The $H_c(T)$ peak was also observed near the Curie point of the amorphous phase. The rising branch of these curves is associated with the suppression of the exchange interaction between the particles. A much higher T_c of the amorphous phase leads to a decrease in H_c in accordance with the predictions of the theories of superparamagnetism.

In [3.397 and 3.398], the phenomenological parameter γ_{am} was introduced, which determines the exchange interaction of the NC through the amorphous phase. In the same place, a model was proposed that predicts a peak on the $H_c(T)$ curve at the Curie temperature of the amorphous phase. It was shown [3.399] that for a parameter $\gamma_{am} > 0.85$ the material behaves as an ensemble of single domain particles that become superparamagnetic near T_c for Al. Nanocrystals (volume fraction 0.15–0.25) located in the amorphous matrix act as inclusions and create the effect of magnetic rigidity [3.400, 3.401].

A further increase in the magnetic characteristics of ferromagnetic two-phase materials can be realized by increasing the spontaneous magnetization of the amorphous phase [3.388, 3.391]. This is due to the dominant role of the stiffness of the exchange interaction of the amorphous phase of A_{am} in the temperature interval below T_c.

The authors of [3.403] investigated the $Fe_{79}Hf_7B_{12}Si_2$ alloy to obtain a microwire with a nanocrystalline structure by the Taylor-Ulitovsky method. It was chosen instead of the nanocrystalline Fe–Zr based alloy because Zr interacts with the glass, and Zr was replaced by Hf. It should be mentioned that even the initial $Fe_{79}Hf_7B_{12}Si_2$ samples have a partially crystalline structure. The average grain size, estimated from the width of the peak from the Debye–Scherrer equation, is about 17 nm. During annealing, the grain size of crystallites increased from 17 to 35 nm after annealing at 873 K. During annealing, a significant improvement in soft magnetic properties occurred at temperatures T_{ann} = 773–873 K at which the grain size D also increased. However, the values of $H_c \approx 600$ A/m and $D \approx 30$ nm for samples annealed at such temperatures (773–873 K) are noticeably higher than the values found in nanocrystalline ribbons of the Finemet type ($H_c \approx 1$ A/m, $D \approx 10$ nm) based on Fe. Figure 3.177 shows the $\frac{H_{c0} - H_{cS}}{H_{cS}}$ dependence (H_{c0} is the coercive force without load, H_{cS} is the coercive force under load, for the initial and annealed microwire) on the the applied mechanical stress σ, measured in $Fe_{79}Hf_7B_{12}Si_2$ samples.

As shown in Fig. 3.177, the growth of H_c with σ was observed in the initial and annealed samples at low annealing temperatures (T_{ann} < 773 K), but at T_{ann} > 773 K there was a decrease in H_c

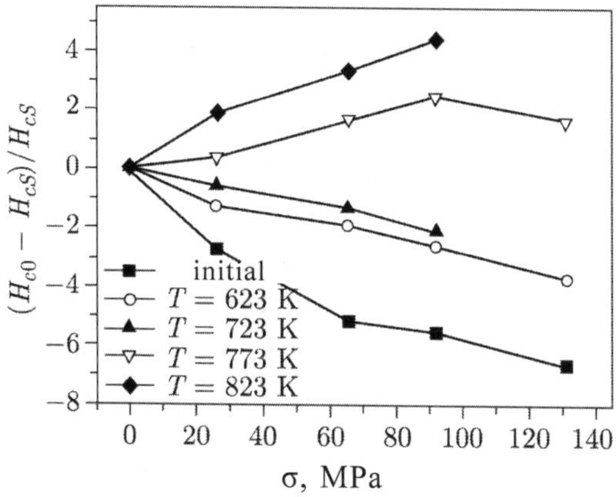

Fig. 3.177. Dependence of the coercive force on the applied mechanical stresses in the initial and annealed state at different temperatures of the $Fe_{79}Hf_7B_{12}Si_2$ microwire.

with increasing σ. This behaviour can be related to the different magnetostrictive character of the samples. It is generally accepted that the effective magnetostriction of saturation in magnetically soft nanocrystalline alloys is determined mainly by the balance of the two contributions, namely, the first contribution arising from the nanocrystalline phase and the second from the residual amorphous matrix. It can be assumed that during annealing, the concentration and distribution of the nanocrystalline and amorphous phases change, which leads to a different dependence of the coercive force on the stresses at different T_{ann}. The grain size plays an important role in explaining the behaviour of the coercive force during annealing. Indeed, there is an increase in the grain size (from 15 to 35 nm) with a simultaneous decrease in the coercive force. In addition, this dependence of the coercive force on the grain size does not correspond to the model of random anisotropy proposed by Herzer for nanocrystalline materials of the Finemet type [3.374].

3.4.2. *Magnetic properties of Fe–Si–Nb–Cu–B alloys (Finemet).*

As we have already noted, systems with a nanocrystalline phase were considered primarily as magnetically hard materials (for example, to create high values of properties in Nd–Fe–B, Pr–Fe–B, Sm–Co, and others). In order to obtain good soft magnetic properties (low coercive force and high magnetic permeability), we sought to realize as large grain sizes as possible. A typical example: electrotechnical steel (Fe–3% Si), the maximum properties of which can be obtained as a result of secondary recrystallisation [3.60]. The situation changed dramatically after Japanese researchers completely accidentally discovered unique soft magnetic parameters in an amorphous-nanocrystalline Fe–Si–B alloy with small additions of copper and niobium, which was later called 'Finemet' [3.402]. Initially, the alloy had the following chemical composition: $Fe_{73.5}Si_{13.5}B_9Nb_3Cu_1$, but later it underwent some changes. A wide variety of alloys of the Fe–Si–B system with small additions of refractory elements (Nb, W, Ta, Zr, Hf, Ti and Mo) and copper, whose composition was located in the immediate vicinity of the classical Finemet [3.402], was studied in detail. In the initial state (after quenching from the melt) they are all amorphous, and the optimum level of properties is achieved after partial crystallisation, as a result of which nanocrystals of the ordered phase of Fe–Si are released in an amorphous matrix. It is important to note that the nanocrystalline phase, reaching a size of about 10–20 nm, does not increase further, which, as it turned out, is due to the hindered diffusion and atmospheres created by the boron, niobium and copper atoms in the amorphous matrix around the growing nanocrystals [3.404], and also due to the retardation of the growth of nanocrystals by ultradispersed precipitates of the metastable boride phase at the interface [3.405]. It is also shown [3.405] that there are copper clusters in the amorphous matrix at the stage preceding nanocrystallisation, which stimulate the separation of FeSi phase nanocrystals arranged on the substrate, as on the substrate, in the order of type DO_3 [3.406]. More details of the structure formation in Finemet-type alloys are given in 3.2.1.

In the classical Finemet, there is no magnetic domain structure in nanocrystals, which in combination with mutual compensation of magnetostriction effects in nanocrystals and in an amorphous matrix leads to the formation of a very low coercive force (5–10 A/m), high initial permeability at ordinary (100 000) and high (10 000) frequencies and small remagnetization losses (200 kW/m^3). An additional positive effect on the properties is also provided by treatment in a magnetic field [3.407] and annealing in an atmosphere containing nitrogen [3.408].

As follows from the previous section, the nature of the magnetic interaction of nanocrystals is crucial for the formation of high magnetic properties in an amorphous–nanocrystalline Finemet alloy. This interaction is reduced or suppressed above the Curie point of the amorphous phase. Since the non-magnetic components of the alloy can be located mainly in the amorphous phase and, therefore, lower its Curie point, it is necessary to closely monitor the composition of the selected alloy.

In soft magnetic materials, their composition and structure should provide the maximum decrease in the magnetocrystalline anisotropy, the main contribution to which is made by the induced anisotropy associated with magnetostrictive deformation. The best materials (including Finemet) have the lowest values of magnetostriction. In the crystallisation annealing of amorphous Finemet quenched from the melt, the saturation magnetostriction λ_s decreases (from 20 to 3×10^{-6}), which can be explained by the small positive λ value for the amorphous phase, by the small negative value for the α-FeSi nanocrystalline phase and compensation of regions $+\lambda$ with regions $-\lambda$. The maximum value of the magnetic anisotropy constant K is observed after annealing at a temperature of 450°C, corresponding to the onset of crystallisation [3.409]. At higher temperatures, K decreases as the distance between the particles of the NC phase decreases and as the interparticle exchange interaction increases.

The saturation magnetization of the Finemet alloy is determined by the reversible rotation of the magnetization vector in accordance with the law

$$M(H) = M_s\left[1 - \frac{a_1}{H} - \frac{a_2}{H^2}\right] + bH^{1/2}, \tag{3.90}$$

where the term a_2/H^2 describes the contribution that follows directly from the model of randomly located axes of anisotropy. It is associated with FeSi nanocrystals. The coefficient a_2 reflects the theoretically predicted effective magnetic anisotropy of the material, while in amorphous alloys it is caused by local stresses and magnetoelastic interaction.

In the opinion of the authors of [3.410], the Finemet compositions proposed in [3.411] are 'padded' with amorphous metalloids, which reduces the magnetic properties and worsens the quality of the ribbon. In this connection, an attempt was made in [3.410] to improve alloys by changing their chemical composition, and also by optimizing the heat treatment regimes. As a result, the Russian analogue of the Finemet alloy was developed, which received the brand 5BDSR.

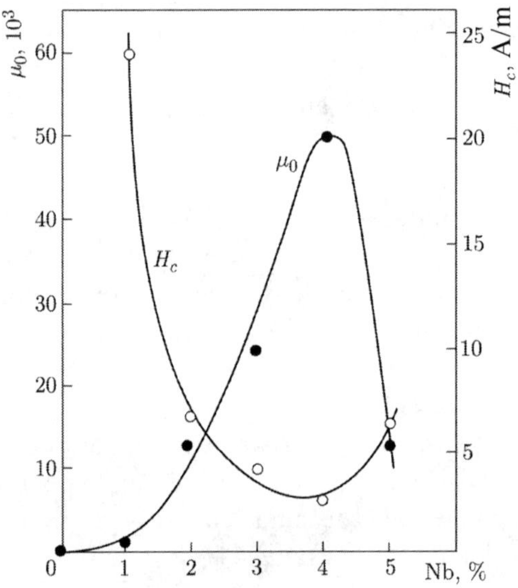

Fig. 3.178. Dependence of magnetic properties on the concentration of niobium in $Fe_{77-x}Nb_xCu_1Si_{16}B_6$ alloys.

Figure 3.178 shows the dependence of the initial magnetic permeability and coercive force H_c, measured in the quasistatic reversal of the magnetization, on the content of niobium in $Fe_{77-x}Nb_xCu_1Si_{16}B_6$ alloys. The maximum of μ_0 and the minimum of H_c on the curves indicate that in order to improve the soft magnetic characteristics, it is necessary to achieve a certain ratio of the components in the phases. In addition, with increasing niobium concentration, the temperature interval for the existence of the necessary amorphous–nanocrystalline structural state increases. This facilitates the choice of the annealing mode to obtain the desired structure and allows it to be carried out at higher temperatures to relieve stresses.

Copper doping, as expected, contributes to the formation of a high density of crystallisation centres and a decrease in the size of critical nuclei, which provides the required nanocrystalline structure [3.404]. In this case, copper alloying embrittles the amorphous ribbon (worsening the technology of its production), therefore, the concentration of copper in the alloy should be minimized. The dependence of μ_0 for copper-free and copper alloys on the annealing temperature is shown in Fig. 3.179. It is seen that the level of properties in the alloy containing copper increases by about an order of magnitude due to the formation of the necessary two-phase

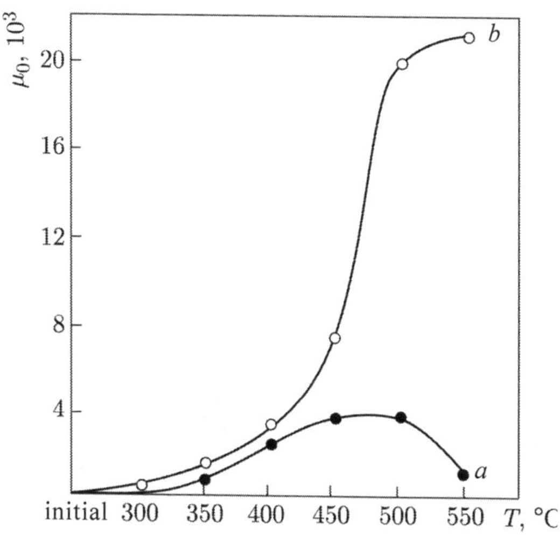

Fig. 3.179. The dependence of the initial magnetic permeability μ_0 of $Fe_{75}Nb_3Si_{16}B_6$ (*a*) and $Fe_{74}Nb_3Cu_1Si_{16}B_6$ (*b*) alloys on annealing temperature; 1 hour exposure.

amorphous–nanocrystalline structure (80% of the nanocrystalline FeSi phase of about 20 nm in size and 20% of the amorphous phase).

When studying $Fe_{73-x}Cu_1Nb_3Si_{13}B_x$ alloys with different boron content, it was found that a high level of properties is achieved in a fairly wide range of concentrations, but the highest values of μ_0 can be obtained only in a narrow range of compositions (Fig. 3.180). According to electron microscopy data, when boron content exceeds 9 at.% after heat treatment at 550°C for 1 hour, in addition to the FeSi phase, undesirable borides are observed [3.412].

The dependence of the properties of nanocrystalline alloys such as $Fe_{89-x}Cu_1Nb_3Si_xB_7$ on the silicon content after various thermal treatments is shown in Fig. 3.181 [3.410]. In alloys with a high silicon content, lower values of H_c can be obtained. In the right-hand corner of Fig. 3.181 there is the dependence of μ_0 after annealing at 550°C. The increase in this characteristic with increasing silicon concentration is due to the fact that near 16–17% Si the saturation magnetostriction changes sign (i.e. $\lambda_S \approx 0$) [3.384]. But with increasing silicon concentration the processability of the alloy deteriorates. Using high silicon alloys it is very difficult to obtain a ductile ribbon.

Figure 3.182 shows the dependence of the saturation magnetostriction λ_S on the annealing temperature for the two compositions

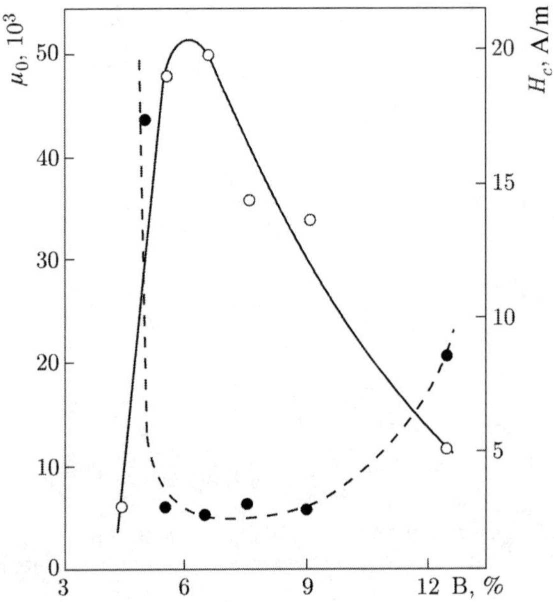

Fig. 3.180. Dependence of magnetic properties on boron concentration in alloys $Fe_{73-x}Cu_1Nb_3Si_{13}B_x$. Measurement of μ_0 at Hc = 0.08 A/m.

$Fe_{74}Cu_1Nb_3Si_{13}B_9$ (*a*) and $Fe_{74}Cu_1Nb_3Si_{16}B_6$ (*b*) described in [3.411]. It can be seen that as a result of annealing the value of λ_S decreases rapidly near the temperature of the onset of nanocrystallisation and passes through a minimum for composition (*a*), and for composition (*b*) passes into the negative region and, with a further increase in temperature, again changes the sign. The observed character of the change in λ_S demonstrates the possibility of making a composition with $\lambda_S \approx 0$, which is less dependent on the annealing regime than alloy *b*, and with a higher level of properties than that of alloy *a*.

The results of the studies carried out in [3.410] were used to determine the composition of the alloy produced under the brand 5BDSR. The material In the form of a ribbon of thickness 20–30 μm and a width of up to 40 mm is intended for the manufacture of magnetic cores of high-frequency pulse transformers, chokes, magnetic amplifiers, current sensors, etc. The normalized properties after thermal processing in accordance with the standard regime (no magnetic field) are characterized by the following values :

Saturation induction B_s, T, not less than 1.2;
Initial magnetic permeability $\bar{\mu}$, not less than 30 000;

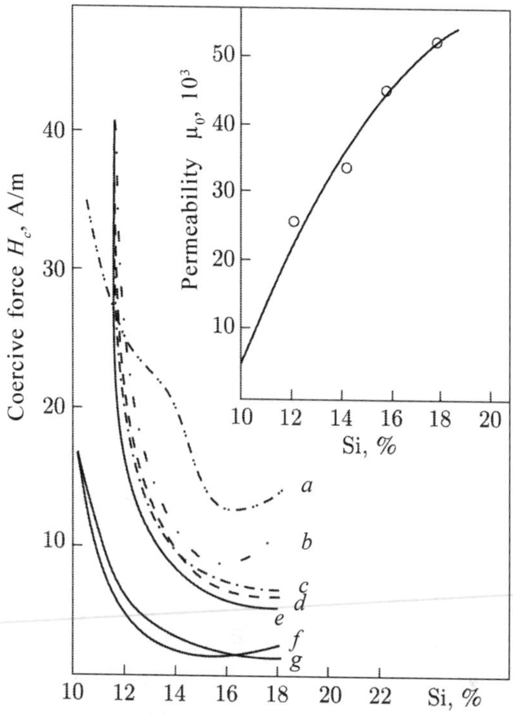

Fig. 3.181. Dependence of the magnetic properties of the $Fe_{89-x}Cu_1Nb_3Si_xB_7$ alloy on the silicon concentration (x = 9; 12; 14; 15.5 and 17.5%) after annealing in a magnetic field of 800 A/m from 400°C (*a*) to 575°C (*g*).

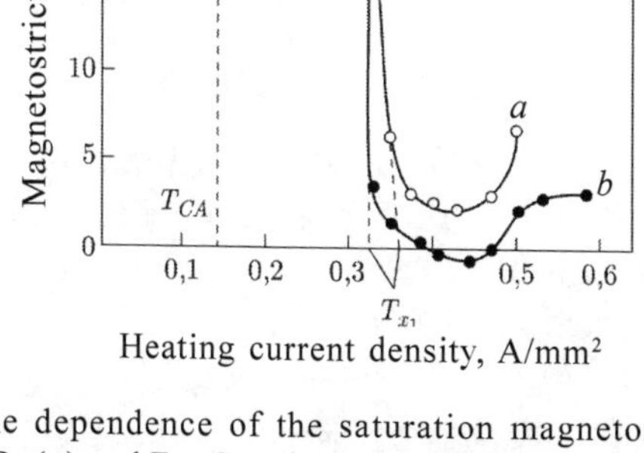

Fig. 3.182. The dependence of the saturation magnetostriction λ_S of the alloys $Fe_{74}Cu_1Nb_3Si_{13}B_9$ (*a*) and $Fe_{74}Cu_1Nb_3Si_{16}B_6$ (*b*) on the current density (on the heating temperature) during annealing (T_{CA} is the Curie temperature of the amorphous state, and T_{x1} is the crystallisation temperature).

Maximum permeability μ_{max}, not less than 100 000;
Coercive force H_c, A/m, not more than 1.6.
The alloy has the following physical properties: density, g/cm^3, 7.6;
Specific electrical resistance, μOhm · m, 1.35;
Hardness according to Vickers HV_{02}, 600.
Curie temperature, °C: in the amorphous state 350; in the nanocrystalline state, 550;
The saturation magnetostriction constant λ_S is less than 10^{-6}.

A rectangular or linear hysteresis loop can be formed by thermomagnetic treatment in a longitudinal or transverse field in magnetic cores made of 5BDSR alloy, respectively (Fig. 3.183). In other words, the heat treatment mode can correct the magnetic parameters of the alloy, depending on its use in a particular product. The frequency dependence of the initial permeability μ after the standard heat treatment is shown in Fig. 3.184.

Comparing the properties of the 5BDSR alloy with the properties of the known high-cobalt coarse-grained alloys, it can be seen that, by the main parameters, it is not inferior to these materials, and the saturation induction is 1.5 times higher in it. The undoubted advantages of the 5BDSR alloy are the relatively low cost, as well as the higher temperature stability of the properties.

In [3.101], the magnetic properties in the amorphous Finemet alloy were analyzed after controlled annealing in the temperature range 540–560°C, the annealing time varied from 0.5 to 2 h.

The dependence of the coercive force H_c on the annealing time at temperatures of 540 and 560°C is shown in Fig. 3.185. In the early stages of heat treatment (0.5 h), annealing at a higher temperature (560°C) reduces H_c to a greater extent. To achieve the same H_c value at 540°C, an annealing of 1 hour is necessary. Further increase in the annealing time leads to the fact that after both annealing temperatures the value of H_c acquires a minimum value of 2.5 A/m (1.5 h), and then slightly increases to 2.7–2.8 A/m (2 h).

Figure 3.186 shows the dependence of the saturation magnetization $4\pi I_s$ of the alloy annealed at temperatures of 540 and 560°C, depending on the duration of heat treatment. Both depend on saturation: at 560°C, starting at 0.5 h ($4\pi I_s$ = 1.27–1.29 T) and at 540°C, starting at 1.5 h ($4\pi I_s$ = 1.20–1.24 T).

In the opinion of the authors of Ref. [3.101], from the point of view of obtaining the optimal magnetic properties (H_c = 2.5 A/m and

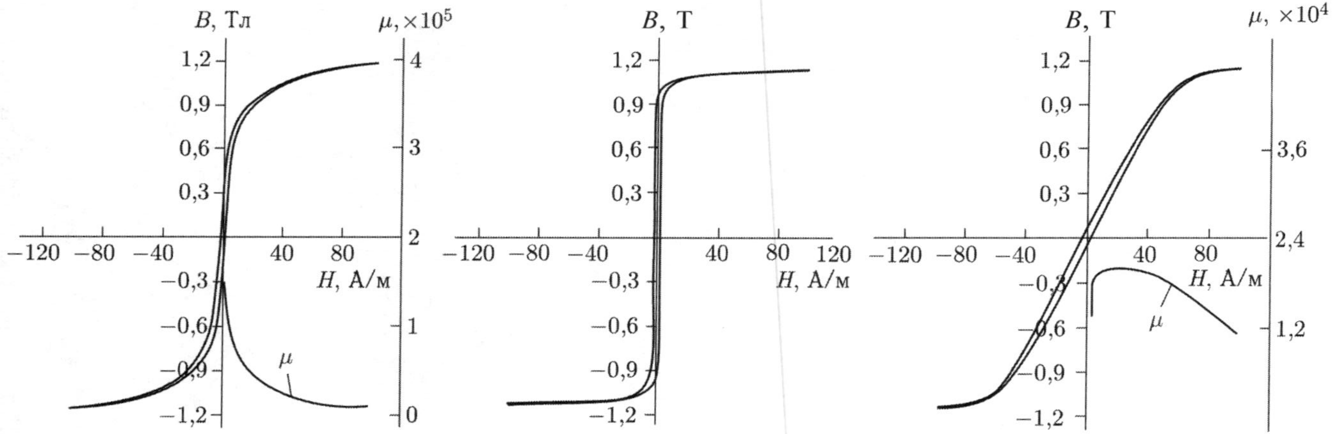

B_{max} = 1,1910 T K_p = 0,24
H_{max} = 80 А/м μ_{max} = 194291
B_r = 0,2810 T $\tilde{\mu}$ = 120000
H_c = 0,96 А/м $H(\mu_{max})$ = 1,12 А/м
a

B_{max} = 1,1306 T
B_r = 0,9730 T
H_c = 1,184 А/м
K_p = 0,86
b

B_{max} = 1,1527 T μ_{max} = 20010
B_r = 0,0622 T $H(\mu_{max})$ = 10,95 А/м
H_c = 2,0 А/м
K_p = 0,054
c

Fig. 3.183. Hysteresis loops of 5BDSP alloy after annealing: *a* – without superposition of the field; *b* and *c* – in a longitudinal and transverse magnetic field.

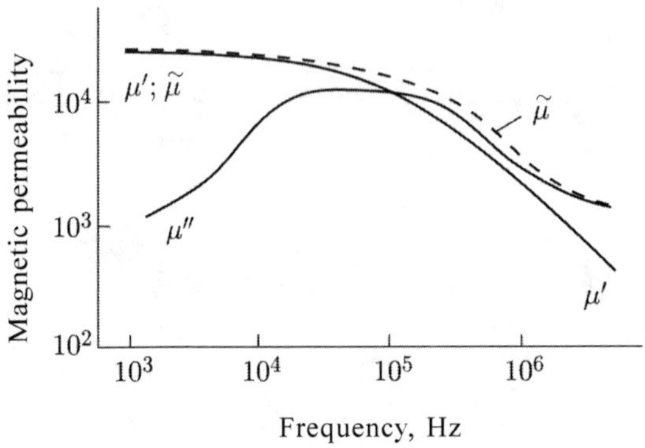

Fig. 3.184. The frequency dependence of the effective magnetic permeability of the 5BDSP alloy in a field of 0.08 A/ m (μ' and μ'' – the real and imaginary components, respectively).

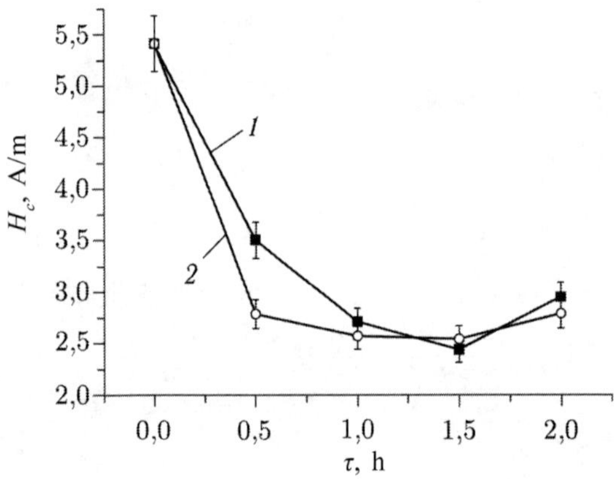

Fig. 3.185. Dependence of the coercive force H_c on holding time τ at annealing temperatures *T*: 1 – 540, 2 – 560°C.

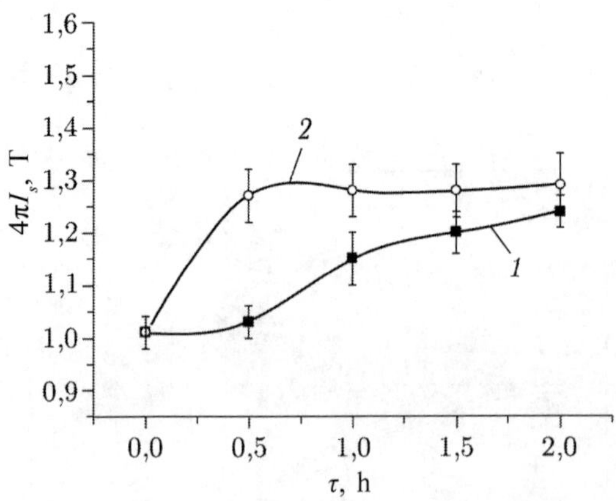

Fig. 3.186. Dependence of saturation magnetization $4\pi I_s$ on holding time τ at annealing temperatures *T*: 1 – 540, 2 – 560°C.

$4\pi I_s$ = 1.24 T), annealing of the $Fe_{73.5}Si_{13.5}B_9Nb_3Cu_1$ alloy at 560°C for 1.5 h is preferred

An interesting correlation of mechanical and magnetic properties was observed in a $Fe_{73.4}Cu_1Nb_{3.1}Si_{13.4}B_{9.1}$ microwire during its nanocrystallisation in the course of annealing: the dependence of the tensile strength σ_y on the annealing temperature correlates with the dependence of the coercive force $H_c(T_{ann})$. As in the case of the dependence $H_c(T_{ann})$, a local minimum of about T_{ann} = 450°C was observed on the $\sigma_y(T_{ann})$ curve. Both H_c and σ_y decreased with T_{ann} at T_{ann} < 450°C. Some growth of both H_c and σ_y was observed at T_{ann} about 575°C. In the end, a noticeable increase in σ_y was observed at T_{ann} > 650°C. Samples annealed at T_{ann} > 700°C were very brittle. At the annealing temperature above 600°C, the nature of the destruction of the samples changed. The metal core became brittle, and the glass coating gave a noticeable contribution to the ultimate strength: the sample broke down immediately after the glass coating was destroyed.

The correlation of the mechanical properties with magnetic behaviour can be associated with structural changes induced by annealing. The precipitation of the second crystalline phase leads to complete crystallisation of the sample, inducing strong internal stresses and changes in the nature of the chemical bonds. This crystallisation process makes the sample brittle [3.413].

3.4.3. Magnetic properties of alloys after MPD

Plastic deformation, especially megaplastic, is a complex process that leads not only to the formation of a deformed solid, but also causes significant changes in the structure and properties of the material itself [3.162]. In particular, deformation stimulates mass transfer and changes in chemical composition at both the macroscale and the microscale levels [3.414]. In turn, the redistribution of the components of the solid solution during plastic deformation can change a number of the physical properties of materials and, in particular, their magnetic properties [3.415].

In [3.416] attention was given in detail to the influence of MPD in the Bridgman chamber and occurring at the same time structural changes in the magnetic properties of a number of amorphous alloys of the metal–metalloid type having a great practical value. Studies were conducted on five amorphous alloys $Ni_{44}Fe_{29}Co_{15}Si_2B_{10}$ (alloy 1), $Fe_{74}Si_{13}B_9Nb_3Cu_1$ (alloy Finemet), $Fe_{57.5}Ni_{25}B_{17.5}$ (alloy 2A), $Fe_{49.5}Ni_{33}B_{17.5}$ (alloy 2B) and $Fe_{70}Cr_{15}B_{15}$ (alloy 3). The magnetic properties were determined with a vibrating magnetometer in a constant magnetic field of up to 720 kA/m.

Figure 3.187 shows, as an example, typical magnetic hysteresis loops obtained for the $Ni_{44}Fe_{29}Co_{15}Si_2B_{10}$ alloy in the initial (after quenching from the melt) (*a*) and treated in a Bridgman (*b*) chamber.

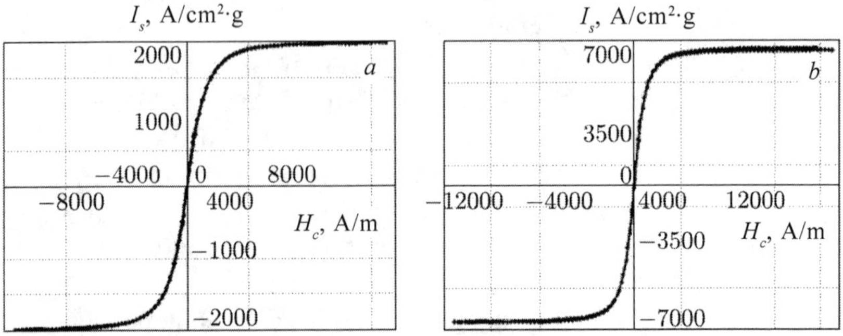

Fig. 3.187. Magnetic hysteresis loops obtained for the $Ni_{44}Fe_{29}Co_{15}Si_2B_{10}$ alloy in the initial state (after melt quenching) (*a*) and treated in a Bridgman chamber ($N = 4$, $T = 293$ K) (*b*).

The magnetic properties in the $Ni_{44}Fe_{29}Co_{15}Si_2B_{10}$ alloy were analyzed, as in [3.145], after MPD at 273 K and 77 K, and also in a partially crystallized state (annealing 380°C, 1 h) after MPD at 273 K. The dependence of I_s on the number of turns N in the Bridgman chamber for the $Ni_{44}Fe_{29}Co_{15}Si_2B_{10}$ alloy is shown in Fig. 3.188.

There is a sharp increase (approximately threefold) in I_s, followed by saturation with increasing N after MPD at 77 K and a sharp maximum of I_s at $N = 1$, followed by a decrease to the initial value after MPD at 293 K.

Characteristically, the maximum value of I_s after MPD at 293 K and the limiting value of I_s for a smooth increase after MPD at 77 K approximately coincide. The $I_s(N)$ dependence for the partially crystallized state of $Ni_{44}Fe_{29}Co_{15}Si_2B_{10}$ alloy after MPD at 293 K qualitatively coincides with the one that corresponds to the initially amorphous state at the same deformation temperature, but is less sharp both at the stage of growth and at the stage of fall. The maximum value of I_s also corresponds to $N = 1$ in this case, but in absolute magnitude is less exactly as much as the volume fraction of the amorphous phase in this state (the volume fractions of the crystalline and amorphous phase after partial crystallisation are determined in [3.145]).

The dependences of $H_c(N)$ for the $Ni_{44}Fe_{29}Co_{15}Si_2B_{10}$ alloy are shown in Fig. 3.189. It can be noted that their character for the initial state (amorphous or amorphous–crystalline) and for different MPD temperatures is qualitatively the same: first a sharp increase in the value of H_c at $N = 0.5$, and then a gradual decrease to values slightly exceeding the original values. As in the case of I_s, the maximum

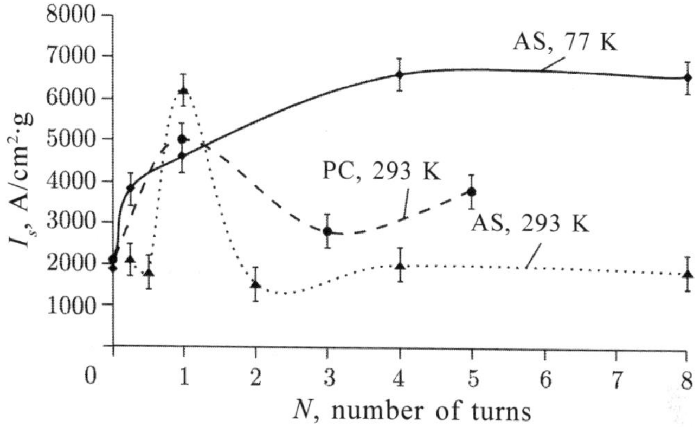

Fig. 3.188. Dependence of I_s on the number of turns N in the Bridgman chamber for the $Ni_{44}Fe_{29}Co_{15}Si_2B_{1}0$ alloy. AS – the initial amorphous state followed by MPD at 77 and 293 K; PC – initial partially crystallized state with subsequent MPD at 293 K.

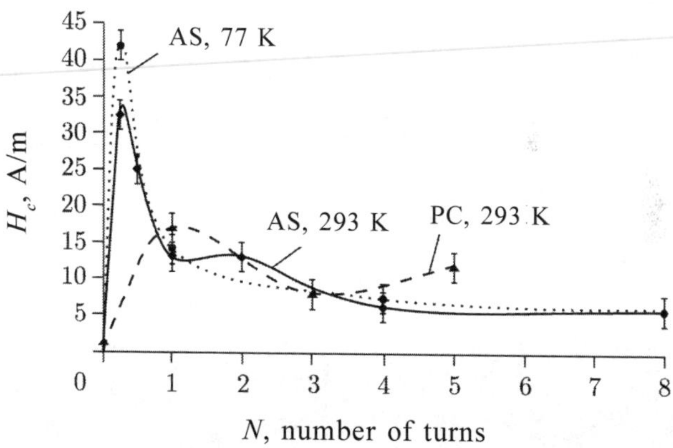

Fig. 3.189. Dependence of H_c on the number of turns in the Bridgman chamber N for $Ni_{44}Fe_{29}Co_{15}Si_2B_{10}$ alloy; The notation of the AS and PC is the same as in Fig. 3.188.

value for the partially crystallized state is noticeably lower, and it in addition corresponds to $N = 1$.

The $I_s(N)$ dependences for the amorphous Finemet alloy after MPD at 77 and 273 K are shown in Fig. 3.190. Here, as in the $Ni_{44}Fe_{29}Co_{15}Si_2B_{10}$ alloy, there is a noticeable increase in magnetization with increasing N, but, firstly, the increase is no more than 40 and 30% and, secondly, the dependences have a clear maximum at $N = 3$ and at $N = 2$ for MPD at 273 and 77 K, respectively.

Figure 3.191 shows $I_s(N)$ curves for $Fe_{57.5}Ni_{25}B_{17.5}$, $Fe_{49.5}Ni_{33}$B17.5, and $Fe_{70}Cr_{15}B_{15}$ alloys after MPD at 273 K.

Their character for different compositions is radically different. If in the alloy $Fe_{57.5}Ni_{25}B_{17.5}$ MPD practically does not affect the magnetization at all values of N, then in the alloy $Fe_{49.5}Ni_{33}B_{17.5}$, an appreciable jump of growth (by 120%) is observed at $N = 1$. In the $Fe_{70}Cr_{15}B_{15}$ alloy, on the contrary, for $N = 1$ there is a sharp drop in the value of I_s (by 250%). The nature of the variation of H_c with increasing N in all the studied alloys is practically the same and corresponds to the dependences shown in Fig. 3.189 for the $Ni_{44}Fe_{29}Co_{15}Si_2B_{10}$ alloy.

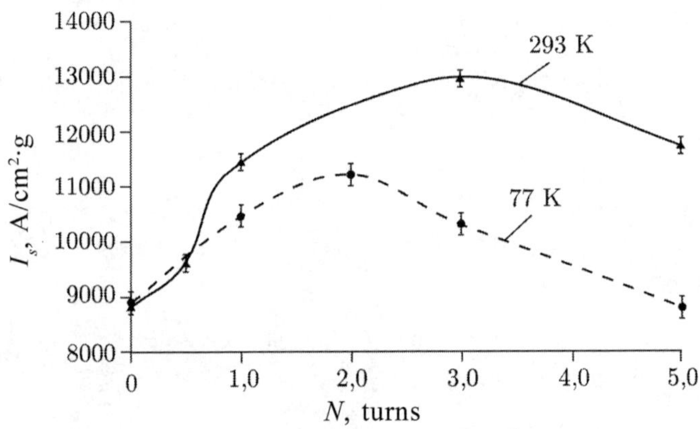

Fig. 3.190. Dependence of I_s on the number of turns N in the Bridgman chamber for the Finemet alloy; MPD at 77 and 293 K.

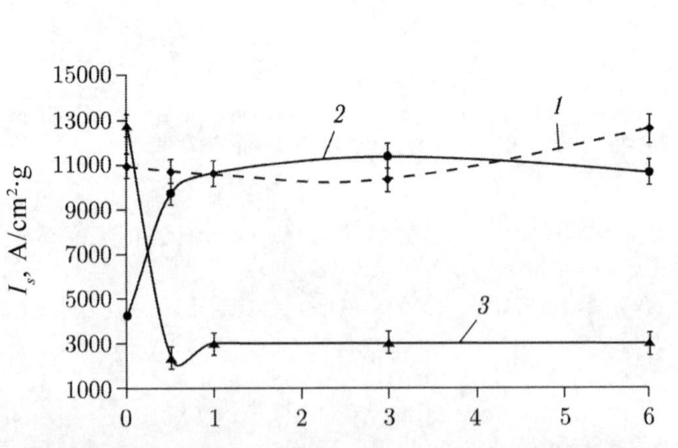

Fig. 3.191. Dependence of I_s on the number of turns N in the Bridgman chamber for $Fe_{57.5}Ni_{25}B_{17.5}$ (1), $Fe_{49.5}Ni_{33}B_{17.5}$ (2), and $Fe_{70}Cr_{15}B_{15}$ (3) alloys; MPD at 293 K.

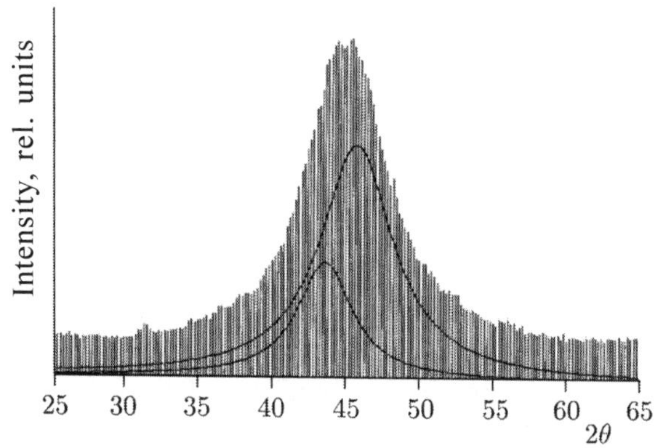

Fig. 3.192. Calculation of the X-ray line profile for amorphous alloy $Ni_{44}Fe_{29}Co_{15}Si_2B_{10}$ after MPD (N = 4, T = 77 K).

The structural state of the alloys corresponding to the maximum values of I_s after MPD is characterized by the presence on the X-ray patterns of an explicit asymmetry of the intensity profile of the main halo corresponding to the amorphous state. At the same time, both X-ray and electron-microscopic studies do not reveal, with rare exception, the existence of nanocrystalline phases in these states. It was possible to show (Fig. 3.192) that in the structural states corresponding to a noticeable increase in the saturation magnetization, the asymmetry of the X-ray halo profile is most likely due to the superposition of two maxima corresponding to different amorphous structures. In other words, the initially homogeneous amorphous matrix is divided under the influence of MPD into two amorphous phases with different chemical composition and, possibly, with different atomic structure. Since the stratification of the matrix into two amorphous phases leads to a noticeable change (increase or decrease) in saturation magnetization, it is reasonable to assume that in the MPD process, regions enriched and, respectively, depleted in ferromagnetic components are formed in the structure.

The following interesting regularity is observed: in the alloy $Ni_{44}Fe_{29}Co_{15}Si_2B_{10}$, where there are three ferromagnetic components (Fe, Ni, and Co), the maximum *Is* increase in MPD processing is 300%; In the $Fe_{49.5}Ni_{33}B_{17.5}$, where there are two ferromagnetic components (Fe and Ni), it is 120%; in the Finemet alloy, where there is one ferromagnetic component, it is 40%, and finally, in the $Fe_{70}Cr_{15}B_{15}$ alloy, where there is one ferromagnetic (Fe) and one antiferromagnetic (Cr) component, a decrease of I_s reaches 250% instead of the increase. Figure 3.193 shows the dependence of the observed effect (ΔI_s) on the number of ferromagnetic components in alloy n. It was assumed that for the $Fe_{70}Cr_{15}B_{15}$ alloy, $n = l_F + l_A = 0$, where l_F and l_A are the numbers of ferromagnetic (Fe) and antiferromagnetic (Cr) components, respectively, they have a conventionally opposite sign. The dependence is close to linear, although its clear physical meaning remains to be determined in the course of further research.

Comparison of the values (ΔI_s) in $Fe_{57.5}Ni_{25}B_{17.5}$ and $Fe_{49.5}Ni_{33}B_{17.5}$ alloys allows us to make an assumption about the local chemical composition of the amorphous phase enriched in ferromagnetic components in the Fe–Ni–B system. We see (Fig. 3.193) that a positive jump (ΔI_s) in the $Fe_{49.5}Ni_{33}B_{17.5}$ alloy leads after the MPD treatment ($N \geq 1$) to the same value of the saturation magnetization as in the $Fe_{57.5}Ni_{25}B_{17.5}$ alloy in the initial state.

At the same time, the $Fe_{57.5}Ni_{25}B_{17.5}$ alloy turned out to be completely indifferent to MPD treatment. Consequently, in the alloys of this system, two amorphous phases turn out to be energetically favourable, one of which, rich in metal components, has a ratio of the ferromagnetic Fe:Ni atoms the same as in the $Ni_{44}Fe_{29}Co_{15}Si_2B_{10}$ alloy, namely, 57.5:25 = 2.3.

Let us briefly discuss the prospects for the practical use of the results obtained. As a result of MPD processing by optimal regimes, we have the opportunity to significantly increase the saturation magnetization – the 'weak link' of amorphous and nanocrystalline magnetically soft alloys – while maintaining very small values of the coercive force. Thus, for an industrial grade product (in Russia - more than a few tens of tons per year) and the Finemet alloy widely used in electronics and instrument engineering, we obtained ΔI_s up to 40%, which in the result leads to a marked improvement in the

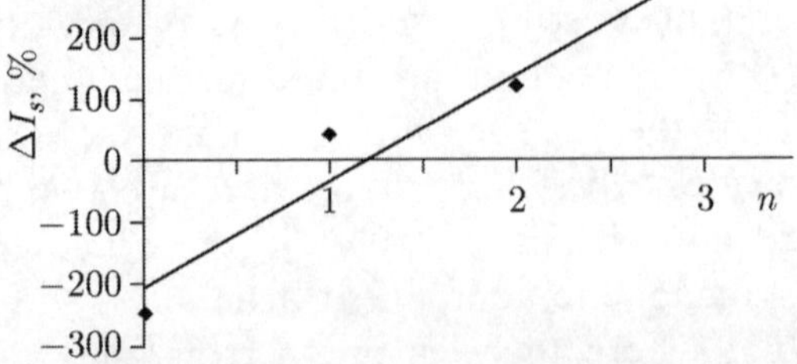

Fig. 3.193. Dependence of the effect of change in the saturation magnetization ΔI_s on the number of ferromagnetic components n in the amorphous alloys studied.

complex of magnetic properties of this alloy after optimum heat treatment conditions, including treatment in a magnetic field. The innovative attractiveness of magnetic alloys that have undergone MPD processing is very high. Inferior to all the Fe–Co(I_s) alloys and amorphous Co(μ)-based alloys in the individual magnetic parameters, the obtained alloys exceed them (including the standard Finemet) in the aggregate of both parameters. This gives grounds for assuming that the results of [3.416] can form the basis for obtaining a new class of promising soft magnetic materials.

3.4.4. Magnetic properties of Fe–M–B–Cu (Nanoperm) and (Fe, Co)–M–B–Cu (Thermoperm) alloys

The following additional requirements can often be applied to magnetically soft alloys [3.417].

1. High combined values of magnetic induction and permeability in constant and variable fields.

2. The ability to retain high magnetic properties at very high frequencies of magnetization reversal, at elevated temperatures, as well as the presence of many other important properties of non-magnetic nature, such as mechanical, corrosive and others.

Unfortunately, alloys of the Finemet type do not exactly satisfy both of the above requirements, since they have insufficiently high saturation magnetization and can not be used at elevated temperatures. The first is due to the presence in the alloy of a large amount of metalloids (more than 20 at.%), which we are forced to introduce in order to obtain the alloy in the amorphous state after quenching from the melt. For the same reason, the Curie point of the amorphous matrix (350°C) is relatively low in Finemet, and this, as we showed above, excludes the formation of high magnetic properties at temperatures above 250–300°C.

For this reason, attempts have been made to obtain alloys with the Finemet effect with improved magnetic characteristics on other systems. In particular, in Japan, soft magnetic alloys Fe–M–B–Cu (M = Zr, Nb, Hf) were developed, which were called Nanoperm [3.418]. These nanocrystalline alloys have been optimized in composition in such a way as to achieve a small magnetostriction coefficient and, as a result, greater permeability. They can be converted during quenching from the melt to an amorphous state at a significantly lower concentration of nonmagnetic elements and, consequently, substantially increase their saturation magnetization while maintaining high permeability. During subsequent annealing, Nanoperm alloys form α-Fe nanocrystals with structural parameters analogous to nanocrystals formed in Finemet.

Figure 3.194 shows the temperature dependences of the magnetization *M*(*T*) of quenched and annealed samples of $Fe_{86}Zr_7Cu_1B_6$ alloy in comparison with the saturation magnetization of pure α-Fe [3.419].

The *M*(*T*) dependences are particularly useful for identifying structural changes associated with the crystallisation process. For the amorphous phase, the Curie point is (333 ± 5) K. After crystallisation, the composite shape of the *M*(*T*) curve reflects the contribution from α-Fe nanocrystals and the intercrystalline amorphous phase with a markedly lower Curie point.

The magnetic moment of the alloy slightly increases during crystallisation due to the displacement of the B and Zr atoms from the nanocrystals into the amorphous matrix, which reduce the magnetic moment. The Curie point of the amorphous phase during crystallisation remains practically unchanged and amounts to only 340 K.

In [3.420], the temperature dependence of the magnetization was measured for $Fe_{93-x}Zr_7B_x$ and $Fe_{93-x}Zr_7B_xCu_2$ alloys (x = 4–14). The temperature dependence shown in Fig. 3.195 is completely analogous to that observed in crystalline invar alloys of the Fe–Ni type.

It is shown [3.421, 3.422] that the increase of soft magnetic properties in the Nanoperm alloys occurs for two reasons: low

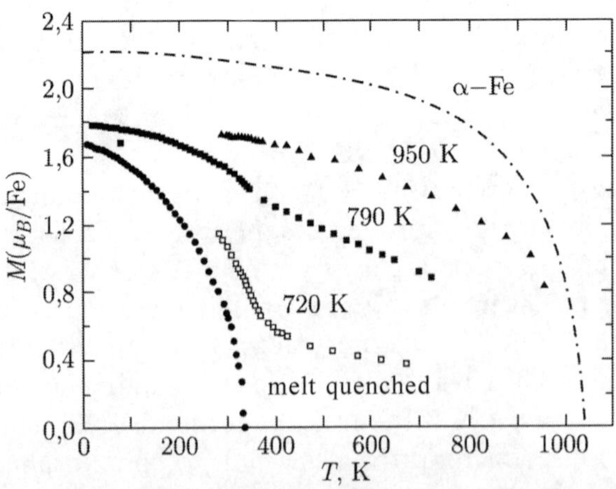

Fig. 3.194. Dependences *M*(*T*) for melt quenched and then annealed at the appropriate temperatures $Fe_{86}Zr_7Cu_1B_6$ alloy as well as for pure α-Fe in the nanocrystalline state [3.419].

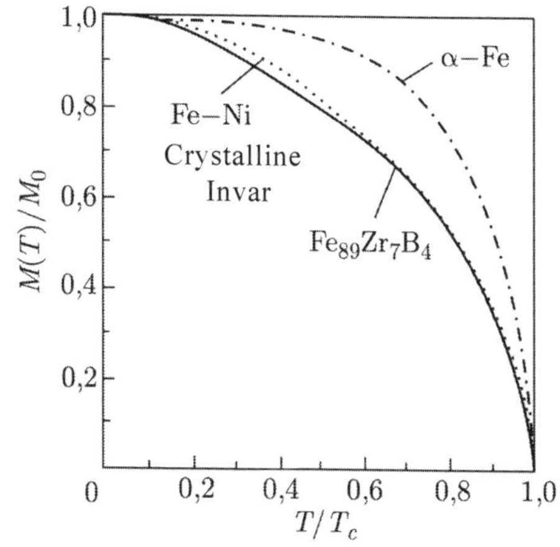

Fig. 3.195. Dependences $M(T)$ for an amorphous–nanocrystalline alloy $Fe_{89}Zr_7B_4$, crystalline Fe–Ni Invar and for α-Fe.

effective anisotropy at the time of crystallisation and a decrease in saturation magnetostriction. The addition of Cu increases the intercrystalline exchange interaction due to a larger number of α-Fe nanocrystals in the interacting unit of volume. This, obviously, reflects the important role of Cu clusters in the process of nucleation of α-Fe particles.

Other amorphous–nanocrystalline Finemet-like alloys are alloys (Fe, Co)–M–B–Cu (M = Nb, Hf or Zr), termed Thermoperm [3.423]. They have increased induction (1.6–2.1 T) in combination with high permeability and high Curie point. In these alloys [3.389, 3.424] nanocrystalline phases are formed on the basis of the BCC superstructure of the type *B*2 (α-FeSi and α′-FeCo) with significantly improved high-temperature magnetic properties in comparison with the Finemet and Nanoperm alloys. Alloys of the Thermoperm type were developed for use as materials with lower permeability, but high induction at high temperatures.

Figure 3.196 shows the frequency dependence of the real and imaginary components of permeability, μ′ and μ″ respectively. The value of μ′ reflects the density of losses due to eddy currents and hysteresis. The maximum permeability for this material is 1800.

The dependence μ″(*T*) has a maximum at a frequency of 20 kHz. The frequency peak is apparently associated with a higher electrical resistivity in nanocrystalline materials, and the losses in the alternating field reflect the behaviour of the domain wall in the viscous continuum. A higher value of ρ (50 mΩ · cm at 300 K) shifts

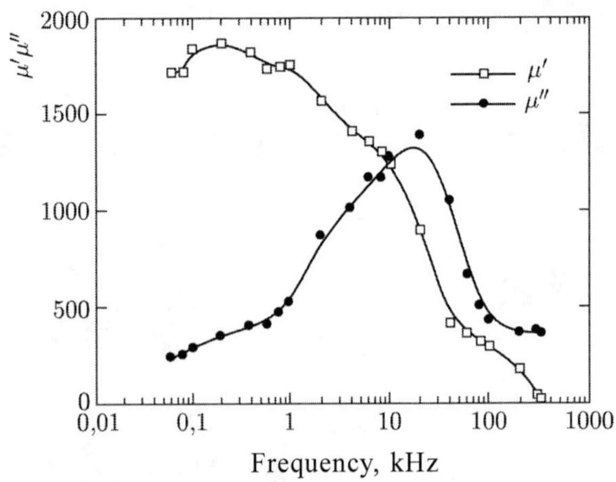

Fig. 3.196. The frequency dependence of the real and imaginary components of the permittivity of the Thermoperm alloy; annealing at 650°C for 1 h [3.389].

the high permeability to higher frequencies, where eddy current losses dominate, including the motion of the domain walls.

In [3.423], $(Fe_{1-x}Co_x)_{88}Hs_7B_4Cu_1$ alloys were investigated as potential candidates for magnetically soft materials for external generators. The materials are used in a high-temperature atmosphere (500–600°C), and they must have induction at a level of 2 T and higher at a temperature of 500°C, and also have thermal stability at 600°C for 5000 h. In addition, the magnetization loss should be below 480 W/kg at 5 kHz and 500°C. The characteristic exchange length for the equiatomic ordered FeCo alloy is 46 nm under the assumption that $A = 1.7 \cdot 10^{-11}$ J/m and $K = 8$ kJ/m³. If we assume a higher value of K (for an alloy with 30% Co $K \sim 25$ kJ/m³), but all other parameters are taken for the equiatomic alloy, this parameter will be 26 nm. Analysis in the framework of the theory of H. Hertzer suggests that a material with a particle size of 30 nm should have $H_c \approx 100$ A/m. This corresponds to a value $H_c \approx 10^{-2}$ A/m, obtained in the experiment for the size of nanocrystals $D \approx 10$ nm, and shows that further refinement of the nanocrystalline phase should lead to even better magnetic properties [3.425].

3.4.5. Magnetic properties of Fe–Nd–B alloys

Among nanomaterials, a separate group consists of magnetic

nanomaterials, which can be used as magnetically hard materials or materials for magnetic recording [3.426, 3.427].

In nanomaterials, the size of structural elements should be so small that nanoscale effects appear, that is, other (in comparison with massive materials) mechanisms of physical phenomena. In the case of magnetic materials, nanosized effects should include: *a*) magnetization reversal of crystallites as single-domain particles; *b*) 'exchange intergrain' interaction; *c*) the formation of a superparamagnetic state.

Single-domain particles are formed at sizes below the critical one for a single-domain state (D_c). The transition from a multi domain to a single-domain state leads to an increase in the coercive force due to a change in the magnetization reversal mechanism. The value of D_c can be calculated if the fundamental magnetic characteristics of the material are known (saturation magnetization I_s, exchange energy A, anisotropy constant K_1) [3.428–3.431]. The calculated value of D_c for α-Fe, characterized by a low magnetocrystallographic anisotropy $K_1 \ll \mu_0 I_s^2$, μ_0 is the magnetic constant), assuming the spherical shape of the particle is about 25 nm, which is in good agreement with the experimental data [3.428]. And for materials with high magnetocrystallographic anisotropy, in the case of the compound $Nd_2Fe_{14}B$ [3.431, 3.432], for barium and strontium ferrite $D_c \sim 0.4$–0.5 μm [3.430], for $SmCo_5$ $Dc \sim 0.7$–0.9 μm [3.432]. It should be noted that for materials with high magnetocrystallographic anisotropy, it is possible to calculate only the value of D_c in the absence of an external magnetic field. When the magnetization is reversed in fields $H < 0$, the value of D_c can be significantly reduced. Therefore, as noted in [3.431], the above value should be considered only as the upper limit of the quantity D_c. The lower limit is determined here (in the field $H = -H_A$, where H_A is the anisotropy field) by the width of the domain boundary, which, for example, in the compound $Nd_2Fe_{14}B$ is 4–7 nm [3.431].

As the particle size decreases noticeably below D_c, the coercive force decreases because of their transition to the superparamagnetic state. Usually superparamagnetism is observed at particle sizes ~ 1 nm [3.429].

In nanopowders, the magnetic interaction between the particles can be neglected. However, in compact nanomaterials, naturally, it is necessary to take into account the magnetic interaction between neighbouring grains, which can significantly affect the magnetic properties.

In particular, it can facilitate a light remagnetization and a decrease in the coercive force [3.433]. At the same time, the magnetic 'exchange interaction' between grains can lead to a significant increase in the remanent magnetization, which is observed in alloys based on the $Nd_2Fe_{14}B$ compound at grain sizes $D < 30$ nm [3.432, 3.434, 3.435].

It should be noted that magnetically hard nanomaterials are produced on an industrial scale. In particular, the production of permanent magnets of non-crystalline alloys of the Nd–Fe–B system is currently about 5 000 tons and, according to experts [3.436], it should increase by more than 2.5 times in the next 5 years. The spectrum of magnetically hard alloys obtained in the nanocrystalline state is quite wide.

Nanocrystalline alloys based on the $Nd_2Fe_{14}B$ phase are the object of numerous studies (see, for example, reviews [3.432, 3.434, 3.435, 3.437]). This is explained by the extensive use of magnets of these alloys in electrical and radio engineering products, computer technology, etc.

The industrial technology for obtaining alloys with a grain size of about 10 nm is quenching from the liquid state. In 1984, General Motors Company patented a method for spinning a melt, which involves injecting a molten alloy stream under excess gas pressure through an opening in the crucible onto the surface of a water-cooled spinning drum (usually made of copper with high thermal conductivity). The rate of solidification of the liquid in this case can reach about 10^5–10^6 K/s, and as a result, ribbons or flakes are formed (due to the high brittleness of the Nd–Fe–B system alloys most often fragmented) with a thickness of 20–50 μm. The process is usually carried out in an inert atmosphere (for example, Ar or He) because of the high chemical activity of neodymium. The structure and magnetic properties of the rapidly quenched alloys of the Nd–Fe–B system depend on the composition of the alloy, the rate of quenching, determined by the speed of rotation of the drum and heat transfer at the point of contact of the melt with the surface of the drum, injection conditions (type and pressure of the inert gas, the size of the crucible's hole, the distance from the crucible to the drum, etc.) and the temperature of the melt. As an example, Fig. 3.197 [3.438] shows the dependence of the coercive force (H_{ci}) of the Nd–Fe–B alloys on the drum spinning speed, where H_{ci} is indicated in oersteds (1 kOe = $(1/4\pi) \cdot 10^6$ A/m ≈ 80 kA/m).

According to typical microstructures obtained during spinning, the alloys of the Nd–Fe–B system can be conditionally divided into three groups. The first one is nanomelts in the 'underquenched' state, formed at low quenching rates: in this case, a significant part of the material consists of a size of 100 nm to 10 μm and higher, which significantly exceeds the optimum grain size necessary for obtaining high properties in a magnetically hard condition.

The second group includes nanoalloys obtained by the so-called 'direct' quenching from the liquid state, with a grain size less than 100 nm, which have the maximum coercive force (Fig. 3.197). As a rule, good magnetic properties can be obtained only in a very narrow range of process parameters, so the 'direct' hardening method is rarely used to produce permanent magnets. The third group includes nanoalloys, obtained by the most common method of 'overquenching', followed by crystallisation annealing in the later stage. 'Overquenching' initially produced alloys many of which are in the amorphous state, so that they have a sufficiently low coercive force. Subsequent annealing leads to the production of nanocrystalline alloys with a characteristic grain size of phases of about 20–50 nm (smaller than the critical size of the single-domain

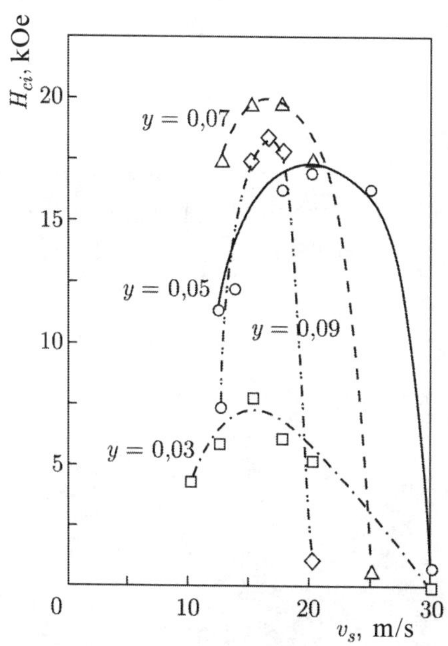

Fig. 3.197. The coercive force H_{ci} of $Nd_{0.15}$ $(Fe_{1-y}B_y)$ as a function of the spinning speed of the drum v_s [3.438].

particle of the $Nd_2Fe_{14}B$ phase), which have high properties in the magnetically solid state.

In Russia, the method of centrifugal melt sputtering has been developed and mastered on an industrial scale to obtain rapidly quenched Nd–Fe–B powders [3.439]. The technology of nanocrystalline materials production by this method includes two main stages: centrifugal spraying of the alloy and crystallisation annealing. In the first stage, an amorphous or amorphous-crystalline alloy is obtained, in the second stage, as a result of its crystallisation, a nanocrystalline state is formed in it. It should be noted that in most cases, this two-stage technology is used to produce magnetically hard nanoalloys, the first operation being rapid quenching, mechanochemical treatment, and vapour deposition.

Fig. 3.198 shows the installation scheme for centrifugal spraying of alloys.

The coarse-grained initial alloy is continuously poured into a water-cooled crucible fixed to the fast-rotating spindle. The melting of the metal in the crucible is provided by an electric arc burning between the metal and a non-consumable tungsten electrode. The melt drops under the action of the centrifugal force break away from the upper edge of the crucible and move with great speed to the cooled screen–crystallizer, which collide with it and quickly hardens. The hardened particles are in the form of thin flakes.

The cooling rate, the size of the droplets and the solidified scales depend primarily on the speed of spinning of the crucible and the screen (fixed on a single spindle), the angle of incidence of the drop on the screen, the temperature of the melt, and other factors. Under

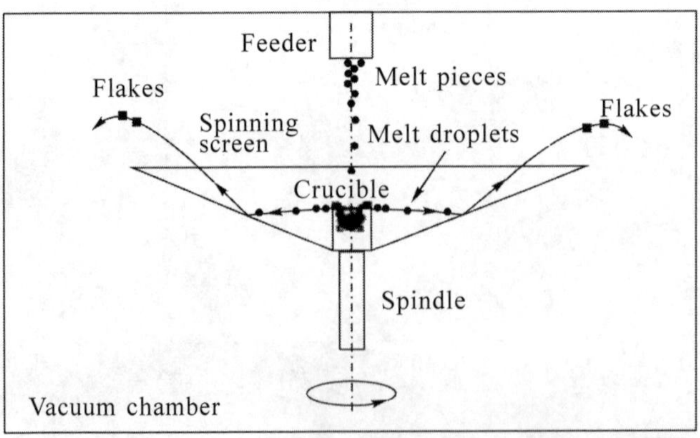

Fig. 3.198. Scheme of installation for centrifugal spraying.

traditional spraying conditions in existing plants, the cooling rate can reach a value of the order of 10^6 K/s, and the solidified flakes have a thickness of 20–40 μm, a width of 0.5–1.0 mm and a length of up to 10–20 mm.

X-ray phase analysis showed that, depending on the spraying regimes and alloy composition, these flakes contain $Nd_2Fe_{14}B$, α-Fe phases, as well as an amorphous component, the amount of which varies from 30 to 100% depending on the composition of the alloy and the spraying regimes [3.440]. The coercive force of the pulverized powders is low. For example, for an alloy with a 30% degree of amorphisation, it was $\mu_0 H_{ci} \approx 0.07$ T. The increase in the degree of amorphisation to 65 and 100% (by volume) leads to a decrease in the value of $\mu_0 H_{ci}$ up to 0.027 and 0.014 T, respectively, which is explained by a fairly easy reversal of the amorphous phase. As a result of crystallisation annealing, the properties of rapidly quenched powders increase noticeably, which is due to the formation of a nanocrystalline structure in them. For illustration, Fig. 3.199 shows an electron micrograph of an annealed sample from a single-phase alloy $Nd_2(Fe_{10.3}Co_{3.7})B$ with an average grain size (25 ± 2) nm [3.441].

Figure 3.200 shows the dependence of the coercive force on the temperature of isochronous annealing (for 5 and 7 min) of a sputtered alloy that is close in composition to $Nd_2Fe_{14}B$. It can be seen that these dependences are described by curves with a maximum.

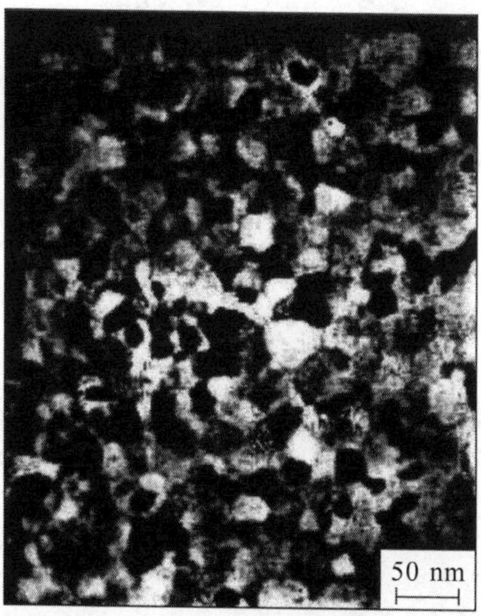

Fig. 3.199. Microstructure of rapidly quenched $Nd_2Fe_{10.3}Co_{3.7}B$.

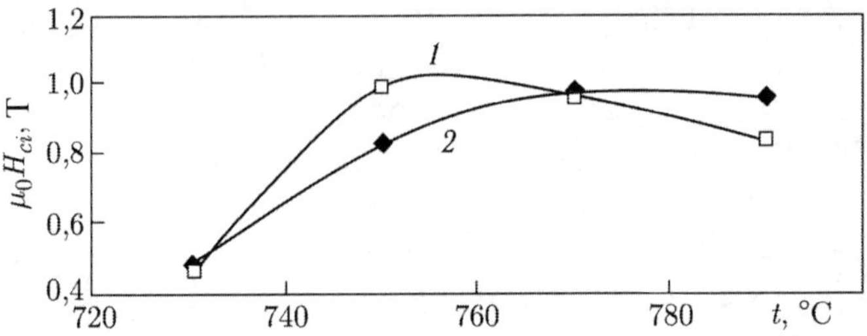

Fig. 3.200. Dependence of the coercive force on the temperature of isochronous annealing: 1 – τ = 5 min; 2 – τ = 7 min.

The same dependence was typical for residual magnetization. The reasons for such an extreme dependence of properties on temperature are structural changes in powders during annealing. It is shown [3.439] that initially low magnetic properties ($\mu_0 H_{ci}$ and $\mu_0 I_r$) are due to the incompleteness of the crystallisation process of the amorphous phase. The best properties are characteristic of the state immediately after the completion of crystallisation.

A subsequent increase in temperature and/or duration of annealing leads to an increase in the crystallite size and a decrease in the $Nd_2Fe_{14}B$ content due to the increase in the amount of neodymium oxide and, accordingly, α-Fe. For example, an increase in temperature of 40°C (compared with the optimal one) leads to an increase in the size of $Nd_2Fe_{14}B$ crystallites from about 40 to 50 nm and an increase in the content of α-Fe and Nd_2O_3 from 7 to 11% and from 2 to 4%, respectively. In this case, the coercive force ($\mu_0 H_{ci}$) decreases from 1.0 to 0.8 T. The alloying of Nd–Fe–B alloys with cobalt has a positive effect on the formation of the nanocrystalline structure and, as a consequence, on their hysteresis properties. In particular, it was shown [3.437] that the phase composition, the size of the phase crystallites in Co-doped alloys in which a high degree of amorphisation (more than 70% (vol.)) was obtained during sputtering, and also the magnetic properties depend only slightly on the annealing temperature (in the range 690–790°C). For example, the powders of the $Nd_{11}Fe_{75.1}Co_{7.1}B_{6.8u7}$ alloy annealed in this temperature range contained the $Nd_2Fe_{14}B$ phase (81 ± 2)%, ⟨D⟩ = 50–60 nm), α-Fe ((12.8 ± 1.7)%, ⟨D⟩ = 30–40 nm), as well as a small amount of Nd_2O_3 oxide (about 2%). In this alloy, after annealing, the $Fe_{23}B_6$ phase (3–5%) was also detected, which is obviously due to an increased boron content (1.2%).

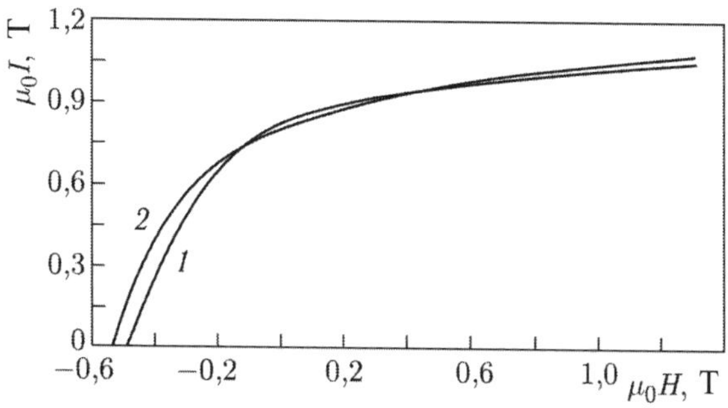

Fig. 3.201. The demagnetization curves of the $Nd_{11}Fe_{75.1}Co_{7.1}B_{6.8}$ alloy annealed at 690 (1) and 810°C (2) for 6 min.

The annealing of alloys doped with Co at a sufficiently low temperature of 690°C for 6 min allowed the crystallisation process to be completely completed. In spite of the multiphase structure, the demagnetization curves of samples from these alloys annealed at different temperatures were smooth (without kinks) and depended weakly on the annealing temperature (Fig. 3.201).

It should be noted that the smooth (without kinks) demagnetization curve in multiphase magnetically hard materials is one of the characteristic features of the so-called nanocomposite 'exchange-coupled' materials (i.e., materials with 'exchange intergranular' interaction). In addition, these materials should have an increased relative remanent magnetization $I_r/I_s > 0.5$ and a high reversibility of the demagnetization curve (in small fields) [3.432, 3.442]. This physical phenomenon associated with the 'exchange intergranular' interaction of nanocrystallites in the material is discussed in many papers (see, for example, the reviews [3.432, 3.434, 3.435]).

The results of a study of the structure and properties of magnetic powders obtained by the centrifugal sputtering method [3.437] are summarized in Fig. 3.202 in the form of the dependence of the coercive force and remanent magnetization on the content of soft magnetic phases (α-Fe and iron borides). It can be seen that all alloys can be divided into two types. Type A – high-coercivity alloys ($\mu_0 H_{ci} > 1$ T) and type B – alloys with an average level of coercive force ($\mu_0 H_{ci}$ = 0.4–0.6 T). Alloys of type A have a chemical composition close to $Nd_2Fe_{14}B$. In these alloys, the $Nd_2Fe_{14}B$ phase content is at least 90%, the rest is α-Fe and a small amount of neodymium oxides (1%). In alloys of type B, the neodymium

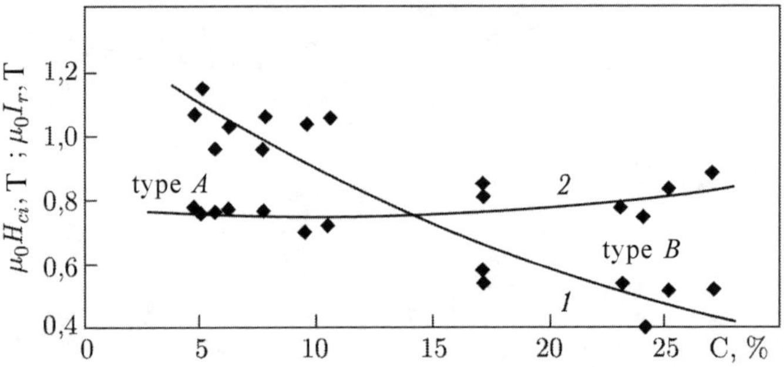

Fig. 3.202. Dependence of the coercive force $\mu_0 H_{ci}$ (1) and the remanent magnetization $\mu_0 I_r$ (2) alloys on the content of soft magnetic phases in them.

content is much less than in the $Nd_2Fe_{14}B$ phase. Alloys of this type contain the $Nd_2Fe_{14}B$ phase (70–80%), α-Fe and borides $Fe_{23}B_6$ or Fe_3B, as well as a small amount of neodymium oxides (1%). With a relatively small coercive force, they have a sufficiently high residual magnetization $\mu_0 I_r$ and energy production $(BH)_{max}$ (up to 80 kJ/m^3).

In recent years, mechanochemical methods have been widely used to produce alloys based on the $Nd_2Fe_{14}B$ compound. Such methods include processing in high-energy mills, intense plastic deformation, and other processes that involve mechanical action on a material that activates the course of phase transformations, often leading to the formation of metastable phases.

Information (see reviews [3.435, 3.438]) on the preparation of nanocrystalline alloys of the Nd–Fe–B system using the processing of powders in planetary high-energy ball mills is available. It is shown that treatment in a high-energy mill initiates the reaction $Nd_2Fe_{14}B \rightarrow$ amorphous phase + α-Fe. The rate of this reaction depends on the energy intensity of the grinding process and at its high values the reaction is completed in a few hours. We note that the above-mentioned character of the breakdown of the $Nd_2Fe_{14}B$ phase is probably due to the fact that according to the phase diagram [3.431] this compound is formed from the peritectic reaction of the annealing phase (with an amorphous structure) and α-Fe.

Crystals of α-Fe, formed during the decay of the $Nd_2Fe_{14}B$ phase, are nanosized (about 10 nm). After completion of this decay, the properties of the powders in the magnetically solid state are low, in particular, the coercive force is $\mu_0 H_{ci} < 0.02$ T. It should be noted that megaplastic torsion deformation (MPTD) also leads to the decay of the phases of $Nd_2Fe_{14}B$ into α-Fe and an amorphous phase, and the decay of the $Nd_2Fe_{14}B$ phase proceeds more with an increase in the degree of deformation [3.437].

Taking into account the positive effect of doping with cobalt, in [3.443] a mixture of $Nd_2Fe_{14}B$ + *X*% Co powders (*X* = 5.7 and 10) was also processed in a high-energy ball mill. To prevent the oxidation of reactive neodymium, grinding was carried out in an Ar atmosphere. As a result of the investigations carried out, it was established that the mechanochemical reaction $Nd_2Fe_{14}B$ + Co → amorphous phase + α-Fe occurs during grinding. After 2 hours of grinding, the reaction is completely complete, with the formed α-Fe crystallites having a size of about 10 nm and a high level of microdeformation of the lattice (about 0.6%). Cobalt dissolves in the phases present in the sample.

To form nanocrystalline alloys, the powders are subjected to crystallisation annealing after grinding in a mill or MPTD.

For example, in [3.443], a powder obtained by grinding a mixture of $Nd_2Fe_{14}B$ + 7% Co was annealed at various temperatures (650 to 770°C) for 6 min. Based on the results of X-ray structural analysis, the annealed alloys contained $Nd_2Fe_{14}B$ and α-Fe phases, as well as a small amount of neodymium oxides NdO and Nd_2O_3. Annealing at a temperature of 650°C for 6 min already led to the completion of the crystallisation process, i.e., the amorphous phase disappeared. The $Nd_2Fe_{14}B$ and α-Fe phases had nanodimensions, with their average dimensions increasing from (23 ± 3) to (36 ± 4) nm for the $Nd_2Fe_{14}B$ phase and from (12 ± 2) up to (31 ± 4) nm – for α-Fe.

Also, when the annealing temperature rose from 650 to 770°C, the amount of the $Nd_2Fe_{14}B$ phase increased from (66 ± 2) to (74 ± 2)% and the amount of α-Fe decreased from (24 ± 2) to (16 ± 1)%. The total amount of neodymium oxides practically did not depend on the annealing temperature and remained almost constant (at the level of 10%). Annealing also led to a sharp decrease in the microdeformation of the $Nd_2Fe_{14}B$ and α-Fe phase lattices. According to the Mössbauer studies, Co is soluble in both α-Fe and $Nd_2Fe_{14}B$, and in the lattice of $Nd_2Fe_{14}B$ the Co atoms preferentially replace Fe atoms in position k_2 [3.443].

The morphology of particles and the crystallites contained in them in alloys obtained by the mechanochemical method was studied by transmission and scanning electron microscopy. It has been found that the powder particles have a relatively equiaxed shape and a size of the order of 0.1–1 μm. As an example, Fig. 3.203 is a dark-field electron microscopic image of a powder particle of about 0.4 μm size obtained by grinding a mixture of $Nd_2Fe_{14}B$ + 7% Co and annealing. A dark field image was obtained by placing the objective diaphragm in the region of the diffraction ring reflections from $Nd_2Fe_{14}B$ and α-Fe observed on the electron diffraction pattern.

At the boundaries of the particle, in the places where it was possible to illuminate it, crystallites of phases $Nd_2Fe_{14}B$ and α-Fe with a size of 25–30 nm are seen, but smaller ones, 10–15 nm in size, are also present. According to X-ray diffraction analysis, the sizes of $Nd_2Fe_{14}B$ and α-Fe crystallites in this sample were (27 ± 3) and (14 ± 2) nm, respectively.

As a result of annealing, due to the formation of a nanocrystalline state, high magnetic properties of the powders are formed. For example, Fig. 3.204 illustrates the change in the magnetic properties in the refined powder $Nd_2Fe_{14}B$ + 7% Co after annealing [3.443].

As can be seen, as the annealing temperature increases, the coercive force increases and the residual magnetization decreases. This is primarily due to the increase in the content of the $Nd_2Fe_{14}B$ phase and the decrease in the amount of α-Fe in the alloy. In this case, naturally, the dependence of the energy product $(BH)_{max}$ on the annealing temperature has an extreme character.

Figure 3.205 shows the dependence of the coercive force of alloys obtained by different methods on the content of magnetically soft phases (α-Fe and iron borides), i.e., in fact, on the phase composition of the alloys.

It can be seen that the coercive force of the powders obtained by mechanochemical methods in the mill is noticeably higher than that

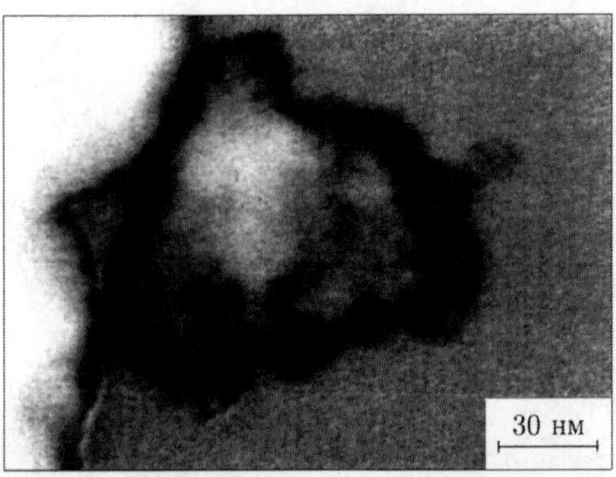

Fig. 3.203. Electron microscopic image of a particle of NdFeCoB powder after annealing at 690°C, ×100 000.

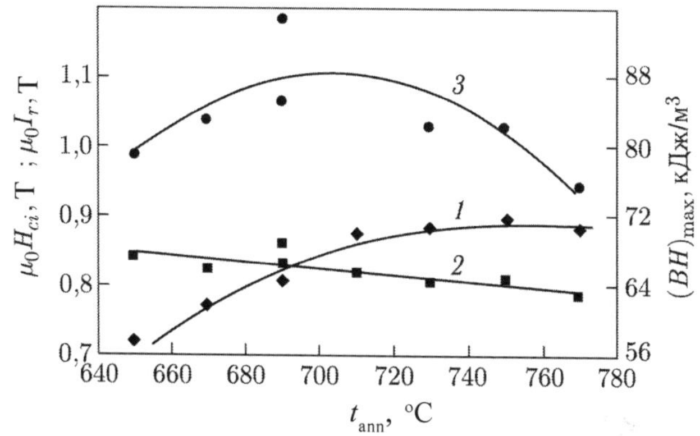

Fig. 3.204. Dependence of the coercive force (1), remanent magnetization (2) and energy product (3) on the annealing temperature of $Nd_2Fe_{14}B$ + 7% Co powder.

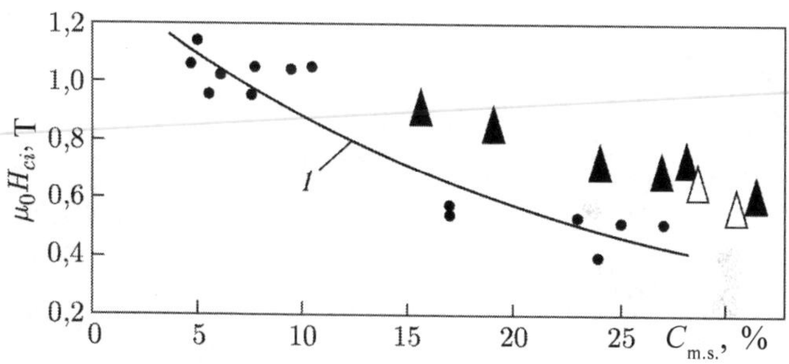

Fig. 3.205. Dependence of the coercive force of alloys obtained by centrifugal sputtering (1) and mechanochemical methods: grinding in a high-energy mill (▲) and MPTD (Δ), on the content of magnetically soft phases.

of powders obtained by centrifugal sputtering and having similar chemical and phase compositions. This is due primarily to the fact that the crystallite sizes of $Nd_2Fe_{14}B$ and α-Fe phases in powders obtained by the mechanochemical method are 1.5–2 times smaller than in powders produced by centrifugal sputtering.

The results of measurements of the magnetic properties of nanocrystalline alloys based on the $Nd_2Fe_{14}B$ compound at room temperature and at 10 K in a pulsed magnetization setup showed that even in a superlarge field of 48 T, which exceeds the anisotropy field of the $Nd_2Fe_{14}B$ compound approximately 7-fold, magnetic saturation is not achieved [3.444]. The absence of magnetic saturation in fields far exceeding the anisotropy field was also observed in nanoalloys of the Fe–O system [3.445] and Fe nanopowders [3.446]. The reasons for this phenomenon need further study.

At the moment, nanocrystalline magnetically hard alloys are not only an object of scientific research, but also are produced on an industrial scale. For example, permanent magnets from rapidly quenched alloys based on the compound $Nd_2Fe_{14}B$ are widely used in products of electrical engineering, acoustics, electronics [3.436]. It should be expected that further investigations of the alloys of the Nd–Fe–Co–B systems will be carried out in two directions. The first is due to the obtaining and investigation of the structure of alloys, which provides a significant increase in magnetic properties, in particular, through the formation of an 'exchange-bound' state. The second is the development of new and improvement of the known processes, which will allow a lower cost of manufactured nanomaterials, or significantly increase their performance (magnetic, strength, etc.) properties.

3.4.6. Magnetic properties of films

Films of magnetically soft alloys are used in the production of miniature high-sensitivity magnetic field sensors of a wide range of applications: magnetic recording systems, communication systems, etc. With increasing recording density and miniaturization of information recording and storage devices, the requirements to the operational properties of the material increase. At present, these requirements are characterized by the following level of properties: high induction of saturation (up to 2 T) in combination with low coercivity (<0.1 Oe) and high magnetic permeability (≫1000) at high frequencies (hundreds of MHz); high electrical resistivity (200 Ω · cm); high wear resistance; thermal stability (up to 600°C), corrosion resistance and the ability to produce material using film technology. In the modern design of the magnetic head, the thickness of the magnetic layer in single- and multilayer cores varies from a few microns to several nanometers, and the core is made using planar technologies (magnetron, ion-plasma spraying, etc.)

It is known that in thin layers a low coercive force and high permeability can be obtained on materials with nanoscale structure [3.447, 3.448]. Such a structure is based on Finemet iron alloys (see 3.4.2), Nanoperm (see 3.4.4) and Heatperm [3.449, 3.450]. The structure of these alloys is formed during the annealing of an amorphous material obtained by quenching from a melt and consists of randomly oriented 10 nm grains of the α-(FeSi) phase and an amorphous phase located in intergranular spaces.

In the early 1990s, the attention of researchers was drawn to another class of film magnetically soft alloys with a nanocrystalline structure belonging to the Fe–Me–X system in chemical composition, where Me is a metal of Group IV or V, for example Ti, Zr , Hf, Nb, Ta; X – nonmetal: N, C, O, B [3.449, 3.451, 3.452]. Films of such alloys are able to provide, in comparison with Finemet alloys, a higher magnetization (B_s = 1.5–1.75 T) in combination with higher values of high-frequency permeability ($\mu_{10\ \mathrm{MHz}}$ = 2000–7000) and are characterized by higher thermal stability. The use of high-energy planar technologies, in particular, magnetron sputtering for the production of films these compositions make it possible to produce ultrathin films, and also in a single technological process to obtain a whole multilayered element of a miniature magnetic sensor design comprising one or more soft magnetic layers. The structure of such films after annealing is nanosized (10–20 nm) crystallites of the ferromagnetic α-Fe phase and dispersed precipitates of the MeX phase along their boundaries. This type of structure is characteristic of dispersed-hardened alloys, which is confirmed by extremely high hardness values (10 GPa) obtained on films of Fe–Ta–N systems [3.454].

Under the conditions of magnetron sputtering, the elements Me and X, for example, Zr and N as alloying elements, are capable of producing films of Fe-based alloys in the amorphous, cluster or nanocrystalline state [3.455, 3.456]. In the process of sputtering or subsequent annealing of the film, a dispersed-hardened structure of the composite type is formed, consisting of an α-Fe ferromagnetic phase with a nanoscale grain ($D_3 \approx$ 10 nm) and a thermodynamically stable, solid non-magnetic interstitial ZrN phase located in the form of highly dispersed particles along the ferrite grain boundaries (Fig. 3.206).

The authors of [3.457] studied the structure and magnetic properties of the $Fe_{78}Zr_{11}N_{11}$ films 0.7 mm thick obtained by frequency reactive magnetron sputtering on glass substrates of the target with the composition Fe–10 at.% Zr, depending on the annealing temperature in the range from 200 to 700°C holding for 1 hour.

Figure 3.207 shows the change in the dependence of the annealing temperature on the effective magnetic anisotropy constant K_{ef} and on the magnetic characteristics H_c and B_s of the films studied [3.458].

Figure 3.207 shows that as the annealing temperature increases, the values of H_c and K_{eff} decrease and at 450–550°C the annealing

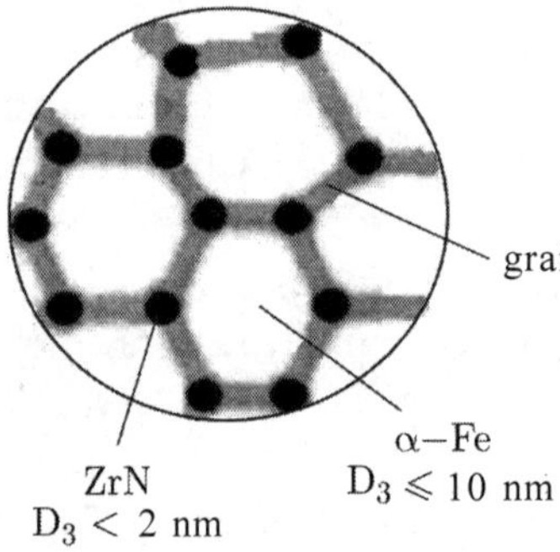

Fig. 3.206. Scheme of the structure of the Fe-ZrN nanocomposite [3.457].

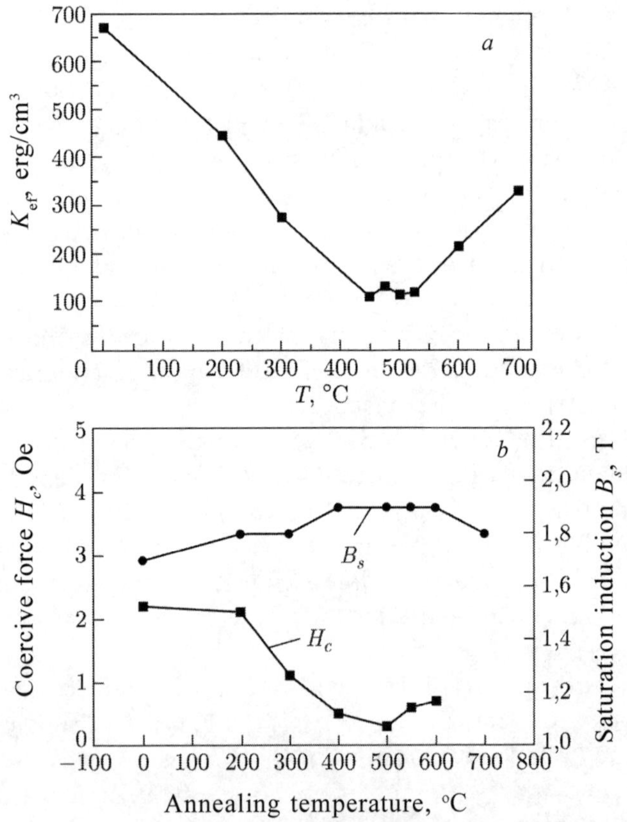

Fig. 3.207. The temperature dependence of the effective magnetic anisotropy constant K_{ef} (*a*) and the magnetic characteristics H_c and B_s (*b*) $Fe_{78}Zr_{11}N_{11}$ films.

temperatures have the minimum values (H_c = 0.3 Oe, K_{ef} = 100 erg/cm^3). At these temperatures, the saturation induction value B_s reaches 1.9 T and decreases to 1.8 T after annealing at 700°C. As the annealing temperature rises above 550 ∘C, the H_c and K_{ef} values increase.

Investigations of the structural state of the films in the initial state and after annealing in the 300–700°C range showed [3.459] that the films in the initial state have a structure consisting of amorphous and two crystalline phases. One of them, the BCC ferromagnetic phase is a strongly supersaturated nitrogen solution based on α-Fe with a grain size of 3–4 nm; The other is the non-magnetic FCC phase Zr_1N_{1-x} with a grain size of 1.5 nm. The quantitative ratio of BCC and FCC phases is 91:9. After annealing, the mutual ratio of the volume fractions of BCC and FCC phases (95:5) varies in the temperature range 400–550°C compared with the initial state, the average size of the BCC-phases does not change. Annealing at 600 and 700°C leads to an increase in the grain size of the BCC phase to 8 and 11, respectively, and to the appearance of an additional FCC Fe_4N phase whose amount after annealing amounts to 20%.

The authors of [3.457] explain the changes in the magnetic properties (Fig. 3.207 *a, b*) as follows. Films are characterized by high values of B_s, since the main structural component is a ferromagnetic BCC phase based on α-Fe. The presence in the initial structure of the amorphous phase and nitrogen in the α-Fe lattice reduces the characteristic value of B_s for α-Fe (2.147 T) to 1.7 T. The decrease in the amount of the amorphous phase as a result of annealing at 550°C leads to an increase in the B_s value of the film to 1.9 T. The decrease in B_s after annealing at 700°C (1.8 T) is due to the appearance of an additional phase Fe_4N with B_s of 1.7 T in the structure.

An explanation of the nature of the changes in the values of H_c and K_{ef} with an increase in the annealing temperature is possible with allowance for the change in the effective constant of induced magnetic anisotropy, the value of which depends on the magnetostriction λ_S and internal stresses σ ($K_{ef} \sim \lambda_S\sigma$), which is confirmed by the structural changes observed after annealing at temperatures up to 500°C [3.459]. The amorphous phase on the basis of Fe with high positive magnetostriction disappears, the nitrogen concentration in the α-Fe lattice decreases, which leads to a change in the magnitude and sign of the magnetostriction: apparently, the stresses arising in films obtained by the magnetron sputtering method decrease. All this leads to a decrease in the value of $\lambda_S\sigma$ and the values of H_c and K_{ef}.

According to the data obtained by the authors, in the investigated films close to the optimal value, the ratio of structural elements is most likely achieved after annealing at 450–550°C, when K_{ef} has a minimum value. After annealing at 600 and 700°C, the grain size of the BCC phase increases, hence the extent of the intergranular boundary regions decreases, and a second magnetic phase Fe_4N appears; all this leads to an increase in λ_S^{ef}, K_{ef} and H_c of the films.

Thus, the authors of [3.457] obtained films thermally stable up to 550°C, which combine high saturation induction (1.9 T) with low values of coercive force (H_c = 0.3 Oe), exceeding in the complex of these properties the well-known magnetically soft films based on Fe (Figure 3.208).

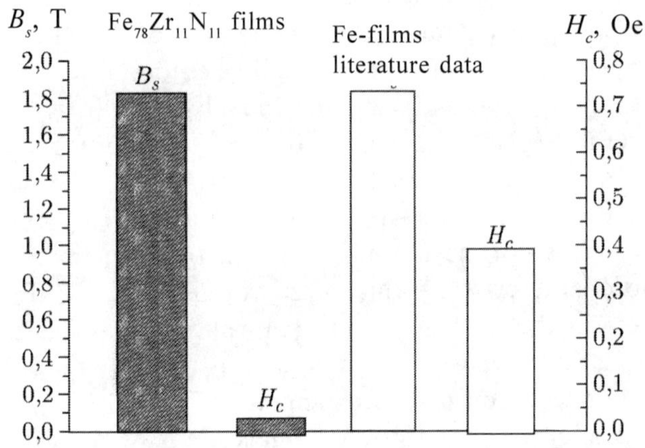

Fig. 3.208. The complex of magnetic properties (H_c and B_c) obtained on $Fe_{78}Zr_{11}N_{11}$ films and on magnetically soft Fe-based films known from publications.

References

3.1. Glezer A.M., Permyakova I.E., Materialovedenie, 2006. No. 9. P. 30–36.
3.2. Lu K., Mater. Sci. Eng. 1996. V16. No. 4. P. 161–122.
3.3. Skakov Yu.A., Phase transformations during heating and isothermal holding in metallic glasses, Itogi Nauki i Tekhniki. Metaoolved. term. obrab. - Moscow, VINITI, 1987, V. 21. P. 53–96.
3.4. Fortov V.E., Extreme states of matter. - Moscow: Fizmatlit, 2009.
3.5. Dobatkin S.V., Lyakishev N.P., Prospects for obtaining and using nanostructured steels. In the collection: Abstracts of the Second All-Russian Conference on Nanomaterials, - Novosibirsk: HTTM SB RAS, 2007. C2007. P. 35–36.
3.6. Segal V.M., et al., The processes of plastic structure formation of metals. - Minsk: Nauka i tekhnika, 1994.
3.7. Valiev R.Z., Aleksandrov I.V., Nanostructured materials obtained by intense plastic deformation. - Moscow: Logos, 2000.
3.8. Golovin Yu.I., Universal principles of natural science. - Tambov, TSU, 2002.
3.9. Valiev R.Z., et al., JOM. 2006. V. 58, No. 4. P. 33–39.
3.10. Large English-Russian dictionary. - Moscow, Russian language, 1988. - T. 2. P. 427.
3.11. Glezer A.M., Izv. RAN, Ser. Fiz., 2007. V. 71, No. 12. P. 1764–1772.
3.12. Shtremel' M.A., Strength of alloys. Part 2. - Moscow: MISiS, 1997.
3.13. Beloshenko V.A., et al., Theory and practice of hydroextrusion. - Kiev: Naukova Dumka, 2007.
3.14. Lesyuk E.A., Alekhin V.P., Formation of nano- and submicrocrystalline structures in instrumental and structural materials and ensuring their thermal stability. - Moscow: MGIU, 2009.
3.15. Gusev A.I., Usp. Fiz. Nauk, 1998. V. 168, vol. 1, P. 55-83.
3.16. Gleiter N., Deformation of Polycrystals: Mechanisms and Microstructure / Eds .: N. Hansen et al. - Roskilde: Riso Nat. Laboratory, 1981.
3.17. Gleiter N., Marquardt P. Z., Metallkud. 1984. V. 75 (4). P. 263.
3.18. Birringer R., et al., Trans. Japan. Inst. Met. 1986. Suppl. 27. P. 43.
3.19. Siegel R.W., Hahn H. , Current Trends in Physics of Materials / Ed. M.Yussouff. - Singapore: World Sci. Publ. Co., 1987, P. 403.
3.20. Gleiter N., Progr. Mater. Sci. 1989. V. 33 (4). P. 233.
3.21. Siegel R.W. J. J., Phys. Chem. Solids. 1994. V. 55 (10). P. 1097.
3.22. Andrievsky R.A., Powder Material Science. - Moscow: Metallurgiya, 1991.
3.23. Matthew M.D., Pechenik A., J. Am. Ceram. Soc. 1991. V. 74 (7). P. 1547.
3.24. Chen D.-J., Maya M. J., Nanostruct. Mater. 1992. V. 2 (3). P. 469.
3.25. Andrievsky R.A., et al., Nanostruct. Mater. 1995. V. 6 (1–4). P. 353.
3.26. Rabe T., Wasche R., Nanostruct. Mater. 1995. V. 6 (1–4). P. 357.
3.27. Kear B.H., Strurr P.R., Nanostruct. Mater. 1995. V. 6 (1–4). P. 227.
3.28. McCandlish L.E., Kear B.H., Kim B.K., Nanostruct. Mater. 1992. V. 1 (1). P. 119.
3.29. Wu L., et al., Proc. 13th Intern. Plansee Seminar. V. 3., Eds . H. Bildstein, R. Eck. - Reutte: Metallwerk Plansee, 1993. P. 667.
3.30. Fang Z., Eason J.W., Proc. 13th Intern. Plansee Seminar. V. 3. Eds. H.Bildstein, R. Eck. - Reutte: Metallwerk Plansee, 1993. P. 625.
3.31. Seegopaul P., McCandlish L. E., Shinneman F.M., Intern. Refe. Met. Hard Mater. 1997. V. 15 (1–3). P. 133.
3.32. Ivanov V.V., et al., DAN SSSR. 1984. Vol. 275 (4). P. 873.
3.33. Ivanov V.V., et al., SFKhT. 1992. Vol. 5 (6). P. 1112.
3.34. Ivanov V.V., et al., Nanostruct. Mater. 1995. V. 6 (1–4). P. 287.

3.35. Ivanov V.V., et al., Fiz. Khim. Obrab. Mater., 1997. V. 3. P. 67.
3.36. Andrievsky R.A., et al., FMM, 1996. V. 81 (1). 137.
3.37. Vavilova V.V., et al., Deformatsiya Razrushenie, 2006. No. 8. P. 17–21.
3.38. Pilipenko V.A. Fast heat treatment in VLSI technology. - Belarusian State University, 2004.
3.39. Vavilova V.V., et al., Deformatsiya Razrushenie, 2006. No. 4. P. 18–22.
3.40. Tagirov R.B., et al. Photodesorption of adsorbed gases in vacuum volumes. In: Physics and vacuum technology. - Kazan': Publishing house of Kazan' University, 1974. - P. 3–11.
3.41. Tagirov R.B., ibid, - Kazan', Publishing house of Kazan' University, 1974. - P. 155-158.
3.42. Lieberman A.B., et al., FTT. 1996. V. 38, No. 5. P. 1596–1597.
3.43. Shtyrkov E.I., et al., Fizika i tekhnika semipoluprovodnikov, 1996. V. 9, No.10. P. 2000–2002.
3.44. Dvurechensky A.V., et al., Pulsed annealing of semiconductor materials. - Moscow, Nauka, 1982.
3.45. Ievlev V.M., Thin films of inorganic materials: mechanism of growth and structure: - Voronezh: Publishing and Polygraphic Centre of Voronezh State University, 2008.
3.46. Dvurechensky A.V., Sorovskii obrazovatel'nyi zhurnal. 2004. V. 8, No. 1. P. 1–7.
3.47. Sidorova G.V., et al., in: Metall-amorphous materials. - Izhevsk: UdGU, 1988. - P. 107–112.
3.48. Felts A. Amorphous and glassy solid bodies. - Moscow: Mir, 1986.
3.49. Andrievsky R.A., Uspekhi Khimii. 1997. Vol. 6, No. 1. P. 57–77.
3.50. Ievlev V.M., et al., Growth and structure of condensed films. - Voronezh: Publishing and Printing Centre of the Voronezh State University, 2000.
3.51. Gotra Z.Yu., Technology of microelectronic devices. - Moscow, Radio i svyaz', 1991.
3.52. Blinkov I.V., Manukhin A.V., Nanodispersed and granulated materials obtained in pulsed plasma. - Moscow: MISIS, 2005.
3.53. Pavlov V.A., FMM, 1989. Vol. 67, No. 5. P. 924–932.
3.54. Stable and metastable equilibrium in metal systems. ed. M.E.Dritz. - Moscow: Nauka, 1985.
3.55. Rapidly quenched metals. ed. B. Kantor, translated from English, ed. A.F. Prokoshin. - Moscow: Metallurgiya, 1983.
3.56. Gridnev S.A., et al., Non-linear phenomena in nano- and microheterogeneous systems. - Moscow: Binom. Laboratory of Knowledge, 2012.
3.57. Yoshizawa Y., et al., J. Appl. Phys. 1988. V. 64, No. 1. P. 6044–6046.
3.58. Inoue A., Mater. Sci. Eng. 1994. A179/180. P. 57–61.
3.59. Choi G., Kim Y., Cho H., Inoue A., Masumoto T., Scr. Met. Mater. 1995. V. 33. P. 1301–1306.
3.60. Livshits B.G., et la., Physical properties of metals and alloys. - Moscow: Metallurgiya, 1980.
3.61. Lu K., Mater. Sci. Eng. Reports. 1996. V. R16. P. 161–221.
3.62. Scott M.G. Crystallization. Amorphous metal alloys. -Moscow, Metallurgiya, 1987. P. 137–164.
3.63. Lu K., et al., J. Non-Cryst. Solids. 1993. V. 156–158. P. 589–593.
3.64. Clavaguera-Mora M.T., et al., Progress in Materials Science. 2002. V. 47. P. 559-619.
3.65. Johnson M.W.A., Mehl K. F.,Trans Am. Inst. Mining. Met. Eng. 1939. V. 135. P. 146.

3.66. Avrami, M., J. Chem. Phys. 1941. V. 9, No. 2. P. 177–184.
3.67. Khonik V.A., et al., J. Appl. Phys. 2000. V. 87, No. 12. P. 8440-8443.
3.68. Burke J., The kinetics of the phase transformation in metals. - London: Pergamon 1965.
3.69. Leonova E.A., et al., Izv. RAN. Ser. Fiz. 2001. Vol. 65, No. 10. P. 1420–1423.
3.70. Leonova E.A., Izv. RAN. Ser. Fiz. 2001. Vol. 65, No. 10. P. 1424–1427.
3.71. Cusido J.A., Isalque A., Phys. Stat. Sol. (a). 1985. V. 90, No. 1. P. 127–133.
3.72. Kester U., Gerold U., Crystallization of metallic glasses. In: Metallic glass. Ionic structure, electron transport and crystallization. ed. G.-J. Günterodt and G. Beck. - Moscow: Mir, 1983. P. 325–371.
3.73. Metastable and nonequilibrium alloys, Ed. Yu.V. Efimova. - Moscow: Metallurgiya, 1988.
3.74. Pustov Yu.A., et al., Zashch. Met., 1999. Vol. 35, No. 6. P. 565–576.
3.75. Zhdanova LI, et al, ibid, 1999. Vol. 35, No. 6. P. 577–580.
3.76. Potapov AP, et al., FMM. 1995. 79, No. 2. P. 51–56.
3.77. Kekalo IB, Loeffler F., FMM, 1989. Vol. 68, No. 2. P. 280–288.
3.78. Betechtin V.I., et al., Pis'ma ZhTF. 1998. Vol. 24, No. 23. P. 58–64.
3.79. Betekhtin V.I., et al., FTT. 2001. Vol. 43, no. 10. P. 1815–1820.
3.80. Koster U., et al., Glastech. Ber. 1983. Bd.K56. P. 584–596.
3.81. Koster U., et al. In: Proc. ISMANAM 94. Grenoble, 1994. - P. 85–88.
3.82. Koster U., Meinhardt J., Mater. Sci. Eng. A. 1994. V. 178, No. 1–2. P. 271–278.
3.83. Ignatieva E.Yu. ,Dissertation, Chernogolovka: IFTT RAN, 2007.
3.84. Abrosimova G.E., et al., JMMM. 1999. V. 203. P. 169–171.
3.85. Abrosimova G.E., et al., FTT. 2006. No. 48, issue. 3. P. 523–528.
3.86. Abrosimova G.E., et al., Mat. Sci. Eng. A. 2007. V. 449–451. P. 485–488.
3.87. Abrosimova G.E., et al., FTT. 2006. No. 48, issue. 1. P. 114–119.
3.88. Inoue A., Kimura H., et al., Novel Nanocrystalline Alloys and Magnetic Nanomaterials: Nanocrystalline, nanquasicrystalline and amorphous Al and Mg alloys, Eds .: B. Cantor, M. J. Goringe, E.Ma. - Institute of Physics Publishing Bristol and Philadelphia, 2005. P. 51.
3.89. Louzguine D., Inoue A. ibid, Philadelphia, 2005. - P. 86.
3.90. Fornell J., et al., Acta Materialia. 2010. V. 58. P. 6256–6266.
3.91. Andrievsky R.A., Usp. Khimii, 2002. 71, No. 10. P. 967–981.
3.92. Glezer A.M., et al., Materialovedenie, 2011. No. 6. P. 18–23.
3.93. Chernyavsky K.S. Stereology in Metallurgy. - Moscow: Metallurgiya, 1977.
3.94. Glezer A.M., et al., Deform, Razrush. Mater., 2010. No. 8. C2010. P. 1–10.
3.95. Yoshizawa Y., Yamauchi K., Mater. Trans. JIM. 1989. V. 31. P. 3324–3326.
3.96. Muller M., et al., JMMM. 1996. V. 157/158. P. 209–210.
3.97. Dugaj P., et al., Mater. Sci. Eng. 1991. V.A133. P. 398–402.
3.98. Yoshizawa Y., Yamauchi K., et al., J. Appl. Phys. 1988. V. 64. P. 6047–6051.
3.99. Yoshizawa Y., Yamauchi K., Mat. Sci. Eng. 1991. V.A133. P. 176–182.
3.100. Maslov V.V., et al., FMM. 2001. V. 91, No. 5. Pp. 47–55.
3.101. Shurygina N.A., et al., Izv. RAN, Ser. Fiz., 2012. V. 76, No. 1. P. 52–59.
3.102. Nemoschkalenko V.V., et al. J. Met. Phys. Adv. Techn. 1998. V. 20, No. 6. P. 22–34.
3.103. Muller M., et al., Z. Metallk. 1991. V. 82, No. 12. P. 895–901.
3.104. Illecova E., et al., Mat. Sci. Eng. 1996. V.A205. P. 166–179.
3.105. Mat'ko I., et al., Mat. Sci. Eng. 1994. V.A179 / A180. P. 557–562.
3.106. Koster U., et al., Mat. Sci. Eng. 1991. V.A133. P. 611–615.
3.107. Hampel G., et al., J. Phys. Condens. Matter. 1992. V. 4. P. 3195–3214.
3.108. Fujinami M., et al., Japan J. Appl. Phys. 1990. V. 29. P. 477–480.

3.109. Chen L.C., Spaepen F., J. Appl. Phys. 1991. V. 69, No. 2. P. 679–688.
3.110. Bakai A.S., Polycluster amorphous structures and their properties. Part 2. - Moscow: TsNIIatominform, 1985.
3.111. Hiraga K., Kohmoto O., Mater. Trans. JIM. 1991. V. 33, No. 9. P. 868–871.
3.112. Chin T.-S., et al., Materials Today. 2009. V. 12, No. 1–2. P. 34–39.
3.113. Iwanabe H., Lu B., et al., J. Appl. Phys. 1999. V. B85. P. 4424–4429.
3.114. Willard M. A., Laughlin D. E. et al., J. Appl. Phys. 1998. V. 84. P. 6773–6780.
3.115. Willard M.A., Huang M-Q., et al., J. Appl. Phys. 1999. V. 85. P. 4421–4429.
3.116. Abrosimova G.E., et al., FTT. 1998. Issue 40, No. 1. P. 10–16.
3.117. Abrosimova G.E., Aronin A.S., FTT. 2008. Issue 50, No. 1. P. 154–158.
3.118. Gutkin M.Yu., Ovidko M.A., Physical mechanics of deformable nanostructures. - St. Petersburg: Janus, 2003.
3.119. Tatjanin E.V., et al., FMM, 1986. V. 62, No. 1. P. 133–137.
3.120. Smirnova N.A., et atl., FMM. 1986. Vol. 61, No. 4. P. 1170–1178.
3.121. Rybin V.V. Large plastic deformations and destruction of metals. - Moscow: Metallurgiya, 1986.
3.122. Firstov S.A.,et al., Izv. VUZ, Fizika, 2002. No. 3. P. 41–48.
3.123. Koneva N.A., Kozlov E.V., Structural levels of plastic deformation and fracture / under. Ed. V.E. Panin. - Novosibirsk, Nauka, 1990, P. 123–186.
3.124. Gapontsev V.L., Kondratiev V.V., DAN. 2002. V. 385, No. 5. P. 684–687.
3.125. Valiev R.Z., Ross. Nanotekhnologii, 2006. V. 1, No.1–2. P. 208–216.
3.126. Bykov V.M., et al., FMM, 1978. V. 45, No. 1. P. 163–169.
3.127. Andrievsky R.A, Fundamentals of nanostructured materials science. Opportunities and problems. - Moscow Binom, Laboratory of Knowledge, 2012..
3.128. Glezer A.M., Izv. VUZ, Fizika, 2008. V. 51, No. 5. P. 36–46.
3.129. Pozdnyakov V.A., Glezer A.M., Izv. RAN, Ser. Fiz.. 2004. V. 68, No. 10. P. 1449-1455.
3.130. Gorelik S.S., et al., Recrystallization of metals and alloys. - Moscow, MISiS, 2005.
3.131. Umanskii Ya.S., et al., Physical basis of metal science. - Moscow: Metallurgizdat. 1949.
3.132. Bokshtein B.S., Diffusion in metals. - Moscow, Metallurgiya, 1978.
3.133. Pozdnyakov V.A., Glezer A.M., FTT. 2002. No. 4. P. 705–710.
3.134. Glezer A.M., Deforma. Razrush. Mater., 2005. No. 2. P. 10-15.
3.135. Kozlov E.V., Fiz. Mezomekh., 2004. V. 7, No. 4. P. 93–113.
3.136. Blinova E.N., Glezer A.M., Materialovedenie, 2005. No. 5. P. 32–39.
3.137. Glezer A.M., Pozdnyakov V.A., Deform. Razrush. Mater., 2005. No. 4. P. 9–15.
3.138. Sherif El-Eskandarany M., et al., Acta Met. 2002. V. 50. P. 1113–1123.
3.139. Glezer A.M., Metlov L.S., Physics of Solid State. V. 52, No. 6. P. 1162–1169.
3.140. Glezer A.M., Izv. RAN, Ser. Fiz., 2007. V. 71, No. 12. P. 1764–1772.
3.141. Brailovski V., et al., Materials Transaction JIM. 2006. V. 47, No. P. 795–804.
3.142. Chen H., et al., Lett. Nature. 1994, V. 367, No. 2. P. 541–543.
3.143. Gunderov D.V., Deform. Razrush. Mater., 2006. No. 4. P. 22-25.
3.144. Glezer A.M., et al., Mat. Sci. Forum. 2008. V. 584–586. P. 227–230.
3.145. Glezer A.M., et al., Izv. RAN, Ser. Fiz., 2009. V. 73, No. 9. 1302–1309.
3.146. Valiev R.Z., et al., JOM. 2006. V. 58, No. 4. P. 33–39.
3.147. Kovneristy Yu.K., et al., Deform. Razrush. Mater., 2008. No. 1. P. 35.
3.148. Donovan P.E., Stobbs W.M., Acta Met. 1981. V. 29, No. 6. P. 1419–1424.
3.149. Gryaznov V.G., et al., Pis'ma ZhTF. 1989. T. 15, No. 2. Pp. 1256-1261.
3.150. Pushin VG, Prokoshkin SD, etc. Alloys of titanium nickelide with shape Part 1. Structure, phase transformations and properties. Ekaterinburg: IFM UrB RAN,

2006.
3.151. Tatyanin E.V.,, FTT. 1997. P. 39. P. 1237–1243.
3.152. Prokoshkin S.D., et al., FMM. 2004. V. 97. P. 84–91.
3.153. Zel'dovich V.I., et al., FMM. 2005. V. 99. P. 90–98.
3.154. Prokoshkin S.D., et al., FMM. 2010. V. 110, No. 3. P. 305–320.
3.155. Gunderev D.V., Electronic scientific journal 'Issledovano v Rossi', 2006. 151.pdf.
3.156. Nosova G.I., et al., Kristallografiya. 2009. V. 54, No. 6. 1111–1119.
3.157. Matveeva N.M., et al., FMM. 1997. 83. P. 82–89.
3.158. Glezer A.M., Izv. RAN, Ser. Fiz., 2007. V. 71, No. 12. P. 1764–1772.
3.159. Beloshenko V.A., et al., Theory and practice of hydroextrusion. - Kiev: Naukova Dumka, 2007.
3.160. Glezer A.M., et al., Izv. RAN, Ser. Fiz., 2010. V. 74, No. 11. P. 1576–1582.
3.161. Glezer A.M., et al., J. Nanoparticle Research. 2003. V. 5. P. 551–560.
3.162. Zhilyaev A.P., Langdon T.G., Prog. Mater. Sci. 2008. V. 53. P. 893–979.
3.163. Ast D.G., Krenitsky D.J. et al., J. Mater. Sci. and Eng. 1980. V. 43. P. 241–236.
3.164. Lavrent'ev E.A., Shabat B.V., Methods of the theory of functions of complex variable. - Moscow: Nauka.
3.165. Timoshenko S.P., Goodyear J. Theory of Elasticity. - Moscow: Nauka, 1979.
3.166. Lykov A.V. Theory of heat conductivity. - Moscow: Vysshaya shkola, 1967.
3.167. Chen H., He Y., et al., Nature, 1994. V. 367, No. 6463. P. 541–543.
3.168. Chen Y.S., J. Appl. Phys. Letters. V.V. 29, No. 6. P. 328–330.
3.169. Bell J.F., Experimental foundations of mechanics of deformable solid bodies. V. 2. Finite deformations. - Moscow: Nauka, 1984.
3.170. Kimura H., et al., et al., J. Appl. Phys. 1982. V. 53, No. 5. P. 3523–3528.
3.171. Zaichenko S.G., et al., Materialovedenie, 2004. No. 9. P. 11–18.
3.172. Cherepanov G.P., Mechanics of brittle fracture. - Moscow: Nauka, 1974.
3.173. Spaepen F., Turnbull D., et al., Scr. Metall. 1974. V. 8, No. 2. P. 563.
3.174. Spaepen F.A., Acta Metall. V. V. 25, No. 3. P. 407–415.
3.175. Wu T.W., Spaepen F., Acta Mertall. 1985. V. 33, No. 11. P. 2185.
3.176. Eyring H., J. Chem. Phys. 1936. V. 4, No. 4. P. 283–291.
3.177. Gleiter H., in: Deformation of Polycrystals: Mechanisms and Microstructure, eds .: N. Hansen et al. - Roskilde: Riso Nat. Laboratory, 1981. P. 15.
3.178. Gleiter H., Marquardt P. Z., Metallkud. 1984. V. 75 (4). P. 263.
3.179. Birringer R., et al., Trans. Japan. Inst. Met. 1986. Suppl. 27. P. 43.
3.180. Siegel R.W., Hahn H., Current Trends in Physics of Materials, Ed. M.Yussouff. -Singapore: World Sci. Publ. Co., 1987. P. 403.
3.181. Gleiter H. Progr. Mater. Sci. 1989. V. 33 (4). P. 233.
3.182. Matthew M.D., Pechenik A., J. Am. Ceram. Soc. 1991. V. 74 (7). P. 1547.
3.183. Chen D.-J., Maya M., J. Nanostruct. Mater. 1992. V. 2 (3). P. 469.
3.184. Ivanov V.V., et al., Nanostruct. Mater. 1995. V. 6 (1–4). P. 287.
3.185. Gleiter H., Nanostruct. Mater. 1995. V. 6 (1–4). P. 3.
3.186. Gleiter H., Nanostruct. Mater. 1992. V. 1 (1). P. 1.
3.187. Gleiter H., Mechanical properties and deformation behavior of materials having ultrafine microstructure, ed. M.A. Nastasi. - Netherlands, Dordrecht: Kluwer Academic Press, 1993. - P. 3.
3.188. Wunderlich W., et al., Scripta Metall. Mater. 1990. V. 24, No. 2. P. 403.
3.189. Thomas G.R., et al., Scripta Metall. Mater. 1990. V. 24, No. 1. P. 201.
3.190. Mutschele T., Kirchheim R., Scripta Met. 1987. V. 21, No. 2. P. 135.
3.191. Schaefer H.E. et al., Nanostruct. Mater. 1992. V. 1, No. 6. P. 523.
3.192. Birringer R., Gleiter H., Encyclopedia of material science and engineering. Suppl.

V.1., Ed. R.W. Cahn. - Oxford: Pergamon press, 1988. - P. 339.
3.193. Zhu X. et al., Phys. Rev. B35. 1987. V. 17. P. 9085.
3.194. Fitzsimmons M. et al., Phys. Rev. B44. 1991. V. 44, No. 6. P. 2452.
3.195. Eastmen J.A. et al., Nanostruct. Mater. 1992. V. 1, No. 1. P. 47.
3.196. Löffler J., et al., Nanostruct. Mater. 1995. V. 6 (5–8). P. 567.
3.197. Weissmüller J., et al., Nanostruct. Mater. 1995. V. 6 (1–4). P. 105.
3.198. Haubold T. et al. Phys. Lett. A. 1989. V. 135 (8–9). P. 461.
3.199. Ishida Y. et al., Nanostruct. Mater. 1995. V. 6 (1–4). P. 115.
3.200. Schlorke N. et al., Nanostruct. Mater. 1995. V. 6 (5–8). P. 593.
3.201. Babanov Yu.A. et al., Nanostruct. Mater. 1995. V. 6 (5–8). P. 601.
3.202. Mitin A.V., et al., 4th Youth Scientific School 'Coherent Optics and Optical Spectroscopy', Kazan, October 26-28, 2000: Collection of articles. - Kazan: KSU, 2000. P. 265–272.
3.203. Mitin A.V., et al., ISVTE-4, ISTFE-12. Collection of works. - Kharkov: NSC KIPT, 2001. P. 443–446.
3. 204. Vavilova V.V., et al., Inorganic Materials. 2003. V. 39, No. 1. P. 72–76.
3.205. Vavilova V.V., Inorganic Materials. 2004. V. 40, No. 2. P. 152–160.
3.206. Vavilova V.V., et al., Inorganic Materials. 2004. V. 40, No. 7. P. 707–715.
3.207. Anosova M.O., et al., Inorganic Materials. 2009. V. 45, No. 9. P. 993–997.
3.208. Hono K., Ping D.H., Materials Characterization. 2000. V. 44, No. 1–2. P. 203–217.
3.209. Pundt A., et al., Z. Phys. B .: Condensed Matter. 1992. V. 87, No. 1. P. 65–72.
3.210. Knyazev Yu.V., Kuz'min Yu.I., Optics and spectroscopy. 2009. V. 10, No. 5. P. 708-712.
3.211. Darinsky B.M., Yudin L.Yu., Izv. RAN, Ser. Fiz., 2010. V. 74, No. 9. P. 1355–1359.
3.212. Yakovlev A.V., et al., Proc. XVII Petersburg readings on the problems of strength, dedication. 90th anniversary of the birth of. Prof. A.N. Orlov, 10-12 April. 2007, St. Petersburg, 2007. Part 1. P. 73–74.
3.213. Ushakov I.V., Vestnik TGU, 2007. V. 12, No. 2. P. 263–266.
3.214. Fedorov V.A., Fund. Probl. Sovrem.Materialoved., 2009. V. 6, No. 2. P. 87–91.
3.215. Fedorov V.A., et al., Issues of modern science and practice. University named after of V.I. Vernadsky. Special issue (36). 2011. P. 74–78.
3.216. Volmer M., Weber A. Z. Phys. Chen. 1926. Bd. 19, No. 3–4. P. 277–301.
3.217. Frank F. C., Van der Merve J.H., Proc. Roy. Soc. A. 1949. V.A198. P. 205–225.
3.218. Frank F. C., Van der Merve J.H., Proc. Roy. Soc. A. 1950. V.A200. P. 125–134.
3.219. Stranski I.N., Krastanow L., Sitz Berl. Akad Wiss. 1938. V. 146. P. 797–807.
3.220. Ievlev V.M., Thin films of inorganic materials: mechanism of growth and structure: Textbook. - Voronezh: CPI VSU, 2008.
3.221. Matthews J.W. in: Physics of thin films. - Moscow: Mir, 1970. - T. 4. P. 167–227.
3.222. Van der Drift A., Phillips Res. Repts. 1967. V. 22. P. 267–276.
3.223. Ievlev V.M.,et al., in: Problems of Solid State Physics. - Voronezh: VSTU, 1973. - P. 142–145.
3.224. Ievlev V.M., et al., FizKhOM. 1975. No. 2. P. 84–87.
3.225. Ievlev V.M., et al, Izv. AN SSSR, Neorg. Mater., AN SSSR. Inorganic materials. 1991. V. 27, No. 3. P. 521–525.
3.226. Smidt F.A., International Materials Reviews. 1990. V. 35, No. 2. P. 61–128.
3.227. Sonnenberg N., et al., J. Appl. Phys. 1993. V. 74 (2). P. 115.
3.228. Isaev A.Yu. Dissertation, Voronezh: VPI, 1999. - 138 p.
3.229. Vermout P., Dekeyser W., Physica. 1959. V. 25, No. 1. P. 53–54.
3.230. Sheftal' N.N., Buzynin A.H., Vestnik MGU. Geologiya. 1972. No. 3. P. 102–104.
3.231. Givargizov E.I. Artificial epitaxy – a promising technology of the elementary base

of microelectronics. - Moscow: Nauka, 1988..
3.232. Gudulin E.A. Directional synthesis of superconductive materials based on REE-barium cuprates. Author's abstract. PhD dissertation. - Moscow, 2003.
3.233. Kukushkin S.A.,FTT. 2007, V. 50. P. 1188–1195.
3.234. Cder H.,et al., J. Appl. Phys. 2005. V. 98. P. 024313 (1–10).
3.235. Aleksandrov L.N., et al., Krystall and Techn. 1969. V. 4. P. 25–30.
3.236. Parmigiani F., et al., Appl. Opt. 1985. V. 24. P. 3335.
3.237. Biegelsen D.K., et al., Appl. Phys. Lett. 1981. V. 38, No. 3. P. 150–152.
3.238. Biegelsen D.K., et al., In Ref. [332]. P. 537-548.
3.239. Voevodin A.A., et al., Mater. Sci. Eng. A. 2008. V. 488. P. 112.
3.241. Chen C.Q., et al., J. Appl. Phys. 2009. V. 105. P. 114314.
3.242. Musil J., et al., J. Vac. Sci. Technol. A. 2010. V. 28. P. 244.
3.243. Blinkov I.V., Manukhin A.V., Nanodispersed and granulated materials obtained in pulsed plasma. - Moscow: MISIS, 2005.
3.244. Voevodin A.A., Zabinski J.S., Compos. Sci. Technol. 2005. V. 65. P. 741.
3.245. Pavesi L., Mater. Today. 2005. V. 8, No. 1. P. 18.
3.246. Tetelbaum D.I., et al., Poverkhnost'. Rent. Sinkhr. Neitr. Issled., 2009. No. 9. P. 1.
3.247. Kogan B.C., et al., In: Structure of metallic films, doped with gases. Central Research Institute of Information and Technical and Economic Research in Atomic Science and Technology: FTI AN USSR. - Kharkiv, 1987.
3.248. Cohen R.W., Abeles B., Phys. Rev. 1968. V. 168, No. 2. P. 444–450.
3.249. Kalinin Yu.E., et al., FizKhOM, 2001. No. 5. P. 14–20.
3.250. Kalinin Yu.E., Persp. Materialy, 2003. No. 3. P. 62–66.
3.251. Bondarev A.V. The atomic structure of amorphous rhenium alloys with transition-metals of the sixth period. Author's abstract. Dissertation, VSTU. Voronezh, 2002.
3.252. Levchenko E.V. Structural models of vitrification of pure metals and metal-metalloid-type systems. Author's abstract. Dissertation, Voronezh: VSTU.
3.253. Kravets V.G.,et al., Fiz. Tverd. Tela, 2003. Vol. 45, no. 8. P. 1456–1462.
3.254. Kalinin Y.E., et al., Ferroelectrics. 2004. V. 307. P. 243–249.
3.255. Barnard Y.A., et al., IEEE Trans. Magn. 1993. V. 29, No. 6. P. 2711–2713.
3.256. Tsoukatos A., et al., IEEE Trans. Magn. 1993. V. 29. No. 6, Pt. 1. P. 2726–2728.
3.257. Avramenko B.A., et al., in: Thin films in optics and electronics.Kharkov International Symposium. (ISTFE-14). - Kharkov: NSC KIPT, 2002. - P. 57–60.
3.258. Daughlon J.M., Chen Y.J., IEEE Trans. Magn. 1993. V. 29, No. 6. P. 2705–2710.
3.259. Rupp G., Van der Berg H.A.M., IEEE Trans. Magn. 1993. V. 29.
3.260. Mouchot J., et al,, IEEE Bill. Magn. 1993. V.29, No. 6. Pt. 1. P. 2732–2734.
3.261. Roschenko S., et ak., FMM, 2000. V. 90, No. 3. P. 58–62.
3.262. Tochitsky T.A., et al., Izv. VUZ, Fizika, 1998. Vol. 41, No. 7. P. 12–17.
3.263. Tochitskii O.A., et al., J. Magn. and Magn. Mater. 2001. V. 224, No. 3. P. 221–232.
3.264. Ievlev V.M., et al., FMM, 1986. Vol. 62. P. 412–413.
3.265. Ievlev V.M., et al., The Physics of Metals and Metallography. 2000. V. 90, No. 2. P. 159–163.
3.266. Ievlev V.M., Bugakov A.B., Vestnik VGTU, Ser. Materialoveenie, 1999. Issue. 1.5. P. 61–68.
3.267. Merkulov G.V. Diffusion-controlled mechanisms of formation of nanocrystalline heterostructures in two-component films with limited mutual solubility. Dissertation, Voronezh: VSTU, 2003.
3.268. Wall I.A., Jankowski A.F., Thin. Sol. Films. 1989. No. 181. P. 313–321.
3.269. Gupta R., et al., Phys. Rev. B. 2003. V. 67, No. 7. P. 075402 / 1-075402 / 7.
3.270. Mikhailov I.F., et al., Fiz. Tverdogo Tela, 1983. Vol. 25, No. 4. P. 1166–1171.

3.271. Palatnik L.C., et al., DAN SSSR. 1977. P. 236. P. 87–90.
3.272. Ievlev V.M., Kushchev S.B., 1979. P. 47. P. 1102–1104.
3.273. Sipatov A.Yu., Epitaxial superlattices and quantum structures of isochalcogenides of lead, tin, europium and ytterbium. Author's abstract. Dissertation, Kharkov: NTU KhPI, 2006..
3.274. Zhigalina O.M., et al., Kristallografiya, 2013. Vol. 58, No. 2. P. 327–336.
3.275. Bannykh O.A., et al., Metally. 2000. No. 2. P. 54–56.
3.276. Zang H.Y., et al., J. Appl. Phys. 1995. V. 77. P. 2811–2813.
3.277. Lu K., ey al., Scr. Metall. Mater. 1990. V. 24. P. 2319–2324.
3.278. Noskova N.I., FMM. 1997. V. 81, No. 1. Pp. 116-121.
3.279. Zielinski P.G., Ast D.G., Acta Metal. 1984. V. 32, No. 3. P. 397–405.
3.280. Samsonov G.V., Vinnitsky I.M. Refractory compounds. - Moscow, Metallurgiya, 1976.
3.281. Diagrams of the state of binary systems (reference book, N.P. Lyakishev). T. 2. - Moscow, Mashinostroenie, 2000.
3.282. Pozdnyakov V.A., Materialovedenie, 2002. No. 11. P. 39.
3.283. Gol'dstein M.I., et al., Metallophysics of high-strength alloys. - Moscow, Metallurgiya, 1986.
3.284. Glezer A.M., Molotilov B.V., Ordering and deformation of ferrous alloys. Moscow, Metallurgiya, 1984..
3.285. Inoue A. et al., Mater. Trans. Japan. Inst. Met. 1994. V. 35. P. 85.
3.286. Golovin Yu.I. Introduction to nanotechnology. - Moscow, Mashinostroenie-1, 2003.
3.287. Nieh T.G., Wadsworth J., Scr. Metall. Mater. 1991. V. 25. No. 4. P. 955–958.
3.288. Shen T.D., et al., Acta Materialia. 2007. V. 55. I. 15. P. 5007–5013.
3.289. Noskova N.I., Deform. Razrush. Mater., 2009. No. 4. Pp. 17–24.
3.290. Tanaka K., Mori T., Acta Metall. 1970. V. 18, No. 8. P. 931–941.
3.291. Mura T., Micromechanics of defects in solids. - Hague: Martinus Nijhoff Publ., 1982.
3.292. Das E.R., Marcikowski M.J., J. Appl. Phys. 1972. V. 43. No. 2. P. 4425–4434. P. 3102–3104.
3.293. Bian Z., et al., 2002. V. 46. P. 407–412.
3.294. Conner R.D., et al., J. Mater. Res. 1999. V. 14. No. 8. P. 3292
3.294. Conner R.D., et al., J. Mater. Res. 1999. V. 14. I. 8. P. 3292–3297.
3.295. Pozdnyakov V.A., FMM. 2004. V. 97, No. 1. P. 9–17.
3.296. Glezer A.M., et al., Deform. Razrush. Mater., 2012. No. 3. P. 1–10.
3.297. Schuh C.A., et al., Acta Mater. 2007. V. 55. P. 4067.
3.298. Zabelin S.F., et al., Vestnik TGU. 2010. V. 15, No. 3. P. 835–836.
3.299. Andrievsky R.A., Khachoyan A.V., Ross. Khim. Zh.,. 2009. V. LIII. No. 2. P. 4.
3.300. Andrievsky R.A., Izv. RAN. Ser. Fiz. 2009. V. 73,No. 9. P. 1290.
3.301. Veprek S., J. Vac. Sci. Technol. A. 1999. V. 17. P. 2401.
3.302. Veprek S., Veprek-Heinman M.G.D., Nanostructured coatings, ed. A. Cavaleiro and D. de Hosson. Trans. from English. ed. R.A.Andrievsky. - Moscow, Tekhnosfera, 2011. - P. 412.
3.303. Igbal M., et al., J. Non-Cryst. Sol. 2008. V. 354. P. 3284.
3.304. Inoue A., Acta Mater. 2000. V. 48. P. 279.
3.305. Qiao J., J. Mater. Sci. Technol. 2013. V. 29, No. 8. P. 685–701.
3.306. Qiao J.W., et al., Metall. Mater. Trans. A. 2011. V. 42, No. 9. P. 2530–2534.
3.307. Greer A. L., et al., Materials Science and Engineering R. 2013. V. 74. P. 71–132.
3.308. Hays C.C., et al., Phys. Rev. Lett. 2000. V. 84. P. 2901–2904.
3.309. Cherepanov G.P., Mechanics of brittle fracture. - Moscow, Nauka, 1974.

3.310. Raye, J., Mathematical Methods in Mechanics of Destruction, Fracture. V. 2. - Moscow, Mir, 1975. - P. 204–335.
3.311. Gutkin M.Yu., Ovidko I.A., FTT. 2010. V. 52. P. 56.
3.312. Kornilov I.I., et al., Titanium nickelide and others alloys with the shape memory effect. - Moscow: Nauka, 1977.
3.313. Schlossmacher P., et al., Mater. Sci. Forum. 2000. V. 327–328. P. 131–134.
3.314. Glezer A.M., et al., Metally. 1998. No. 4. P. 45–47.
3.315. Golovin Yu.I., In: Perspective materials. Structure and methods of research. Textbook, ed. D.L. Merson. - TSU, MISiS, 2006. - Pp. 89–244.
3.316. Glezer A.M., et al., Deform. Razrush. Mater., 2011. No. 11. Pp. 1–8.
3.317. Gorelik S.S., et al., Radioactive and electron-optical analysis. - Moscow: MISiS, 2002.
3.318. Malygin G.A., FTT. 1995. P. 37. P. 2281–2292.
3.319. Firstov S.A., Rogul' T.G., Deform. Razrush. Mater., 2011. No. 5. P. 1–7.
3.320. Firstov S.A., Rogul' T.G., Theoretical (ultimate hardness). Reports of the National Academy of Science of Ukraine. 2007. No. 4. Pp. 110–114.
3.321. Ke M., et al., Nanostruct. Mater. 1995. V. 5. P. 689–697.
3.322. Pozdnyakov V.A., Glezer A.M., Doklady RAN, 2002. V. 384, No. 2. P. 177–180.
3.323. Li Y. S., et al., Acta Mater. 2008. V. 56. P. 230.
3.324. Lu L. et al., Science. 2004. V. 304. P. 422.
3.325. Zhang X. et al., Acta Mater. 2004. V. 52. P. 995.
3.326. Shen Y.F., et al., Scr. Mater. 2005. V. 52. P. 989.
3.327. Beygelzimer Ya.E., Mechanics of Materials, 2004. V. 17. P. 753.
3.328. Beygelzimer Ya.E., et al., Helical extrusion as the process of accumulation of deformations. - Donetsk, TE AN, 2003..
3.329. Kopylov V.I., Chuvildeev V.N., Metally, 2004. No. 1. P. 22.3297.
3.295. Pozdnyakov V.A., FMM. 2004. V. 97, No. 1. P. 9–17.
3.296. Glezer A.M., et al., Deform. Razrush. Mater., 2012. No. 3. P. 1–10.
3.297. Schuh C.A., et al., Acta Mater. 2007. V. 55. P. 4067.
3.298. Zabelin S.F., et al., Vestnik TGU. 2010. V. 15, No. 3. P. 835–836.
3.299. Andrievsky R.A., Khachoyan A.V., Ross. Khim. Zh. 2009. V. LIII. No. 2. P. 4.
3.300. Andrievsky R.A., Izv. RAN. Ser. Fiz. 2009. V. 73, No. 9. P. 1290.
3.301. Veprek S., J. Vac. Sci. Technol. A. 1999. V. 17. P. 2401.
3.302. Veprek S., Veprek-Heinman M.G.D., Trans. from English. ed. R.A. Andrievsky. - Moscow, Tekhnosfera, 2011.
3.303. Igbal M., et al., J. Non-Cryst. Sol. 2008. V. 354. P. 3284.
3.304. Inoue A., Acta Mater. 2000. V. 48. P. 279.
3.305. Qiao J., J. Mater. Sci. Technol. 2013. V. 29, No. 8. P. 685–701.
3.306. Qiao J.W., et al., Metall. Mater. Trans. A. 2011. V. 42, No. 9. P. 2530–2534.
3.307. Greer A.L., et al., Materials Science and Engineering R. 2013. V74. P. 71–132.
3.308. Hays C.C., et al., Phys. Rev. Lett. 2000. V. 84. P. 2901–2904.
3.309. Cherepanov G.P., Mechanics of brittle fracture. - Moscow, Nauka, 1974.
3.310. Raye J., Mathematical Methods in Mechanics of Destruction, Fracture. V. 2. - Moscow: Mir, 1975. - P. 204–335.
3.311. Gutkin M.Yu., Ovidko I.A., FTT. 2010. V. 52. P. 56.
3.312. Kornilov I.I., et al., Titanium nickelide and others alloys with the shape memory effect. - Moscow: Nauka, 1977.
3.313. Schlossmacher P., et al., Mater. Sci. Forum. 2000. V. 327-328. P. 131-134.
3.314. Glezer A.M., et al., Metally. 1998. No. 4. Pp. 45-47.
3.315. Golovin Yu.I., Perspective materials. Structure and methods of research. Textbook.

Ed. D.L. Merson. - TSU, MISiS, 2006. - P. 89–244.
3.316. Glezer A.M., et al., Deform. Razrush. Mater., 2011. No. 11. P. 1–8.
3.317. Gorelik S.S., et al., Radiographic and electron-optical analysis. - Moscow: MISiS, 2002..
3.318. Malygin G.A., FTT. 1995. P. 37. P. 2281–2292.
3.319. Firstov S.A., Rogul' T.G., 2011. No. 5. P. 1–7.
3.320. Firstov S.A., Rogul' T.G., Theoretical (ultimate hardness), Reports of the National Academy of Science of Ukraine. 2007. No. 4. P. 110–114.
3.321. Ke M.,et al., Nanostruct. Mater. 1995. V. 5. P. 689–697.
3.322. Pozdnyakov V.A., Glezer A.M., Doklady RAN, 2002. V. 384, No. 2. P. 177–180.
3.323. Li Y. S., et al., Acta Mater. 2008. V. 56. P. 230.
3.324. Lu L., et al., Science. 2004. V. 304. P. 422.
3.325. Zhang X. et al., Acta Mater. 2004. V. 52. P. 995.
3.326. Shen Y. F. et al., Scr. Mater. 2005. V. 52. P. 989.
3.327. Beygelzimer Ya.E., Mechanics of Materials. 2004. V.˙17. P.˙753.
3.328. Beygelzimer Ya.E.,et al., Helical extrusion is the process of accumulation of deformations. - Donetsk: company TE AN, 2003.
3.329. Kopylov V.I., Chuvildeev V.N., Metally. 2004. No. 1. P. 22.
3.330. Chuvildeev V.N. Nonequilibrium grain boundaries in metals. Theory and experiment. - Moscow: Fizmatlit, 2004.
3.331. Kopylov V. I., Chuvildeev V.N. The Limit of Grain Refinement during ECAP (Deformation // Severe Plastic Deformation, Ed. by A. Burhanettin. - New York: Nova Science Publishers, 2006..
3.332. Khonik B.A., Proc. of the IV International School-Conference "Micromechanisms of plasticity, fracture and related phenomena ". - Tambov: TSU, 2007.
3.333. Bobrov O.P., et al., FTT. 2004. V. 46, No. 10. P. 1801.
3.334. Malygin G.A., FTT. 2001. P. 43, No. 10. P. 1832.
3.335. Malygin G.A., FTT. 2002. P. 44, No. 7. P. 1249.
3.336. Malygin G.A., FTT. 2002. P. 44, No. 11. C. 1979.
3.337. Malygin G.A., Usp. Fiz. Nauka, 1999. Vol. 169, No. 9. P. 979.
3.338. Metlov L.S., Vest. Donetsk Univ., Ser. A. Estestv. Nauki, 2006. Issue 1. P. 269.
3.339. Metlov L.S., Vest. Donetsk Univ., Ser. A. Estestv. Nauki, 2006. Issue 2. P. 169.
3.340. Metlov L.S. Deform. Razrush. Mater., 2007. No. 2. P.˙40.
3.341. Metlov L.S. MFiNT. 2007. No. 2. 40.
3.342. Metlov L.S., Proc. of the IV international- conference "Micromechanisms of plasticity, destruction and accompanying phenomena ". - Tambov: TSU, 2007
3.343. Rubi J.M., Gadomski A., Physica A. 2003. V. 326. P. 333.
3.344. Santamaria-Holek I., et al., http://arxiv.org/abs/cond-mat / 0409362,2004.
3.345. Metlov L.S. Vest. Donetsk Univ., Ser. A. Estestv. Nauki, 2007. Issue. 1. P. 169.
3.346. Slobodskoy M.I., et al., Matemat. modelirovanie sistem i protsessov. 1995. No. 3. P. 88.
3.347. Slobodskoy M.I., Popov L.E., *ibid*, 1997. No. 5. P. 105.
3.348. Popov L.E., et al., *ibid.* 1999. No. 7. P. 67.
3.349. Puspesheva S.I., et al., *ibid*, 2002. No. 10. P.˙92.
3.350. Ievlev V.M., et al., Vestnik TSU. 2010. V. 15, No. 3. P. 927–928.
3.351. Viswanath B., Scripta Materialia. 2007. V. 57. P. 361–364.
3.352. Belousov O.K., et al., Deform. Razrush. Mater., 2007. No. 6. P. 10–18.
3.353. Vavilova V.V., et al., Deform. Razrush. Mater., 2007. No. 9. P. 29–32.
3.354. Vavilova V.V., Deform. Razrush. Mater., 2006. No. 4. P. 18–22.

3.355. Veprek S., et al., Thin solid films. 2005. V. 476. P. 1–29.
3.356. Andrievsky R.A., Izv. RAN, Ser. Fiz., 2009. V. 73, No. 9. P. 1290–1294.
3.357. Voevodin A.A., Zabinski J.S., Diam. Rel. Mater. 1998. V. 71. P. 463–467.
3.358. Andrievsky R.A. Fundamentals of nanostructured materials science. Opportunities and problems. - M .: Binom. Laboratory of Knowledge, 2012. 252 p.
3.359. Musil J., et al., J. Vac. Sci. Technol. A. 2010. V. 28. P. 244–249.
3.360. Chen C.Q., et al., J. Appl. Phys. 2009. V. 105. P. 114314.
3.361. Korotaev A.D., et al., Fiz. Mezomekhanika. 2005. Issues 8, No. 5. P. 103–116.
3.362. Yamauchi K., Yoshizawa Y., Nanostruct. Mater. 1995. V. 6, No. 1–4. P. 247–256.
3.363. Suryanarayana C. Nanophase materials, Int. Mater. Rev. 1995. V. 40. P. 41–64.
3.364. Gusev A.I., Usp. Fiz. Nauk, 1998. V. 168. P. 29–58.
3.365. Kronmuller H., Nanostruct. Mater. 1995. V. 6. P. 157–168.
3.366. Ustinov V.V., Kravtsov E.A., In: Nanostructured Materials: Science and Technology, eds. G.-M. Chow, N. I.Noskova. Dordrecht: Kluwer Acad. Publ., 1998. P. 441-456.
3.367. Becker J.I., J. Appl. Phys. 1984. V. 55, No. 6. P. 2067–2072.
3.368. Glazer A.A., FMM, 1979. Vol. 48, No. 6. P. 1165–1172.
3.369. Drozdova M.A., et al., FMM. 1989. Vol. 67, No. 5. P. 896–901.
3.370. Yagodkin Yu.D., Izv. VUZ, Chernaya metallurgiya, 2007. No. 1. Pp. 37–46.
3.371. Herzer G., Hilzinger H.R., J. Mag. Mag. Mat. 1986. V. 62, No. 2–3. P. 143–151.
3.372. Herzer G., Hilzinger H.R., Physica Scripta. 1989. V. 39. P. 639–642.
3.373. Herzer G., IEEE. Trans. Mag. 1989. V. 25. P. 3327–3329.
3.374. Herzer G., IEEE.Trans. Mag., 1990. V. 26. P. 1397–1402.
3.375. Herzer G., J. Mag. Mag. Mat. 1992. V. 112, No. 1–3. P. 258–262.
3.376. Herzer G., Warlimont H., Nanostruct. Mat. 1992. V. 1, No. 3. P. 263–268.
3.377. Coey J.M.D., Rare-earth iron permanent magnets. Oxford: Oxford Science Publications: Clarendon Press, 1996.
3.378. Kneller E.F., Hawig R., IEEE. Trans. Mag. 199l. V. 27. P. 3588–3560.
3.379. Skomski, R., Coey J.M.D., Phys. Rev. V. 1993. V. 48. P. 15812–15816.
3.380. McHenry M.E.,et al., Recent advances in the chemistry and physics of fullerenes and related materials. eds. K.M. Kadish, R. S. Ruoff. PV 96-10. ECS Symposium Proceedings. Penmington. NJ. 1996. P. 703.
3.381. Host J. J., et al., J. Mat. Res. 1997. V. 12, No. 5. P. 1268–1273.
3.382. Bertotti G., et al., Mat. Sci. Eng. A. 1997. V. 226–228. P. 603–613.
3.383. Herzer G., Mat. Sci. Eng. A. 1991. V. 133. P. 1.
3.384. Herzer G., Scr. Metall. Mater. 1995. V. 33, No. 10–11. P. 1741–1756.
3.385. Hoffmann H., Fujii T., J. Mag. Mag. Mat. 199. V. 128. P. 395–400.
3.386. Fujii Y., et al., J. Appl. Phys. 1991. V. 70. P. 6241–6243.
3.387. Tsuei C.C., et al. J. Appl. Phys. 1966. V. 37. P. 435.
3.388. Turgut Z., et al., J. Appl. Phys. 1997. V. 81. P. 4039–4041.
3.389. Inoue A., et al., J. Appl. Phys. 1998. V. 83. P. 6326–6332.
3.390. Gallagher K.A., et al., J. Appl. Phys. 1999. V. 85. P. 5130–5132.
3.391. Turgut Z., et al., J. Appl. Phys. 1998. V. 83. P. 6468–6470.
3.392. Turgut Z., et al., J. Appl. Phys. 1999. V. 85, No. 8. P. 4406–4408.
3.393. Varga L.K., et al., Mat. Sci. Eng. A. 1994. V. 179–180. P. 567–571.
3.394. Stolloff N.S., Mater. Res. Soc. Proc. 1985. V. 39. P. 3–12.
3.395. Suzuki K., et la., Mat. Trans. JIM. 1990. V. 31. P. 743–746.
3.396. Smith C.H., IEEE. Trans. Mag. 1982. V. 18. P. 1376–1381.
3.397. Hernando A., Kulik T., Phys. Rev. V. 1994. V. 49. P. 7064–7067.
3.398. Hernando A., Navarro I., Nanophase materials / Eds. G.C. Hadjipanyis, R.W. Siegel.

- Netherlands: Kluwer, 1994.
3.399. Hernando A., et al., Phys. Rev. B. 1995. V. 51. P. 3281–3284.
3.400. Gomez-Polo C., et al., Phys. Rev. B. 1996. V. 53. P. 3392–3397.
3.401. Malkinski L., Slawska-Waniewska A., J. Mag. Mag. Mat. 1996. V. 160. P. 273–279.
3.402. McHenry M.E., et al., Prog. Mater. Sci. 1999. V. 44. P. 291–433.
3.403. Garcia C., et al., Physica B. 2008. V. V. 403. P. 286–288.
3.404. Hono K., et al., Acta Met. Mater. 1992. V. 40. P. 2137–12144.
3.405. Glezer A.M., Kirienko V.I., Metally, 1998. No. 2. Pp. 44–48.
3.406. Ayers J.D., et al., J. Mater. Sci. 1995. V. 30. P. 4492–4498.
3.407. Zusman A.I., Artsishevsky M.A., Thermomagnetic treatment of iron-nickel alloys. - Moscow, Metallurgiya, 1984.
3.408. Grognet S., Atmani H., Teilett J. Nanocrystallization by nitriding treatment of FeSiB-based amorphous ribbon, UMR. 6634. CNRS. 19 p.
3.409. Gonzalez J., J. Appl. Phys. 1994. V. 76. P. 1131–1134.
3.410. Sadchikov V.V., et al., Stal'. 1997. No. 11. P. 58–61.
3.411. Yoshizawa Y., Yamauchi K., Mat. Sci. Engin. 1991. V.A133. P. 176–182.
3.412. Makarov V.A., FMM, 1991. No. 9. Pp. 139-149.
3.413. Zhukova V., et al., J.Magn. Magn. Mater. 2002. V. 249. P. 79–84.
3.414. Deryagin A.I., et al., FMM 2008. 106. No. 3. P. 301–311.
3.415. Inoue A., Acta Mater. 2000. V. 48. P. 279-286.
3.416. Glezer A.M., et al., Izv. RAN, Ser. Fiz., 2009. V. 73, No. 9. P. 1310–1314.
3.417. Kekalo I.B., Samarin B.A., Physical metallurgy of precision alloys. Alloys with special magnetic properties. - Moscow, Metallurgiya, 1989.
3.418. Kojima A., et al., Mat Sci. Eng. 1994. V.A179–A180. P. 511–515.
3.419. Gorria P., et al., IEEE. Trans. Mag. 1993. V. 29. P. 2682–2684.
3.420. Suzuki K., et al., J. Appl. Phys. 1991. V. 70. P. 6232–6237.
3.421. Garcia-Tello P., et al., IEEE. Trans. Mag. 1997. V. 33. P. 3919–3921.
3.422. Garcia Tello P., et al., J. Appl. Phys. 1998. V. 83. P. 6338–6340.
3.423. Iwanabe H., et al., J. Appl. Phys. 1999. V. 85. P. 4424–4426.
3.424. Kim K.S., Yu S.C., J. Appl. Phys. 1997. V. 81. P. 4649–4652.
3.425. Inoue A., Zhang T. et al., J. Appl. Phys. 1998. V. 83. P. 6326–6332.
3.426. Pooh C., Owen F., Nanotechnology: Trans. with English. Ed. Yu.I. Golovin. - Moscow: Tekhnosfera, 2004.
3.427. New materials, Ed. Yu.S. Karabasova. - Moscow: MISiS, 2002..
3.428. Kondorskiy E.I., Izv. AN SSSR. Ser. Fiz., 1978. Vol. 42, No. 8. P. 1638–1645.
3.429. Nepijko S.A., Physical properties of small metal particles. -Kiev: Naukova Dumka, 1985.
3.430. Rezlescu L., et al., J. Magn. Magn. Mater. 1999. V. 193. P. 288–290.
3.431. Buschow K.H. J., New permanent magnet materials. - North-Holland, Amsterdam, 1986.
3.432. Hadjipanayis G.C., J. Magn. Magn. Mater. 1999. V. 200. P. 373–391.
3.433. Kekalo I.B., Samarin B.A., Physical Metallurgy of precision alloys. Alloys with special magnetic properties. - Moscow, Metallurgiya, 1989.
3.434. Savchenko A.G., Proc. II Rusian-Japanese seminar:Perspective Technologies, Materials and Equipment of Solid-State Electronic Components ». 2004. - Moscow, MISA Publishing. - P. 280–332.
3.435. Yagodkin Yu.D., et al., Izv, VUZ, Chernaya metallurgiya. 2004. No. 9. P. 28–38.
3.436. Luo Y., Proc. 18th Intern. Workshop High performance magnets and their application, 29 Aug. - 2 Sept. 2004. Annecy, France, 2004. P. 28–39.

3.437. Yagodkin Yu. D., Izv. VUZ. Chernaya metallurgiya. 2007. No. 1. P. 37–46.
3.438. Croat J.J., et al., Appl. Phys. Lett. 1984. V. 44. P. 148–149.
3.439. Glebov V.A., et al., Perspekt. Materialy. 2003. No. 2. P. 55–60.
3.440. Minakova S.M., et al., Zavod. Lab., 2004. V. 70, No. 8. P. 34–37.
3.441. Melsheimer A., et al., J. Magn. Magn. Mater. 1999. V. 202. P. 458–464.
3.442. Bernard J., et al., J. Magn. Magn. Mater. 2000. V. 219. P. 186–198.
3.443. Minakova S.M., et al., MITOM. 2005. No. 10. P. 29–32.
3.444. Minakova S.M., et al., Proc. XV Int. Conf. by permanent magnets. Suzdal' (19-23 Sept. 2005). - Moscow 2005. - P. 42.
3.445. Yagodkin Yu.D., et al., Perspekt. Materialy. 2002. No. 4. P. 70–73.
3.446. Tanimoto N., et al., Scripta mater. 2000. V. 42. P. 961–966.
3.447. Yoshizawa Y., et al., J. Appl. Phys. 1988. V. 64. P. 6044–6046.
3.448. Herzer G., IEEE Transactions on magnetics. 1990. V. 26, No. 5. P. 1397–1402.
3.449. Yamauchi K., Yoshizawa Y., Nanostructured Mater. 1995. V. 6. P. 247–254.
3.450. McHenry M.E., Laughlin D.E., Acta Mater. 2000. V. 48. P. 223–238.
3.451. Nago K., et al, IEEE Transactions J. on Magn. in Japan. 1992. V. 7, No. 2. P. 119-127.
3.452. Hasegawa N., Saito M., JMMM 103. 1992. P. 274–284.
3.453. Makito A., Hayakawa Y., Sci. Eng. 1993. V.A181 / A182. P. 1020-1024.
3.454. Isiwata N., J. Magn. Soc. Japan. 1994. V. 18. P. 744–749.
3.455. Sheftel' E.N., Bannykh O.A., Metally. 2006. No. 5. P. 33–39.
3.456. Iskhakov RS,, FMM. 1995. V. 79. No.. 6. P. 122–135.
3.457. Sheftel' E.N., et al., The Baikov Institute of Metallurgy and Materials Science of the Russian Academy of Sciences - 70 years Scientific works by Academician K.A. Solntseva. - Moscow, Interkontakt Nauka, 2008.
4.458. Sheftel' E.N., et al., Pis'ma v ZhTF. 2007. V. 33, No. 20. P. 64–72.
3.459. Sheftel' E.N., et al., FMM. 2008. V. 105, No. 5. P. 1–6.

Index

A

alloy
- 5BDSR 408, 410, 412
- $Al_{85}Y_4Nd_5Co_2$ 182
- $Co_{70}Fe_5Si_{15}B_{10}$ 91, 96
- $Co_{75}Fe_5Si_{13}$C7 36
- $Co_{85}Nb_{7.5}Ti_{7.5}$ 94
- $Co_{86}Nb_7Zr_{3.5}Mo_{3.5}Cr_{1.5}$ 94
- Co–Fe–Cr– Si 64
- Co–Fe–P–B 96
- Co–Mn 89
- Elinvar 95
- $Fe_5Co_{70}Si_{15}B_{10}$ 122
- $Fe_5Co_{75}Si_{13}C_7$ 34
- $Fe_{27}Ni_{53}P_{14}B_6$ 77
- $Fe_{36}Co_{36}B_{19.2}Si_{4.8}Nb_4$ 182
- $Fe_{40}Ni_{38}Mo_4B_{18}$ 71
- $Fe_{40}Ni_{40}B_{20}$ 82, 83
- $Fe_{40}Ni_{40}P_{14}B_6$ 20, 34, 36, 73, 78, 188, 267
- $Fe_{50}Ni_{33}B_{17}$ 191, 199, 200, 201, 202, 203, 205, 218, 219, 322, 325, 326, 327, 328, 329, 330, 333, 335, 336
- $Fe_{58}Ni_{25}B_{17}$ 191, 194, 195, 196, 197, 198, 200, 201, 320, 321, 325, 326, 335, 336, 339, 340, 346, 352, 353
- $Fe_{65.5}Ni_{17}P_{17.5}$ 40
- $Fe_{65}Ni_{17}P_{18}$ 45
- $Fe_{70}Cr_{15}B_{15}$ 184, 185, 187, 188, 189, 191, 192, 193, 194, 199, 210, 236, 316, 317, 318, 319, 320, 333, 373, 375, 376, 377, 378, 379, 415, 418, 419, 420
- $Fe_{73.5}Cu_1Nb_3Si_{13.5}B_9$ 122
- $Fe_{73.5}Si_{13.5}B_9Cu_1Nb_3$ 211, 213
- $Fe_{73.5}Si1_{3.5}B9Nb_3Cu_1$ 205, 206, 209, 210, 215, 216, 217, 331, 332, 334, 339, 340, 406, 415
- $Fe_{73}P_{18}Mn_5V_4$ 391
- $Fe_{75}Ni_2B13Si_{10}$ 81
- $Fe_{77}Ni_1Si_9B_{13}$ 36, 37, 38
- $Fe_{80}P_{12.5}C_{7.5}$ 91
- $Fe_{80}P_{13}C_7$ 91
- $Fe_{80-x}Ni_xSi_6B_{14}$ 74
- $Fe_{81}Si_7B_{12}$ 122
- $Fe_{82.5}B_{17.5}$ 18, 75
- $Fe_{82}B_{18}$ 59, 61, 65, 66, 95

$Fe_{83}B_{17}$ 79, 93
$Fe_{83-x}Ni_xB_{17}$ 191
$Fe_{84}B_{16}$ 35, 70
Fe–B 32, 61, 62, 63, 66, 67, 68, 72, 73, 74, 76, 77, 78, 80, 93
Fe–B–Ce 63, 68, 73
Fe–Co 89, 93, 96
FeCo 143, 218, 423, 424
Fe–Co–Zr–B 145
Fe–Cr–B 373, 374, 376, 379
Fe–Ni 32, 34, 38, 39, 44, 78, 80, 89, 91, 93
Fe–Ni–B 219, 240, 420
Fe–Ni–P 38, 39, 44, 78, 80
Fe–P–C–Ga–Si–Cu 401
Fe–Si–B–Nb–Cu 218, 274, 275, 397
$Fe_xNi_{80-x}B_{20}$ 33
$Fe_xNi_{80-x}P_{14}B_6$ 33
Finemet 72, 205, 211, 212, 213, 214, 218, 236, 379, 401, 404, 405, 406, 407, 408, 412, 415, 418, 420, 421, 422, 423, 437
Heatperm 437
Invar 93
nanoalloy 108
nanocrystalline 103, 104, 105, 106, 107, 108, 109, 110, 111, 112, 113, 114, 115, 116, 117, 119, 120, 121, 122, 124, 125, 126, 127, 128, 129, 130, 131, 132, 133, 134, 135, 136, 139
Nanoperm 218, 421, 422, 423, 437
Nd–Fe–B 406, 426, 427, 428, 430, 432
Ni44Fe29Co15B10Si2 201, 204, 205, 206, 207, 208, 218, 219, 322, 325, 327, 328, 329, 333, 335, 336
$Ni_{44}Fe_{29}Co_{15}Si_2B_{10}$ 236, 237, 238, 242, 243, 244, 379, 380, 415, 416, 417, 418, 419, 420
$Ni_{55}Pd_{35}P_{10}$ 65
$Ni_{70}Mo_{10}B_{20}$ 179, 180, 181
$Ni_{70}Mo_{10}P_{20}$ 178, 180, 181
Ni–Co 89
Ni–Fe–Co–Si–B 219, 236
Ni–Mn 89
Ni–Mo–B 218, 219, 220
$Pd_{78}Cu_6Si_6$ 35
$Pd_{78}Cu_6Si_{16}$ 60
Pd–Cu–Ni–P 276, 389
Tb–Co 93
Thermoperm 218, 421, 423
$Ti_{49}Ni_{24}Cu_{24}B_3$ 360, 362, 363, 364, 367, 368, 369, 370, 371, 372
$Ti_{50}Co_{50}$ 93
$Ti_{50}Ni_{25}Cu_{25}$ 247, 248, 249, 250, 251, 252, 253, 254, 255, 257, 259, 359, 360, 361, 362, 367, 368, 369, 371, 372
Vitreloy 1 356, 357
WC–Co 107, 108
Zr–Cu–Al–Ni–Nb 356
Zr–Ti–Cu–Ni–Al 276, 389

amorphisation 1, 2, 4, 14, 33
anisotropy
  effective anisotropy 399, 400, 401, 402, 423
  equiaxial anisotropy 401
  magnetocrystalline anisotropy 400, 407
  magnetocrystallographic anisotropy 425
Avrami index 212

B

Bridgman chamber 109, 148, 149, 226, 236, 247, 248, 251, 254, 257, 261, 270, 379, 415, 416, 417, 418

C

cathode sputtering 163, 165
creep
  Coble creep 229
crystallisation
  eutectic crystallisation 172
  polymorphous crystallisation 172
  primary crystallisation 172
  surface crystallisation 177

D

deformation
  severe plastic deformation 105, 106, 109, 111, 112, 113, 115, 116, 131, 132
ductile–brittle transition 61, 71, 78, 83

E

effect
  high-temperature hardening effect 38
  magnetostriction effect 92
  ΔE-effect 92, 94, 95
equal-channel angular pressing 109, 111
equation
  Arrhenius equation 174
  Debye–Scherrer equation 404
  Ginzburg–Landau 383
  Griffith-type energy balance equation 58
  Irwin equation 30
  Johnson–Mehl–Avrami equation 174
  Kissinger equation 174
  Landau–Khalatnikov equations 382

F

factor
  Debye–Waller factor 112
ferrimagnetics 84, 86, 87, 97
ferrimagnetism 84, 86, 87

ferromagnetics 84
ferromagnetism 84, 85, 86, 87, 88, 89
function
  radial distribution function 6, 7, 8, 9, 10, 13, 20, 76
  reduced interference function 8

G

grain boundary microsliding 359
grain boundary microslip 127

H

Hall–Petch ratio 120, 122, 123, 124, 126
high-temperature gas extrusion (HTGE) 107
high-temperature superconductors 289
hot isostatic pressing 106

I

ion-beam assisted deposition 289
ionised cluster beam deposition 163
ion plasma sputtering 166

M

magnetoresistance 302, 303, 304
magnetron sputtering 167, 168, 286, 288, 295, 298, 299, 313, 392, 437, 440
megaplastic deformation 146, 147, 150, 151, 222, 271, 380
megaplastic torsion deformation 433
metal-organic chemical vapour deposition 163
method
  Bridgman-Stockbarger method 297
  Taylor-Ulitovsky method 404
model
  Bernal model 11
  Eyring model 270
  Gilman dislocation model 47
  Hertzer model 401, 403
  Spaepen free volume model 15
  Spaepen–Taub model 49
molecular beam epitaxy 163
moment
  magnetic moment 84, 86, 88, 89, 90, 93
MPD 146, 147, 148, 222, 223, 224, 225, 226, 227, 228, 229, 230, 231, 232, 233, 234, 235, 236, 238, 239, 240, 241, 242, 243, 244, 245, 246, 247, 248, 250, 251, 252, 253, 254, 255, 256, 257, 258, 259, 260, 261, 264, 267, 270, 271, 379, 380, 381, 382, 385, 387, 388, 389, 415, 416, 417, 418, 419, 420, 421

N

nanocrystals 108, 110, 112, 114, 116, 117, 120, 121, 122, 123, 124, 126, 127, 128, 129, 130, 134, 136

nanoindentation 355, 389, 393, 394
nanopowder 106, 108, 131, 132

P

paramagnetism
  Curie paramagnetism 133
  Pauli paramagnetism 133
Peierls barrier 232, 233
polyclusters 21
polyhedrons
  Bernal polyhedrons 11
  Voronoi polyhedrons 11
pulsed photon irradiation 390

R

ratio
  Hall–Petch ratio 143, 337, 339, 359, 368, 369, 372
relation
  Hall–Petch relation 119, 122, 124, 125, 128, 129, 337, 339, 352
remagnetisation 90

S

spin nonet 135
structural relaxation 14, 15, 19, 20, 21, 24, 35, 37, 38, 39, 50, 51, 70, 76, 78, 79, 91, 98
superparamagnetism 404, 426
surface hardening ultrasonic treatment 150

T

temperature
  blocking temperature 135
  Curie temperature 14, 77, 85, 88, 89, 95, 96
  equicohesion temperature 41, 71, 73
  glass transition temperature 1, 3, 36, 42
thermal spraying 164
treatment
  pulsed photon treatment 160, 161

Z

zero ferromagnetics 89